ECOSYSTEM RESPONSE MODELLING IN THE MURRAY-DARLING BASIN

ECOSYSTEM RESPONSE MODELLING IN THE MURRAY-DARLING BASIN

NEIL SAINTILAN AND IAN OVERTON

CSIRO PUBLISHING

National Library of Australia Cataloguing-in-Publication entry

Ecosystem response modelling in the Murray-Darling Basin/
editor, Neil Saintilan, Ian Overton.

9780643096134 (pbk)

Includes index.
Bibliography.

Wetland management – Darling River Watershed (Qld. and N.S.W.)
Wetland management – Murray River Watershed (N.S.W.–S. Aust.)
Darling River Watershed (Qld. and N.S.W.) – Environmental conditions.
Murray River Watershed (N.S.W.–S. Aust.) – Environmental conditions.

Saintilan, Neil
Overton, Ian.
333.9180994

Published by
CSIRO PUBLISHING
36 Gardiner Road, Clayton VIC 3168
Private Bag 10, Clayton South VIC 3169
Australia

Telephone: [+613] 9545 8555
Local call: 1300 788 000 (Australia only)
Fax: +61 3 9662 7555
Email: csiropublishing@csiro.au
Web site: www.publishing.csiro.au

Front cover: Monoman Creek on the Chowilla Floodplain in South Australia (photo by Ian Overton)

Set in Adobe Minion 10/12 and Stone Sans
Edited by Adrienne de Kretser
Cover and text design by James Kelly
Typeset by Desktop Concepts Pty Ltd, Melbourne
Index by Russell Brooks
Printed by Ingram Lightning Source

CSIRO PUBLISHING publishes and distributes scientific, technical and health science books and journals from Australia to a worldwide audience and conducts these activities autonomously from the research activities of the Commonwealth Scientific and Industrial Research Organisation (CSIRO).

The research in this book was supported by the following organisations:

CSIRO PUBLISHING

Australian Government
Water for the Future

Department of
Environment, Climate Change and Water NSW

Feb26_RP_ILS

Foreword

The National Water Initiative (NWI) is Australia's enduring blueprint for water reform. Under this initiative, Australian, state and territory governments have made commitments to return all currently overallocated and overused groundwater and surface water systems to environmentally sustainable levels of extraction. Meeting these commitments will require rigorous scientific knowledge of the links between flow regimes and ecosystem responses.

Predictive ecological modelling is a new area of aquatic science in Australia. The relationships between flow regimes and ecosystem responses are complex and difficult to represent in models. Nevertheless, predictive ecological models will need to continue to be developed if aquatic ecology is to fulfil the required role in water planning and management processes. These models will support both water planning, to guide the development of hydrologic regimes that allocate sufficient water to achieve environmental outcomes, and water management, to ensure that water is delivered in ways that maximise the health of rivers and wetlands. These models are needed to take the guess work out of optimising our most valuable natural resource.

The Murray-Darling Basin is the water system where this predictive capability is most needed in Australia. Not only is the pressure to optimise the allocation of water between consumptive and environmental uses greatest in the Basin, but there is more information on ecosystem responses to flows in this region than in any other part of Australia. This book is timely, given the current efforts to establish a sustainable diversion limit and a plan for water use in the Basin.

This publication is the result of contributions by Australia's leading ecological scientists and gives a comprehensive overview of ecosystem modelling techniques and outcomes in the Murray-Darling Basin in a way that is accessible to stakeholders. This work strongly supports the goals of the National Water Initiative.

Ken Matthews AO
Chair and Chief Executive Officer
National Water Commission

Australian Government
Water for the Future

Contents

Preface

The Murray-Darling Basin covers an area of over 1 000 000 square kilometres, one-seventh of Australia. It supports over 30 000 wetlands, including many that are of international significance. The Basin also produces 70% of Australia's irrigated crops and is home to over 2 million people.

The most recent report of the Sustainable Rivers Audit (MDBC 2008a) painted a bleak picture of riverine health in the Murray-Darling Basin. Of 23 valleys, one was reported as being in good condition (the Paroo), two in moderate condition, seven in poor condition and 13 in very poor condition. It is unsurprising that the iconic wetland sites of the basin are similarly affected by declining water availability. The Murray-Darling Basin Commission reported in 2008 on the condition of the six icon sites along the River Murray: the Barmah-Millewa Forest; Gunbower-Koondrook-Perricoota; Hattah Lakes; Chowilla Floodplain and Lindsay and Wallpolla Islands; Lower Lakes, Coorong and Murray Mouth; and the River Channel (MDBC 2008b). Few waterbirds were supported across the sites and most of the floodplain wetlands were dry. The river red gum and black box communities were in decline and there was a reduction in the diversity of understorey vegetation (MDBC 2008b). Flows at the Murray Mouth have reduced by more than 60%, with flow ceasing there 40% of the time – an increase from 1% under natural conditions (CSIRO 2008). As a result, the Coorong has become hypersaline and is in danger of irreversible change.

A similar pattern of decline is evident in the Ramsar-listed wetlands of the northern basin. At Narran Lakes, approximately half of the river red gums and river coobas have died (Thoms *et al.* 2007). The core wetland of the Macquarie Marshes, the Northern Nature Reserve, has seen the loss of approximately three-quarters of the Cumbungi and water couch within a decade; over the same period the wetland has seen the invasion of drought-tolerant chenopod shrubs (Bowen and Simpson 2008). In the Gwydir wetlands, three-quarters of the water couch/spike rush association and over half of the *Bulbochenus* was lost between 1996 and 2005 (Bowen *et al.* 2008).

Dam releases for environmental purposes (environmental flows) have been an important tool for wetland management in Australia for several decades. The effectiveness of environmental flows has been limited not only because of the proportion of water available but also due to a lack of strategic purpose in many situations. For much of the early 1990s the management of environmental water was 'essentially ad hoc' (Walker *et al.* 1995). As recently as 2000 environmental flow management could be described as being based on 'best informed guesses' (Thoms *et al.* 2000).

A more scientifically grounded strategy for environmental flow management is warranted, given the staggering levels of government investment in environmental water over the past decade. These initiatives include the $26.4 million NSW Wetland Recovery Plan (2006), the $173.27 million Rivers Environmental Restoration Program (incorporating the NSW River-Bank program) (2007), the $700 million The Living Murray restoration program and the $12.9 billion National Plan (2008), which includes $3.1 billion for water purchase in the Murray-Darling Basin.

Much of this funding has been provided by the Commonwealth government, which is assuming a greater role in the management of water within the Murray-Darling Basin. The *Water Act* (2007) provided a legislative basis for the establishment of the Murray-Darling Basin Authority, which will develop and implement a basin plan, including an environmental watering plan. Water-sharing plans previously developed by state jurisdictions will, upon their expiry, require Commonwealth endorsement for their renewal. This provides a junction for a Commonwealth review of water-sharing arrangements, and conformity to the objectives of a basin-wide environmental watering plan is likely to be one criterion of acceptance.

Quantitatively based ecosystem response models will be an important tool with which the jurisdictions will be able to model the ecological implications of proposed watering arrangements. Ecosystem response models therefore have an important role in guiding water planning. They have the potential to model the condition of ecosystems under a range of management scenarios, including the confounding influence of climate change. Specifically, the roles of ecosystem response modelling can include:

- linking water volumes and timing to ecological outcomes;
- interactive testing of management and climate scenarios;
- improvement of our understanding of ecosystem processes;
- identification of the key drivers in the system, and optimal strategies;
- identification of environmental opportunity costs;
- a framework for monitoring;
- scientific rigour through repeatable analysis and transparent decision-making;
- a strong influence on those involved;
- a method for communicating complex ecosystem drivers, processes and responses; and
- a framework for integrating the best science.

This book is a compilation of papers presented at the Ecosystem Response Modelling in the Murray-Darling Basin conference held in Sydney, Australia, in November 2008. Papers were invited from researchers and managers operating within the Murray-Darling Basin in the areas of ecosystem response modelling for environmental water management. The purpose of the conference, and this book, was to expose ecosystem response modelling issues and capabilities to water managers and researchers during a time of significant drought, declining ecosystem condition and major water reform in Australia. The limited resource and significant environmental decline have led to complex water allocation issues with numerous stakeholders. Improvements in identifying equitable options and reducing the risk in managing resources for these outcomes can be greatly increased with quantitative, repeatable and open decision support systems, including ecosystem response models.

The book is divided into three sections. Part I covers three areas of general interest to ecosystem response modelling and will be applicable to researchers and managers across the world. These chapters include fundamental principles of ecosystem response modelling and issues of spatial and temporal scales, along with decision-making with sparse data.

Part II contains seven chapters on the northern part of the Murray-Darling Basin – the Darling Basin – and includes ecosystem response modelling of the Gwydir Wetlands, the Macquarie Marshes and the border rivers. The chapters include modelling of waterbirds, fish, trophic structures and riparian vegetation.

Part III contains 13 chapters on the southern portion of the basin – the Murray Basin – and includes ecosystem response modelling of the Lachlan River, the Murrumbidgee, the Murray and the Goulburn. It includes ecosystem response modelling of the Lowbidgee floodplain, The Living Murray icon sites at Barmah, Chowilla floodplain and Lindsay and Wallpolla Islands and the Coorong. The chapters discuss modelling of birds, fish, frogs, riparian vegetation, aquatic vegetation and invasive plants, as well as modelling whole of ecosystems using holistic methods.

The final chapter explores the lessons learnt and future issues for ecosystem response modelling in the Murray-Darling Basin. Ecosystem response modelling as applied to environmental water allocation is a rapidly advancing science across many disciplinary fronts. While this book was being produced, the Commonwealth purchased $650 million in water entitlements for environmental purposes in the Murray-Darling Basin. The application of this water across the basin provides a fresh imperative for the application of models, and their further calibration with the results of targeted monitoring.

References

Bowen S and Simpson S (2008). Changes in extent and condition of the vegetation communities of the Macquarie Marshes floodplain 1991–2008. Draft Macquarie Marshes Adaptive Environmental Management Plan. Water for the Future, NSW Dept of Environment and Climate Change.

Bowen S, Simpson S and McCosker R (2008). Changes in extent and condition of floodplain wetland vegetation communities in the Gwydir wetland and floodplain. NSW Dept of Environment and Climate Change, Sydney.

CSIRO (2008). Water availability in the Murray-Darling Basin. Report from the CSIRO to the Australian Government. CSIRO Water for a Healthy Country, Canberra.

MDBC (2008a). Murray-Darling Basin Rivers: Ecosystem Health Check, 2004–2007. Sustainable Rivers Audit Report 1. Murray-Darling Basin Commission, Canberra.

MDBC (2008b). The Living Murray. Icon Site Condition Report, October 2008. Murray-Darling Basin Commission, Canberra.

Thoms M, Suter P, Roberts J, Koehn J, Jones G, Hillman T and Close A (2000). Report of the River Murray Scientific Panel on Environmental Flows: River Murray – Dartmouth to Wellington and the Lower Darling River. Murray-Darling Basin Commission, Canberra.

Thoms M, Capon S, James C, Padgham M and Rayburg S (2007). The Narran Ecosystem Project: the response of a terminal wetland system to variable wetting and drying. Final report to the Murray-Darling Basin Commission. MDBC Publication No. 40/08.

Walker KF, Sheldon F and Puckeridge JT (1995). A perspective on dryland river ecosystems. *Regulated Rivers: Research and Management* **11(1)**, 85–104.

Neil Saintilan and Ian Overton
January 2010

Acknowledgments

The research within this book was made possible by the funding support of several agencies and programs, too numerous to mention here. Particular recognition is made, however, of the contribution of the Commonwealth Government's Water Smart Australia program, which funded the Rivers Environmental Restoration Program, a major initiative in ecological response modelling in the northern Murray-Darling Basin. The Murray-Darling Basin Authority, through its previous incarnation as the Murray-Darling Basin Commission, funded many of the southern basin projects. The CSIRO Water for a Healthy Country Flagship has supported the CSIRO River Murray projects.

All chapter contributions were submitted to independent review by two expert referees prior to acceptance. We thank the following experts who contributed their time in assessing and improving the contributions:

- Angela Arthington, Griffith University;
- Lex Cogle, Murray-Darling Basin Authority;
- Peter Cottingham, Consultant;
- Susan Cuddy, Water for a Healthy Country Flagship, CSIRO;
- Jocelyn Dela Cruz, NSW Department of Environment, Climate Change and Water;
- Tanya Doody, Water for a Healthy Country Flagship, CSIRO;
- Neil Foster, NSW Department of Environment, Climate Change and Water;
- George Ganf, University of Adelaide;
- Ben Gawne, Murray-Darling Freshwater Research Centre;
- Mike Geddes, University of Adelaide;
- Geoff Gordon, NSW Department of Environment, Climate Change and Water;
- Ivor Growns, NSW Department of Environment, Climate Change and Water;
- Barry Hart, Emeritus Professor, Monash University;
- Tony Herbert, SA Murray-Darling Basin NRM Board;
- David Hunter, NSW Department of Environment, Climate Change and Water;
- Jeff Kelleway, NSW Department of Environment, Climate Change and Water;
- Joanne Ling, NSW Department of Environment, Climate Change and Water;
- Rebecca Lester, Flinders University;
- Heather Malan, University of Cape Town, South Africa;
- Nick Marsh, Consultant; Yorb Pty Ltd
- Alex Meehan, Murray-Darling Basin Authority;
- Lucy Nairn, University of New South Wales;
- Gina Newton, Department of Environment, Water, Heritage and the Arts;
- Rod Oliver, Water for a Healthy Country Flagship, CSIRO;
- Ian Overton, Water for a Healthy Country Flagship, CSIRO;
- Naomi Rea, Charles Darwin University;
- Mike Reid, University of New England;
- Jane Roberts, Consultant;

- Darren Ryder, University of New England;
- Neil Saintilan, NSW Department of Environment, Climate Change and Water;
- Mike Stewardson, University of Melbourne;
- Kate Stokes, Water for a Healthy Country Flagship, CSIRO;
- Martin Thoms, University of New England;
- Keith Ward, Goulburn Broken Catchment Management Authority;
- Angus Webb, University of Melbourne;
- Paul Wettin, Consultant.

Finally, we thank Briana Melideo, Adrienne de Kretser and the team at CSIRO Publishing for their support, and Frank Saintilan for his careful reading of the manuscript.

List of contributors

Michelle Akeroyd Murray-Darling Basin Authority, 51 Allara St, Canberra, ACT, Australia

Angela Arthington Australian Rivers Institute, Griffith University, Nathan, QLD 4111; eWater Co-operative Research Centre, Australia

Stephen Balcombe Australian Rivers Institute, Griffith University, Nathan, QLD 4111; eWater Co-operative Research Centre, Australia

Sharon Bowen Rivers and Wetlands Unit, NSW Department of Environment, Climate Change and Water, PO Box A290, Sydney South, NSW 1232, Australia

Kate Brandis Australian Wetlands and Rivers Centre, School of Biological, Earth and Environmental Sciences, University of NSW, NSW 2052, Australia

Shahadat Chowdhury NSW Department of Environment, Climate Change and Water, PO Box 3720, Parramatta, NSW 2124, Australia

Lex Cogle Murray-Darling Basin Authority, 51 Allara St, Canberra, ACT, Australia

Matthew Colloff CSIRO Entomology, GPO Box 1700, Canberra, ACT 2601, Australia

Peter Cottingham Peter Cottingham & Associates; eWater CRC, Melbourne, 56 Como St, Alphington, Vic. 3078, Australia

Steve Cox Rivers and Wetlands Unit, NSW Department of Environment, Climate Change and Water, PO Box A290, Sydney South, NSW 1232, Australia

David Crook Arthur Rylah Institute for Environmental Research; eWater Co-operative Research Centre, Australia

Susan Cuddy CSIRO Land and Water, Canberra, ACT, Australia

Louisa Davis Qld Department of Environment and Resource Management, 120 Meiers Rd, Indooroopilly, QLD 4068; eWater Co-operative Research Centre, Australia

Tanya Doody Water for a Healthy Country, CSIRO Land and Water, Waite Rd, Urrbrae, SA 5064, Australia

Patrick Driver NSW Department of Environment, Climate Change and Water; Cooperative Research Centre for eWater, PO Box 53, Orange, NSW 2800, Australia

Peter Fairweather School of Biological Sciences, Flinders University, GPO Box 2100, Adelaide, SA 5001, Australia

George Ganf Ecology and Evolutionary Biology, University of Adelaide, SA 5005, Australia

Tahir Hameed NSW Department of Environment, Climate Change and Water, PO Box 3720, Parramatta, NSW 2124, Australia

Elizabeth Heagney School of Biological, Earth and Environmental Science, University of New South Wales, Sydney, NSW 2052, Australia

Terry Hillman Consultant, 567 Brown St, Lavington, NSW 2611, Australia

David Hohnberg Murray-Darling Basin Authority, 51 Allara St, Canberra, ACT, Australia

Joel Huey eWater Co-operative Research Centre; Australian Rivers Institute, Griffith University, Nathan, QLD 4111

Simon Hunter Rivers and Wetlands Unit, NSW Department of Environment, Climate Change and Water, PO Box A290, Sydney South, NSW 1232, Australia

Sarah Imgraben Rivers and Wetlands Unit, NSW Department of Environment, Climate Change and Water, PO Box A290, Sydney South, NSW 1232, Australia

Tony Jakeman Integrated Catchment Assessment and Management Centre, Fenner School of Environment and Society, Australian National University, Canberra, ACT, Australia

Kim Jenkins Australian Wetlands and Rivers Centre, School of Biological, Earth and Environmental Sciences, University of NSW, NSW 2052, Australia

Jeff Kelleway NSW Department of Environment, Climate Change and Water, PO Box A290, Sydney South, NSW 1232, Australia

Richard Kingsford Australian Wetlands and Rivers Centre, School of Biological, Earth and Environmental Sciences, University of NSW, NSW 2052, Australia

Tsuyoshi Kobayashi NSW Department of Environment, Climate Change and Water, PO Box A290, Sydney South, NSW 1232, Australia

Jong Lee Murray-Darling Basin Authority, 51 Allara St, Canberra, ACT, Australia

Rebecca Lester School of Biological Sciences, Flinders University, GPO Box 2100, Adelaide, SA 5001, Australia

Joanne Ling Rivers and Wetlands Unit, Scientific Supporting Division, NSW Department of Environment, Climate Change and Water, PO Box A290, Sydney South, NSW 1232, Australia

Stuart Little Murray-Darling Basin Authority, 51 Allara St, Canberra, ACT, Australia

Jaye Lobegeiger Qld Department of Environment and Resource Management, 120 Meiers Rd, Indooroopilly, QLD 4068; eWater Co-operative Research Centre

Zygmunt Lorenz CSIRO Land and Water, Private Bag 2, Glen Osmond, SA 5064, Australia

Yi Lu Rivers and Wetlands Unit, NSW Department of Environment, Climate Change and Water, PO Box A290, Sydney South, NSW 1232, Australia

Mike Maher NSW Department of Environment, Climate Change and Water, PO Box A290, Sydney South, NSW 1232, Australia

Nick Marsh Managing Director, Yorb Pty Ltd, 906 Sandgate Rd, Clayfield, QLD 4011

Jonathan Marshall Qld Department of Environment and Resource Management, 120 Meiers Rd, Indooroopilly, QLD 4068; eWater Co-operative Research Centre

Debashish Mazumder Australian Nuclear Science and Technology Organisation, PMB 1, Lucas Heights, Menai, NSW 2234, Australia

Wendy Merritt Integrated Catchment Assessment and Management Centre, Fenner School of Environment and Society, Australian National University, Canberra, ACT, Australia

Lucy Nairn Australian Wetlands and Rivers Centre, School of Biological, Earth and Environmental Sciences, University of NSW, NSW 2052, Australia

Rod Oliver CSIRO Land and Water, Private Bag 2, Glen Osmond, SA 5064, Australia

Martin O'Rourke NSW Department of Environment, Climate Change and Water, PO Box 550, Tamworth, NSW 2340, Australia

Ian Overton Water for a Healthy Country, CSIRO Land and Water, Urrbrae, South Australia

Dave Penton Water for a Healthy Country, CSIRO Land and Water, GPO Box 1666, Canberra, ACT 2601, Australia

Carmel Pollino Integrated Catchment Assessment and Management Centre, Fenner School of Environment and Society, Australian National University, Canberra, ACT, Australia

Sue Powell Integrated Catchment Assessment and Management Centre, Fenner School of Environment and Society, Australian National University, Canberra, ACT, Australia

Tim Ralph Rivers and Wetlands Unit, NSW Department of Environment, Climate Change and Water, PO Box A290, Sydney South, NSW 1232, Australia

Thomas Rayner Australian Wetlands and Rivers Centre, School of Biological, Earth and Environmental Sciences, University of NSW, NSW 2052, Australia

Jane Roberts PO Box 6191, O'Connor, ACT 2602, Australia

Kerrylee Rogers Rivers and Wetlands Unit, NSW Department of Environment, Climate Change and Water, PO Box A290, Sydney South, NSW 1232, Australia

Ian Rutherfurd 8 Nicholson St, East Melbourne, Vic. 3002, Australia

Neil Saintilan Rivers and Wetlands Unit, NSW Department of Environment, Climate Change and Water, PO Box 3720, Parramatta, NSW 2124, Australia

Mustak Shaikh NSW Department of Environment, Climate Change and Water, PO Box 3720, Parramatta, NSW 2124, Australia

Shannon Simpson Rivers and Wetlands Unit, NSW Department of Environment, Climate Change and Water, PO Box A290, Sydney South, NSW 1232, Australia

Neil Sims Water for a Healthy Country, CSIRO Land and Water, Private Bag 10, Clayton South, Vic. 3169, Australia

Jennifer Spencer Rivers and Wetlands Unit, NSW Department of Environment, Climate Change and Water, PO Box A290, Sydney South, NSW 1232, Australia

David Sternberg Australian Rivers Institute, Griffith University, Nathan, QLD 4111, Australia

Michael Stewardson eWater CRC; Department of Civil and Environmental Engineering, The University of Melbourne, Parkville, Vic. 3010, Australia

Kate Stokes CSIRO Entomology, GPO Box 1700, Canberra, ACT 2601, Australia

Jody Swirepik Murray-Darling Basin Authority, 51 Allara St, Canberra, ACT, Australia

Rachael Thomas Rivers and Wetlands Unit, NSW Department of Environment, Climate Change and Water, PO Box A290, Sydney South, NSW 1232; Australian Wetland and Rivers Centre, School of Biological, Earth and Environmental Sciences, University of New South Wales, NSW 2052, Australia

Martin Thoms Riverine Landscapes Research Lab, University of New England, Armidale, Australia

Keith Ward Goulburn Broken Catchment Management Authority, Shepparton, Victoria, Australia

Paula Ward Goulburn Broken Catchment Management Authority, Shepparton, Victoria, Australia

Skye Wassens Institute of Land, Water and Society, School of Environmental Sciences, Charles Sturt University, PO Box 789, Albury, NSW 2640, Australia

Robyn Watts Institute for Land, Water and Society, School of Environmental Sciences, Charles Sturt University, PO Box 789, Albury, NSW 2640, Australia

Angus Webb eWater CRC; Department of Resource Management and Geography, The University of Melbourne, Parkville, Vic. 3010, Australia

Li Wen Rivers and Wetlands Unit, Scientific Supporting Division, NSW Department of Environment, Climate Change and Water, PO Box A290, Sydney South, NSW 1232, Australia

Sean White Ecology and Evolutionary Biology, University of Adelaide, SA 5005, Australia

Glenn Wilson Cotton Catchment Communities Cooperative Research Centre; Ecosystem Management, School of Environmental and Rural Science, University of New England, Armidale, NSW 2351, Australia

PART I

ISSUES IN ECOSYSTEM RESPONSE MODELLING

Chapter 1

Ecological modelling to support natural resource management

Nick Marsh and Susan Cuddy

Introduction

Ecological response models are important tools for predicting the impacts of management actions on the natural environment. Yet developing them is difficult because natural resource management decisions are usually based on achieving multiple ecological outcomes from combinations of management strategies, and because the models exist in a wide range of incompatible forms. This chapter presents a simple taxonomy for describing ecological models based on their purpose, not their numerical method. The focus of our interest then moves from the promotion of a single numerical approach, to how to combine models of different numerical form. We present a partial solution to this later in the chapter.

To demonstrate the difficulty of applying models of different numerical forms to a single natural resource management (NRM) problem, consider the problem of informing a water manager of the relative ecological benefits of different water delivery scenarios for a wetland, reported as the health of vegetation, birds and fish (such as that adopted under the River Murray icon sites Living Murray Program (MDBA 2009)). Evaluating each scenario requires an assessment of its impact on the reporting measures (vegetation, birds, fish). Each of these reporting measures could comprise five to 10 key species (as used in the Murray Flow Assessment Tool (Young *et al.* 2003)), with each species possibly having more than one life-history stage that is critically influenced by water. If we consider just five key species, each with two life-history stages across the three reporting measures, then reporting for that wetland requires 30 predictions (potentially 30 models and/or submodels) for each water delivery scenario. The model predictions then need to be aggregated in some way to provide scores for each of the reporting measures, and perhaps an overall score across reporting measures. Traditionally, such a computation would require that each of the 30 models be of a predefined and complementary numerical form so that the aggregation of their results is meaningful. The Murray Flow Assessment Tool is a good example – its models are of a similar numerical form with their outputs scaled to the same index (0–1), making them comparable and easily aggregated. This approach successfully manages the computational dilemma of handling many predictions simultaneously, however, the representation of the ecosystem processes is forced to comply to a single numerical model form which may not be the most appropriate for all elements being modelled.

Ecological modelling has not reached a level of consistency in modelling form where there is a single adopted numerical approach which is accepted as being adequate for all elements of the ecosystem. Nor does it need to. There are many alternative (ecological) model forms that can be used to estimate essentially the same element, and may be appropriate given constraints such as access to data to construct the model, modelling expertise and the target question to be addressed by the model. While embracing the multitude of model forms that can be brought to bear on a single NRM problem (e.g. expert-driven habitat suitability models, data-driven regression models, population models that include predator–prey interactions) does mean that we need to develop mechanisms to integrate these different model forms, the alternative of restricting models to a single model form is neither acceptable nor practical. Such an approach restricts the capacity for ecological prediction to only that research which has presented its findings in the chosen model form, and consequently reduces the potential ecological understanding that can be used to inform NRM.

There is a significant body of research into methods for model integration and aggregation, though not confined to nor focused on ecological modelling, and we describe an approach to ecological model integration later in this chapter. To inform this later discussion, the next section presents a taxonomy that distinguishes different approaches to ecological response modelling.

A taxonomy for ecological response models

In examining the range of approaches employed in modelling ecological response, it is useful to consider just what constitutes ecological response modelling. Ecological response modelling is a subset of modelling that seeks to predict change of state as a consequence of one or more external stimuli. In the case of ecological modelling, stimuli are commonly forcing functions such as changes in water levels, flows or temperature, which are the local- or habitat-scale results of management strategies (e.g. nutrient or sediment supply, land use change, water conservation measures). The ecosystem component of interest is thus conceptualised in relation to the forcing functions. These are in turn linked to the management strategies of interest, together with response measures/indicators that are meaningful for interpretation of the results.

To provide a basis for describing ecological modelling approaches, consider a simple system of categorising and classifying ecological models – a simple model taxonomy. For a taxonomy to be effective it must work for known examples and not be a prescriptive description of a particular (biased) view of what a model should be, nor of how it should be constructed. The distinction between groups or subgroups of the taxonomy must be meaningful and useful in the communication of models, not simply for the sake of further differentiation. Where possible, a taxonomy should be as simple as possible while still being useful.

Our ecological model taxonomy (which could equally be applied to any modelling approach) has three elements:

- purpose or focus ecosystem component;
- model abstraction;
- computational form.

With knowledge of these three elements we are able to conceive how models can interact and be combined for reporting in an ecological modelling framework (Table 1.1). In a 10-step iterative approach for model construction Jakeman *et al.* (2006) commence with essentially these three steps, followed by steps to gauge model performance. The approach that is used to construct a model can also be adopted to describe it.

Each of the taxonomy elements is expanded below (and summarised in Figure 1.1) with examples of modelling approaches, followed by a description of recent advances in modelling platforms that integrate these modelling approaches.

Table 1.1: Ecological response model taxonomy

1. Purpose: reporting the response of an ecosystem component	
Description	The thing/s that you want to predict a response in and report as output.
Inputs	Management decision to define important assets.
Example	'To assess the value of an artificial flood we wish to model the likely river red gum response.' The ecosystem component is river red gum.
2. Abstraction: conceptualisation of the key elements and how they interact	
Description	A description of the process of simplification of the natural world to allow it to be a fit-for-purpose model. Model abstraction would often include a diagram 'conceptual model'. Abstraction includes reference to the ecosystem component. It is often constrained by known data limitations and is sometimes constrained to a numerical form.
Inputs	System understanding of the key processes which influence the ecosystem component. Explicit consideration of the spatial and temporal extent of the model which will address the ecosystem component and will inform the selection of the appropriate computational form.
Example	'We don't know the precise model form, but we have observed that where river red gums are permanently inundated or flooded less frequently than every 10 years, there is a high mortality. Similarly, we have observed a larger response after winter floods than summer floods and for longer-duration floods than shorter-duration floods. We will construct a relationship which compares flood frequency, where that flood frequency is determined by applying a local wetland hydraulic model to convert main channel flows into a spatial extent of flooding. This model will allow comparison between scenarios based on the prevalence of conditions within a known suitable range, rather than a prediction of river red gum condition.'
3. Computational form: model implementation	
Description	The numerical approach or approaches used to implement the model abstraction. This may include the combination of several submodels of different form.
Inputs	Mathematical ability and available input data.
Example	The numerical approach is to summarise a daily time series of the percentage of the wetland inundated under each flow scenario. The magnitude (what spatial extent), duration (how long they last), timing (what time of year they occur) and frequency (period between) of floods is compared between scenarios.

Purpose: reporting the response of an ecosystem component

There is no correct ecosystem component at which to target model development. The target ecosystem component is largely a function of the model purpose defined by the NRM options, further constrained by the availability of relevant data and expert knowledge. For example, a fully integrated food web model may be the preferred endpoint to inform the management decision, however, the actual model may be a simple one of habitat availability for a single life-history stage of a single species, just because the errors inherent in trying to represent a complex system about which we have an incomplete understanding far outweigh the value of developing a full food web model. A simple model about a small element of the system often has to suffice. The selection of the target output for the model (e.g. the performance and/or response indicators) should be developed through dialogue with the model stakeholders, as it is critical that they be meaningful and useful to the potential model users (Johnson 2009).

This step of defining the scope and objectives of the modelling exercise (Jakeman *et al.* 2006) allows model developers to distinguish between research and application. It is important to make a distinction between models which are developed solely to improve our understanding, such as statistical models constructed to maximise the amount of variance in the environmental data explained by the model (explanatory models), and models which are developed to inform a management decision. In the latter, the model design should be focused on those input variables which are subject to alteration through NRM and predict a future state (predictive models). Explanatory models can form the basis of subsequent analysis used to develop predictive models, however 'explanatory power' over data used to create a model and the 'predictive accuracy' when that model is extended beyond the input data range are two distinct qualities (Shmueli and Koppius 2009).

Drivers of change: scenarios

From a modelling perspective, the forcing functions and stressors described earlier are commonly termed 'scenarios' or decision variables. Their state is provided as an input to the model conceptualisation, such as a time series of flow past a point, a land use map or a set of dam release rules. A typical approach in ecosystem response modelling is to express the policies and actions that natural resource managers can implement as scenarios, then model the likely response of the system to those scenarios.

The choice of scenarios and their formulation directly affects the model conceptualisation.

Model abstraction: conceptualising the real world

This section provides an overview of some of the common approaches to conceptualising the system of interest, from the perspectives of:

- habitat;
- primary productivity;
- mass balance;
- community or ecosystem population.

Habitat

Where a specific physical habitat requirement of a life-history stage of a single species is known, the quantification of the change in available habitat between scenarios has been used in many applications as a powerful quantitative tool. A widely used process is the Instream Flow Incremental Methodology (IFIM) and its software package PHABSIM (physical habitat simulation), which was developed for defining the hydraulic habitat requirements of salmonids in northern American streams (Bovee *et al.* 1998). Field observations of depth and velocity at the point of observing a salmonid are combined to construct a graph relating depth and velocity of flow to habitat suitability. A hydraulic model of a reach of stream can be used to predict the velocity and depth distribution for a given discharge, and this can be converted to a distribution of suitable habitat for different discharges. Constructing this for several discharges creates a curve which allows the transformation of a time series of flow into a time series of suitable habitat.

Habitat: connectivity

A critical link in an ecosystem is the periodic (or continuous) spatial connections between different geographic locations. For example, after a prolonged drought where many river waterholes have dried up, the flowing of the river and subsequent connection of remaining refuge waterholes is a critical determinant in defining the population (or even existence) of a given fish. The need for connectivity can be seen as a subset of the physical habitat requirements approach described above, although the focus is usually on the infrequent connection events (such as their frequency and duration) as opposed to the IFIM approach, which is usually focused on measures of central tendency (the average weighted usable area of habitat).

Primary productivity

The basic premise of focusing ecological modelling on primary productivity is that ecosystems are critically dependent on the lowest trophic levels and the system can be conceptualised at this level. An example is the role of carbon and nutrients leached from the floodplain in the functioning of river systems. Complex biogeochemical interactions occur under different flooding regimes; for example, in a river red gum forest inundation initially promotes a burst of microbial activity in the soil, which eventually becomes anoxic under an extended period of inundation in warmer months (Howitt *et al.* 2007). These primary productivity activities cycle nutrients and influence water quality, which is critical for the survival of higher-level organisms.

Mass balance

Another family of models seek to balance mass or, more usually, energy. Good examples are commercial fisheries; the use of 'Ecopath' for constructing mass balance based food web models to describe the average state of fisheries systems is described well in Pauly *et al.* (2000). A difficulty in mass balance or energy balance models is the ability to test and calibrate them; this applies to most ecological models. This is partially compensated for by undertaking sensitivity analysis to quantify their uncertainty. To do this, each of the model's input parameters is varied through its likely range of values. The effect on model results of changing each parameter's value provides some guide to the importance of that parameter.

Community or ecosystem models: population models

The term 'ecological modelling' is commonly associated with the ability to predict population. There are many approaches to population modelling, which include a mass balance approach (described earlier) or a life-history based approach where the probability of transition from one life-history stage to another is described using a stage-classified matrix and age classes can be tracked (Cochran and Ellner 1992). A simpler approach which also focuses on life-history stage transitions are vital attribute style models, which focus on the likelihood of and unsuccessful transition from one life-history stage to another – the predicted result is the likelihood of a local extinction rather than a growing population (Roberts 1999). Vital attribute models can be described as a life-history version of presence–absence models.

Presence–absence models

Presence–absence models tend to be based on the provision of all known necessary physical requirements (e.g. physical habitat elements) to predict a likelihood of the organisms being present. Logistic regression is commonly used as it can accommodate both continuous and binary variables. Presence–absence models can also take the form of decision trees, which essentially encompass a series of nested 'if' statements that ensure all the required habitat elements occur in the appropriate temporal and spatial order to allow conditions to be suitable for the organism to be present.

Size or age-structured population models

In this family of models, the population is usually divided into subunits of size class or age class, and the relative interactions of these classes are modelled separately. An example is a model of fish populations which includes submodels for each life-history stage of spawn, juvenile and adult. Such models require a detailed understanding of the population dynamics. Perhaps their most common use is to build a robust understanding of how the population changes over time, and to investigate how these dynamics may be affected by environmental triggers such as changes in temperature.

Individual-based population models

The modelling of a population is conducted by modelling the growth of individual elements of the population and applying that over larger spatial areas. This approach works where there is little interaction between individuals, or the form of the interaction does not change over time. This is particularly relevant for vegetation modelling. We may consider, for example, the growth requirements of river red gum then consider how those requirements are met through time at many locations across the landscape. The result is landscape-scale representation of response, but the model is not developed with any specific interaction between the trees.

Scale

While the conceptualisation of how to adequately represent the ecosystem component may take many forms, ecological models have some notion of both spatial and temporal scale. The spatial and temporal scale of a model should form an explicit part of the model abstraction. There are three distinct time scales to consider:

- intergenerational or long-term suitability of a site;
- within-generation or suitable recruitment conditions;
- short-term disturbance, such as a flood last week.

Models that are driven by time series input, such as environmental flow response models, should explicitly deal with these time scales. The appeal of habitat-based modelling is that we can focus on the long-term suitability, and whether average conditions are likely to be suitable. However, short-term planning (one to two years) requires a measure of condition in order to avoid a particular outcome (e.g. another two years without water will result in the death of river red gum). Predictions of condition require knowledge of current state and the ability to infer response from within-generation (e.g. recruitment opportunity to infer age structure) and short-term disturbances, such as prolonged survival of river red gum due to partial flooding events.

Similarly, we could consider three distinct spatial scales:

- landscape (macro) – e.g. whether there is a potential recolonisation source, where it is and how the ecosystem component connects to it;
- habitat range (meso) – e.g. the average water depth range for a pool or wetland;
- daily range (micro) – e.g. the hydraulic conditions around a preferred snag habitat.

For the flow-driven response modelling presented in this book, the spatial scale is usually set (e.g. a river reach or a wetland) and the flow-driven input (flow past a given point or into a given wetland) lends itself to predicting responses to temporally varying conditions alone. However, future modelling to support NRM must consider both spatial and temporal interventions, to address questions such as what combinations of altering flow and reintroducing habitat will give the best ecological response.

Computational form: implementing the model

Once you have conceptualised your model with regard to the issues, scenarios, appropriate indicators and relevant model abstraction, the next step is to identify an appropriate computational model form that suits the model formulation and the available data. In practice, the decision also reflects availability of computing expertise, time constraints, the nature and urgency of the enquiry, and personal bias.

Jakeman *et al.* (2006) provide a useful list of different model forms. We have reinterpreted this classification as the computational forms relate to ecological modelling, with a particular emphasis on predicting flow-related responses.

1 **Empirical** data-driven statistical models, with structures chosen primarily for their versatility and assuming little in advance. Empirical models are often referred to as 'black boxes' because there is little assumed process understanding, other than variable selection. Empirical models are commonly used to help describe a system – reviewing the variance explained by different elements of a fitted model can infer some level of causality and the explanation of these associations helps to describe the system. Common model forms are data-mined clusters, parametric and non-parametric time series models, regressions and their generalisations such as autoregressive moving-average exogenous models, generalised linear modelling, generalised additive modelling, power laws and neural networks.
2 **Deterministic**, process-based or specific theory based models: These model forms are driven by a specific understanding of how the system works and are common in environmental physics and economics. This model form can also include empirical elements where the data are available. For example, the sediment generation component of a catchment model might be based on an empirical relationship, but how the sediment is transported to the stream and routed through the river network may be based on deterministic equations.
3 **Stochastic** models may be based on deterministic or empirical model forms, but the computational process embraces the stochasticity of the input data or model process, allowing for a range of possible outcomes, where those outcomes are governed by a random element. This is usually done by adding a random component to the input data (e.g. state-space models and hidden Markov models).
4 **Conceptual** models. The model abstraction process described above is often manifested in a pictorial diagram describing the system understanding, often termed a conceptual model (or mind map). The links between elements in a conceptual model can be prescribed functional relationships (to allow predictive capacity); empirical and deterministic models may be described in this way. In contrast, Bayesian (decision) networks formulate conceptual models based on the cause–effect interaction between model elements. They are commonly expressed as discrete or continuous probability distributions.
5 **Rule-based** models, e.g. expert systems and decision trees, include heuristic models which are most often captured as qualitative (linguistic) statements of the form 'If X then Y', where X is a large number of conditions that can be tested and Y is the consequence of those conditions being true. Rule-based models can be empirical, such as those developed from a classification tree approach, or deterministic, such as those generated by expert-based judgement of how a system functions.
6 **Agent-based** models are constructed by combining discrete submodels to observe emergent behaviour. For example, a fish population model may be coupled with an algal growth model to allow interaction between the model elements. The responding interaction between these models may generate a response which is not directly predicted by considering the models in isolation.

Applying the taxonomy to NRM

Consider the above taxonomy applied to some simplified models of the response of vegetation to flow (Table 1.2). All three of the models in Table 1.2 have a similar purpose, but different

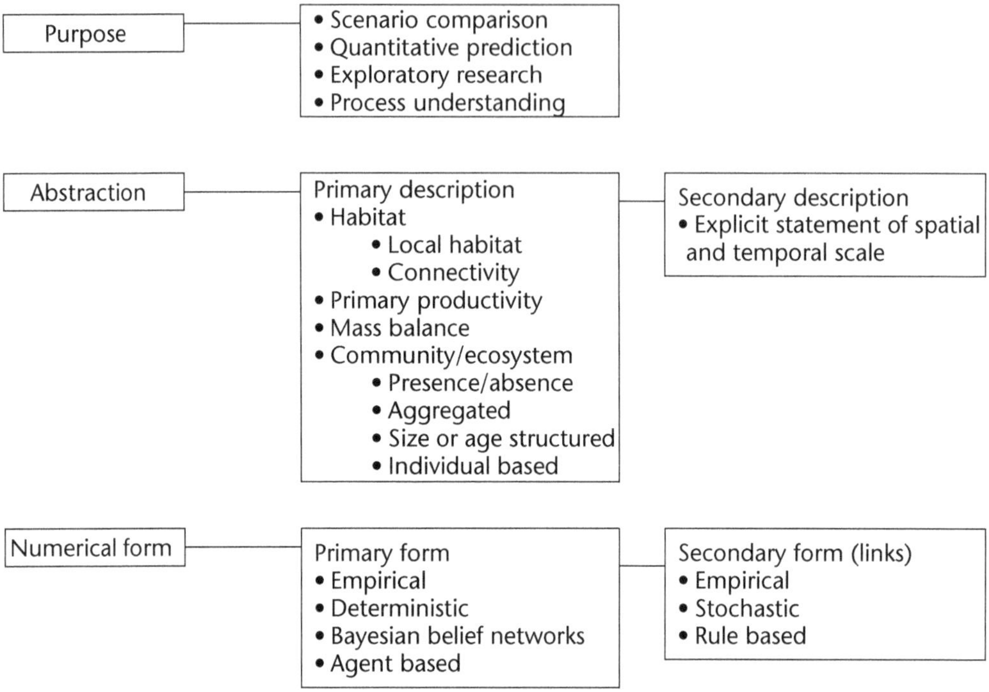

Figure 1.1: Ecological model taxonomy with suggested subcategories to describe a model.

abstractions and numerical form. This leads to the second issue highlighted in the introduction: how do we integrate models with a complementary purpose but different computational form?

The need for integration

The complex nature of NRM requires an ability to draw together knowledge of condition and possible consequences across ecological, social and economic dimensions. The most favourable environmental outcomes are often constrained, an optimal solution is difficult to identify and trade-offs are usually required. This necessitates an integrated decision-making framework. Where decision-making is supported by models, this needs to be supported by an integrated modelling framework. The following section describes the qualities that such an integrated modelling framework needs to provide, using the Ecological Modeller (eMod) software package developed by the eWater Cooperative Research Centre as an example of a partial solution. The literature includes many other examples based on a common numerical model form, such as the general ecosystem model (GEM) (Fitz *et al.* 1996), modular systems such as ECOBAS and LHEM, and tailored systems such as the Australian Environmental Flows Decision Support System (EFDSS) (Young 1999).

An integration framework

eMod is structured around a library of quantitative models to predict ecological response, represented as population change, or habitat response to single or multiple time-variant drivers such as flow or water quality generated by alternate NRM options (Marsh *et al.* 2007). Models are constructed for ecosystem components (species or ecosystem processes) and their attributes, such as life-history stages.

Table 1.2: Example applications of the ecological response model taxonomy for some simplified vegetation response models

1. The similarity to natural flow can be used to assess the likelihood of change in plant assemblage	
Purpose	Scenario comparison: To differentiate between alternative flow scenarios.
Abstraction	Habitat availability: 'The suite of plant species present in a wetland exists in response to the particular water regime that has historically prevailed in that wetland' (McCosker 1998). Measures of deviation of the water regime from the natural case can be used as a measure for comparing alternative flow scenarios.
Numerical form	Deterministic: Analysis of flow data by conversion to summary metrics such as flood frequency, duration and timing where those metrics are derived from expert judgment of likely import components of the flow.
2. Preferred inundation for river red gum	
Purpose	Scenario comparison and quantitative prediction: To differentiate between flow scenarios and to prescribe an optimal scenario.
Abstraction	Habitat availability: Different plant species are present at different places on the floodplain, which can be characterised by different wetting regimes. Therefore, based on a given wetting regime, we can predict the presence or absence of a given species. For example, high-quality river red gum have been observed where the minimum flood duration is one month and the maximum is 18 months, and this occurs seven years in 10 during the June–November period (Green *et al.* 1997).
Numerical form	Empirical: Analysis of flow data by conversion to summary metrics such as flood frequency, duration and timing, where those metrics are derived from a classification of flow events favouring different vegetation classes.
3. Preferred inundation for a single life-history stage (recruitment potential) for a single species	
Purpose	Scenario comparison and quantitative prediction: To differentiate between flow scenarios and to prescribe an optimal scenario.
Abstraction	Ecosystem model: individual-based. The long-term survival of a given species is dependent on a water regime which allows successful recruitment, which in turn requires successful germination. From laboratory experiments we can determine the proportion of successful germinations from different wetting and drying cycles for a given species.
Numerical form	Deterministic: with empirical relationships to quantify relationships. By relating the proportion of successful germinations for different wetting and drying cycles, we can construct a series of empirical relationships which allow us to predict the flow regimes that will provide the opportunities for successful germination and thus recruitment.

The models of ecosystem components require inputs (Figure 1.2) that are a time series of the habitat attributes or stressors for that ecosystem component. For example, a model for a fish may define a particular flow condition for spawning, hence the input would be a flow time series. Any number of different input scenarios can be considered simultaneously. The models are run and the results from all models are aggregated into a score for the scenario, to allow comparison between scenarios.

This general approach is similar to other integrated environmental modelling such as the generalised ecosystem model, ECOBAS and EFDSS. The difference of eMod is that the numerical form of the available model is less restrictive. The only constraints for the model form are

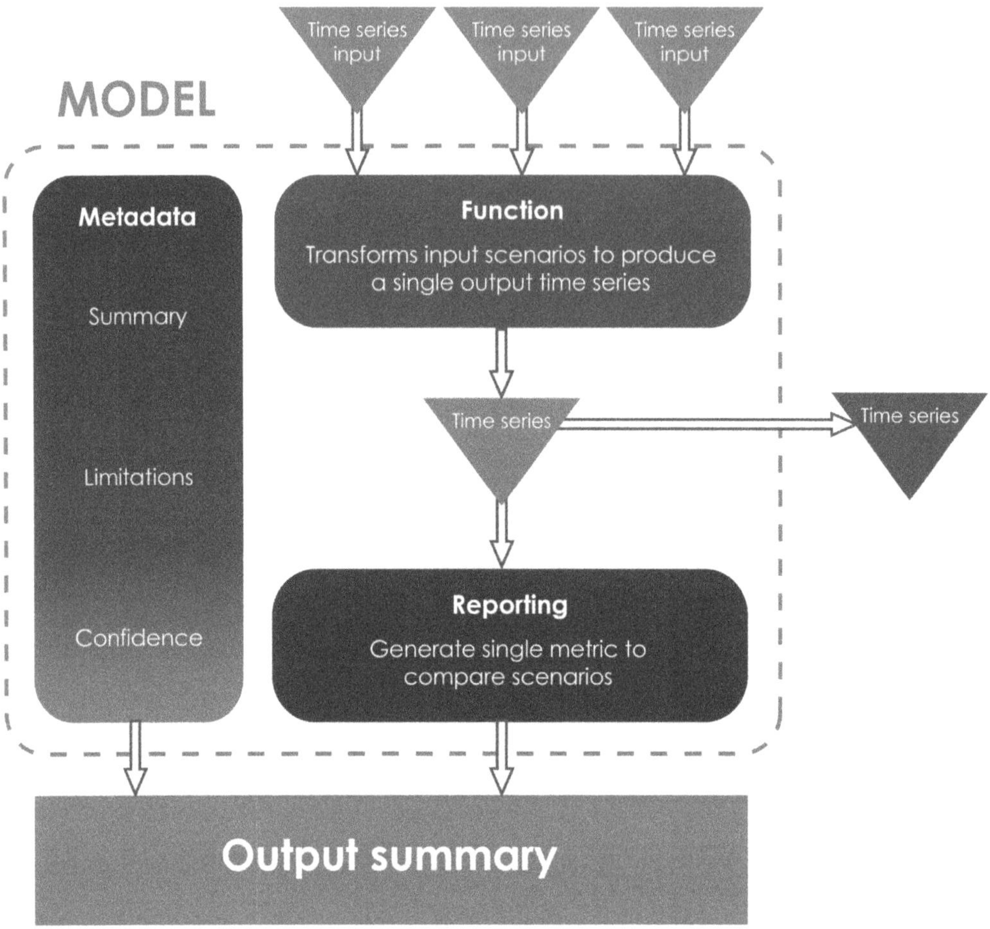

Figure 1.2: An ecological response model (dashed area) has an input of time series that represent key habitat attributes or stressors, meta information to quantify who produced it and the confidence in the model, a function for transforming and combining the input time series into a single output time series (usually a reflection of habitat availability through time), and a method of summarising that time series to produce a single key metric to allow comparison across alternative scenarios.

that they must rely on input data as time series, and produce an output which can be summarised as a score that can be meaningfully compared between scenarios. This comparative analysis for use in scenario assessment means that the absolute predictions are less important than the relative changes between scenarios. The alternative modelling forms, such as stochastic and Bayesian network models, generally produce distributions as outputs. They do not conform to the requirement of an eMod model without further reducing the distribution to a summary value such as a measure of central tendency (mean/median) or distribution (e.g. interquartile range).

The eMod framework provides an interface for entering three numerical functions commonly used in ecological modelling studies to transform input time series to produce an output time series: rating or habitat preference curves; hydraulic rules such as water depth and velocity constraints; and formula-style functions for creating complicated functions and conditional statements. In addition, eMod allows a 'plug-in' so that highbred model forms, uncommon data formats or executable files of models created elsewhere can be handled.

Conclusions

There are many numerical forms of ecological modelling. We have resisted the temptation to provide a long and detailed list of the nuances and limitations of each method, because a good model is one which fits its purpose well. Generally, the simplest implementation that can meet this criteria is best. Rather than consider models according to their numerical form, many of which overlap, describing models primarily by their purpose will increase their opportunity to contribute to NRM. We have presented a simple taxonomy based on the model purpose, its abstraction or how the model simplifies the real world and on its computational form. To adopt an approach whereby the numerical model form is of secondary importance, we need a mechanism that integrates the results of models of different form to inform NRM. We have presented a partial solution whereby the input and output of a model are constrained to a given form but the underlying computational approach can be chosen to suit the purpose, understanding, available data and skills of the model developer.

References

Bovee KD, Lamb BL, Bartholow JM, Stalnaker CB, Taylor J and Henricksen J (1998). *Stream Habitat Analysis using the Instream Flow Incremental Methodology.* USGS.

Cochran M and Ellner S (1992). Simple methods for calculating age-based life history parameters for stage-structured populations. *Ecological Monographs* **62(3)**, 345–364.

Fitz HC, DeBellevue E, Costanza R, Boumans R, Maxwell T, Wainger L and Sklar F (1996). Development of a general ecosystem model for a range of scales and ecosystems. *Ecological Modelling* **88 (1/3)**, 263–295.

Green D, Shaikh M, Maini N, Cross H and Slaven J (1997). *Assessment of Environmental Flow Needs for the Lower Darling River.* NSW Dept of Land and Water Conservation Centre for Natural Resources, Sydney.

Howitt J, Baldwin D, Rees G and Williams J (2007). Modelling blackwater: predicting water quality during flooding of lowland river forests. *Ecological Modelling* **203**, 229–242.

Jackson L, Trebitz A and Cottingham K (2000). An introduction to the practice of ecological modelling. *Bioscience* **V50(8)**, 694–706.

Jakeman A, Letcher R and Norton J (2006). Ten iterative steps in development and evaluation of environmental models. *Environmental Modelling and Software* **21**, 602–614.

Johnson M (2009). Participation and perceptions of watershed modelling. *Society and Natural Resources* **22**, 79–87.

Marsh N, Arene S and Ogden R (2007). Predicting ecological responses for improved stream management. In *Proceedings of the 5th Australian Stream Management Conference. Australian Rivers: Making a Difference.* May 2007, Charles Sturt University, Albury. (Eds A Wilson, R Dehaan, R Watts, K Page, K Bowmer and A Curtis) pp. 247–253.

McCosker R (1998). Methods addressing the flow requirements of wetland, riparian and floodplain vegetation. In *Comparative Evaluation of Environmental Flow Assessment Techniques: Review of Methods.* (Eds AH Arthington and JM Zalucki). LWRRDC Occasional Paper 27/98.

MDBA (2009). The Living Murray Program website. Murray-Darling Basin Authority. Available at http://www.thelivingmurray.mdbc.gov.au. Accessed 15 February 2009.

Pauly D, Christensen V and Walters C (2000). Ecopath, ecosim, and ecospace as tools for evaluating ecosystem impact of fisheries. *ICES Journal of Marine Science* **57**, 697–706.

Refsgaard J, Van der Sluijs J, Holberg A and Vanrolleghem P (2007). Uncertainty in the environmental modelling process: a framework and guidance. *Environmental Modelling and Software* **22**, 1543–1556.

Roberts D (1999) Modelling forest dynamics with vital attributes and fuzzy systems theory. *Ecological Modelling* **90**, 161–173.

Shmueli G and Koppius O (2009). The challenge of prediction in information systems research. Robert H. Smith School Research Paper No. RHS 06-058. Available at SSRN: http://ssrn.com/abstract=1112893. Accessed 10 April 2009.

Young WJ, Cuddy SM, Marston FM and Farley TFN (1999). Environmental Flows Decision Support System – a prototype. CSIRO Land and Water. Technical Report 11/99. CSIRO, Canberra.

Young W, Scott A, Cuddy S and Rennie B (2003). Murray flow assessment tool: a technical description. Client Report, 2003. CSIRO Land and Water, Canberra.

Chapter 2

Measuring ecosystem responses to flow across temporal and spatial scales

Richard T. Kingsford, Kate J. Brandis, Kim M. Jenkins, Lucy C. Nairn and Thomas S. Rayner

Introduction

Flow regime – encompassing magnitude, frequency of occurrence, duration, timing or predictability and rate of change – is the ultimate determinant of the responses of aquatic organisms (Poff *et al.* 1997; Bunn and Arthington 2002; Nilsson and Svedmark 2002). In naturally flowing rivers with little river regulation, organisms respond to the flow regime across the spectrum of temporal and spatial scales, determined by climate and geomorphology (Naiman *et al.* 2002). Patterns of temporal variation ultimately dictate the biotic distribution, abundance and ecological processes of dependent ecosystems – the natural flow-regime paradigm (Richter *et al.* 1996; Poff *et al.* 1997; Lytle and Poff 2004). With river regulation, the flow regime changes significantly, resulting in altered and dampened ecosystem and organism responses. As many of the world's rivers and their dependent wetlands are significantly modified by dams and river regulation (Lemly *et al.* 2000; Vörösmarty *et al.* 2003; Nilsson *et al.* 2005), there are clear priorities to determine ways of better managing rivers for the environment.

Ecological impacts of river regulation are pervasive and widespread across Australia's rivers, predominantly in the south-east of the continent and particularly within the Murray-Darling Basin (Kingsford 2000; Arthington and Pusey 2003). This has affected many plants and animals (Boulton and Brock 1999) and basic ecosystem processes (Walker *et al.* 1995; Ellis *et al.* 1999; Valett *et al.* 2005; Jenkins and Boulton 2007). River regulation and extraction change flows (Bunn and Arthington 2002), typically reducing the frequency of medium and large flows and quantity downriver. The extent of the impact depends on how much water is extracted and its location. This has significantly reduced the inundation frequency and extent of many wetlands (Kingsford 2000; Kingsford and Thomas 2004). As a result, there is widespread recognition of the poor ecological health of Australia's Murray-Darling Basin rivers and wetlands.

There is clear recognition that rivers and their dependent wetlands are legitimate users of water, demanding an environmental flow (Naiman *et al.* 2002; Arthington *et al.* 2006). Governments are now increasing the environmental share of a river's flow, given that much of it is diverted. The Water for the Future program of the Australian Government across the Murray-Darling Basin aims to purchase $3.1 billion worth of water through irrigation licences, one of the largest rehabilitation efforts for rivers in the world (http://www.environment.gov.au/water/action/index.html). Given the social impact of such a policy on the irrigation industry, it is essential to demonstrate good environmental management and outcomes in use of this water. Within the context of wide-ranging effects of river regulation across entire catchments, rehabilitation efforts with increased environmental flows are a significant challenge for management and science. Defined targets will be required, potentially resulting in trade-offs between environmental assets. For example, some parts of a river may not receive environmental flows if a high-priority ecological asset (e.g. Ramsar wetland) on that river takes precedence in demand for water. In all regulated ecosystems, priority ecosystems for management need to be identified and managers need to use ecological indicators to measure the success or otherwise of environmental flow use, ideally applicable at the largest spatial scale.

The environmental flow of a regulated river needs to be managed for the considerable complexity of ecological interactions downstream. In this chapter, we focus on the management of environmental flows in regulated rivers because these are the priority for active management of water usually held in dams. Our approach can equally be used to determine potential impacts of increased river regulation or extraction. Understanding the degree of change to a river's flow and its effect on ecosystems is critical, particularly how changed flow affects surrogate organisms of ecosystem health at different scales. It remains a major constraint on progress in environmental flow management (Richter *et al.* 1996; Bunn and Arthington 2002).

There are two critical objectives for managing a major wetland in a regulated river:

- determine the past and current wetland ecosystem extents and their water needs so that it is clear what can be conserved;
- maximise the environmental outcomes, given the available environmental water.

Large floodplain wetlands usually show the extent of the natural inundation pattern, evident in the boundaries of long-lived perennial flood-dependent trees, except where areas have been cleared for agriculture. These trees are often dead or in poor health (Armstrong 2008). For example, the lower Murrumbidgee wetland still has some black box trees *Eucalyptus largiflorens* alive within forests of dead trees that reflect more than 100 years of river regulation (Kingsford and Thomas 2004). Given a clear definition of which areas can no longer be managed for, the next objective is to maximise environmental outcomes from the remaining flow that goes to the environment. River managers define the component share that goes to the environment as the environmental share, but this represents only the contemporary environmental flow left after regulation and extraction – the remnant environmental flow. In natural free-flowing rivers, all flow has some environmental effect.

Much of the scientific literature on environmental flows and the 200+ different methods available (Tharme 2003) has taken a predominantly hydrological perspective to defining elements of the flow regime for management (Bunn and Arthington 2002; Arthington *et al.* 2006), primarily because empirical data for flow–ecology responses remains poor. Scientific and management consensus that environmental flows need to mimic natural flow variability (magnitude, frequency, timing, duration, rate of change) (Arthington *et al.* 2006) are not always practical because they do not explicitly acknowledge inevitable trade-offs. Further, such approaches do not necessarily link well to ecological outcomes of flow management in terms of species survival, reproduction and recruitment or ecosystem processes in a clear measurable way. A related limitation is that there is usually little spatially explicit connection with the flows and their delivery, which is fundamental for managers.

Establishment of an adaptive environmental management framework (*sensu* Biggs and Rogers 2003; Arthington *et al.* 2006) allow clear identification of objectives relevant to the management of a particular wetland. Once identified, key indicators (e.g. food web links, inundation patterns, fish diversity, waterbird breeding) can be selected and monitored to determine when some threshold is exceeded that might cause irreversible change (Naiman *et al.* 2002) or as a target for rehabilitation. These biological indicators could report on the

Table 2.1: Hierarchical model of structure for rivers, their dependent floodplain wetlands and organism

Scale	Space (m)	Spatial feature	Time	Biological response
Micro	$<10^0$	Individual plants, rocks	Hours/days	Colonisation, migration, feeding behaviour
	10^0–10^1	Snags, plants, riffle zones	Weeks	Reproduction
Meso	10^1–10^2	Habitats, e.g. waterholes, lakes, vegetation types	Months and/ or seasons	Migration, reproduction, feeding behaviour
	10^2–10^5	Sections or reaches: upstream, downstream	1 year	Reproduction
Macro	10^5–10^6	River basin	Decade	Changes in community structure, fluxes of nutrients and energy
	10^6–10^7	Drainage catchment	Century	Life-history survival for long-lived species
	$>10^7$	Continental, intercontinental	Evolutionary	Speciation and extinction, migration

Modified from Walker *et al.* (1995) and Stanley *et al.* (1997).

effectiveness of management of environmental flows to a wetland in a regulated system. If flow–ecology relationships are known and can be modelled, impacts of changing climate or increased environmental flows can also be assessed, allowing forecasting of ecological consequences (Naiman *et al.* 2002).

Despite increased understanding of how flow characteristics vary in rivers across the globe, including highly variable dryland rivers (Puckridge *et al.* 1998; Peel *et al.* 2001; Young and Kingsford 2006), flow–ecology relationships tend to be studied through the lens of a particular species or group of organisms. This simplifies analysis by avoiding the complexity of multiple interactions among organisms. Adding more focus species considerably increases the information demands – species are affected differently by flow regime and may also vary with life-history stages of growth, survival, reproduction or recruitment. Fully understanding how each species copes with naturally variable flow regimes over different temporal and spatial scales is challenging, if not impossible. Yet managers need to integrate this information for rehabilitation of regulated rivers and their dependent aquatic ecosystems, to measure outcomes of their management.

Our aim is to show how management of environmental flows in regulated rivers can be measured, by examining and understanding flow–ecology relationships of a range of organisms and ecosystem processes, across different temporal and spatial scales relevant to environmental flow management and the organisms (Table 2.1). Just knowing how different organisms respond is not sufficient. We need to be able to measure their response, integrate this within an adaptive management context and ultimately predict response in terms of different management regimes or scenarios. We use six major categories of organisms spanning the range of temporal and spatial scales: micro-organisms, plants, fish, waterbirds, other vertebrates and ecological processes (primary and secondary production). We examine how empirical flow–ecology relationships can be determined for different organisms within this context by:

- focusing on the different scales that relate to the range of organisms;
- the management priorities in linking flow to ecological responses;
- how these may be measured at different temporal and spatial scales.

Biotic scales for communities and organisms

The effects of environmental flows vary with the organism and ecological process, from the micro to macro temporal and spatial scales (Table 2.1; Naiman *et al.* 2002). Responses of different organisms will be scale-dependent, but even fine-scale processes depend on broad-scale factors, such as geomorphology and inundation history. These can then be aggregated to the broad scale by spatial modelling. Ultimately this is important, as rehabilitation efforts using environmental flows must be large-scale (Pinay *et al.* 2002), co-operative and implemented through adaptive management frameworks (Poff *et al.* 2003). A challenge is to identify organism responses to flow that can be interpreted at large spatial scales. We have adapted the predominantly geomorphological or fluvial scales (Walker *et al.* 1995; Stanley *et al.* 1997) to reflect scales of organism response.

Micro scales vary from less than 1 m to 10 m and extend from hours to weeks, particularly representing micro and macro-invertebrates, bacteria and some plant communities (e.g. algae) that can rapidly change at these scales. At the meso scale, waterholes, lakes, floodplain vegetation communities and sections of rivers can be defined and temporal changes occur at monthly and annual scales. These capture the direct responses (migration, reproduction and feeding) of fish, waterbirds, frogs and other vertebrates to environmental flows as well as broad patterns of microscopic biota. Finally, the macro scale includes whole catchments and river basins, continents and, for some migratory bird species, intercontinental scales over decades or centuries and longer (Table 2.1). These scales are particularly relevant to waterbirds, but also to fish during large flood events. Such scales are useful conceptually, but in practice organisms interact with many scales.

All organisms respond to or are influenced by temporal and spatial biotic and abiotic factors (Ward *et al.* 1999), operating at micro, meso and macro scales (*sensu* Walker *et al.* 1995; Stanley *et al.* 1997; Table 2.1). For example, waterbirds may nest and roost on individual wetlands, forage for food at several wetlands across a region and migrate seasonally at continental scales to locate suitable habitats (Kingsford and Norman 2002; Roshier *et al.* 2008). Such complexity can confound experimental investigations and exacerbate the already difficult task of isolating species' flow-responses from background ecological variability. This is particularly true when multiple taxa and/or life-history stages are considered together, as ecological communities. Even within a community, patterns and processes of response differ. For example, some waterbirds nest in large colonies while many other species are more secretive, with breeding confined to pairs or where females are primarily responsible for parental care (Kingsford and Norman 2002). As a result, *a priori* identification of biotic components that exhibit their strongest responses to flows at a specific scale can substantially boost investigative power and provide opportunities for predictive power in management of environmental flows. We discuss three scales equally applicable in spatial and temporal dimensions and link them to key responses in groups of aquatic biota (Table 2.1). Such scales have no agreed status but are useful in terms of investigation of environmental flows.

Micro-scale responses occur in habitat patches (<10 m, e.g. fallen timber, plant stands) or within hours, days or weeks. Seedbanks germinate and create heterogeneous mosaics of seedling patches at this scale (Brock *et al.* 2006; Capon and Brock 2006) while micro-organisms such as protozoans, micro-invertebrates (e.g. cladocerans) and low-order plants (e.g. algae) respond by germinating, hatching or growing within hours to days of inundation (Casanova and Brock 1999; Brock *et al.* 2003; Jenkins and Boulton 2003; Porter 2007). Aquatic photosynthesis, respiration and nutrient cycling are triggered when floodwater inundates the stored organic matter in floodplain sediments. Building rapidly, micro-organism numbers and rates of ecosystem metabolism and nutrient turnover fuel the boom in higher-order biota in arid-zone river systems (Bunn *et al.* 2006; Kingsford *et al.* 2006). At the micro-habitat scale, this rapid response to floods is extremely patchy, influenced by the presence of trees or macrophytes, water depth (topography), flow, water quality, duration and flood history (Jenkins and Boulton 2007; Morris 2008). Similarly, micro-habitats influence the short-term movements of high-level organisms. For example, large wading waterbirds feed on the edge of wetlands among emergent macrophytes where invertebrates are available, while pink-eared ducks *Malacorhynchus membranaceus* revolve around each other in pairs with their bills in water as they feed along the concentrated plankton seiches that form in inland lakes. Snags, macrophytes, gravel substrates and patches of slow- versus fast-flowing water form micro-habitats that also influence fish movement.

At meso scales (10–100 000 m, months to a year), the water regime within a floodplain wetland varies with some areas flooding frequently (every year) and others flooding moderately (once every three to 10 years) while high areas flood less frequently (one in 10–20 years). At this scale, clear patterns in vegetation communities emerge with the strong influence of water regime (Brock *et al.* 2006; Capon and Brock 2006). Reeds *Phragmites australis* fringe rivers and can form large beds in terminal wetlands (e.g. Macquarie Marshes, Cumbung Swamp), as does water couch *Paspalum distichum* (Roberts and Marston 1999). On the floodplain, lignum *Muehlenbeckia florulenta* can form extensive stands, providing essential habitat for colonially nesting waterbirds when inundated. Eucalypts (e.g. *Eucalyptus camaldulensis, E. largiflorens, E. coolabah*) dominate floodplain woodlands of the Murray-Darling Basin, with understorey species varying locally (Beadle 1981). These woodlands provide essential habitat to a wide range of fauna and contribute large amounts of organic matter to rivers and wetlands following inundation (Briggs *et al.* 1985).

Different habitats (e.g. lagoons, floodplain, creeks) influence invertebrate and fish communities (Manatunge *et al.* 2000; Sheldon *et al.* 2002; Arthington *et al.* 2005; Jenkins and Boulton 2007). These become connected in time and space at the meso scale by floodwaters that allow fish movements or transport of invertebrates that augment *in situ* recruitment (Jenkins and Boulton 2003; Hughes 2007). When floodwaters arrive, adult fish may move upstream from perennial waterholes into newly inundated habitats to feed or spawn (Arthington *et al.* 2005; Balcombe *et al.* 2005). Newly hatched recruits feed on the abundance and diversity of resources. Freshwater turtles exhibit similar behaviour and life-history traits with expanding habitat and inundation (Kingsford *et al.* 2006). Temporal and habitat factors also interact. For example, invertebrate communities may differ between upstream and downstream reaches due to shifts in season during slow-moving floods (Jenkins and Boulton 2003), while seasonal fluctuations in habitat and food resources can affect the composition of fish communities (Balcombe *et al.* 2005).

Other vertebrate life-histories vary with species across meso scales, with some species completely reliant on permanent water while others are floodplain dwellers. The platypus *Ornithorhynchus anatinus* and water rat *Hydromys chrysogaster* are obligate aquatic mammals of the Murray-Darling Basin rivers. Both species require perennial water and so may be less reliant on floodplain interactions than other species. There are also spatial differences. In the reptiles, for example, turtles may require permanent water while some snakes (e.g. red-bellied black snake, *Pseudechis porphyriacus*) occupy the riparian edge of rivers. Use of wetland habitats by waterbird species can vary among species, but an individual species' use of habitat may also vary with life-history. Some specialist species use particular habitats (e.g. bitterns in reed beds) while many waterbirds (e.g. ducks, piscivores) feed on open water lagoons where foods are accessible (Kingsford and Norman 2002). Specific breeding requirements can also demand particular vegetation characteristics. For example, many colonial waterbird species (e.g. straw-necked ibis *Threskiornis spinicollis*) show high fidelity to breeding areas with dense vegetation, using them each year (Kingsford and Auld 2005). At other times, these birds may use all types of wetlands. Breeding requirement for waterbirds, whether they are colonial or solitary breeders, can be quite specific in terms of habitat, with inundation duration also critically important. From a temporal perspective, waterbirds require flows lasting several months to recruit successfully. Flows which are too short may cause abandonment of eggs or chicks (Kingsford and Auld 2005). Some amphibian species inhabit the floodplains and temporary pools supplied by rivers and creeks, with a life-history that is primarily governed by short-term flooding, particularly in more arid regions. During times of flooding or after widespread rain, large numbers of frogs can emerge from desiccated cocoons (Lee and Mercer 1967; Rayner *et al.* unpubl. data), reproduce and recruit rapidly (Sullivan 1989; Predavec and Dickman 1993). Frog species have varying responses, dependent on the type of wetland and fine-scale habitat attributes. They may also be dependent on mosaics of habitat (Ch. 14, this volume).

Macro-scale responses are widespread and long-lasting, often covering entire catchments and sometimes beyond for waterbirds (Roshier *et al.* 2001, 2008), and spanning several decades. For micro-organisms, critical nutrients and flows provided by upstream processes ensure reproduction, survival and recruitment. Some of these organisms may even move between catchments, transported by waterbirds (Green *et al.* 2008). Freshwater turtles and their biology seem to be intricately tied to the semi-permanent waterholes of desert rivers (e.g. Cooper Creek) but the presence and age of turtles in waterholes across a river may operate at a macro scale. More permanent waterholes are occupied by the older turtles, possibly reflecting relative refugia permanence. Drying of waterholes may cause significant mortality in populations of species that seldom venture onto the expansive floodplains (Georges and Adams 1996; White 2002; Kingsford *et al.* 2006). At the scale of the river, these waterholes may be particularly important, emphasising the importance of not letting them dry out.

Unlike many micro-organisms, vertebrates and many tree species are long-lived and reliant on periodic flooding and drying processes at the macro scale (Fig. 2.1). Within the life-span of most vertebrates, a series of flooding and dry periods would respectively favour reproduction and recruitment and mortality. Some coolibah and black box form similar aged stands of mature trees that ring the edges of small depressions, sometimes forming contours that could indicate recruitment responses to different flood events at the macro scale. Despite their short adult life-cycles, populations of dormant micro-organisms (egg and seedbanks) are also influenced by the cycle of floods and droughts over these long time periods (Fig. 2.1; Jenkins and Boulton 2007).

Changes in flow regime affect organisms at all three scales, in temporal and spatial dimensions. Potentially there is more resilience at the micro scale because of the small areas and rapid generation times. River regulation effects are often most obvious at the macro scale (Kingsford 2000; Nilsson *et al.* 2005), but affect all fine temporal and spatial scales, dampening ecosystem response. Changes to flow regimes and management of environmental flows will necessarily effect changes in the life-histories of the range of organisms and their biotic and abiotic interactions, reflecting the magnitude of regulation. For example, growth rings of large trees provide a sessile record of environmental conditions at a particular point in the landscape, while the health and community structure of flood-dependent forests may convey the ecological legacy of past flow management decisions. Similarly, dormant seeds and eggs decline when dry periods extend beyond 10 years (Capon and Brock 2006; Jenkins and Boulton 2007). Declines in waterbird abundance and species richness in key wetlands may affect broad regional patterns (Kingsford and Thomas 1995, 2004; Kingsford *et al.* 2004; Kingsford and Porter 2009). The effects of reduced breeding opportunities on the relatively few wetlands that support colonial waterbirds (Kingsford and Johnson 1998; Leslie 2001; Kingsford and Auld 2005) may affect species' survival. At evolutionary scales, significant changes caused by climate

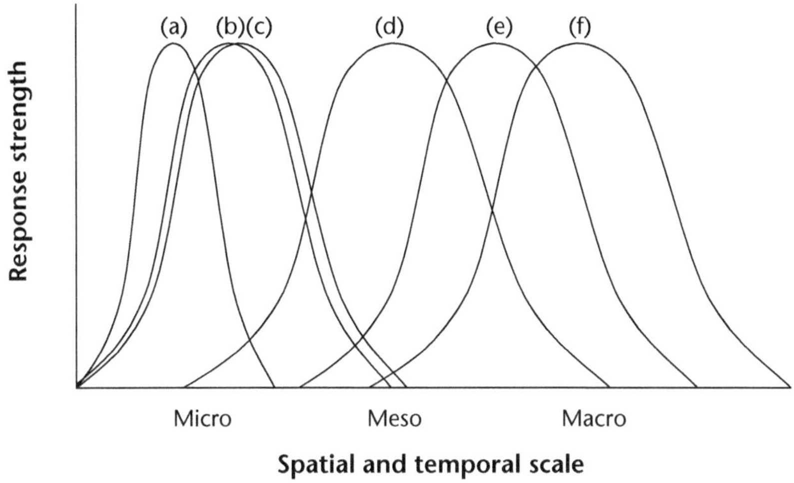

Figure 2.1: Hypothetical representation of the responses of the biotic components of river and floodplain ecosystems to flow regime across micro, meso and macro spatial and temporal scales. Each biotic component typically responds more strongly over a narrow subset of scales: (a) algae germination; (b) invertebrate recolonisation and emergence; (c) aquatic macrophytes; (d) fish recruitment and movement; (e) waterbird breeding and other vertebrate responses; (f) perennial flood-dependent plant species. River regulation and extraction will dampen responses, which will be further mediated by interactions with other ecosystem stressors such as deterioration of catchment health, riparian degradation and loss of connectivity.

or river regulation are likely to be registered in evolutionary processes with changes to species' distribution, potential local extinctions and effects on migration pathways.

Linking organism responses to flow

Knowledge of how organisms and ecological processes respond to flow has increased significantly (Boulton and Brock 1999). Much of our understanding revolves around the micro-temporal dimensions of responses and micro- and meso-spatial scales (when and where species breed, recruit, move or germinate). This growing knowledge contrasts with our relatively poor understanding of explicit spatial frameworks which recognise interactions of landscape elements. Yet the connectivity of riverine ecosystems across meso- to macro-temporal and spatial scales is fundamental to ecological responses (Ward *et al.* 1999; Tockner *et al.* 1999) and of particular importance for management. Understanding how organisms respond to natural flows in unregulated rivers is useful; it can inform management of regulated rivers, particularly in terms of establishing rehabilitation targets.

Managers need to know how much water is required by their aquatic ecosystem, including its components or assets. In highly regulated rivers where much of the flow is stored in dams and has to pass control structures such as weirs, the river has been highly modified and there is no choice but to actively manage the environmental share of a river's flow. The challenge is increasing as governments buy back more water, currently stored in the dams, for the environment. This environmental water has to be managed and released at certain times and in quantities tied to environmental outcomes. The only restrictions are often the volume of environmental flow available and the capacity to release water (e.g. outlet valve capacity on the dam, effects on other users). The decisions on timing and quantity of the release will translate into frequency and duration of inundation for ecosystems downstream.

The position of the wetland manager, responsible for looking after an aquatic ecosystem reliant on environmental flows from a regulated river, can be unenviable. The wetland may be a remnant of the original floodplain (e.g. Macquarie Marshes, Lowbidgee, Gwydir) but the manager may have statutory responsibilities to protect the aquatic assets of the system. There are many wetland protected areas in the major floodplains of the Murray-Darling Basin rivers. A critical knowledge gap is predicting how environmental flows (including the allocated environmental share, dam spills and other unregulated flows) affect ecosystem response in the managed wetland. There are two essential pieces of information for the manager: knowledge of the spatial arrangement of organisms and their responses to flow and subsequent flooding in all its temporal and spatial complexity. In a perfect world, the manager would know how each organism will respond and could integrate those responses to ensure the viability of the entire ecosystem over short- and long-term scales. In reality, the manager has to identify a few key variables to measure, such as flagship or umbrella species and ecosystem responses. The use of flagship or umbrella species is useful but not without dangers (Simberloff 1998). As major floodplains and their organisms are products of their flow history, usually before river regulation, a key question is whether a particular flagship species or group is still viable and can produce self-sustaining populations, given the current environmental flow share.

For example, there may simply not be enough environmental water in a river to sustain a river red gum forest because a substantial component of the flow in the river is now diverted for human use, such as irrigation. There is little point in managers wasting their time trying to manage for the original extent of this organism – there is insufficient environmental water. The challenge is to determine how much of particular communities can be managed for. A management response would be to point out that the particular red gum forest will not survive, or it might survive but not reproduce, given the current environmental water share. If the viability criteria are met, the manager needs to decide how to best manage the available environmental flow share. The importance of understanding how organism scales affect the decision about whether an organism is viable or, given it is, how to manage for it, is critical. This can be important at different temporal and spatial scales for all organisms. It may be just as important to know whether an *E. coolabah* forest will survive the next 100 years as whether sufficient macro-invertebrate species reproduce in response to a single flood. The former provides organic material and habitat for many organisms, the latter may contribute to the food base for many different vertebrates and invertebrates.

The challenge for managers of environmental flows is to integrate information across scales. Understanding flow history, current flows and inundation patterns is essential. The manager needs to know the extent of a system that can be managed for, given current environmental flow. As many long-lived species, such as floodplain eucalypts, may still be extant it may be necessary to compare natural and regulated flow regimes to make decisions about viability. The manager also needs to know the current flow and inundation patterns. Spatial modelling of flow and inundation linked to organism responses at different temporal and spatial scales can guide the manager. This would necessitate linking flow and inundation modelling (Ren *et al.* 2009) to build an explicit spatial model as well as a temporal model using inundation data (Ch. 9, this volume). Spatial integration of organism responses linked to flow and inundation modelling would allow modelling of different responses to flow by indicators or key organisms identified as the focus for monitoring. Most importantly, this spatial modelling would allow testing of different scenarios of increased or decreased environmental flows and the likely long-term impacts on different organisms. Such scenario testing would allow managers a way to assess different options for release of an environmental flow. This could involve assessment of different flow release options (rules-based versus actively pulsed flows) and the opportunity costs, in terms of storage of flows, for delayed environmental flow release with increased volume.

Measuring responses to flows at different scales

Flow governs biotic interactions at different spatial and temporal scales. Flow regimes dampened by river regulation, diversions, earthworks or climate change will still produce biotic responses that reflect the quantum of change. Conversely, rehabilitation efforts involving environmental flow require some understanding and measurement of organism responses, particularly if management is to learn from the process of environmental flow delivery. This is essential. Environmental flows need to be managed because they are held in storage. Decision frameworks are needed to govern the release of the water to produce environmental outcomes. This much is usually agreed, but the measurement of environmental outcomes is seldom implemented at the appropriate organism scales that allow for measurement of an ecosystem response. This does not mean everything has to be measured.

Environmental flows produce responses over micro and meso temporal scales. These include raising river levels and inundating wetlands for weeks to months, which result in biotic responses. Measurement of responses can include short-term localised studies of habitat use combined with long-term programs to determine changes in population structure. Identifying and addressing clear questions is critical (Lake 2001; Palmer and Bernhardt 2005). This must be followed by evaluation of success or failure, with explicit thresholds for management (Biggs and Rogers 2003). We have outlined vital issues, for key processes and each biotic group, that managers need to know to evaluate the ecological response to flows at different scales. We describe an array of monitoring and experimental approaches to measure the targeted responses and highlight knowledge gaps that hinder management decisions on the location, timing and magnitude of environmental water allocations.

Processes

Processes refer to rates of basic ecosystem functions such as decomposition, photosynthesis, recruitment and migration (Boulton and Brock 1999). These define the supply and movement of energy through aquatic food webs and underpin the boom and bust cycles in arid-zone floodplain wetlands (Boon 2006; Bunn *et al.* 2006). Their relevance to environmental flows lies in their capacity to synthesise impacts across entire ecosystems. For example, an environmental flow in winter may trigger slower rates of primary production and decomposition than in spring–summer, resulting in few nutrients to support microbes, invertebrates and waterbirds. In heterotrophic wetlands, high rates of oxygen consumption by micro-organisms create anaerobic (black water) conditions that slow rates of decomposition (Gehrke *et al.* 1995; Vallet *et al.* 2005) and adversely affect microbes (Boon 2006), invertebrates (Benke 2001) and fish (Gehrke *et al.* 1995). Understanding the role of ecosystem processes could avoid negative impacts on biota when designing and allocating the timing, frequency and duration of environmental flows.

Ecosystem processes in wetlands are relatively easy to measure. Metabolism (primary production and community respiration) can be measured using the change in dissolved oxygen over time. It provides information about wetland function across relatively broad spatial scales (Bunn *et al.* 2006; Jenkins 2006), but has also been adapted for experiments at fine spatial scales (Bunn *et al.* 2003; Fellows *et al.* 2006). Decomposition can be estimated via experiments using leaf-packs placed in creeks and floodplain habitats before and after floods to examine how the breakdown rates of a range of organic matter sources vary in response to environmental flows (Boulton and Boon 1991). Isotopic labels can provide information on the assimilation of these different sources of carbon into aquatic food webs (Whitcraft *et al.* 2008). Experiments that unravel the mechanisms linking different components of aquatic food webs are lacking for arid-zone wetlands.

When floodwaters inundate the dry floodplain of arid-zone wetlands, rates of carbon metabolism are orders of magnitude higher than those recorded in main channel habitats during dry periods (Bunn *et al.* 2006). Booms in fish and waterbird recruitment follow these high rates of decomposition, photosynthesis and invertebrate colonisation (Kingsford *et al.* 1999; Puckridge *et al.* 2000; Jenkins and Boulton 2003). At micro scales, BACI experiments can assess if environmental flows and floodplain inundation increase the rates of metabolism, migration and recruitment. These data can help explain observed structural patterns in invertebrate, waterbird, fish and vegetation communities and enable predictions on viability. For example, the poor condition of fish communities in the regulated Campaspe River was reflected in low rates of recruitment in native species compared to those in a nearby unregulated river (Humphries *et al.* 1999). Experiments that incorporate meso-scale factors such as vegetation community, wetland type or flood history (i.e. time since last flood) can provide insights into the variable response of processes to environmental flows. For example, the loss of carbon as dry periods extend may fuel varying rates of metabolism (Jenkins *et al.* 2005) that affect higher-order biota after an environmental flow.

Invertebrates and microbes

Although the response of micro-organisms to environmental flows is interesting for their inherent and biodiversity values, their role in aquatic food webs is critical. Primarily, managers want to know whether inundated wetlands are colonised at densities that will support waterbirds and fish, ideally through breeding events. For example, densities of microinvertebrates (35–250 mm, particularly cladocerans) below 100 L^{-1} do not support recruiting fish (King 2004). We know little of density thresholds of micro-invertebrates that sustain filter-feeding waterbirds such as pink-eared ducks, nor the larger macro-invertebrates (>250 mm) that are eaten by wading waterbirds, ducks and large-bodied adult fish (Boulton *et al.* 2006). Experiments and/or sampling are needed to assess the condition of waterbirds and fish through maintenance and breeding phases in relation to availability of varying prey densities.

To measure if micro-invertebrate responses match vertebrate food requirements, organisms need to be sampled within two weeks of inundation and at monthly intervals for the duration of the waterbird or fish recruitment season (up to five months). There is considerable knowledge of short-term (days to weeks, see Tan and Shiel 1993; Jenkins and Boulton 2003) invertebrate responses to flooding, but little of how the initial pulses vary across a flood event (Jenkins *et al.* 2005). The spatial scale of monitoring should ideally match the foraging locations of the predators, although knowledge of principal habitats used by predators is poor. There are relevant questions at multiple spatial scales. The micro-habitats of foraging waterbirds and fish could be sampled for densities and diversity of micro-organisms (e.g. shallow littoral habitats, around snags, macrophytes). At a meso scale, the key objective should be to determine the relative value of different feeding habitats such as lagoons, channels, red gum floodplain, open water, couch floodplain and reed beds, in supporting varying sized populations of waterbird, fish, amphibian or reptile communities. There is little understanding in Australia of relationships between micro-organisms and wetland and vegetation type.

At a macro-habitat scale, the challenge is understanding how invertebrate communities respond across an entire river catchment with varying habitats, lentic versus lotic flows and water regimes. Density and diversity of invertebrates available to sustain waterbirds and fish falls in microcosm studies as the time extends between flooding (Boulton and Lloyd 1992; Jenkins and Boulton 2007). In line with this relationship, wetlands that naturally flooded frequently (e.g. the Macquarie Marshes) support micro-invertebrate communities with higher densities and diversity than lakes on the Darling River floodplain that flood less often (Jenkins and Boulton 2003; Morris 2008). Invertebrate densities are also lower in regulated channels than in temporary channels and floodplain (Jenkins *et al.* 2008), likely explaining why water-bird populations decline when temporary wetlands are permanently inundated (Kingsford *et al.* 2004). Managers need to predict and measure these responses when allocating water to wetlands that have experienced different flow regimes across meso- and macro-time scales to maximise the productivity of wetlands. Experimental approaches could be used to assess if colonisation pathways are operating for micro-organisms transported by floodwaters at large spatial scales (Jenkins and Boulton 2003). Although we do not know all the scales across which micro-organisms connect in a system, genetic techniques offer the potential to identify these patterns (Hughes 2007).

Fish

Freshwater fish are found in almost all riverine and wetland environments. This ubiquity, combined with an absolute dependence on aquatic habitats, an inability to senesce through dry periods and strong trophic links with other taxa, makes fish a unique biotic component for the measurement of ecosystem responses to flows. Fish are particularly useful for investigating meso-scale spatial and micro- to meso-scale temporal dynamics (Table 2.1), as these can be readily quantified through field surveys and experiments. For example, analyses of fish community structure may reveal recent patterns of hydrological connectivity (Arthington *et al.* 2005). At macro scales (e.g. entire catchments) fish–flow relationships can be confounded by factors such as longitudinal gradients, seasonal fluctuations in environmental conditions and the presence of instream barriers to fish movement (Growns 2008). Genetic analysis has unravelled patterns of connectivity in fish populations that relate to flow at multiple scales (Hughes 2007).

Most freshwater fish species are relatively widespread in their distributions, enabling comparison of flow responses within and between rivers. In the Murray-Darling Basin, for example, data from unregulated subcatchments provide a good reference framework for investigations in adjacent regulated subcatchments (Gehrke and Harris 2001; Davies *et al.* 2008). Wherever fish populations exist, they respond strongly to increased flows by exploiting newly inundated habitats (Rayner *et al.* 2009). Large floodplains are particularly productive habitats. For example, on Cooper Creek, more than 90% of native species (10 of 11) were found on the floodplain, including larval fish of six species which exhibited significant growth in these productive habitats (Balcombe *et al.* 2007). Flooding also shifted the diets of adult fish (Balcombe *et al.* 2005) and allowed for biomass transfer between floodplains and deep waterholes, known to be important refuge sites for fish during dry periods (Arthington *et al.* 2005; Balcombe *et al.* 2006).

In regulated river systems, fish responses to flows are dampened by the interactive effects of reduced health of the fish community and exacerbation of extreme conditions (particularly droughts) by water extraction and anthropogenic habitat modification (Gehrke and Harris 2001). Humphries *et al.* (1999) hypothesised that native Australian fish species may recruit successfully in non-flood years, provided within-channel flow pulses stimulate spawning and provide suitable habitats for the retention and growth of juveniles. However, this research was primarily in the River Murray, which is relatively degraded compared to systems such as Cooper Creek. This raises the question of whether such dynamics represent only a subset of the total potential flow–response spectrum. Even in highly regulated rivers, there may be strong recruitment of some native fish species (King *et al.* 2009). Similarly, we demonstrated dispersal of fish from isolated refugia and evidence of recruitment in most species following a 23 000 ML environmental flow in the Macquarie Marshes, although the specific roles of floodplain resources (habitats and foods) were difficult to elucidate (Rayner *et al.* 2009). In the unregulated Cooper Creek, fish recruit during low flows (A. Kerezy, *pers. comm.*) but only during floods do populations boom (Puckridge *et al.* 2000). Appropriate management of low flows and floods, using environmental allocations, will be critical to the recovery of fish communities in regulated rivers.

Fish are a strong indicator of ecosystem responses to flows, with the maintenance of natural flow regimes central to their conservation (Arthington *et al.* 2006). While evidence linking fish to floodplains in unregulated systems is emerging, these links require more attention in regulated systems. For example, the inundation of shallow floodplain areas over meso-scale time frames (weeks to months) could disproportionately advantage alien species if conditions remain unsuitable for more sensitive native species (Rayner *et al.* 2009). We need to develop threshold criteria to guide the delivery of environmental flows to priority habitats (e.g. floodplain wetlands) for native fish species in regulated rivers, supported by experimental approaches at micro to meso scales.

Fish biologists often conceptualise the factors determining fish community structure as a series of hierarchical filters which limit the available species pool at increasingly fine spatial and temporal scales (Poff *et al.* 1997; Chessman 2006). For example, macro-scale catchment characteristics (e.g. presence of instream barriers to fish movement) may determine the species present at a site, before local habitat conditions determine relative abundance (Pusey *et al.* 1995; Pusey and Kennard 1996). This model provides a useful framework for designing flow–response monitoring programs because it allows a focus on a specific spatial or temporal scale, while remaining cognisant of factors operating at other scales. Three key investigative techniques are relevant to the micro, meso and macro scales: laboratory or field-based experiments, fine- and broad-scale monitoring programs, and predictive modelling.

Laboratory experiments provide information on the specific mechanisms governing fish responses to flows, such as swimming behaviour (Mallen-Cooper 1994; Lyon *et al.* 2007), water quality tolerances and feeding, growth and age relationships (Anderson *et al.* 1992). *In situ* manipulative and Before-After-Control-Impact experiments provide a rigorous foundation for fine-scale monitoring programs by extending work conducted in the lab (Schlosser 1998). Ideal control sites are seldom available, so less-structured comparative approaches are needed (Gehrke *et al.* 1999). For example, Astles *et al.* (2003) constructed pens adjacent to the Macquarie River, downstream from Burrendong Dam, to examine the potential impacts of cold water pollution on native fish. They used the unregulated Bogan River as a reference.

Monitoring programs can measure responses of fish communities to environmental changes including improved flow delivery and changes in water quality, habitat structure and food availability. They can measure biotic interactions that show the community structure, growth, feeding and recruitment responses of fish to environmental flows. Meso-scale monitoring programs provide cost-effective snapshots of condition which can be repeated using standardised methodology (Davies *et al.* 2008), typically over 10–40 sites, and related to individual flow events. Where appropriate, techniques requiring additional equipment and expertise can be added, for example otolith microchemistry and stable isotope analysis (Dufour and Gerdeaux 2001; Bunn *et al.* 2003), radio-tracking (Crook 2004; O'Connor *et al.* 2005) and larvae surveys (King *et al.* 2003a). This scale is also appropriate for community-supported research, such as fish tagging by local fishing clubs.

At the macro scale, broad long-term programs can be used to identify fish population responses to scenario-based environmental flow management changes (King *et al.* 2003b). This may involve linking a series of environmental flows over several years to response variables such as native:pest species ratio or the recovery of a threatened species. At large scales, predictions of declines in extant wild stocks provide information for complementary management strategies (e.g. restocking) and examine potential impacts of climate change.

Vegetation

At a macro scale, woodlands decline or terrestrial plants invade formerly aquatic habitat; at a micro scale individual dying trees shift community structure and soil–nutrient dynamics. Plants (e.g. algae) whose life-histories occupy fine temporal and spatial scales respond immediately to flow, in contrast to the long slow decline in populations of black box when inundation intervals stretch from every five to 12 or more years (Jolly 1996; Sharley and Huggan 1995).

Techniques for measuring plant responses to environmental flows range from remote sensing of vegetation cover, health and vigour (Cunningham *et al.* 2006; Stone and Haywood 2006) to measuring water potential in individual trees (Bacon *et al.* 1993). Changes in plant distribution over macro scales can be mapped using historic and current aerial photography, or by comparing maps made at different times and interrogating these in relation to flow regimes. Remote sensing needs to be related to field-based data for calibration and accuracy assessment (Stone and Haywood 2006; Ch. 9, this volume).

Field data inform on micro and meso responses to flow, such as changes in local distribution and vegetation health, including the regenerative potential and demographics of populations. Repeated sampling following inundation provides insight into succession, capturing temporal variation in the response of seedbanks and standing vegetation. This allows a more accurate assessment of the success of an environmental flow release. For instance, macrophyte seeds may be triggered to germinate by a short flood event, with a response seen within weeks, but a succession of small floods may deplete the seedbank by not providing enough water to allow flowering and setting of seed (Brock *et al.* 2003; Brock and Rogers 1998). Similarly, if the management goal is to maintain the overstorey of river red gum, small floods may be sufficient in the short term but longer periods of repeated flooding may be necessary for successful recruitment (George *et al.* 2005; Roberts and Marston 2000).

An experimental approach can help predict responses of vegetation to flows at micro scales. Flooding seedbank samples can provide insight into the likely effects of different flow parameters, including depth and duration of inundation (Capon and Brock 2006; Casanova and Brock 2000). Similarly, growth experiments can determine the optimum water regimes for different plant taxa. These can be performed by simulating floods in a laboratory or greenhouse setting or by measuring the response of individuals or communities to controlled flooding in the field (Bacon *et al.* 1993). Combined methods in the laboratory, field sampling and remote sensing can all be used to determine effects of environmental flow on vegetation response.

Waterbirds

Waterbirds use the landscape at a range of different spatial and temporal scales. They use individual wetlands for nesting and roosting sites, they forage regionally moving between wetlands, and at a continental scale they travel long distances in search of suitable habitat (McKilligan 1975; Roshier *et al.* 2001, 2008). Temporal variability in their use of wetlands is high and linked to river flows and flooding (Kingsford and Auld 2005). When conditions are suitable waterbirds, particularly colonially breeding waterbirds, can respond in spectacular ways with tens to hundreds of thousands of individuals forming a single breeding colony (Kingsford *et al.* 2008). While the inundation of a wetland may occur on a temporal meso scale (Fig. 2.1), it provides a critical opportunity for breeding and the long-term (macro-scale) persistence of waterbird populations.

In addition to the temporal and spatial scales at which waterbirds use habitats in the landscape, different species of waterbirds use specific habitats at different intensities. For breeding habitats, some species nest individually or in small groups (cormorants, ducks), other species nest in colonies of hundreds of thousands of individuals (ibis, pelicans) and some species nest in small groups or in colonies depending upon conditions (spoonbills). During non-breeding periods colonial waterbirds disperse and use habitats at a lower intensity. The intensity at which waterbirds use the habitat is linked to the life traits of the species.

Drivers of waterbird response include size of the flood, duration of inundation, depth of inundation, availability of suitable nesting habitats and food resources. Measuring these drivers can help identify their role in influencing waterbird breeding success. Flooding can occur at a range of spatial scales from meso to macro scale, while duration of inundation is measured on a temporal meso scale.

Measuring waterbird response can be achieved at a range of temporal scales from the micro to the macro. At the micro temporal scale waterbird response can be measured through reproductive success studies and examining chick survival rates on a daily basis (Mayfield 1975). At the meso temporal scale, colony success can be assessed in relation to changes in inundation characteristics (Kingsford *et al.* 2008) and location and populations can be measured on an annual basis (Porter *et al.* 2006). The continued surveying of waterbird populations at the temporal meso scale can provide a measure of population fluctuations at the macro scale (Porter *et al.* 2006).

Habitat use by waterbirds can be also measured on a spatial scale from the micro scale to the macro. Movements of individual birds can be tracked using satellite tracking (Roshier and Asmus 2009) and bird banding (ABBBS 2009). Tracking bird movements can provide information on their movements at micro, meso and macro spatial scales.

Other vertebrates

Life-histories of organisms and their resilience are first approached by determining the distribution and abundance of populations and their ability to successfully reproduce and recruit. The information for reptiles, amphibians, birds other than waterbirds and mammals remains poor (Kingsford *et al.* 2006). The first and most important objective is to define the distribution and abundance of a species. Current distribution and abundance of groups of species could be effectively measured. Standard weight and length measurements are important as they may be correlated to age and habitat suitability. Greater understanding of the relative importance of different aquatic habitats (at the meso scale) to a species is also critical, particularly for long-lived species such as frogs. Understanding the extent of movements will help define the framework of habitat that is important for the management of environmental flows. Categorising different habitats on the basis of their flow regime can help identify whether some vertebrate species depend on particular habitats more than others. Linking habitat use of different vertebrates to an understanding of different floodplains and their flow regimes is essential to further understanding of the management of environmental flows and likely successful outcomes. A pragmatic approach would be to identify target species or groups and determine their distribution and abundance in relation to flow regime and inundation patterns, and possibly infer flow–ecology relationships. More detailed studies could follow.

Conclusions

The ecological health of rivers and their dependent wetlands is in decline all over the world. Increasingly, communities are demanding that such systems be rehabilitated, by the recovery of water that is currently extracted. Within the Murray-Darling Basin, a current policy and management focus is the buy-back of some irrigation licences to increase the environmental flow to high conservation value wetlands. The challenge for managers is to manage environmental flow strategically. This demands informed decision-making about the best timing and quantity for release. Governments have generally opted for mimicking natural flows by passing flows through the dam wall according to a simplistic set of rules which largely ignore the complex interactions and functional dependencies of organisms (Arthington *et al.* 2006). This often means that flows remain within the main channel of the river, with little opportunity to inundate wetlands which are often the target for environmental flows. Major wetlands require large-scale inundation events, managed or not. It may actually be detrimental to use a set of rules, and not the most effective way to use environmental water (Kingsford and Auld 2005). The challenge for the science of environmental flows is that a flow must not only be good and useful but also be used (Naiman *et al.* 2002). This demands considerable engagement by scientists and managers, where managers can see and measure the results of their flow management and predict future scenarios in the inevitable dilemma of making trade-offs in space (different environmental assets) or time (different years).

Acknowledgments

We thank Angela Arthington for her incisive constructive comments on an early draft, which considerably improved the chapter. Funding support for the research discussed in this chapter comes from the University of New South Wales, the Australian Research Council (LP0453694 and LP0884160), the NSW Environmental Trust (RD0103 and RD009), Land and Water Australia, Natural Heritage Trust, Department of Environment and Climate Change and Department of Water and Energy.

References

ABBBS (2009). Australian Bird and Bat Banding Scheme. Dept of Environment, Water, Heritage, Environment and the Arts. Available at http://www.environment.gov.au/biodiversity/science/abbbs/index.html. Accessed 25 February 2009.

Anderson JR, Morison AK and Ray DJ (1992). Age and growth of Murray cod, *Maccullochella peelii* (Perciformes: Percichthyidae), in the lower Murray-Darling Basin, Australia, from thin-sectioned otoliths. *Marine and Freshwater Research* **43**, 983–1013.

Armstrong J (2008). The effect of river regulation on red gum (*Eucalyptus camaldulensis*) swamps in the Booligal Wetlands, Lachlan River. Honours thesis. University of New South Wales, Sydney.

Arthington AH and Pusey BJ (2003). Flow restoration and protection in Australian rivers. *River Research and Applications* **19**, 377–395.

Arthington AH, Balcombe SR, Wilson GA, Thoms MC and Marshall J (2005). Spatial and temporal variation in fish-assemblage structure in isolated waterholes during the 2001 dry season of an arid-zone floodplain river, Cooper Creek, Australia. *Marine and Freshwater Research* **56**, 25–35.

Arthington AH, Bunn SE, LeRoy Poff N and Naiman RJ (2006). The challenge of providing environmental flow rules to sustain river ecosystems. *Ecological Applications* **16**, 1311–1318.

Astles KL, Winstanley RK, Harris JH and Gehrke PC (2003). Experimental study of the effects of cold water pollution on native fish. NSW Fisheries Office of Conservation, Sydney.

Bacon PE, Stone C, Binns DL, Leslie DJ and Edwards DW (1993). Relationships between water availability and *Eucalyptus camaldulensis* growth in a riparian forest. *Journal of Hydrology* **150**, 541–561.

Balcombe SR, Bunn SE, McKenzie-Smith FJ and Davies PM (2005). Variability of fish diets between dry and flood periods in an arid zone floodplain river. *Journal of Fish Biology* **67**, 1552–1567.

Balcombe SR, Arthington AH, Foster ND, Thoms MC, Wilson GG and Bunn SE (2006). Fish assemblages of an Australian dryland river: abundance, assemblage structure and recruitment patterns in the Warrego River, Murray-Darling Basin. *Marine and Freshwater Research* **57**, 619–633.

Balcombe SR, Bunn SE, Arthington AH, Fawcett JH, McKenzie-Smith FJ and Wright A (2007). Fish larvae, growth and biomass relationships in an Australian arid zone river: links between floodplains and waterholes. *Freshwater Biology* **52**, 2385–2398.

Beadle NCW (1981). *The Vegetation of Australia*. Cambridge University Press, Cambridge.

Benke AC (2001). Importance of flood regime to invertebrate habitat in an unregulated river-floodplain ecosystem. *Journal of the North American Benthological Society* **20**, 225–240.

Biggs HC and Rogers KH (2003). An adaptive system to link science, monitoring, and management in practice. In *The Kruger Experience. Ecology and Management of Savanna Heterogeneity.* (Eds JT du Toit, KH Rogers and HC Biggs) pp. 59–80. Island Press, Washington DC.

Boon PI (2006). Biogeochemistry and bacterial ecology of hydrologically dynamic wetlands. In *Ecology of Freshwater and Estuarine Wetlands.* (Eds DP Batzer and RR Shavitz) pp. 115–176. University of California Press, Berkeley.

Boulton AJ and Boon PI (1991). A review of methodology used to measure leaf litter decomposition in lotic environments: time to turn over an old leaf. *Australian Journal of Marine and Freshwater Research* **42**, 1–43.

Boulton AJ and Brock MA (1999). *Australian Freshwater Ecology: Processes and Management.* Gleneagles Publishing, Adelaide.

Boulton AJ and Lloyd LN (1992). Flooding frequency and invertebrate emergence from dry floodplain sediments of the River Murray, Australia. *Regulated Rivers: Research and Management* **7**, 137–151.

Boulton AJ, Sheldon F and Jenkins KM (2006). Natural disturbance and aquatic invertebrates in desert rivers. In *Ecology of Desert Rivers.* (Ed. RT Kingsford) pp. 133–153. Cambridge University Press, Cambridge.

Briggs SV, Maher MT and Carpenter SM (1985). Limnological studies of waterfowl habitat in south-western New South Wales. I. Water chemistry. *Australian Journal of Marine and Freshwater Research* **36**, 59–67.

Brock MA and Rogers KH (1998). The regeneration potential of the seed bank of an ephemeral floodplain in South Africa. *Aquatic Botany* **61**, 123–135.

Brock MA, Nielsen DL, Shiel RJ, Green JD and Langley JD (2003). Drought and aquatic community resilience: the role of eggs and seeds in sediments of temporary wetlands. *Freshwater Biology* **48**, 1–12.

Brock MA, Capon SJ and Porter JL (2006). Disturbance of plant communities dependent on desert rivers. In Ecology of Desert Rivers. (Ed. RT Kingsford) pp. 100–132. Cambridge University Press, Cambridge.

Bunn SE and Arthington AH (2002). Basic principles and ecological consequences of altered flow regimes for aquatic biodiversity. *Environmental Management* **30**, 492–507.

Bunn SE, Davies PM and Winning M (2003). Sources of organic carbon supporting the food web of an arid zone floodplain river. *Freshwater Biology* **48**, 619–635.

Bunn SE, Balcombe SR, Davies PM, Fellows CS and McKenzie-Smith FJ (2006). Aquatic productivity and food webs of desert river ecosystems. In *Ecology of Desert Rivers.* (Ed. RT Kingsford) pp. 76–99. Cambridge University Press, Cambridge.

Capon SJ and Brock MA (2006). Flooding, soil seed bank dynamics and vegetation resilience of a hydrologically variable desert floodplain. *Freshwater Biology* **51**, 206–223.

Casanova MT and Brock MA (1999). Life histories of charophytes from permanent and temporary wetlands in eastern Australia. *Australian Journal of Botany* **47**, 383–397.

Casanova MT and Brock MA (2000). How do the depth, duration and frequency of flooding influence the establishment of wetland plant communities? *Plant Ecology* **147**, 237–250.

Chessman B (2006). Prediction of riverine fish assemblages through the concept of environmental filters. *Marine and Freshwater Research* **57**, 601–609.

Crook DA (2004). Is the home range concept compatible with the movements of two species of lowland river fish? *Journal of Animal Ecology* **73**, 353–366.

Cunningham SC, MacNally R, White M, Read J, Baker PJ, Thomson J and Griffioen P (2006). Mapping the current condition of river red gum (*Eucalyptus camaldulensis* Dehnh.) stands along the Victoria Murray River floodplain. Report to the Northern Victorian Catchment Management Authorities/Dept of Sustainability and Environment, Melbourne.

Davies P, Harris JH, Hillman TJ and Walker KF (2008). SRA Report 1: Report on the Ecological Health of Rivers in the Murray-Darling Basin, 2004–2007. Prepared by the Independent Sustainable Rivers Audit Group for the Murray-Darling Basin Ministerial Council. Murray-Darling Basin Commission, Canberra.

Dufour E and Gerdeaux D (2001). Contribution of stable isotopes to fish ecological studies. *Cybium* **25**, 369–382.

Ellis LM, Molles Jr MC and Crawford CS (1999). Influence of experimental flooding on litter dynamics in a Rio Grande riparian forest, New Mexico. *Restoration Ecology* **7**, 193–204.

Fellows CS, Clapcott JE, Udy JW, Bunn SE, Harch BD, Smith MJ and Davies PM (2006). Benthic metabolism as an indicator of stream ecosystem health. *Hydrobiologia* **572**, 71–87.

Gehrke PC and Harris JH (2001). Regional-scale effects of flow regulation on lowland riverine fish communities in New South Wales, Australia. *Regulated Rivers: Research and Management* **17**, 369–391.

Gehrke PC, Brown P, Schiller CB, Moffatt DB and Bruce AM (1995). River regulation and fish communities in the Murray-Darling River system, Australia. *Regulated Rivers: Research and Management* **11**, 363–375.

Gehrke PC, Astles KL and Harris JH (1999). Within-catchment effects of flow alteration on fish assemblages in the Hawkesbury-Nepean River system Australia. *Regulated Rivers: Research and Management* **15**, 181–198.

George AK, Walker KF and Lewis ML (2005). Population status of Eucalypt trees on the River Murray floodplain, South Australia. *River Research and Applications* **21**, 271–282.

Georges A and Adams M (1996). Electrophoretic delineation of species boundaries within the short-necked chelid turtles of Australia. *Zoological Journal of the Linnean Society, London* **118**, 241–260.

Green AJ, Jenkins KM, Bell D, Morris PJ and Kingsford RT (2008). The potential role of waterbirds in dispersing invertebrates and plants in arid Australia. *Freshwater Biology* **53**, 380–392.

Growns I (2008). The influence of changes to river hydrology on freshwater fish in regulated rivers of the Murray-Darling basin. *Hydrobiologia* **596**, 203–211.

Hughes JM (2007). Constraints on recovery: using molecular methods to study connectivity of aquatic biota in rivers and streams. *Freshwater Biology* **52**, 616–631.

Humphries P, King AJ and Koehn JD (1999). Fish, flows and floodplains: links between freshwater fishes and their environment in the Murray-Darling River system, Australia. *Environmental Biology of Fishes* **56**, 129–151.

Jenkins KM (2006). Links between flow, aquatic productivity and diversity: setting rehabilitation targets for regulated floodplain wetlands. Report to the State Wetland Advisory Committee. University of New England, Armidale, NSW.

Jenkins KM and Boulton AJ (2003). Connectivity in a dryland river: short-term aquatic microinvertebrate recruitment following floodplain inundation. *Ecology* **84**, 2708–2723.

Jenkins KM and Boulton AJ (2007). Detecting impacts and setting restoration targets in arid-zone rivers: aquatic microinvertebrate responses to reduced floodplain inundation. *Journal of Applied Ecology* **44**, 823–832.

Jenkins KM, Boulton AJ and Ryder DS (2005). A common parched future? Research and management of Australian arid-zone floodplain wetlands. *Hydrobiologia* **552**, 57–73.

Jenkins KM, Wolfenden BJ, Davies JN, Boulton AJ and Rolls R (2008). Macroinvertebrate and fish communities of the Macquarie Marshes: responses to low flows and floods. Report to the Environmental Trust. University of New South Wales, Sydney.

Jolly ID (1996). The effects of river management on the hydrology and hydroecology of arid and semi-arid floodplains. In *Floodplain Processes*. (Eds MG Anderson, DE Walling and PD Bates) pp. 577–609. John Wiley and Sons, London.

King AJ (2004). Density and distribution of potential prey for larval fish in the main channel of a floodplain river: pelagic versus epibenthic meiofauna. *River Research and Applications* **20**, 883–897.

King AJ, Humphries P and Lake PS (2003a). Fish recruitment on floodplains: the roles of patterns of flooding and life history characteristics. *Canadian Journal of Fisheries and Aquatic Sciences* **60**, 773–786.

King J, Brown C and Sabet H (2003b). A scenario-based holistic approach to environmental flow assessments for rivers. *Regulated Rivers: Research and Management* **19**, 619–639.

King AJ, Ward KA, O'Connor P, Green D, Tonkin Z and Mahoney J (2009). Adaptive management of an environmental watering event to enhance native fish spawning and recruitment. *Freshwater Biology*. DOI 10.1111/j.1365-2427.2009.02178.x.

Kingsford RT (2000). Ecological impacts of dams, water diversions and river management on floodplain wetlands in Australia. *Austral Ecology* **25**, 109–127.

Kingsford RT and Auld KM (2005). Waterbird breeding and environmental flow management in the Macquarie Marshes, arid Australia. *River Research and Applications* **21**, 187–200.

Kingsford RT and Johnson WJ (1998). The impact of water diversions on colonially nesting waterbirds in the Macquarie Marshes of arid Australia. *Colonial Waterbirds* **21,** 159–179.

Kingsford RL and Norman FI (2002). Australian waterbirds: products of the continent's ecology. *Emu* **102**, 47–69.

Kingsford RT and Porter JL (2009). Monitoring waterbird populations with aerial surveys: what have we learnt? *Wildlife Research* **36**, 29–40.

Kingsford RT and Thomas RF (1995). The Macquarie Marshes in arid Australia and their waterbirds: a 50-year history of decline. *Environmental Management* **19**, 867–878.

Kingsford RT and Thomas RF (2004). Destruction of wetlands and waterbird populations by dams and irrigation on the Murrumbidgee River in arid Australia. *Environmental Management* **34**, 383–396.

Kingsford RT, Curtin AL and Porter J (1999). Water flows on Cooper Creek in arid Australia determine 'boom' and 'bust' periods for waterbirds. *Biological Conservation* **88**, 231–248.

Kingsford RT, Jenkins KM and Porter JL (2004). Imposed hydrological stability on lakes in arid Australia and effects on waterbirds. *Ecology* **85**, 2478–2492.

Kingsford RT, Georges A and Unmack PJ (2006). Vertebrates of desert rivers: meeting the challenges of temporal and spatial unpredictability. In *Ecology of Desert Rivers*. (Ed. RT Kingsford) pp. 154–200. Cambridge University Press, Cambridge)

Kingsford RT, Brandis K and Porter J (2008). Waterbird response to flooding in the northern Murray-Darling Basin 2008. Final report to the Murray-Darling Basin Authority. University of New South Wales, Sydney.

Lake PS (2001). On the maturing of restoration: linking ecological research and restoration. *Ecological Management and Restoration* **2**, 110–115.

Lee AK and Mercer AH (1967). Cocoon surrounding desert-adapted frogs. *Science* **159**, 87–88.

Lemly AD, Kingsford RT and Thompson JR (2000). Irrigated agriculture and wildlife conservation: conflict on a global scale. *Environmental Management* **25**, 485–512.

Leslie DJ (2001). Effect of river management on colonially nesting waterbirds in the Barmah-Millewa forest, south-eastern Australia. *Regulated Rivers: Research and Management* **17**, 21–36.

Lyon J, Ryan TJ and Scroggie MP (2007). Effects of temperature on the fast-start swimming performance of an Australian freshwater fish. *Ecology of Freshwater Fish* **17**, 184–188.

Lytle DH and Poff NL (2004). Adaptation to natural flow regimes. *Trends in Ecology and Evolution* **19**, 94–100.

Mallen-Cooper M (1994). Swimming ability of adult golden perch, *Macquaria ambigua* (Percichthyidae), and adult silver perch, *Bidyanus bidyanus* (Teraponidae), in an experimental vertical-slot fishway. *Marine and Freshwater Research* **45**, 191–198.

Manatunge JM, Asaeda T and Priyadarshana T (2000). The influence of structural complexity on fish–zooplankton interactions: a study using artificial submerged macrophytes. *Environmental Biology of Fishes* **58**, 425–438.

Mayfield HF (1975). Suggestions for calculating nest success. *Wilson Bulletin* **87**, 456–466.

McKilligan NG (1975). Breeding and movements of the straw-necked ibis in Australia. *Emu* **75**, 199–212.

Morris PM (2008). Microinvertebrate community response to changing water regimes in the Macquarie Marshes, NSW, Australia. PhD thesis. University of New England, Armidale.

Naiman RJ, Bunn SE, Nilsson C, Petts GE, Pinay G and Thompson LC (2002). Legitimising fluvial ecosystems as users of water: an overview. *Environmental Management* **30**, 455–467.

Nilsson C and Svedmark M (2002). Basic principles and ecological consequences of changing water regimes: riparian plant communities. *Environmental Management* **30**, 468–480.

Nilsson C, Reidy CA, Dynesius M and Revenga C (2005). Fragmentation and flow regulation of the world's large river systems. *Science* **308**, 405–408.

O'Connor JP, O'Mahony DJ and O'Mahony JM (2005). Movements of *Macquaria ambigua* in the Murray River, south-eastern Australia. *Journal of Fish Biology* **66**, 392–403.

Palmer MA and Bernhardt ES (2005). Standards for ecologically successful river restoration. *Journal of Applied Ecology* **42**, 208–217.

Peel MC, McMahon TA, Finlayson BL and Watson FGR (2001). Identification and explanation of continental differences in the variability of annual runoff. *Journal of Hydrology* **250**, 224–240.

Pinay G, Clément JC and Naiman RJ (2002). Basic principles and ecological consequences of changing water regimes: riparian plant communities. *Environmental Management* **30**, 468–480.

Poff LN, Allan JD, Bain MB, Karr JR, Prestegaard KL, Richter BD, Sparks RE and Stromberg JC (1997). The natural flow regime: a paradigm for river conservation and restoration. *BioScience* **47**, 769–784.

Poff NL, Allan JD, Palmer MA, Hart DD, Richter BD, Arthington AH, Rogers KH, Meyer JL and Stanford JA (2003). River flows and water wars: emerging science for environmental decision making. *Frontiers in Ecology and the Environment* **1**, 298–306.

Porter JL (2007). Contrasting emergence patterns of *Lamprothamnium macropogon* (Characeae, Charophyceae) and *Ruppia tuberosa* (Potamogetonaceae) from arid-zone saline wetlands in Australia. *Charophytes* **1**, 19–27.

Porter JL, Kingsford RT and Hunter SJ (2006). Aerial surveys of wetland birds in eastern Australia, October 2003–2005. NSW Dept of Environment and Conservation, Sydney. Occasional Paper No. 37.

Predavec M and Dickman CR (1993). Ecology of desert frogs: a study from southwestern Queensland. In *Herpetology in Australia*. (Eds D Lunney and D Ayres) pp. 159–169. Royal Zoological Society of NSW, Sydney.

Puckridge JT, Sheldon F, Walker KF and Boulton AJ (1998). Flow variability and the ecology of arid zone rivers. *Marine and Freshwater Research* **49**, 55–72.

Puckridge JT, Walker KF and Costelloe JF (2000). Hydrological persistence and the ecology of dryland rivers. *Regulated Rivers: Research and Management* **16**, 385–402.

Pusey BJ and Kennard MJ (1996). Species richness and geographical variation in assemblage structure of the freshwater fish fauna of the Wet Tropics region of northern Queensland. *Marine and Freshwater Research* **47**, 563–573.

Pusey BJ, Arthington AH and Read MG (1995). Species richness and spatial variation in fish assemblage structure in two rivers of the Wet Tropics of northern Queensland, Australia. *Environmental Biology of Fishes* **42**, 181–199.

Rayner TS, Jenkins KM and Kingsford RT (2009). Small environmental flows, drought and the role of refugia for freshwater fish in the Macquarie Marshes, arid Australia. *Ecohydrology* **2**, 440–453.

Ren S, Kingsford RT and Thomas RF (2009). Modelling flow to and inundation of the Macquarie Marshes in arid Australia. *Environmetrics*. http://www3.interscience.wiley.com/cgi.bin/fulltext/122489750/PDFSTART

Richter BD, Baumgartner JV, Powell J and Braun DP (1996). A method for assessing hydrologic alteration within ecosystems. *Conservation Biology* **10**, 1–12.

Roberts J and Marston F (2000). Water regime of wetland and floodplain plants in the Murray-Darling Basin: a source book of ecological knowledge. CSIRO Land and Water, Canberra. Technical Report 30/00.

Roshier DA and Asmus MW (2009). Use of satellite telemetry on small-bodied waterfowl in Australia. *Marine and Freshwater Research* **60**, 299–305.

Roshier DA, Robertson AI, Kingsford RT and Green DG (2001). Continental-scale interactions with temporary resources may explain the paradox of large populations of desert waterbirds in Australia. *Landscape Ecology* **16**, 547–556.

Roshier D, Asmus M and Klaassen M (2008). What drives long-distance movements in the nomadic grey teal *Anas gracilis* in Australia? *Ibis* **150**, 474–484.

Schlosser IJ (1998). Fish recruitment, dispersal, and trophic interactions in a heterogeneous lotic environment. *Oecologia* **113**, 260–268.

Sharley T and Huggan C (1995). Chowilla Resource Management Plan. Murray-Darling Basin Commission's Chowilla Working Group. Murray-Darling Basin Commission, Canberra.

Sheldon F, Boulton AJ and Puckridge JT (2002). Conservation value of variable connectivity: aquatic invertebrate assemblages of channel and floodplain habitats of a central Australian arid-zone river, Cooper Creek. *Biological Conservation* **103**, 13–31.

Simberloff D (1998). Flagships, umbrellas, and keystones: is single-species management passé in the landscape era? *Biological Conservation* **83**, 247–257.

Stanley EH, Fisher SG and Grimm NB (1997). Ecosystem expansion and contraction in streams. *BioScience* **47**, 427–435.

Stone C and Haywood A (2006). Assessing canopy health of native eucalypt forests. *Ecological Management and Restoration* **7**, 24–30.
Sullivan BK (1989). Desert environments and the structure of anuran mating systems. *Journal of Arid Environments* **17**, 175–183.
Tan LW and Shiel RJ (1993). Responses of billabong rotifer communities to inundation. *Hydrobiologia* **255/256**, 361–369.
Tharme RE (2003). A global perspective on environmental flow assessment: emerging trends in the development and application of environmental flow methodologies for rivers. *River Research and Applications* **19**, 397–442.
Thomas RF, Bowen S, Simpson SL, Cox SJ, Sims NC, Hunter SJ and Lu Y (2010). Inundation response of vegetation communities of the Macquarie Marshes in semi-arid Australia. In *Ecosystem Response Modelling in the Murray-Darling Basin.* (Eds N Saintilan and I Overton) pp. 137–150. CSIRO Publishing, Melbourne.
Tockner K, Schiemer F, Baumgartner C, Kum G, Weigand E, Zweimüller I and Ward JV (1999). The Danube Restoration Project: species diversity patterns across connectivity gradients in the floodplain system. *Regulated Rivers: Research and Management* **15**, 245–258.
Valett HM, Baker MA, Morrice JA, Crawford CS, Molles MC, Dahm CN, Moyer DL, Thibault JR and Ellis LM (2005). Biogeochemical and metabolic responses to the flood pulse in a semiarid floodplain. *Ecology* **86**, 220–234.
Vörösmarty CJ, Meybeck M, Fekete B, Sharma K, Green P and Syvitksi J (2003). Anthropogenic sediment retention: major global-scale impact from the population of registered impoundments. *Global and Planetary Change* **39**, 169–190.
Walker KF, Sheldon F and Puckridge JT (1995). A perspective on dryland river ecosystems. *Regulated Rivers: Research and Management* **11**, 85–104.
Ward JV, Tockner K and Schiemer F (1999). Biodiversity of floodplain river ecosystems: ecotones and connectivity. *Regulated Rivers: Research and Management* **15**, 125–139.
Wassens S (2010). Flooding regimes for frogs in lowland rivers of the Murray-Darling Basin. In *Ecosystem Response Modelling in the Murray-Darling Basin.* (Eds N Saintilan and I Overton) pp. 213–228. CSIRO Publishing, Melbourne.
Whitcraft LA, Talley D and Crooks JA (2008). Utilization of invasive tamarisk by salt marsh consumers. *Oecologia* **158**, 259–272.
White M (2002). The Cooper Creek turtle persisting under pressure: a study in arid Australia. Honours thesis. University of Canberra, Canberra.
Young WJ and Kingsford RT (2006). Flow variability in large unregulated dryland rivers. In *Ecology of Desert Rivers.* (Ed. RT Kingsford) pp. 11–46. Cambridge University Press, Cambridge.

Chapter 3

Modelling ecological responses to flow alteration: making the most of existing data and knowledge

Michael Stewardson and Angus Webb

Introduction

Quantitative predictions of ecological responses to flow alterations are fundamental to sound environmental water planning and delivery. The cost of providing environmental water is considerable, with billions of dollars being invested in water recovery for the Murray-Darling Basin alone (the federal government committed to spending $3 billion to purchase water for the environment as part of the National Water Plan announced in April 2008). Similarly, using water held in environmental water accounts can affect water held in storages and may appear to affect other users of the resource. Environmental water managers need models to make transparent and defensible decisions.

Efforts to model quantitative responses to flow alteration began over 25 years ago. The Physical Habitat Simulation Model (PHABSIM) is used in the Instream Flow Incremental Method (IFIM) to represent changes in micro-habitat availability with flow modifications (Milhous *et al.* 1989; Bovee 1982). Although widely criticised, it has been used in its original or modified forms for many years and in many countries (Arthington *et al.* 1992; Blazkova 1994; Gippel and Stewardson 1995; Gore *et al.* 1991; Johnson *et al.* 1993; Jowett 1989; King and Tharme 1993; Saccardo *et al.* 1994; Saltveit and Heggenes 1991; Souchon *et al.* 1989). Most environmental flow methods require some form of quantitative response model (Arthington *et al.* 2006). However, despite over two decades of research we still have little capacity to model quantitative ecological responses to flow alteration. In most cases, environmental flow methods rely on models that predict the effects of flow alteration on physical habitat conditions. Consequent changes in the biota are normally assumed – the so-called 'field of dreams' approach to restoration planning (Palmer *et al.* 1997). Often, these predictions of biotic response are not underpinned by strong scientific evidence.

This chapter discusses possible methods for modelling ecological responses to flow alteration. Our interest is specifically rivers whose flows have been altered by water resource development or other catchment change. This is in contrast to many studies that use statistical models to relate indices of ecological condition to site-to-site variations in flow regime without considering the level of flow modifications. We argue that it is unreasonable to infer impact of flow alterations from such studies alone. Our methods have also been developed for application to monitoring data collected over a wide geographic range and over a range of flow conditions. The models aim to be as generally applicable as possible. In contrast, most flow–response investigations have concentrated on more spatially and temporally restricted case studies (Hancock and Boulton 2005; Siebentritt *et al.* 2004). The methods we outline below are applicable to such studies, but the resulting models will have less generality than those produced from wider-scale data.

We discuss three stages in the development of quantitative response models.

- The first stage provides a method for testing postulated causal mechanisms of ecological response against available evidence. These tests inform sensible choices regarding the processes that should be represented in a model. In most investigations, this stage of model development is given cursory treatment, relying on the experience and judgment of the scientific team. However, such processes are subject to bias and overconfidence (Morgan *et al.* 1990). We argue that a systematic review of existing knowledge can reduce such influences when identifying the causal mechanisms that underpin a model.
- The key task in developing a quantitative (process-based) predictive model is rendering the causal mechanisms in mathematical form. Deciding on the mathematical structure of a numerical model is rarely discussed in the literature but is central to building a robust and realistic model. The literature shows a bias towards model testing and application rather than addressing the subjective decisions made during model development. The second part of this chapter discusses formulation of models for predicting responses to flow alterations.
- We conclude the chapter with a discussion of model parameter estimation. There are many approaches that can be taken. We highlight the benefits of hierarchical models formulated within the Bayesian statistical framework. By taking a hierarchical approach to data collected across many sites, parameter estimates for individual sites are more robust than they would be had the sites been analysed separately. We argue that this approach makes best use of the often-scant quantitative data, and provides a valid mechanism for extrapolating the results of empirically parameterised models to new sites or rivers that were not included in the original analysis.

Identifying response mechanisms

A conceptual response model is essentially a tool for documenting and communicating an individual or collective understanding of causal pathways in the system under investigation. An influence diagram (Fig. 3.1) is the most common means of depicting a conceptual model, but they can also be expressed purely as text or in a variety of tabular and graphical forms. Quantitative models seek to represent the causal pathways in mathematical form and conceptualising these pathways is a key step in model development. The question is how to go about developing the conceptual model of response to flow.

We begin by arguing that the conceptual model should be made explicit during the model development process. Our experience is that it is not uncommon for model developers to simply propose a model in mathematical form without first documenting the causal mechanisms they believe are (and are not!) fundamental to the responses they are modelling. However, this approach lacks the openness required for constructive scientific review, and lacks repeatability. The discipline of explicitly articulating a conceptual model forces the modeller and/or other stakeholders to express their beliefs concerning causal mechanisms, and facilitates the process of model revision during development (Burgman 2005). Failure to document and justify the conceptual model increases the likelihood of expert bias remaining undetected, and considerable modelling effort might be wasted in attempting to model processes that do not influence ecological responses of interest.

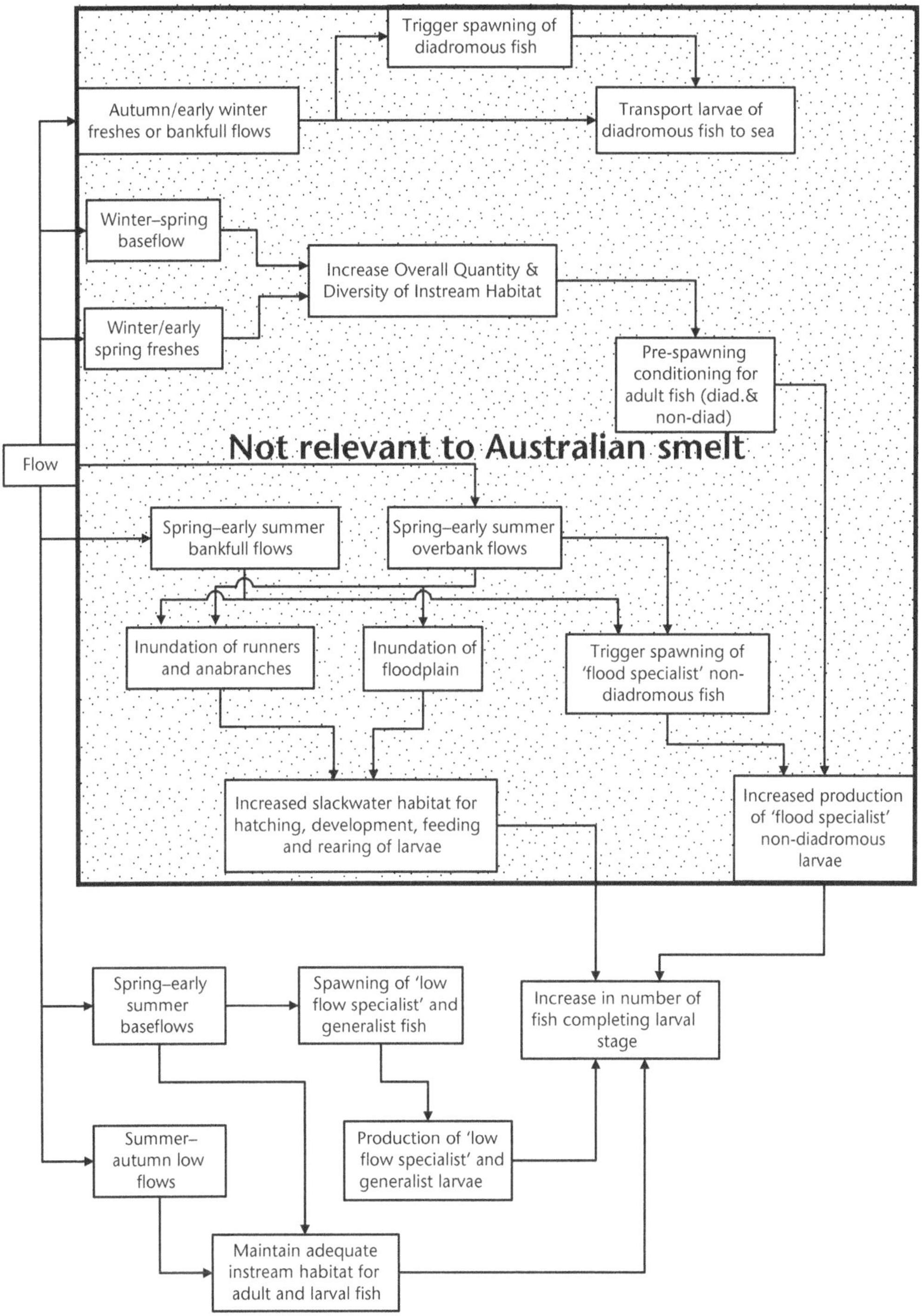

Figure 3.1: The generic conceptual model of fish responses to flow alteration developed by Chee *et al.* (2006), highlighting only the portions deemed relevant to the analysis of flow regime on Australian smelt. Summer and autumn low flows create slow water habitat for larval and juvenile fish, allowing spawning and production, which results in increased numbers of recruits and greater abundance.

While there is considerable speculation about the key mechanism by which flow alteration affects river ecosystems in both the grey and scientific literature, evidence of the importance of these mechanisms is often fragmented. This speculation has been driven by a need to plan environmental flows in the face of limited scientific knowledge. Experts are frequently asked to make judgments concerning the key effects of flow alteration. Their conceptual model of how a river responds to flow alteration is fundamental to these judgments. In cases where this has led to proposals for reallocating water resources from human to environmental use, the expert judgments are often questioned by those adversely affected (Gordon 1995; Marohasy 2003). The expert-based approach to developing environmental flows is vulnerable to political and legal challenges, particularly where there is no basis for resolving differences in opposing expert views (Norris *et al.* 2005).

We argue the need for more robust testing of conceptual models against the available scientific evidence. Such tests should help reduce differences of scientific opinion and improve the credibility of knowledge used in planning environmental flows. There is a substantial and rapidly growing body of scientific knowledge that might be used for such tests. Our analysis of the literature (based on key word searches in Web of Science and review of abstracts) shows that scientific papers providing evidence of the environmental effects of flow alteration grew from a total of 17 in all years prior to 1990, to 190 in 2005, with a total of 1491 papers published prior to mid-2006 when our survey was conducted (Fig. 3.2). It is incumbent upon scientists to make better use of this literature than has been the case in the past.

One promising approach to assessment of causal mechanisms in conceptual models is the use of causal criteria analysis. Causal criteria represent a broad range of logical devices to test hypotheses against experimental evidence. The evolution of causal criteria has a long history in epidemiology, beginning in the late 19th century (Susser 1991). Koch's famous postulates are an early form of causal criteria that were used to identify pathogens. The causal criteria were assembled in their modern form in a landmark 1964 report prepared by an advisory

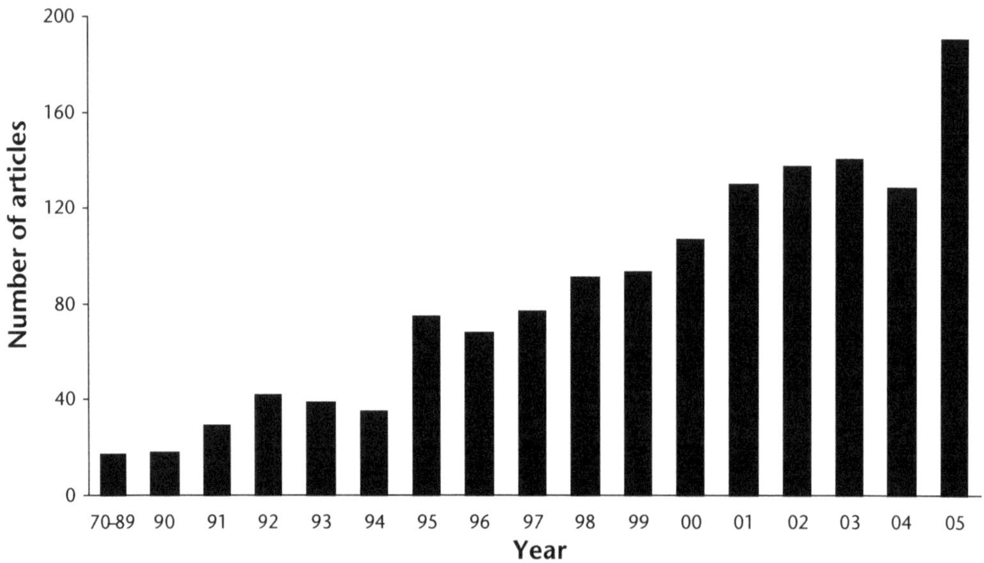

Figure 3.2: Numbers of studies identified per year that deal with environmental responses to flow alteration. We used keyword combination searches in the Web of Science (Thomson ISI) and reviewed all abstracts returned by the searches. Search details are available from the authors on request. Note left-most bar is all years from 1970–89. The search located no articles in years prior to 1970.

committee to the US Surgeon General on the health effects of smoking (USDHEW 1964). The committee recognised that 'Statistical methods alone cannot establish the proof of a causal relationship in an association. The causal significance of an association is a matter of judgment which goes beyond any statement of probability.' There have since been a number of refinements to the causal criteria suggested for use in epidemiology (Hill 1965; Susser 1991).

These causal criteria have also been proposed for identifying human impacts on the environment (Downes *et al.* 2002; Norris *et al.* 2005; USEPA 2000), although there are few case study applications as yet (Adams 2005; Fabricus and De'ath 2004; Norris *et al.* 2005). A group of researchers in the eWater CRC (including both current authors) are developing products to facilitate the use of causal criteria for testing causality in environmental studies. They are focusing on the three causal criteria that are emerging as the most useful for testing causality in environmental systems.

- **Strength of association** is the apparent size of effect associated with the cause. This might be measured by the differences in the effect between samples with and without the cause. A strong association is indicative of a causal relationship. However, weak associations do not necessarily mean there is no causal link.
- **Consistency of association.** If an association between the effects and putative cause has been observed in a variety of situations using a variety of techniques, there is more likely to be a cause and effect relationship. Hill (1965) places more weight on similar results reached in quite different ways, compared to repeated results using the same methods or populations. As with strength of association, the lack of consistent associations does not necessarily mean there is no cause–effect relationship.
- **Coherence of association** refers to the observation of patterns in the association that are consistent with current theory and knowledge about the (possible) causal link. The most obvious example is the existence of a coherent dose–response relation, where increasing size of effect is associated with a more severe application of the cause.

The eWater CRC is developing tools to collate evidence from the literature and catalogue this evidence against the causal criteria, to build a case for causality. The causal criteria tools (www.toolkit.net.au) consist of the Causal Criteria Evidence Database (CCED), which is an on-line searchable database containing attributes and results of individual studies that deal with cause and effect in environmental systems, and the Causal Criteria Assessment Software (CCAS), which uses data from CCED to test for causality. Norris *et al.* (2005) provide an early case study of this approach. Other case studies are under way.

Model formulation

Eventually, the models must be used to make predictions or test hypotheses. Descriptive variables must be replaced with explicit state variables and causal links must be replaced with equations, creating a numerical model. How do we use conceptual models to develop robust and defensible numerical models? This is an under-researched area in the literature. A number of authors have tackled parts of the process in great depth, but we could find no publications that proceeded through the entire process. This lack of direction is somewhat concerning, given the importance of asking the correct questions within any modelling study. In this section, we outline our approach.

What type of model? Complexity and inclusiveness

The first task lies in defining the purpose of the numerical model. Models exist along a gradient of complexity – the simplest might be a statistical model statement designed to test a hypothesis implicit within one part of the conceptual model, through to complex 'ecosystem' models, where all parts of the conceptual model are converted to state variables and equation-based linkages, with the ultimate purpose of creating an analogue of the system under study. Enormous amounts have been written about the numerical evaluation of models at either end of this spectrum. The simpler hypothesis-testing models are covered in texts on statistical inference (Quinn and Keough 2002; Underwood 1997), while ecosystem models are covered by texts on ecological modelling (Gillman and Hails 1997; Jørgensen 1986). Fewer texts acknowledge that many ecological models will lie somewhere along the gradient (Chatfield 1995; Hilborn and Mangel 1997), and indeed that the choice of complexity is a critical part of the modelling process. Chatfield (1995), in particular, stresses the need to specify the objectives of the model before defining structure, and that the challenge for any model-builder is to build a model of appropriate complexity to capture components of perceived importance.

Where the conceptual model comes into this

A number of authors recommend methods for choosing the most appropriate numerical model from a number of contenders (Burnham and Anderson 2002; Gillman and Hails 1997; Hilborn and Mangel 1997). However, these are mostly data-driven choices based on a trade-off between model fit and complexity (i.e. model parsimony). Burnham and Anderson (2002) go so far as to say that only 'plausible' models should be considered within the pool of potential models from which the eventual 'best' model is identified. However, they provide no guidance on how a model may be defined as plausible. Gillman and Hails (1997) recommend reducing the number of explanatory variables in a model via statistical tests until the 'simplest realistic model' is left, but do not define how the choice of 'realistic' is to be made. We believe that a robust conceptual model can be used to immediately restrict the set of potential models to those that are plausible and realistic. If the conceptual model contains no causal linkage between a particular environmental variable and the modelling endpoint, there is no need to consider that variable within any set of potential models. Using the conceptual model as part of this decision process explicitly details the choice of state variables and processes represented within potential models. Moreover, use of the conceptual model in this fashion forces us to acknowledge that choice of a numerical model is more than simply a question of model fit to the data (Chatfield 1995); it forces us to choose a numerical model that is consistent with our background knowledge of the system. Consideration of the conceptual basis of a numeric model also provides a means for choosing between numerical models of approximately equal fit to the data (Chatfield 1995).

Functional form and data transformations

A graphical conceptual model or influence diagram (Fig. 3.1) will be of less assistance in defining the functional form of relationships among model components, and whether (and how) data should be transformed. In some cases, there will be a sound conceptual basis for choosing data transformations. In Webb *et al.* (2010), we scaled summer discharge volumes by the recommended summer low-flow discharge because we believed this would provide an estimate of slow-flow habitat that was less dependent on channel size in different reaches. In other cases, transformations will be necessary to better meet the assumptions of the model (Quinn and Keough 2002) and some choices may be made using data-driven model choice methods (Burnham and Anderson 2002). Concerning functional forms within models, previous research may provide guidance. Jørgensen (1986) provides tables of functional forms to link state variables. However, that text concentrates on lake eutrophication models – systems that are far better understood than the majority of flow–ecology relationships. More often, in defining the functional form for representing links between flow and ecological response, we will be relying on empirical data and/or expert judgment. Jørgensen (1986) refers to such empirically driven models as 'black box', revealing his clear preference for functional forms based on previously elucidated processes ('white box'). He notes that any empirically derived relationship applies solely to the system for which it was determined. We take some issue with this stand, and argue that a hierarchical approach to data analysis provides a method to use the model beyond the systems for which it was developed.

For example, Webb *et al.* (2010) analysed the effect of flow regulation on the abundance of Australian smelt (*Retropinna semoni*) in the Thomson River, Victoria. The reasoning behind the majority of choices that led to the final model will not be repeated here, but it is worth highlighting the role of the conceptual model in the process. We had previously constructed a generic conceptual model of fish responses to flow regulation (Chee *et al.* 2006). Using this model as a basis, we determined the parts of the flow regime likely to affect Australian smelt. Using published information (King 2004; Milton and Arthington 1985; Pusey *et al.* 2004), we determined that large parts of the conceptual model were inapplicable to that species. However, we formed the working hypothesis that that the number of fish recruiting to the adult population would be a function of the amount of slow-flow habitat in the river over the summer period, and that the amount of habitat is a function of the provision of discharges that are sufficient to provide habitat, but not so great that the habitat becomes too disturbed. Figure 3.1 shows the generic model for fish responses (Chee *et al.* 2006), highlighting only the portions relevant to the analysis of smelt abundance.

Hierarchical modelling of flow–ecology responses

We argue that any attempt at modelling flow–ecology responses must make the fullest possible use of available data and evidence. This is perhaps most important for the quantification of numerical models with field data. However, there is a difficulty with using data collected at different sites/reaches/rivers – hereafter referred to as sampling units. All sampling units will differ to some extent, and we may reasonably expect that ecological response to flow alteration may also differ among sampling units. This logical inference may lead us to consider data from smaller and smaller numbers of sampling units when attempting to quantify flow–response relationships. However, such division of a data set will lead to models informed by few data points, and hence with great uncertainty. Such models will also lack generality, as they have been parameterised using highly localised data. How similar must sampling units be, before they can be considered as replicates of the same environment and hence be combined in the same analysis?

One approach is to use ecologically based classifications to define groups of sampling units that are expected to behave similarly (Arthington *et al.* 2006). A number of such classifications exist for flow regimes (Hughes and James 1989; Poff 1996; Poff and Ward 1989; Snelder *et al.* 2005). However, many existing classifications have been constructed for other purposes and it may be risky to assume the classification will be relevant to the question at hand. The use of a classification also means that the modelling exercise starts with an assumption of difference among groups of sampling units, rather than allowing the data to determine whether such a difference exists. Thus, it may reduce the opportunity for maximal use of existing information in quantifying the flow–response relationship.

A hierarchical approach to data analysis (*sensu* Gelman and Hill 2007) provides a rigorous and mathematically consistent approach for combining data from different sampling units when it is justified by the data collected. It also allows data from seemingly dissimilar sampling units to be combined when we can formulate a model of the processes that lead to dissimilarity of such units. As we show below, combining data from multiple sampling units has advantages for ecological inference, predictive ability of models and the ability to extrapolate model results to new sites.

In a hierarchical analysis, sampling units are considered to be exchangeable. This implies that parameter values of interest (e.g. a regression slope) for the different sampling units are drawn from a larger distribution (Gelman *et al.* 2004). For example, the rate at which salinity decreases with increasing discharge in lowland rivers will vary somewhat, but all these rates may be considered as being drawn from a single distribution of possible rates. The parameters of this distribution (e.g. mean and variance) are known as hyperparameters; they provide an estimate of how the sampling unit-level parameter value may be expected to vary across the entire population of sampling units (not just those actually sampled).

The estimates of parameter values for the individual sampling units are also affected when the data are analysed hierarchically. The means of individual estimates will be drawn towards the global mean of all estimates, and uncertainties around individual estimates will be reduced. These properties are respectively known as 'shrinkage' and 'borrowing strength'. This leads to more robust sampling unit-scale estimates of parameter values, and thereby increased inferential strength and more precise predictions from the model.

To illustrate, Webb *et al.* (2010) examined differences in the abundances of fish (Australian smelt) with changes in summer flows in the Thomson River, Victoria, Australia. When river reaches were examined individually, there was some tendency towards higher than recommended summer flows being associated with reduced smelt abundance. One reach showed the opposite tendency, however, and uncertainties around estimates were such that it was difficult to infer any effect. When the reaches were included in a single hierarchical model, the precision of estimates for all reaches was increased and the central tendencies were shrunk together such that the analysis provided some evidence of a river-scale reduction in fish abundances during years of higher summer flows (Fig. 3.3).

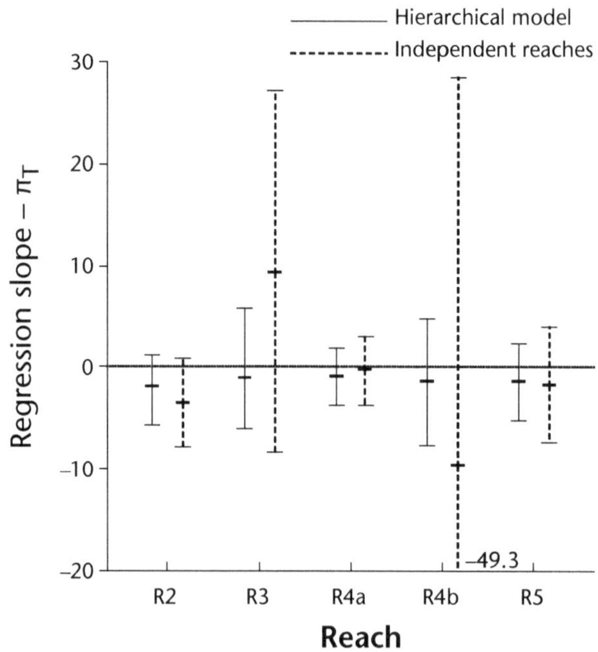

Figure 3.3: Estimates of the effect of summer flows on abundance of Australian smelt across five reaches (R2, R3, R4a, R4b, R5) of the Thomson River. Bars show median and 95% credible intervals for the estimate of the regression slope that quantifies the relationship between summer flows and fish abundance for hierarchical and non-hierarchical (independent reaches) models. Reproduced from Webb *et al.* (2010) with permission (Wiley-Blackwell Publishing).

Sometimes, we may have *a priori* reasons to expect parameter values to vary among sampling units due to known differences. For the salinity example above, it is reasonable to expect that the rate of change of salinity with increasing flow may vary as a function of the distance from the top of the catchment, as water will tend to become saltier further down a river. In such a case, the concept of exchangeability may be extended to that of 'conditional exchangeability'. For this example, the rate of change of salinity with increasing flow is expected to be drawn from a common distribution conditional upon the sampling unit's distance from the top of the catchment. Conditional exchangeability means that data from seemingly dissimilar sampling units can be analysed together, as long as a plausible covariate model can be formulated. The effect of the covariate on individual sampling units is accounted for within the analysis, and the magnitude of the covariate effect calculated – something that can lead to important subsidiary results in itself.

In Webb and King (2009), we used a hierarchical model to show temporal declines in stream ecological condition around Melbourne, Australia, despite having few data points at any individual site. A covariate model for the hyperparameters showed that the rate of degradation increased with the level of urbanisation of the catchment. In this case, the covariate model led to more robust estimates of the rate of degradation at each site, and to the new finding that urbanisation affected the rate of decline.

Hierarchical models are easiest to formulate and implement within the Bayesian statistical framework. The flexibility of Bayesian analysis means that it is possible to build the complex models often demanded by ecological data (Clark 2005). Bayesian analyses are very good at handling and representing uncertainty, something that is critical given our level of uncertainty in flow–ecology relationships.

The concepts of exchangeability and conditional exchangeability mean that it is possible to use more data in quantifying flow–ecology relationships. This leads to better inferential strength at the level of the sampling unit, and allows us to quantify effects and test hypotheses across sampling units. Use of a hierarchical approach also provides a valid basis for extrapolating the results of a model to sampling units for which no data have been collected, something that has been generally frowned upon for empirical models (Jørgensen 1986). We undertake such extrapolation anyway, with the use of 'representative' sites in environmental flow assessments (DNRE 2002). Hardly a publication exists where the authors do not attempt to discuss the implications of empirical findings for environments well beyond those sampled. From the point of view of a hierarchical model, a new sampling unit can be expected to have a parameter value taken from the distribution defined by the previously quantified hyperparameters, and it is possible to use the model to produce estimates of parameter values for non-sampled units. Appropriately, these estimates will have more uncertainty around them than those for sampling units where data have been collected. If a covariate relationship describes the distribution of parameter values, then information on that covariate for the non-sampled unit will improve the precision of the estimate, although not to the level of those for which data were collected.

Conclusions

Most environmental flow models evaluate changes in physical habitat rather than biological consequences, and there is a great need for quantitative ecological response models to underpin environmental water planning and operation. However, there has been limited success in building quantitative models of ecological responses to flow alteration. Instead, expectations of ecological response have been based on expert opinion. There is a need to develop better approaches for developing flow–response models, and we see three elements of the model development process that can be improved.

We believe that the available evidence for causal pathways should be used to test the processes in conceptual models of ecological response. Modelling efforts should target those processes for which there is stronger evidence of their importance. Causal criteria and associated tools being developed by eWater CRC to apply these criteria offer a practical approach to use existing evidence to test causal links between flow alteration and ecological change.

A numerical model – anything from a simple statistical hypothesis test to a complex ecosystem model – is simply a mathematically explicit form of a conceptual hypothesis (Hilborn and Mangel 1997). While we have noted that much has been written about different types of numerical models, and choosing between them, there has been a failure to link these choices to consideration of a sound conceptual model. In making this link, we obtain guidance in our choices and a better theoretical justification of the numerical models used to explore the links between flow and ecological response.

Given the often-quoted paucity of data for quantifying flow–ecology relationships, we believe that any attempt at model quantification should include as much data as possible. Hierarchical models allow us to test sequentially whether:

- effects are similar enough across sampling units to employ a simple hierarchical model;
- effects vary between sampling units, but that the difference can be accounted for with a covariate model;
- whether subsets of the data should be tested separately because of differences among sampling units that we cannot explain (Webb *et al.* 2010).

The classification-based approaches outlined above only consider the third (and least powerful) of these options. Incorporation of data from multiple sampling units within a hierarchical analysis improves inferential strength, provides more robust estimates of effects for individual sampling units, improves the precision of predictions of future effects, and allows estimates of effects for sampling units for which no data have been collected.

Finally, there are a growing number of monitoring programs designed to investigate effects of flow alterations, some of which are published in this book. These studies are generally focused on individual rivers or sites and involve a short period of environmental monitoring (Webb *et al.* in press). The primary aim is usually to detect whether an anticipated ecological response to changes in the flow regime actually occurs. The studies are rarely intended to establish predictive models, either site-specific or generally applicable across many rivers. Larger-scale co-ordinated approaches to investigating the effects of flow alteration are relatively rare, but programs such as IMEF in New South Wales (Chessman and Jones 2001) and VEFMAP in Victoria (Chee *et al.* 2006; Cottingham *et al.* 2005) have begun to fill this gap. VEFMAP has gone furthest in this regard, establishing co-ordinated monitoring and analysis of data across multiple rivers. Large-scale co-ordinated programs have efficiencies of scale and great advantages in terms of the strength of inference concerning the effects of flow alteration. The data generated by such programs can be used in the development of generally applicable quantitative response models, rather than being restricted to answering case-study specific questions. Thus, while the methods we describe may be applicable to single-site and short-term investigations, we believe they will be of most use in the consideration of larger data sets. Without doubt, the establishment of large-scale co-ordinated programs presents considerable technical and logistic difficulties (Webb *et al.* in press) but programs such as VEFMAP (Chee *et al.* 2006; Cottingham *et al.* 2005) will provide better analysis of the effectiveness of environmental flow programs, along with an ability to refine predictive models for future rounds of flow recommendations. It will also mean we can make better use of the scarce scientific resources, particularly the limited number of scientists with expertise in this field.

References

Adams SM (2005). Assessing cause and effect of multiple stressors on marine systems. *Marine Pollution Bulletin* **51**, 649–657.

Arthington AH, Bycroft BM and Conrick DL (1992). Environmental study: Barker-Barambah Creek, Vol. 1, Executive Summary. Centre for Catchment and In-stream Research, Griffith University, Griffith, Queensland.

Arthington AH, Bunn SE, Poff NL and Naiman RJ (2006). The challenge of providing environmental flow rules to sustain river ecosystems. *Ecological Applications* **16**, 1311–1318.

Blazkova S (1994). Physical habitat simulation in the Czech Republic. In *Proceedings of the 6th International Symposium on Regulated Rivers.* Ceske Budejovice, Czech Republic.

Bovee KD (1982). A guide to stream habitat assessment using the Instream Flow Incremental Methodology. Instream Flow Information Paper 12. US Fish and Wildlife Service, Fort Collins, Colorado.

Burgman MA (2005). *Risks and Decisions for Conservation and Environmental Management.* Cambridge University Press, New York.

Burnham KP and Anderson DR (2002). *Model Selection and Multimodel Inference.* Springer, New York.

Chatfield C (1995). *Problem Solving: A Statistician's Guide.* Chapman and Hall, London.

Chee YE, Webb JA, Stewardson MJ and Cottingham P (2006). Victorian Environmental Flows Monitoring and Assessment Program: monitoring and evaluation of environmental flow releases in the Broken River. University of Melbourne/eWater CRC, Melbourne. Available at http://tinyurl.com/lltjma.

Chessman BC and Jones H (2001). Integrated monitoring of environmental flows: design report. NSW Dept of Land and Water Conservation, Sydney.

Clark JS (2005) Why environmental scientists are becoming Bayesians. *Ecology Letters* **8**, 2–14.

Cottingham P, Stewardson M and Webb A (2005). Victorian Environmental Flows Monitoring and Assessment Program. Stage 1: Statewide Framework. CRC for Freshwater Ecology/ CRC for Catchment Hydrology, Melbourne. Available at http://tinyurl.com/2feys3.

DNRE (2002). The FLOWS method: a method for determining environmental water requirements in Victoria. Victorian Dept of Natural Resources and Environment, Melbourne. Available at http://tinyurl.com/yws84m.

Downes BJ, Barmuta LA, Fairweather PG, Faith DP, Keough MJ, Lake PS, Mapstone BD and Quinn GQ (2002). *Monitoring Ecological Impacts: Concepts and Practice in Flowing Waters.* Cambridge University Press: Cambridge.

Fabricus KE and De'ath G (2004). Identifying ecological change and its causes: a case study of coral reefs. *Ecological Applications* **14**, 1448–1465.

Gelman A, Rubin JB, Stern HS and Rubin DB (2004). *Bayesian Data Analysis.* Chapman and Hall/CRC, Boca Raton.

Gelman A and Hill J (2007). *Data Analysis using Regression and Multilevel/hierarchical Models.* Cambridge University Press, Cambridge.

Gillman M and Hails R (1997). *An Introduction to Ecological Modelling: Putting Practice into Theory.* Blackwell Science, Oxford.

Gippel CJ and Stewardson MJ (1995). Development of an environmental flow management strategy for the Thomson River, Victoria, Australia. *Regulated Rivers: Research and Management* **10**, 121–135.

Gordon ND (1995). Summary of Technical Testimony in the Colorado Water Division 1 Trial. Stream Systems Technology Center, USDA Forest Service, Rocky Mountain Forest and Range Experiment Station, Fort Collins, Colorado.

Gore JA, King JM and Hamman KCD (1991). Application of the instream flow incremental methodology to southern African rivers: protecting endemic fish of the Olifants River. *Water South Africa* **17**, 225–236.

Hancock PJ and Boulton AJ (2005). The effects of an environmental flow release on water quality in the hyporheic zone of the Hunter River, Australia. *Hydrobiologia* **552**, 75–85.

Hilborn R and Mangel M (1997). *The Ecological Detective: Confronting Models with Data.* Princeton University Press, Princeton, NJ.

Hill AB (1965). The environment and disease: association or causation? *Proceedings of the Royal Society of Medicine* **58**, 295–300.

Hughes JMR and James B (1989). A hydrological regionalisation of streams in Victoria, Australia, with implications for stream ecology. *Australian Journal of Marine and Freshwater Research* **40**, 303–326.

Johnson IW, Elliot CRN, Gustard A, Armitage PD, Ladle M, Dawson FH and Beaumont WRC (1993). *Ecologically Acceptable Flows: Assessment of Instream Flow Incremental Methodology.* Institute of Hydrology, National Environment Research Council, Wallingford, UK.

Jørgensen SE (1986). *Fundamentals of Ecological Modelling.* Elsevier, Amsterdam.

Jowett IG (1989). RYHABSIM Computer Manual. Freshwater Fisheries Centre, Riccarton, NZ.

King AJ (2004). Ontogenetic patterns of habitat use by fishes within the main channel of an Australian floodplain river. *Journal of Fish Biology* **65**, 1582–1603.

King JM and Tharme RE (1993). Assessment of the Instream Flow Incremental Methodology and Initial Development of Alternate Instream Flow Methodologies for South Africa. Freshwater Research Unit, University of Cape Town, Cape Town, South Africa.

Marohasy J (2003). Myth and the Murray: measuring the real state of the river environment. *IPA Backgrounder* **15(5).**

Milhous RT, Updike MA and Schneider DM (1989). Physical habitat simulation system reference manual. Version II. Instream Flow Information Paper 26. *Biological Reports* **89(16).** US Fish and Wildlife Service, Fort Collins, Colorado.

Milton DA and Arthington AH (1985). Reproductive strategy and growth of the Australian smelt, *Retropinna semoni* (Weber) (Pisces, Retropinnidae), and the Olive Perchlet, *Ambassis nigripinnis* (De-Vis) (Pisces, Ambassidae), in Brisbane, south-eastern Queensland. *Australian Journal of Marine and Freshwater Research* **36**, 329–341.

Morgan MG, Henrion M and Small M (1990). *Uncertainty: A Guide to Dealing with Uncertainty in Quantitative Risk and Policy Analysis.* Cambridge University Press, Cambridge.

Norris R and Liston P *et al.* (2005). Multiple lines and levels of evidence for eetecting ecological responses to management intervention. In *Proceedings of the 4th Australian Stream Management Conference.* 19–22 October. Launceston, Tasmania. (Eds ID Rutherfurd, I Wiszniewski, MJ Askey-Doran and R Glazik). Tasmanian Dept of Primary Industries, Water and Environment.

Palmer MA, Ambrose RF and Poff NL (1997). Ecological theory and community restoration ecology. *Restoration Ecology* **5**, 291–300.

Poff NL (1996). A hydrogeography of unregulated streams in the United States and an examination of scale-dependance in some hydrological descriptors. *Freshwater Biology* **36**, 71–91.

Poff NL and Ward JV (1989). Implications of streamflow variability and predictability for lotic community structure: a regional analysis of streamflow patterns. *Canadian Journal of Fisheries and Aquatic Science* **46**, 1805–1818.

Pusey BJ, Kennard MJ and Arthington AH (2004). *Freshwater Fishes of North-eastern Australia.* CSIRO Publishing, Melbourne.

Quinn GP and Keough MJ (2002). *Experimental Design and Analysis for Biologists.* Cambridge University Press, Cambridge.

Saccardo I, Vitali R, Guzzi L, Annoni P and Stigliano P (1994). First application of methodologies to define a proper minimum discharge for the preservation of aquatic species in Italian Pre-Alps. In *Proceedings of the 1st International Symposium on Habitat Hydraulics.* Trondheim, Norway. pp. 261–267.

Saltveit SJ and Heggenes J (1991). En konsekvensvurdering av reguleringsvirkninger pa laks og orret i Gjengedalsvassdraget, Sogn og Fjordane. Zoological Museum, University of Oslo, Oslo, Norway.

Siebentritt MA, Ganf GG and Walker KF (2004). Effects of an enhanced flood on riparian plants of the River Murray, South Australia. *River Research and Applications* **20**, 765–774.

Snelder TH, Biggs BJF and Woods RA (2005). Improved eco-hydrological classification of rivers. *River Research and Applications* **21**, 609–628.

Souchon Y, Trocherie F, Fragnoud E and Lacombe C (1989). Instream flow incremental methodology: applicability and new developments. *Revue des Sciences de L'eau* **2**, 807–830.

Susser M (1991). What is cause and how do we know one? A grammar for pragmatic epidemiology. *American Journal of Epidemiology* **133**, 635–647.Underwood AJ (1997). Experiments in Ecology: Their Logical Design and Interpretation using Analysis of Variance. Cambridge University Press, New York.

USDHEW (1964). Smoking and Health. Report of the Advisory Committee to the Surgeon General of the Public Health Service. US Dept of Health Education and Welfare, Washington DC.

USEPA (2000). Stressor Identification Guidance Document. USEPA Report EPA/822/B-00/025. US Environment Protection Agency, Washington DC.

Webb JA and King EL (2009). A Bayesian hierarchical trend analysis finds strong evidence for large-scale temporal declines in stream ecological condition around Melbourne, Australia. *Ecography* **32**, 215–225.

Webb JA, Stewardson MJ and Koster WM (2010). Detecting ecological responses to flow variation using Bayesian hierarchical models. *Freshwater Biology* **55**, 108–126.

Webb JA, Stewardson MJ, Chee YE, Schreiber ESG, Sharpe AK and Jensz MC (in press). Negotiating the turbulent boundary: the challenges of building a science-management collaboration for landscape-scale monitoring of environmental flows. *Marine and Freshwater Research.*

PART II

ECOSYSTEM RESPONSE MODELLING IN THE NORTHERN BASIN

Introduction: Darling Basin

The northern part of the Murray-Darling Basin covers the drainage area of the Darling River. It encompasses an area of approximately 640 000 square kilometres. The end of the basin system is the confluence of the Darling River with the River Murray, although the basin is characterised by many terminal wetlands.

The northern basin has extremely variable annual river flows, ranging from 1% to 1000% of the long-term mean. Periods of zero flow in most rivers can extend to months and, in the drier parts, to years. River flows in the Darling Basin are not strongly seasonal. However, northern rivers generally flood in the summer from monsoonal rains, while southern rivers generally have high river flows from winter which extend to the following autumn. The rivers of the northern basin are less regulated than the southern rivers, with many rivers such as the Paroo and Warrego relatively unchanged from pre-European settlement. Future climate change scenarios suggest a reduction of 10% in available water, resulting in less flooding and longer dry periods.

The basin is home to many wetlands of international and national significance, including the Ramsar-listed Macquarie Marshes, the Gwydir wetlands, the Narran Lakes and the Paroo wetlands.

Modelling flows and ecological responses in this part of the basin is challenging, due to the high variability of river flows, the rapid changes in flood conditions and the large floodplain and wetland complexes inundated during flood events.

Chapter 4

Science and management challenges for the wetlands of the northern Murray-Darling Basin

Neil Saintilan, Joanne Ling and Mike Maher

Introduction

There is no question that mistakes have been made in the past in the allocation of water between the environment and other users in the Murray-Darling Basin. To some extent these mistakes were made in ignorance of the volume and timing of water required to sustain floodplain wetlands and their values. To rectify this, the Murray Flow Assessment Tool (MFAT) (Young *et al.* 2003) was developed as a decision support tool to guide the implementation of the Living Murray program along the River Murray corridor. MFAT contains empirical flow–response models for a range of ecological components, and provides the basis for integrating information on hydrological inputs with ecological response. However, the need for more detailed floodplain hydrodynamic/hydrological modelling as a basis for wetland decision support has been recognised by several authors (Davies and Acreman 2003; Ch. 16, this volume).

This chapter describes a co-ordinated research effort to improve the knowledge underpinning environmental flow management for the significant wetlands of the northern Murray-Darling Basin. Several wetlands in the northern basin have been recognised as nationally and internationally significant, and their prominence and management issues form strong analogies to the network of icon sites along the River Murray corridor. A program of research, funded under Subprogram II of Rivers Environmental Restoration Program, is developing decision support systems underpinned by hydrological and ecosystem response models for some of these significant wetlands.

Northern basin wetlands and their management challenges

The combined effects of water abstraction and prolonged drought have taken their toll on the wetlands of the northern Murray-Darling Basin, with reduced inundation frequency and extent leading to declining ecological condition. This has become obvious in the Ramsar-listed wetlands of the basin, where management attention has been greatest and the monitoring effort most thorough (Kingsford 2000; Thoms *et al.* 2007).

Within the northern basin, several wetlands are of particular significance for the breeding of colonially nesting and other waterbirds: the Narran Lakes, the Paroo, the Gwydir Wetlands, the Macquarie Marshes, the lower Lachlan and the lower Murrumbidgee (or Lowbidgee) floodplain. In each of these sites historical records show the nesting of several hundred thousand breeding pairs (Kingsford and Thomas 2004; Kingsford and Auld 2005; Thoms *et al.* 2007).

Of these, the Paroo is the least affected by water regulation and the natural values have been largely preserved (Kingsford 1999). The Paroo responded well to flooding in the summer of 2007/08 (Kingsford, *pers. comm.*). The wetlands within the regulated valleys have fared less well, as outlined below.

Narran Lakes

The Narran Lakes are the northern-most of the Ramsar-listed wetlands in the New South Wales portion of the Murray-Darling Basin (Fig. 4.1/colour plate, p. 301) and are largely dependent on cross-border inflows from Queensland. Water development in Queensland has seen the frequency of moderate-sized floods into Narran decline, particularly since 1992 (Thoms *et al.* 2007). Vegetation health has suffered, with surveys indicating high morality among river red gums (54%), river coobas (45%) and coolibahs (38%) (Thoms *et al.* 2007).

The summer of 2007/08 saw the first significant inflows into the Narran Lakes in five years, and a much-publicised breeding event of straw-necked ibis. In the course of eight weeks between mid-December 2007 and mid-February 2008, 24 GL passed the Narran Park gauge, flooding most of the northern lake. A colony of 30 000 breeding pairs of straw-necked ibis were noted by early February, which grew to over 50 000 nests by mid-March (Kingsford *et al.* 2008).

It became obvious by mid-March 2008 that nest abandonment was a real possibility, with water levels in the northern lake falling by nearly 2 cm per day. The Murray-Darling Basin Commission intervened to provide $2 million in funding for the purchase of water from private storages on the Narran River. Peter Terrill (Senior Wetlands and Rivers Conservation Officer, DECCW) coordinated the delivery of 8100 ML that sustained the breeding event a further five weeks until late April, by which time over 19 000 chicks had fledged. However, many nests were abandoned (Kingsford *et al.* 2008).

Never before in the historical records had birds bred with such low flows into the Narran Lakes. Previous ecosystem response models had suggested that a flow of 100 GL at Wilby Wilby was required as a trigger for a major ibis breeding event (Terrill 2008; Kingsford *et al.* 2008). Perhaps this was prompted by the extraordinary length of time since the last major breeding event in the basin, two years previously, when more than 20 000 breeding pairs nested at Telegraph Swamp in the Lowbidgee and 15 000 pairs bred on Merrowie Creek in the lower Lachlan. An alternative hypothesis is that the event was triggered by flows within the catchment which, were they allowed to enter the lakes, would have been sufficient to sustain water levels for the duration of the event.

While the establishment of a large breeding colony between Clear and Back Lakes is heartening, only one quarter of the Narran system was inundated in the 2007/08 event (Thoms, *pers. comm.*) and the condition of the broader Narran ecosystem continues to decline. Recent (post-flood) surveys have indicated that although lignum condition in the breeding areas responded positively to inundation, the condition of lignum elsewhere on the floodplain is poor, a factor that potentially limits the size of breeding colonies in larger floods (Capon 2009).

Macquarie Marshes

The Macquarie Marshes covers approximately 110 000 ha of semi-permanent and ephemeral floodplain vegetation downstream of Warren. As early as the 1900s, the Macquarie Marshes were designated a game reserve and parts of the marshes, including the northern and southern nature reserves, were listed as wetlands of international significance under the Ramsar convention in 1985 (Masman and Johnstone 2000).

Notwithstanding the international recognition of the value of the Macquarie Marshes, a succession of low-inflow years have taken their toll. Vegetation community mapping co-ordinated by the NSW Department of Environment, Climate Change and Water under the Wetland Recovery Plan has demonstrated a gradual 'terrestrialisation' of vegetation. Between 1991 and 2008, cumbungi declined in the Northern Nature Reserve by 75%, water couch by 87%, coolibah by 97% and river red gum with wetland understorey by 22%. In the same period, drought-tolerant chenopod shrubland, virtually absent from the reserve in 1991, increased to over 7000 ha (Bowen and Simpson 2008a). The annual waterbird survey of eastern Australia has documented a decline in the populations of waterbirds in the northern reserve, from more than 30 000 birds in the early 1980s to 10 birds counted in the 2008 survey (Kingsford and Porter 2008). A small egret rookery in 2008 (2000 breeding pairs) was the first breeding of colonial waterbirds in the marshes in eight years.

Gwydir Wetlands

The terminal wetlands of the Gwydir and Gingham channels were included on the Ramsar estate in 1999. Historically, the Gwydir Wetlands have been perceived as one of the most important waterbird breeding sites in New South Wales, with early reports of hundreds of thousands of breeding pairs (Morse 1922). A flood in 1998 of 192 000 ha prompted one of the largest bird breeding events in the Murray-Darling Basin in recent decades, with estimates of 150 000–183 000 breeding pairs of straw-necked ibis, 27 000 pairs of intermediate egret, over 69 000 pairs of nankeen night heron and 38 000 pairs of Australian white ibis (DECCW 2008).

Recent surveys of the Gwydir Wetlands under the Wetland Recovery Plan (Bowen *et al.* 2008) have noted a shift to terrestrial vegetation. The condition of river red gum has declined in many parts of the Gwydir Wetlands, due to extended drought and the abstraction of water for irrigated agriculture. Added to the complexity of vegetation dynamics in the Gwydir is the widespread clearance of woody wetland vegetation between 1996 and 2005. Over this period 45% of coolibah woodland, 75% of the watercouch/spike rush/tussock rush association, 58% of marsh clubrush and 34% of the river cooba/lignum shrubland has been lost. The invasive groundcover lippia spread over the same period to cover more than 4500 ha (Bowen *et al.* 2008).

Recent surveys have shown few birds in the Gwydir, with only 1007 waterbirds of 37 species identified in aerial and on-ground surveys across the wetlands in spring 2007 (DECCW 2008; Ch. 7, this volume). In 2008, after modest environmental flows, numbers of waterbirds increased modestly. Most increases were dominated by ducks occupying water storages. Few wading birds were identified.

Lowbidgee

Yanga National Park was purchased by the NSW government in 2005 for what was, at the time, the largest sum ever paid for an addition to the reserve system. Yanga contains the largest contiguous stand of river red gum in the lower Murrumbidgee floodplain, along with semi-permanent lakes and billabongs that support significant populations of the endangered southern bell frog.

Recent assessment and modelling of hydrological changes in the Murrumbidgee River have demonstrated declines in the frequency of medium to high flows that would normally have inundated the ephemeral wetland vegetation of Yanga National Park, with 1970 emerging as a turning point (Wen 2009). Indeed, average peak flows in November, the month of historically highest flow, have decreased by as much as 90%.

The reduction in flow over recent decades is reflected in the changing condition of river red gum stands, mapped by McCosker (2008) and analysed by Wen *et al.* (Ch. 15, this volume). While stands lining the river have maintained good condition, higher or less frequently inundated stands have substantially declined in health. The reduction in flow size and frequency has also taken a toll on the southern bell frog, with the local extinction of some populations (Ch. 14, this volume).

Implications

This brief survey of the significant wetlands of the northern basin demonstrates the ubiquity of the problems caused by the combined effects of prolonged drought and water abstraction. While there is no evidence of the problems of salinisation and acid sulphate leachate that compound the problems of the wetlands of the River Murray corridor, the northern basin wetlands are in a critical state of decline. Certainly, it cannot be argued that the northern basin is in better condition than the south. Nor does moving a greater quantum of water from the northern basin to the Murray corridor necessarily represent a wise strategy for achieving basin-wide conservation outcomes. As Bill Johnson succinctly noted, 'There is a Coorong story at the end of every river in the Murray-Darling Basin'.

Ecosystem response modelling and the management of environmental water

The protracted drought conditions of the past few years have exposed inadequacies in the balance between environmental and extractive water use. Contributing to these mistakes has been a lack of detailed information on the water requirements of large wetland systems, and the hydrological inputs required to sustain them. It was rare that the specific ecological consequences of individual decisions on water allocation were flagged at the time they were made, either because the ecosystem response models were not fully developed or because the decision-making process did not fully incorporate consideration of environmental costs.

Modelling the response of key ecological assets is necessary if we are to appreciate the opportunity costs of extractive water use. Individuals and committees charged with making decisions about the allocation of water or the use of environmental water need to be aware, not only of the benefits of a water use strategy, but the immediate and long-term costs to the environment of those decisions, both individually and collectively. The extremely dry conditions since 2000 and the water-use strategies adopted have led to widespread ecosystem collapse in the wetlands of the northern Murray-Darling Basin.

To avoid the mistakes of the past, we recommend three principles that should underpin management of significant wetland complexes.

First, there should be clear articulation of specific objectives for wetland management, in terms of a desired ecological extent and condition of the wetland within naturally wet and dry phases. These objectives should be quantitatively defined where possible, setting a benchmark against which our environmental watering strategies can be judged. Environmental objectives need to be formulated at three spatial scales: the scale of the wetland, the scale of the catchment and the scale of the basin. The Adaptive Environmental Management Plans being developed for the Macquarie Marshes and Gwydir Wetlands will provide a useful case study in the development of objectives for two significant wetlands in the northern basin. The basin plan being developed by the Murray-Darling Basin Authority provides the opportunity for an overarching plan within which wetland and valley-specific plans can be nested.

Second, strategies to meet those objectives need to be defined on the basis of the best available ecosystem response modelling. These models should provide the capacity to compare alternative water use strategies in terms of the likely fulfilment of objectives defined for the ecological system, both in the short and long terms. Ideally, we would have decision support tools developed that could identify and document the risk for habitat decline or population impacts associated with individual decisions or longer-term water-use strategies.

At a higher level, similar decision support tools need to be developed at the scale of the valley and basin, where higher-level objectives will inform objectives set at the scale of the wetland, and transparent decisions can be made about water at these scales. For example, the opportunistic nature of bird breeding may require objectives for the frequency and size of bird breeding events set at a broader spatial scale (e.g. the basin) to have primacy over similar objectives at finer spatial scales. This might, for example, justify the diversion of resources to large spontaneous events such as recently occurred at Narran Lakes during the 2008 flood. A nested set of objectives over basin, valley and wetland scales would lend weight to such a strategy.

Finally, monitoring should complete the adaptive management loop between the formulation of objectives and strategies and their implementation. Certainly the immediate outcomes of a watering event require documentation, and contribute to the improvement of hydrological and ecosystem response models. Importantly, though, the longer-term trends in the extent and condition of the resource, in comparison to model forecasts and the objectives set for the wetland, need to be continually documented. Too often monitoring has demonstrated small victories in the benefits of environmental water when overall we have been comprehensively losing the war.

Subprogram II of Rivers Environmental Restoration Program

Background and objectives

The water planning framework in the regulated rivers of the northern Murray-Darling Basin was largely in place by the turn of the century. These plans were negotiated using hydrology as a common language at a time when environmental needs had not been clearly articulated in hydrological terms and when the current protracted drought had not been anticipated. The uniquely dry condition of the past eight years has brought into sharp focus the over-allocation of water in the basin and the vulnerability of rivers and wetlands to ecological collapse.

The response of the NSW and Commonwealth governments has been to enter the water market to establish a reserve of strategic environmental water. This water could be used to supplement water made available under the water-sharing plans without the compulsory re-acquisition of entitlements. Funding the acquisition of entitlements from willing sellers is the main purpose of the $172 million Rivers Environmental Restoration Program.

The RERP was jointly funded by the NSW government ($101 million) and Commonwealth government ($71 million) under the Water Smart Australia program, commencing in 2007. The program has the overarching objective of improving the ecological conditions of significant ecological assets in five valleys: the Macquarie, Gwydir, Narran, lower Lachlan and lower Murrumbidgee. This is achieved through four subprograms, comprising the four key components of management intervention:

- subprogram I – acquisition and management of water for environmental purposes in NSW river valleys is the primary activity under the RERP;
- subprogram II – better use of environmental water (science);
- subprogram III – better delivery of environmental water (infrastructure);
- subprogram IV – benefits of environmental water on private land (community engagement).

Key elements of Subprogram II

The ability to specify environmental water requirements of key wetland assets in terms of inundation patterns (volumes, duration, season or timing) is critical to informing the next

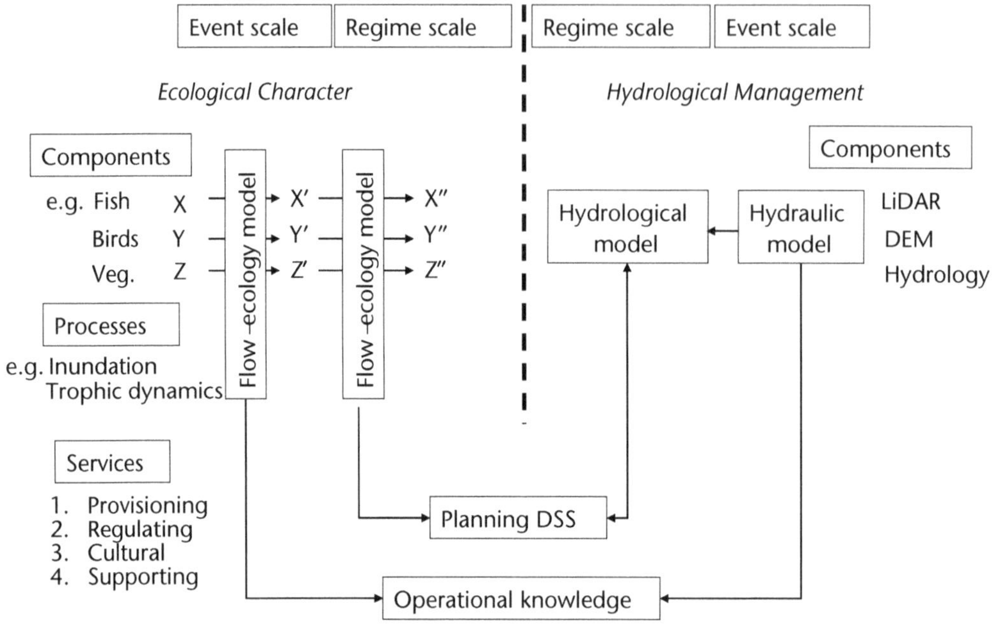

Figure 4.2: Conceptual model of the objectives of Subprogram II.

round of water planning, as well as effectively managing wetlands from year to year (Fig. 4.2). The objectives of Subprogram II, the science components of RERP, are therefore:

- improving our understanding of the water needs of stressed systems;
- demonstrating the ecological outcomes of environmental water management;
- increasing our capacity for adaptive management and efficient use of water for the environment.

In its simplest form, an ecosystem response model combines the insights of a hydrology model and a model of the ecological character of a wetland (Fig. 4.2) in response to flow. Subprogram II is developing decision support tools as a front end to ecosystem response models for the Macquarie Marshes, Gwydir Wetlands, Narran Lakes, lower Lachlan and Lowbidgee wetlands.

Several projects contribute to the development of a wetland hydrology model. The approach of Subprogram II is to combine the insights of historic inundation mapping (Ch. 9, this volume) and detailed hydrodynamic models based on the physical characteristics of the wetland (Figs 4.3, 4.4).

Improved hydrodynamic modelling is being made possible through the acquisition of LiDAR-based elevation models covering 1.2 million ha of significant wetland (including bare earth and canopy elevation models), improved gauging in wetlands and the detailed survey of soil properties including hydrological conductivity. These efforts will extend the capacity of river hydrology models, such as IQQM, to forecast changes in floodplain wetland hydrology over time scales relevant to water-sharing plans and changes in climate.

The relationships between hydrology and ecology in these wetlands are being clarified through time-series mapping of vegetation condition, in time scales ranging from the individual flow event to interdecadal sequences. Experimental studies into the response of birds, fish, amphibians and invertebrates to environmental flows are helping to define the effect of managed flows on the diversity, abundance and trophic dependencies of aquatic organisms

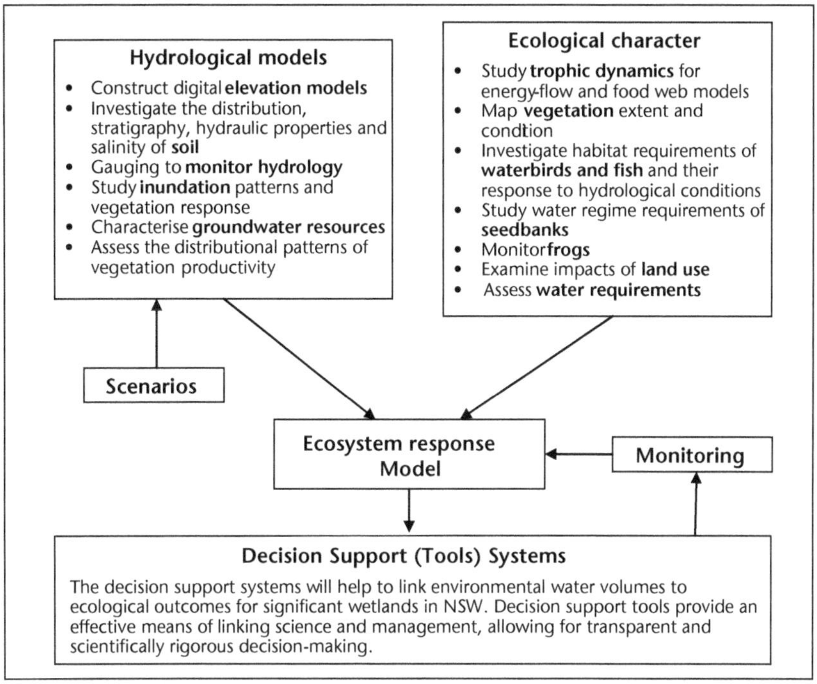

Figure 4.3: Modelling components of Subprogram II.

(Chs 6 and 7, this volume). All these relationships are being incorporated into decision support tools that will give wetland managers ready access to ecosystem response models to help them make scientifically rigorous and transparent decisions in the context of wetland and environmental water planning (Ch. 8, this volume; Fig. 4.5).

Outstanding science questions

Subprogram II will deliver improved wetland hydrology and ecological response models by 2011, in time for application in the development of water-sharing plans when many of these are

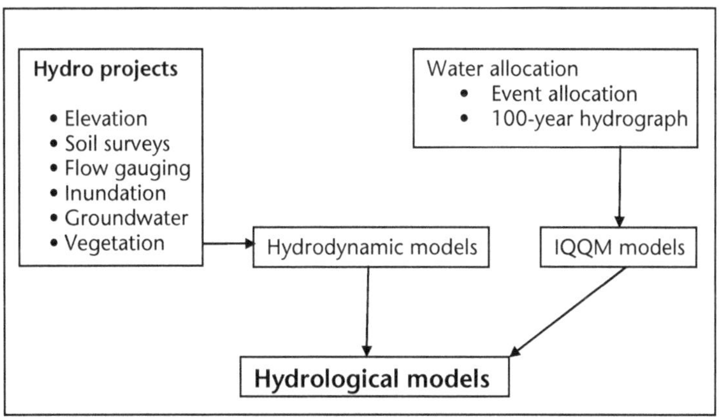

Figure 4.4: Components of the hydrological models constructed for significant wetlands under Subprogram II.

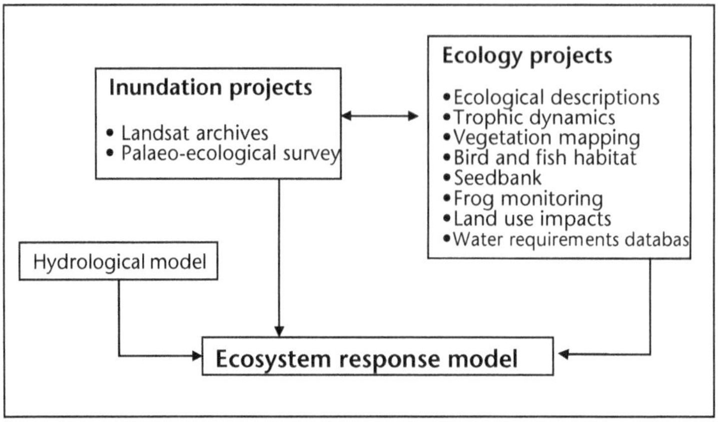

Figure 4.5: Components of the ecosystem response models constructed for significant wetlands under Subprogram II.

revised in 2013/14. The program has provided a useful means of integrating best available science in a decision support framework, but several significant knowledge gaps constrain our capacity to plan for water use over larger temporal and spatial scales. Here we identify three: improved wetland inventory at the valley and basin scales, the potential impacts of climate change, and the temporal and spatial scales governing recruitment of organisms to wetlands.

Need for improved inventory

Wetland scientists in the US recently celebrated the 25th anniversary of their National Wetlands Inventory. This inventory provides a unifying framework for forecasting the impacts of climate change, the enforcement of the 'no-net-loss' policy and systematic conservation planning for wetlands in the US. A national wetlands inventory is still an aspirational goal in Australia, where wetland mapping is incomplete and inconsistent between jurisdictional boundaries. The danger of our focus on icon sites is that the remaining 99% of wetlands are forgotten and degraded. As a national priority, the mapping and inventory framework developed by the Queensland Wetlands Program should be adopted and funded nationally.

Climate change

Global climate models, downscaled to regional projections, have indicated lower water availability in many of the rivers and wetlands of the Murray-Darling Basin (Dunlop and Brown 2008). The work of the CSIRO Murray-Darling Basin Sustainable Yields project has translated these projections into valley-specific forecasts of water availability (CSIRO 2008).

The current extended dry period has commonly been interpreted in the context of these projections as indicative of broader trends of climatic regime shift, the new reality of hydrological conditions in the valley. As a result, wetland managers are scaling back expectations of what might be achieved in wetlands, and in some cases are considering hard engineering solutions that might preserve smaller wetland areas at the expense of the wider historic floodplain. These decisions may be irrevocable.

The phases of the *El Niño* southern oscillation, and more broadly the Pacific Decadal Oscillation (PDO) and the Indian Ocean Dipole, influence climatic and hydrological variability over large sections of the Murray-Darling Basin during summer and winter respectively (Verdon *et al.* 2004; Verdon and Franks 2005). Moderate to high-level flooding has occurred in some catchments when the positive phase of *El Niño* has coincided with the negative phase of the PDO (Ralph 2008), and a periodicity of the PDO as suggested by some historic and palaeolimnological data. Declines in autumn rainfall, that account for most of the rainfall decline in south-east Australia, are less closely linked to these oceanic drivers and are more probably related to greenhouse gas-induced climate change (Murphy and Timbal 2008).

It would be helpful if models projecting hydrological variability reconciled the forecasts of the downscaled global climate models with the insights of palaeo-ecological research into interdecadal variability of rainfall within the basin. This would provide wetland managers with a better sense of the relative significance of natural variability and human-induced climate change (and their interaction) over time scales relevant to water planning.

Recruitment dynamics and system resilience

The rotifer *Lecane shieli* was recently discovered in the Macquarie Marshes (Kobayashi *et al.* 2008). It is the only occurrence of this species in Australia and was previously identified only in Thailand and Turkey. It is an introduced species, not through human agency but more probably by migratory birds using the east Asian–Australasian Flyway. The observation is illustrative of the variety of recruitment vectors and the spatial scales over which they operate, a distinguishing feature of wetlands.

Biological diversity at the scale of the individual wetland depends on the rates of recruitment and local extinction. There are a number of factors that influence extinction rates, the primary one being drying of the wetland (Brock 1986; Boulton and Brock 1999). In the highly variable climatic conditions of inland rivers, organisms have adapted to periodic drying though a range of mechanisms from setting seeds to dispersion (Boulton and Brock 1999).

The presence of *Lecane shieli* in the Macquarie Marshes points to the importance of birds as a vector of genetic material. The affinities in wetland vegetation between coastal and inland environments and along the great flyways are further testament to the importance of birds as agents of recruitment. We still know little about the spatial scales over which colonially nesting waterbirds move in search of habitat, or how protracted drought influences population sizes.

In the end, the decline of biological diversity in the wetlands of the Murray-Darling Basin may not be driven by local extinction so much as by the failure of recruitment mechanisms, including the collapse of bird populations nationally and, for migratory species, internationally. Basin-wide watering strategies should focus on the support of basin-wide populations.

Conclusions

The years between 2001 and 2009 have seen a widespread natural experiment into the resilience of wetlands within the Murray-Darling Basin. Research questions concerning the thresholds for condition and survival of a range of ecosystem components are being answered. The predictable results have been the loss of wetland vegetation, the encroachment of semi-arid vegetation and the failure of waterbird breeding throughout the basin.

The mistakes made over the past decade should inform the development of our ecosystem response models. We now have, for example, ample evidence from which to construct the flood recurrence intervals required to sustain floodplain wetland vegetation communities. We also have the capacity to develop better hydrological models of large floodplain wetlands, made possible through advances in critical data acquisition and modelling.

Such has been the decline of these wetland systems that the focus of coming decades will have to be rehabilitation rather than maintenance. It is unclear what will emerge from the long dormancy of the wetlands, whether the original condition or something quite different. The development of models that can predict the response of key ecological assets and their testing is an iterative approach, expressed through management decisions and the monitoring of outcomes. Hopefully, transparent decisions based on scientifically rigorous research will lead to improvements in the condition of significant wetlands in the northern Murray-Darling Basin.

Acknowledgments

The work of Subprogram II of Rivers Environmental Restoration Program is funded through the Water for the Future Program of the Commonwealth government of Australia.

References

Boulton AJ and Brock MA (1999). *Australian Freshwater Ecology: Processes and Management.* Gleneagles Publishing, Adelaide.

Bowen S and Simpson S (2008a). Changes in extent and condition of the vegetation communities of the Macquarie Marshes Floodplain 1991–2008. Draft Macquarie Marshes Adaptive Environmental Plan. Australian Government Water for the Future.

Bowen S and Simpson S (2008b). Changes in extent and condition of vegetation communities in the Gwydir Wetlands and Floodplain 1996–2005 (Draft). NSW Dept of Environment and Climate Change/Australian Government Water for the Future.

Bowen S, Simpson S and McCosker R (2008). Changes in extent and condition of floodplain wetland vegetation communities in the Gwydir Wetland and floodplain. NSW Dept of Environment and Climate Change, Sydney.

Brock MA (1986). Adaptations to fluctuation rather than to extremes of environmental parameters. In *Limnology in Australia.* (Eds P De Deckker and WD Williams) pp. 131–140. CSIRO Publishing/Junk, Melbourne/Dordecht.

Capon S (2009). Vegetation response to flooding, Narran Lakes, 2008–2009. Draft Report to the NSW Dept of Environment and Climate Change. Riverine Research Laboratory, University of Canberra.

CSIRO (2007). Water availability in the Gwydir. Report to the Australian Government from the CSIRO Murray-Darling Basin Sustainable Yields Project. CSIRO, Australia.

CSIRO (2008). Water availability in the Macquarie-Castlereagh. Report to the Australian Government from the CSIRO Murray-Darling Basin Sustainable Yields Project. CSIRO, Australia.

Davies P and Acreman M (2003). Review of the interim report of the SRP, the Murray Flow Assessment Tool and the SRP-REG scientific process for the Murray-Darling Basin Commission.19 September 2003.

DECCW (2008). Final report on the Gwydir waterbird and fish habitat study. NSW Wetland Recovery Program. Rivers and Wetlands Unit, NSW Department of Environment, Climate Change and Water.

Dunlop M and Brown PR (2008). Implication of climate change for Australia's national reserve system: a preliminary assessment. CSIRO Sustainable Ecosystems report to Dept of Climate Change and Dept of the Environment, Water, Heritage and the Arts, Canberra.

Franks SW and Kuczera G (2002). Flood frequency analysis: evidence and implications of secular climate variability, New South Wales, *Water Resources Research* **38(5)**, 1062.

Kelleway J, Mazumder D, Wilson G and Kobayashi T (2010). Using isotopic techniques to assess trophic structure in northern Murray-Darling Basin wetlands. In *Ecosystem Response Modelling in the Murray-Darling Basin.* (Eds N Saintilan and I Overton) pp. 85–102. CSIRO Publishing, Melbourne.

Kingsford RT (1999). A free-flowing river: the ecology of the Paroo River. National Parks and Wildlife Service, Sydney.

Kingsford RT (2000). Review: ecological impacts of dams, water diversions and river management on floodplain wetlands in Australia. *Austral Ecology* **25**, 105–127.

Kingsford RT and Auld KM (2005). Waterbird breeding and environmental flow management in the Macquarie Marshes, arid Australia. *River Research and Applications* **21**, 187–200.

Kingsford RT and Porter JL (2008). Aerial survey of wetland birds in eastern Australia: October 2007. Annual Summary Report. University of New South Wales, Sydney.

Kingsford RT and Thomas RF (2004). Destruction of wetlands and waterbird populations by dams and irrigation on the Murrumbidgee River in arid Australia. *Environmental Management* **34**, 383–396.

Kingsford RT, Brandis K and Porter JL (2008). Waterbird response to flooding in the northern Murray-Darling Basin 2008. School of Biological, Environmental and Earth Sciences, University of New South Wales, Sydney.

Kobayashi T, Shiel RJ and Segers H (2008). First record of the rotifer *Lecare shieli* Segers & Sanoamuang, 1994 from Australia. *Zoologist* **34(2)**, 181–183.

Masman K and Johnstone M (2000). *Reedbed Country: The Story of the Macquarie Marshes.* Macquarie Marshes Catchment Management Committee.

McCosker R (2008). Yanga Vegetation Mapping: Historical Community Extent and Condition. Technical report to NSW Dept of Environment and Climate Change, Sydney.

Morse FC (1922). Egret and glossy ibis rookeries. *Emu* **22**, 36–39.

Murphy BF and Timbal B (2008). A review of recent climate variability and climate change in southeastern Australia. *International Journal of Climatology* **28(7),** 859–879.

Ralph TJ (2008). Channel breakdown and floodplain wetland morphodynamics in the Macquarie Marshes. PhD thesis. Dept of Physical Geography, Macquarie University, Sydney.

Terrill P (2008). Colonial waterbird breeding, Narran Lakes Nature Reserve, Summer–Autumn 2008. Status Report, 27 March 2008. NSW Dept of Environment and Climate Change.

Thomas RF, Bowen S, Simpson SL, Cox SJ, Sims NC, Hunter SJ and Lu Y (2009). Inundation response of vegetation communities of the Macquarie Marshes in semi-arid Australia. In *Ecosystem Response Modelling in the Murray-Darling Basin.* (Eds N Saintilan and I Overton) pp. 137–150. CSIRO Publishing, Melbourne.

Thoms M, Capon S, James C, Padgham M and Rayburg S (2007). The Narran Ecosystem Project: the response of a terminal wetland system to variable wetting and drying. Final report to the Murray-Darling Basin Commission. MDBC Publication No. 40/08.

Verdon DC and Franks SW (2005). Indian Ocean sea surface temperature variability and winter rainfall: eastern Australia. *Water Resources Research* **41**, W09413

Verdon DC, Wyatt AM, Kiem AS and Franks SW (2004). Multidecadal variability of rainfall and streamflow: eastern Australia. *Water Resources Research* **40**, W10201.

Wassens S (2010). Flooding regimes for frogs in lowland rivers of the Murray-Darling Basin. In *Ecosystem Response Modelling in the Murray-Darling Basin.* (Eds N Saintilan and I Overton) pp. 213–228. CSIRO Publishing, Melbourne.

Wen L (2009). Reconstruction of natural flow in a regulated system, the Murrumbidgee River, Australia, using time series analysis. *Journal of Hydrology* **364(3–4),** 216–226.

Wen L, Saintilan N, Rogers K and Ling J (2010). Linking river red gum condition to hydrological change at Yanga National Park. In *Ecosystem Response Modelling in the Murray-Darling Basin.* (Eds N Saintilan and I Overton) pp. 229–242. CSIRO Publishing, Melbourne.

Wilson GG, Spencer JA and Heagney EC (2010). Responses of fish and waterbirds to flow variability in the Gwydir wetlands. In *Ecosystem Response Modelling in the Murray-Darling Basin*. (Eds N Saintilan and I Overton) pp. 103–118. CSIRO Publishing, Melbourne.

Young WJ, Scott AC, Cuddy SM and Rennie BA (2003). Murray Flow Assessment Tool: a technical description. Client Report 2003. CSIRO Land and Water, Canberra.

Chapter 5

Comparing fish biomass models based on biophysical factors in two northern Murray-Darling Basin rivers: a cautionary tale

Stephen Balcombe, Joel Huey, Jaye Lobegeiger, Jonathan Marshall, Angela Arthington, Louisa Davis, David Sternberg and Martin Thoms

Introduction

The Queensland–upper NSW rivers of the northern Murray-Darling Basin are typical lowland rivers in that they generally flow through lowland (<300 m above sea level) areas (Harris and Gehrke 1997). As lowland rivers they are characteristically linked to their floodplains and many sections are dominated by sinuous anabranching and distributary systems (Thoms and Sheldon 2000). Lowland floodplain regions are geomorphologically diverse as a result of their lateral instability and contain features such as anabranches, billabongs, cut-offs, backwaters, scroll-swales and depression basins. The northern Murray-Darling Basin rivers are best described, however, as dryland rivers as much of their length flows through arid and semi-arid landscapes (<500 mm p.a.) and they thus experience large-scale evaporative water losses. The more easterly rivers of the northern Murray-Darling Basin tend to contribute the greatest run-off to the Murray-Darling Basin as they drain the western margins of the Great Dividing Range (Thoms and Sheldon 2000). These rivers include the Macintyre and Condamine-Balonne in southern Queensland and the Gwydir, Namoi, Castlereagh and Macquarie Rivers in northern NSW. In contrast, the westerly rivers originate in the more arid landscapes of south-west Queensland (Warrego and Paroo Rivers) and tend to contribute less to annual Murray-Darling Basin flows, although they can produce significant inputs during major floods.

The dryland rivers of the northern basin are characterised by highly variable flow regimes, driven by the variable climatic conditions in the region (Medeiros and Arthington 2008). In dryland rivers, flow is irregular and infrequent, with the main channel of catchments dominated by isolated refugial waterholes for most of the year (Knighton and Nanson 1994). These waterholes support aquatic and terrestrial biota (and supply water for stock, irrigation and human consumption) during dry periods, allowing populations to persist in a highly variable and harsh environment (Morton *et al.* 1995). During the summer period, wetter conditions prevail, generating channel flows and the occasional widespread flood which connect the otherwise isolated waterholes. This connectivity generates the opportunity for biota to recolonise flooded waterholes and utilise new resources (Balcombe *et al.* 2007).

The combination of highly dynamic hydrological regimes and geomorphological variability leads to a productive yet spatially and temporally variable suite of fish assemblages across these northern basin rivers (Gehrke *et al.* 1995; Balcombe *et al.* 2006; Lintermans 2007). Some of the more widely distributed and abundant fish species in these rivers include the medium- to large-bodied native species such as yellowbelly (*Macquaria ambigua*, Richardson, 1845), eel-tail catfish (*Tandanus tandanus*, Mitchell, 1838) and bony bream (*Nematolosa erebi*, Günther, 1868) and the exotic species such as goldfish (*Carassius auratus*, L., 1758) and common carp (*Cyprinus carpio*, L., 1758).

With the rest of the basin, the northern rivers have been affected to varying degrees by water resource development (Arthington 1996; Thoms *et al.* 2005). Fish stocks in particular have been reduced in the Murray-Darling Basin largely due to river regulation (MDBC 2004; Lintermans 2007). Future efforts for the rehabilitation of fish stocks, and hence management of rivers with altered flow regimes, require a thorough understanding of the drivers of fish population dynamics. This chapter presents a case study modelling the spatial patterns in fish biomass (of two common species) across two northern basin rivers. These models are compared for each species across the two rivers to demonstrate the applicability of generic models of fish productivity in the northern basin.

Comparing fish biomass models in two upper Murray-Darling Basin rivers

Factors that affect fish assemblages and abundance patterns in floodplain rivers operate at multiple spatial and temporal scales (Scheurer *et al.* 2003; Kingsford 2006). The influence of processes at each spatial and temporal scale must be considered to effectively manage species inhabiting variable environments (Fausch *et al.* 2002). Landscape-scale processes form and maintain river morphology and refuge habitats; fine-scale habitat characteristics regulate the functioning of local populations (Labbe and Fausch 2000). In the headwater rivers of the Murray-Darling Basin, the links between the physical setting, landscape characteristics (e.g. floodplain size and frequency of inundation) and the various levels of productivity (primary – algae, tertiary – predatory fish) are not well understood (Walker *et al.* 1995; Medeiros and Arthington 2008). Efforts to maintain or restore populations of iconic and threatened biota (e.g. fish) will be strengthened by sound understanding and predictive models of the influence of various drivers on species abundance and biomass (Fausch *et al.* 2002).

Broadly speaking, drivers of fish productivity can be divided into two different categories – indirect and direct. Indirect factors are those that drive productivity at lower levels of the food web (e.g. primary producers and herbivores) and hypothetically cascade upwards to drive productivity at higher trophic levels (Power 1992). For example, benthic and planktonic metabolism can be limited by low light levels (retarding photosynthesis) and by temperature and nutrient availability (Bunn *et al.* 2006a; Burford *et al.* 2008; Fellows *et al.* 2009). Direct factors influencing fish biomass include those that are important for determining the recruitment and mortality of fish populations (Welcomme 2001; Welcomme *et al.* 2006). These include aspects of habitat structure and availability at several spatial scales – the floodplain landscape, waterhole scale and local habitat scale (Puckridge *et al.* 2000; Knight and Arthington 2008).

Flow is another major driver of productivity in riverine ecosystems, shaping ecological responses in dryland rivers (Puckridge *et al.* 2000; Kingsford 2006). From an indirect perspective, the timing of flow events determines which physical characteristics shape ecological responses (Humphries *et al.* 1999; King *et al.* 2003). For example, it is proposed that, soon after a flood, the impact of the local biophysical template is tempered and that factors linked to connectivity among waterholes (Amoros and Bornette 2002; Sheldon *et al.* 2002) and catchment-scale processes such as floodplain inundation (Welcomme *et al.* 2006; Kingsford 2006) are important for determining ecological processes within waterholes. As the period of time since inundation increases and waterholes progressively isolate, localised factors increase in importance and dominate ecological processes within waterholes (Arthington *et al.* 2005). From a direct perspective, flow plays a dominant role in determining recruitment in many Australian freshwater fish, with increases in spawning usually coinciding with wetter months (Pusey *et al.* 2004; Balcombe and Arthington 2009).

In the dryland rivers of western Queensland, freshwater fish occupy the secondary and tertiary levels of the aquatic food web. In the headwater streams of the Murray-Darling Basin, species such as bony bream (*Nematolosa erebi*) graze upon sources of primary production (algae, macrophytes) then serve as food, along with other herbivorous species such as macroinvertebrates, for higher-order predators such as yellowbelly (*Macquaria ambigua*) (Balcombe *et al.* 2005; Sternberg *et al.* 2008). If we accept the bottom-up model of productivity regulation in dryland rivers and assume that other stressors to the system are equal across waterholes, we should observe higher fish biomass in waterholes that exhibit higher primary production (algal biomass). For example, Fellows *et al.* (2009) found some evidence for a relationship between rates of benthic primary production and fish abundances in Cooper Creek (Lake Eyre Basin) and the Warrego River (Murray-Darling Basin).

This chapter is concerned with the direct factors influencing fish biomass, particularly aspects of habitat structure and availability at several spatial scales – the landscape, waterhole and local habitat scales – and at two times of year with different flow conditions (no flow and post flow). It explores the possibility that similar predictive habitat–biomass relationships can be developed for the two catchments, the Moonie and the Warrego in the northern Murray-Darling Basin, and if not, why the relationships might differ.

This chapter aims to:

1 predict fish production in Moonie River waterholes using a suite of physical habitat variables measured largely at one spatial scale (whole waterhole);
2 predict fish production in waterholes in the Warrego River using a similar suite of physical habitat variables;
3 determine whether Warrego models are improved by including physical variables from a number of scales (floodplain, whole-waterhole, within-waterhole);
4 consider the transferability of these models between catchments.

Background to the Moonie and Warrego rivers

We chose to compare two catchments in the northern Murray-Darling Basin, the Moonie and Warrego River catchments (Fig. 5.1, Table 5.1). The Moonie catchment drains approximately 14 870 km^2 and runs for 575 km across the Queensland–NSW border into the Barwon River. The system has a semi-arid climate with an average rainfall of 500–600 mm yr^{-1} with considerable annual variation in temperature and interannual variation in rainfall (Sternberg *et al.* 2008). The river flows for only 35% of the time at Nindigully gauge (Gauge No. 417201B), and exists for much of the year as a series of unconnected waterholes (Sternberg *et al.* 2008). We chose 14 waterholes in this system distributed along the length of the main channel from the headwaters to the lower regions of the catchment (Fig. 5.1). For more information regarding site locations and catchment details, see Sternberg *et al.* (2008).

Table 5.1: Biophysical variables and relevant scales of measurement for waterholes in the Moonie and Warrego River catchments

Moonie catchment	Scale	Warrego catchment	Scale
Straight-line distance to nearest waterhole (SLD)	LAND	Total floodplain width (TFW)	LAND
Turbidity (TUR)	WHOLE	Effective floodplain width (EFW)	LAND
Area (A)	WHOLE	Channel distance to nearest waterhole (CD)	LAND
Depth (D)	WHOLE	Straight-line distance to nearest waterhole (SLD)*	LAND
Width (W)	WHOLE	Bifurcation ratio (BR)	LAND
Wetted perimeter (WP)	WHOLE	Number of channels (NOC)	LAND
Cross-sectional area (CSA)	WHOLE	Turbidity (TUR)*	WHOLE
Persistence (PERS)	WHOLE	Width (W)*	WHOLE
Maximum depth (MAXD)	WHOLE	Area (A)*	WHOLE
Volume (V)	WHOLE	Cross-sectional area (CSA)*	WHOLE
Perimeter (PER)	WHOLE	Width to depth ratio (WD)*	WHOLE
Length (L)	WHOLE	Wetted perimeter (WP)*	WHOLE
Planktonic volume (PLV)	WHOLE	Shape index (SI)	WHOLE
Benthic area (BENA)	WHOLE	Fetch length (FL)	WHOLE
Snags (SN)	WITHIN	Perimeter (P)	WHOLE
		Length (L)*	WHOLE
		Circularity (C)	WHOLE
		Elongation ratio (ER)	WHOLE
		Length to width ratio (LW)	WHOLE
		Volume (V)*	WHOLE
		Hydraulic radius (HR)	WHOLE
		Depth of cross-section (DCS)	WHOLE
		Mid-channel bars (MCB)	WHOLE
		Backwater (BAW)	WHOLE
		Off-take channels (OC)	WHOLE
		Bench 0–1/3A (B1)	WHOLE
		Bench 1/3–2/3 (B2)	WHOLE
		Bench 2/3–3/3 (B3)	WHOLE
		Side bars (SB)	WHOLE
		Backwater (BAW)	WHOLE
		Anabranches (AN)	WHOLE
		Bed and bank complexity (BBC)	WHOLE
		Eroding banks (EB)	WITHIN
		Snags (SN)*	WITHIN
		Anabranches (AN)	WITHIN
		Boulders (BOU)	WITHIN
		Fringing vegetation (FV)	WITHIN
		Overhanging vegetation (OV)	WITHIN

LAND = landscape, WHOLE = whole waterhole, WITHIN = within waterhole.
* = indicates variables used in reduced dataset for some regression analyses.

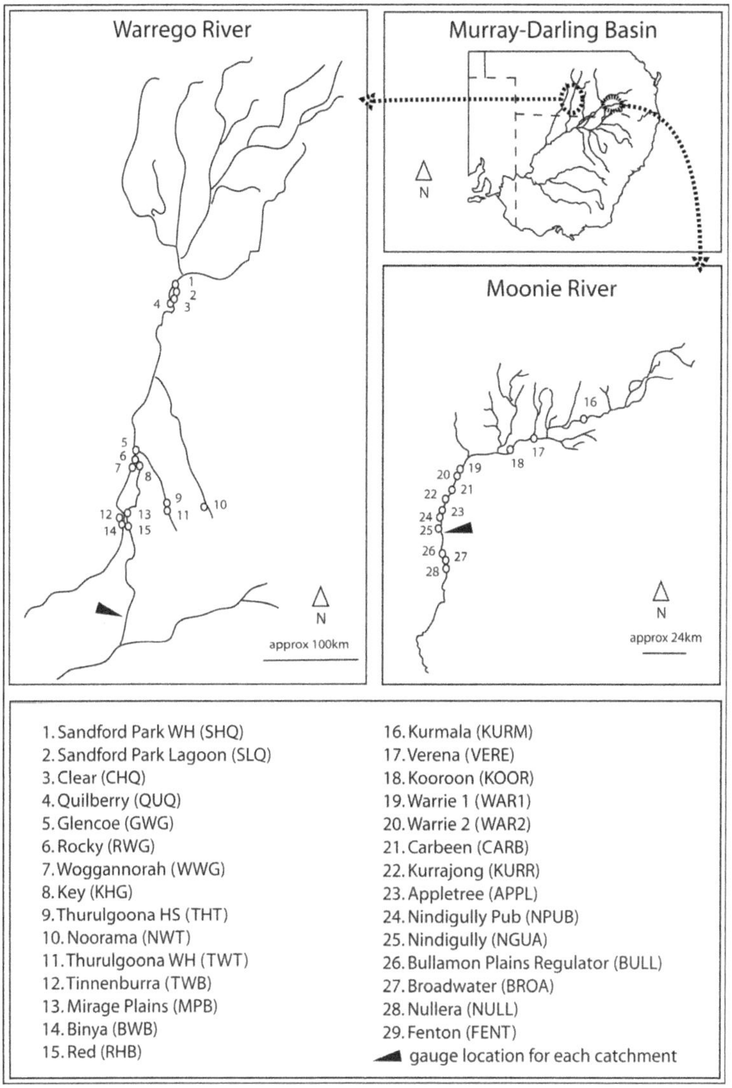

Figure 5.1: Map of study area and sites (and site abbreviations) in the Warrego and Moonie River catchments.

The much larger Warrego catchment covers approximately 70 000 km^2, with the vast majority (85%) located north of the Queensland–NSW border. The long-term average annual rainfall for this system is 450 mm yr^{-1}. However, this rainfall is highly variable and moderately seasonal, usually falling in the warmer months between December and March. Hydrology in the Warrego catchment is equally variable: peak discharges exceed 690 000 ML day^{-1} and the longest recorded period of no-flow reached 22 months in the 1950s at Ford's Bridge, NSW (Gauge No. 23001). Morphologically, the Warrego catchment ranges from constricted headwaters in narrow valleys to more sinuous and anastomosing channel structures in the lower parts of the catchment. This morphology gives rise to permanent and semi-permanent channel waterholes which provide significant drought refuge for aquatic and terrestrial biota. Fifteen waterholes in four reaches (Quilberry, Glencoe, Binya and Thurulgoona) were chosen for this study. More information regarding these waterholes and the Warrego catchment can be found in Balcombe *et al.* (2006).

The study species

For this study, the productivity (biomass per unit area) of two species was modelled. These species were selected to represent different trophic levels in the food web. Bony bream is a moderately sized clupeid and arguably the most widespread dryland fish species in Australia (Pusey *et al.* 2004). Typically, the species is associated with relatively deep slow-flowing habitats, but displays minimal association with waterhole microhabitat structure. Bony bream display strong tolerance of a range of environmental extremes (Ly 2007), an opportunistic reproductive strategy (Balcombe *et al.* 2007) and dietary plasticity (Sternberg *et al.* 2008). Given these adaptations, they often achieve high biomass in isolated waterhole environments. The species is primarily a detritivore/algivore, consuming variable amounts of microbenthic algae, and therefore may respond significantly to changes in primary production (Medeiros and Arthington 2008).

Yellowbelly is a moderate to large percichthyid commonly growing to 500 mm and less than 5 kg. It is often associated with deep slow-flowing pools, large woody debris and overhanging vegetation (Pusey *et al.* 2004). Yellowbelly are widespread and abundant in isolated waterholes of Australian dryland rivers due to a suite of evolutionary adaptations that allow them to cope with the range of environmental conditions present in these types of habitats (Arthington *et al.* 2005; Balcombe *et al* 2006). Females may reabsorb ripe ovaries during extended low flows, presumably to satisfy energy requirements. This indiscriminate carnivore sits at the highest trophic position in isolated waterholes of dryland rivers and therefore is likely to show a strong response to the cascading effects of limited bottom-up production in these environments.

Fish sampling

Fish in both catchments were sampled according to the methods outlined in Balcombe *et al.* (2006). Up to three fyke nets were set overnight at each site for approximately 19 hours, with a wing width of approximately 10 m. For the Warrego data, the wet weight of each individual was measured at the site. For the Moonie, wet weights were not measured at the time of sampling. Instead, the wet weight was derived from the standard length of the individuals using length to weight curves. These curves were generated for each species using data from later field trips in the same catchment at similar times of the year (yellowbelly: post-flow, $y = 0.0132x^2 - 1.19774x + 76.582$, $R^2 = 0.9721$; no-flow, $y = 0.0103x^2 - 1.1888x + 36.424$, $R^2 = 0.9739$; bony bream: post-flow, $y = 0.011x^2 - 1.4971x + 52.36$, $R^2 = 0.9665$; no-flow, $y = 0.012x^2 - 1.8939x + 82.556$, $R^2 = 0.9484$). For each site and sampling trip, biomass data were standardised to 19 hours and 10 m fyke wing width, and averaged across replicate nets within waterholes. The data represented catch per unit effort (CPUE) and catches were comparable across space and time.

Fish biomass was sampled from waterholes in the Moonie catchment in February, May and September 2006. The February 2006 trip occurred immediately after a 356 ML day^{-1} flow (417201B, Moonie River at Nindigully, 29 January 2006; Fig. 5.2), with no further flow events occurring until December 2006, three months after the September 2006 trip. The three trips were labelled 'summer, post-flow', 'autumn, post-flow' and 'spring, no-flow'.

In the Moonie River, fish production was highly variable through time. This was particularly so for bony bream, whose peak production was measured in the summer post-flow time period (744 g CPUE), falling to 184 g CPUE by the spring no-flow sampling period (Table 5.2). Peak yellowbelly production was also recorded in the summer post-flow period, but its production fluctuated less through time than that of bony bream. Across sites within any one time period, the greatest variation in yellowbelly production occurred in the summer period, as

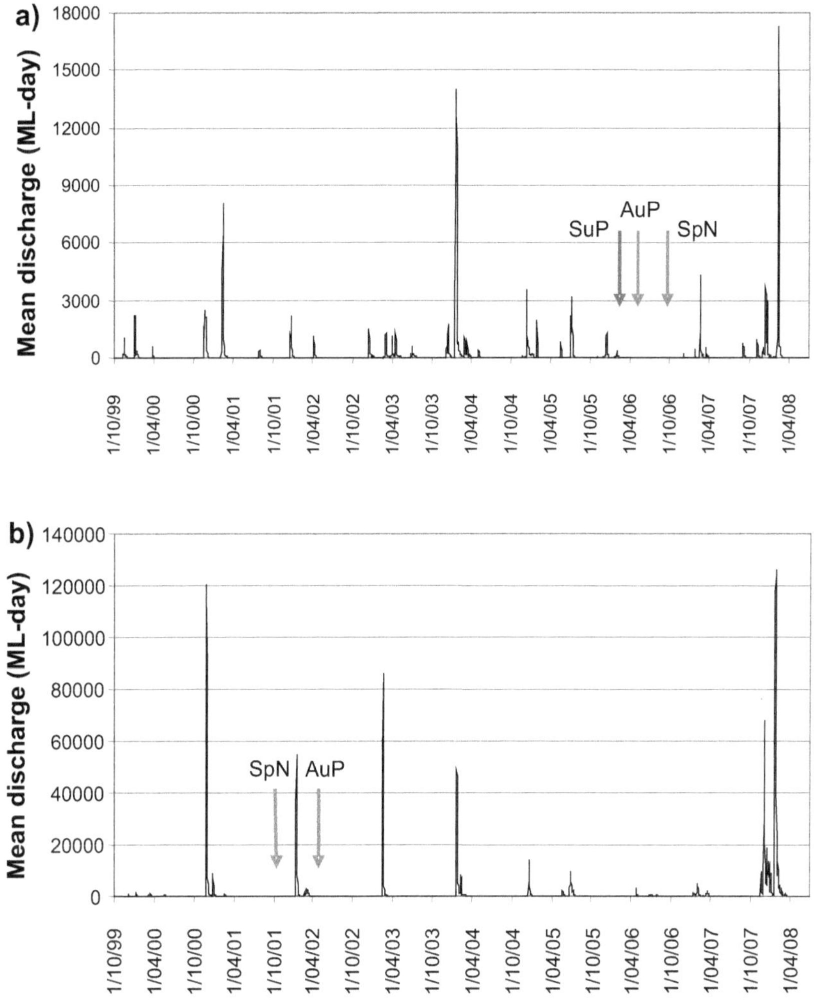

Figure 5.2: Daily discharge for (a) the Moonie River at Nindigully and (b) the Warrego River at Cunnamulla Weir. Arrows indicate individual sampling occasions. SpN = spring no-flow, AuP = autumn post-flow, SuP = summer post-flow.

demonstrated by the high production at Bullamon Plains Regulator and Fenton, intermediate production at sites such as Carbeen and Kurmala, then the majority of sites with very low production (Fig. 5.3a). There was also reasonably high variability in spring yellowbelly production but less fluctuation among sites in autumn. Bony bream production was also highly variably among sites and times, with the most variability evident in summer and autumn (Fig. 5.3b).

In the Warrego catchment, sampling occurred in October 2001 and April 2002. The October 2001 trip occurred after approximately 10 months without a significant flow in the system. In contrast, the April 2002 trip occurred approximately three months after a major flood peak (54 000 ML day^{-1} at Cunnamulla, 423202C, 14 January 2002; Fig. 5.2b). The October 2001 trip has been labelled 'spring, no-flow' and the April 2002 trip has been labelled 'autumn, post-flow'.

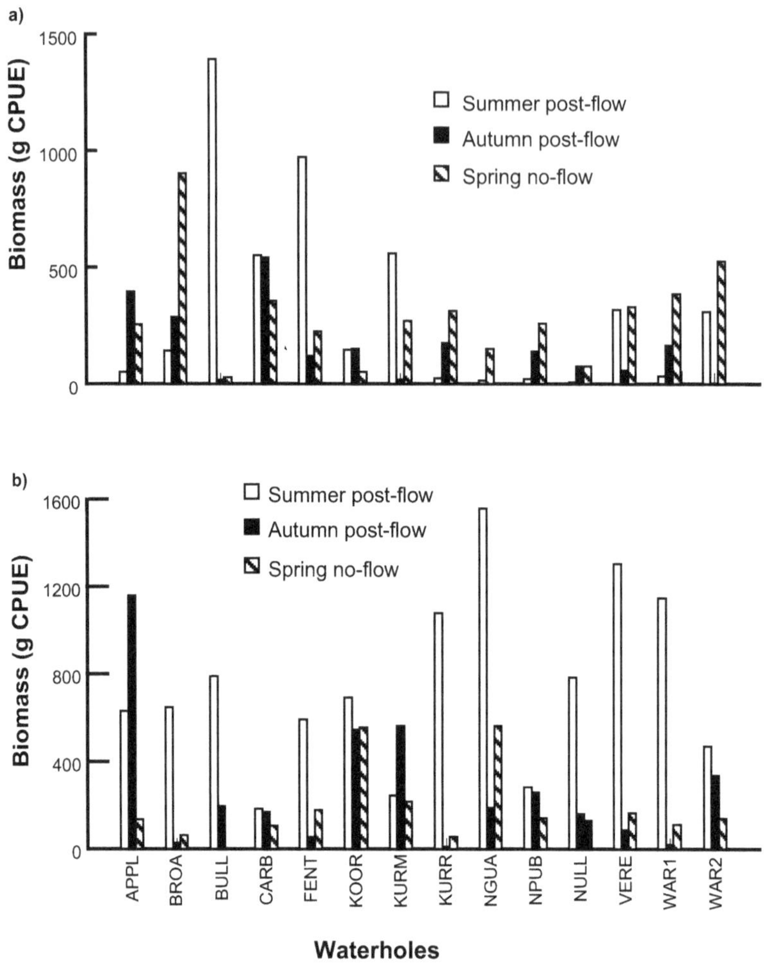

Figure 5.3: Fish biomass on three sampling occasions in 14 waterholes of the Moonie River catchment for (a) yellowbelly and (b) bony bream.

In the Warrego River, total yellowbelly and bony bream production were less temporally variable than in the Moonie River, with only a slight increase in yellowbelly and decrease in bony bream from spring no-flow to summer post-flow periods (Table 5.2). There was some variability among sites within each sampling period for both species (Fig. 5.4). There were also inconsistencies between the two sampling periods in relation to the site with peak yellowbelly production. In the 'spring, no-flow' period this was mostly reflected in Clear, Quilberry, Sandford Park WH and Tinnenburra, while in the 'autumn post-flow' period yellowbelly production was dominated by Noorama and, to a much lesser extent, Key and Thurulgoona HS (Fig. 5.4a). Similarly, bony bream production was largely dominated by five sites in spring (Key, Mirage Plains, Rocky, Tinnenburra and Wogganorah) and four in autumn with only one (Mirage Plains) also dominant in spring (Fig. 5.4b).

Physical variables

Variables that represent the physical template were estimated at three different spatial scales: landscape scale, whole-waterhole scale and within-waterhole scale. Landscape-scale factors

Table 5.2: Summary biophysical data for the Moonie and Warrego River catchments for all sampling occasions where measured

Sampling period	Factor	Mean	Range	CV (%)
Moonie biophysical				
All times	Depth	2.07	1.32–2.87	25
All times	Area x 10^3	65	10.8–266	106
All times	Cross-sectional area	34.9	16.2–60	34
All times	Width	26.5	18.6–35.2	19
All times	Length	2521	588–8673	92
All times	Volume x 10^3	77.4	9–224	89
Warrego biophysical				
All times	Depth	4.89	2.05–5.20	35
All times	Area x 10^3	111.5	3.2–315	770
All times	Cross-sectional area	174.7	79.9–358.6	50
All times	Width	61.1	4.19–96.9	37
All times	Length	3126	48.4–22413	177
All times	Volume x 10^3	522	14.4–1068	73
Moonie fish biomass				
All times	Yellowbelly biomass	260	0–1393	112
All times	Bony bream biomass	399	3.7–1557	99
Summer post-flow	Yellowbelly biomass	325	7.5–1393	129
Summer post-flow	Bony bream biomass	744	184–1557	55
Autumn post-flow	Yellowbelly biomass	169	18.5–543	88
Autumn post-flow	Bony bream biomass	270	10.3–1159	115
Spring no-flow	Yellowbelly biomass	285	0–902	82
Spring no-flow	Bony bream biomass	184	3.7–563	91
Warrego fish biomass				
All times	Yellowbelly biomass	797	0–6453	152
All times	Bony bream biomass	1438	0–3751	89
Spring no-flow	Yellowbelly biomass	764	0–1960	84
Spring no-flow	Bony bream biomass	1684	0–3751	90
Autumn post-flow	Yellowbelly biomass	827	0–6453	194
Autumn post-flow	Bony bream biomass	1207	0–2988	83

describe the extent of the floodplain, connectivity between habitat patches (i.e. waterholes) and the form of the river channel. Whole-waterhole scale factors relate to the size and shape of a waterhole, how long it retains water and its light environment. Within-waterhole scale factors describe substrate and morphological complexity and habitat availability. The importance of factors at each scale varies through time, depending on flow regime and biological response.

For the Moonie River, the majority of physical habitat variables were measured at the entire-waterhole scale (Table 5.1). The bathymetry of each waterhole was profiled to create digital elevation models in ArcMap (version 9.2). Measurements such as surface area, volume and benthic area were calculated from the digital elevation models in ArcMap (version 9.2), using the 3D Analyst extension. Cross-sections were placed through the models at random

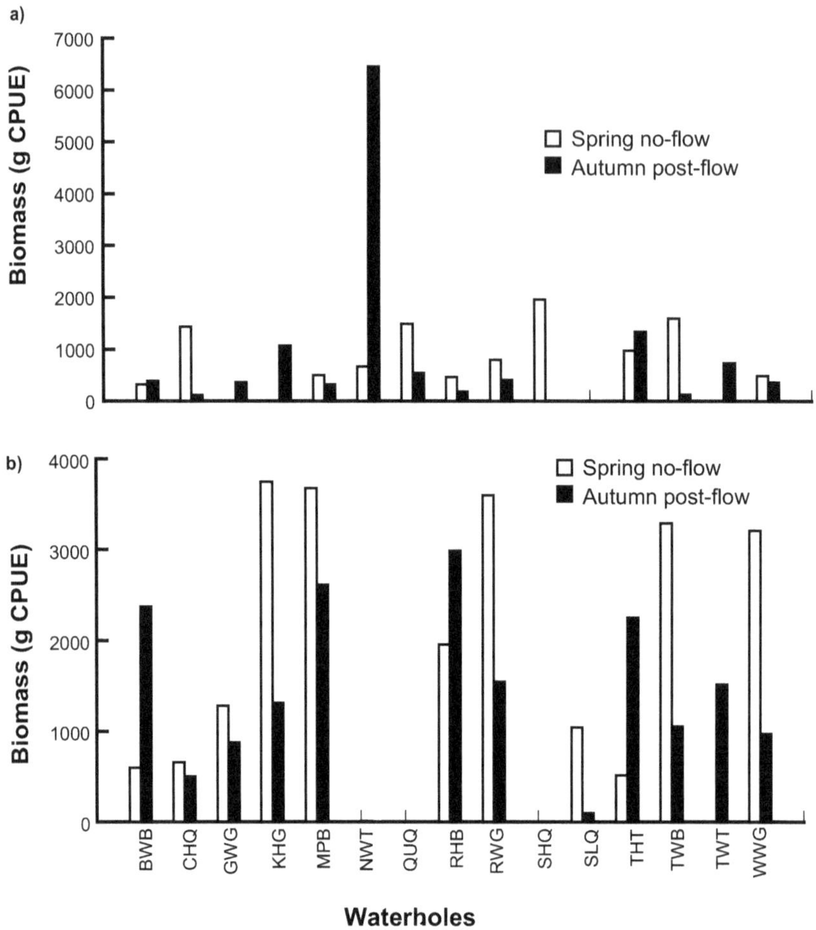

Figure 5.4: Fish biomass on two sampling occasions in 15 waterholes of the Warrego River catchment for (a) yellowbelly and (b) bony bream.

intervals and used to calculate measures such as cross-sectional area, wetted perimeter and average depth. Turbidity was measured *in situ* and waterhole persistence time was modelled (J. Lobegeiger, unpubl. data). A single landscape-scale variable (straight-line distance to nearest sampled waterhole) was measured from maps and a within-waterhole variable (snags) was estimated *in situ* as % cover along a representative transect.

Physical habitat variables from all three scales (landscape, whole-waterhole, within-waterhole) were measured in the Warrego River (Table 5.1). Calculations were largely derived from on-ground physical surveys and *in situ* measurements, with some desktop mapping. For more information on the methods employed in the Warrego, refer to Balcombe *et al.* (2006). To compare the models for each catchment, a reduced physical habitat dataset was derived for the Warrego analyses, containing only measures that were comparable to those collected for the Moonie.

Comparisons of basic physical measures of waterholes from the two study catchments reveals that Moonie river waterholes are generally shorter, shallower and narrower than Warrego River waterholes (Table 5.2). These differences in waterhole size are reflected in gross comparisons of fish production between the two rivers (Fig. 5.1, Table 5.2). Across all sampling times yellowbelly production was approximately three and bony bream 3.5 times greater in the Warrego catchment than in the Moonie catchment.

Regression modelling

Relationships between fish biomass and physical variables were investigated using interactive forward-stepping multiple linear regression (Systat 11, 2006). Variable entry was set at $P < 0.15$ and variables were only retained in the final models if they were significant ($p < 0.05$) or where they added greater than 5% to the overall model's explanatory power. Residual plots of final models were examined to ensure they met the assumption of multiple regression.

To enable comparisons between the two rivers and the validity of using various scales of habitat in the models, three groups of analyses were undertaken. These were regression analysis on the Moonie River fish/habitat, then the Warrego River fish/reduced habitat and finally the Warrego River fish/full habitat dataset.

Prior to running models all data were screened to meet the assumptions of regression. Subsequently, it was necessary to transform some variables. Where transformation was needed, $Log_e (x + 1)$ was found to be suitable for the physical habitat variables, while $X^{0.5}$ was suitable for some of the fish biomass variables. We checked for co-linearity among biophysical variables prior to running models. Where two or more variables were correlated (at $r > 0.75$), only one variable was inserted into the model. To ensure that we had not rejected the wrong variable, alternative models were run with rejected variables. The final variable retained (if significant) was determined by the complete model fit.

Analysis of fish production in relation to physical variables in the Moonie River generally failed to produce any significant relationships (Tables 5.3 and 5.4). Across both species and three sampling times, only one significant fish production/physical habitat model was produced. In the summer post-flow period, bony bream production was weakly (30% of variation explained) associated with maximum waterhole depth. In contrast to the Moonie River regression models, a comparative analysis in the Warrego River revealed that fish production could be predicted from a suite of waterhole physical characteristics. This was demonstrated in the analysis of fish biomass in the Warrego River using the reduced physical dataset, which produced highly predictable production/physical models for both species across the two sampling times. These models predicted 56–81% of the variation in fish production. The parameters consistently selected in final models were depth, snags and wetted perimeter. Two of these factors were opposite in sign between the two sampling times (Table 5.3), with snags positively associated with yellowbelly biomass in spring but negatively associated in autumn. Similarly, bony bream biomass was negatively associated with wetted perimeter in spring and positively in autumn.

When using the full biophysical dataset in the Warrego River, yellowbelly and bony bream biomass models were more predictable than with the reduced dataset (Table 5.3). These models predicted 70–97% of the variation in fish biomass on the two sampling occasions. The suite of biophysical parameters contributing to individual models was not consistent for the two species or the two times. There were, however, some parameters that did consistently appear as significant individual factors in models. These included depth, snags, cross-sectional area, mid-channel bars and turbidity. For any final model there was always a suite of factors made up of parameters from at least two spatial scales (landscape, whole waterhole or within waterhole). The signs of some factors were inconsistent between the two sampling times, as with the reduced dataset. For example, depth was positively associated with bony bream biomass in spring and negatively associated in autumn.

Table 5.3: Multiple regression analyses testing the relationship between fish biomass and biophysical relationships in the Warrego and Moonie river catchments

Regression equation, F_{ratio}	**Adjusted R^2**	**P**
Moonie summer post-flow		
YB = no significant relationships		
BB = -0.43MaxD + 7.9, F1,12 = 6.5	0.3	0.03
Moonie autumn post-flow		
YB = no significant relationships		
BB= no significant relationships		
Moonie spring no-flow		
YB = no significant relationships		
BB= no significant relationships		
Warrego spring no-flow reduced dataset		
YB0.5 = 4.8D – 75.2SN + 48.3, F2,11 = 9.4	0.56	0.004
BB = –0.000035V + 0.932 WP +112, F2,11 = 17.62	0.72	<0.0001
Warrego autumn post-flow reduced dataset		
YB0.5 = 3.3D + 60.0SN + 0.04TUR -21.0, F2,11 = 12.3	0.71	0.001
BB = 0.003A + 445D – 2396LnCSA – 17 WP + 11830, F4,10 = 16.3	0.81	<0.0001
Warrego spring no-flow full dataset		
YB0.5=6.0D – 31.6MCB – 34.3EB –101SN +72.9, F4,9 = 8.523	0.7	0.004
BB0.5 = -12.6D + 48.1B2 – 12.0LnSLD – 29.1LnCSA + 86.2LnBAW – 0.03TUR + 236, F4,9 = 73.1	0.97	<0.0001
Warrego autumn post-flow full dataset		
YB0.5 = 54.0MCB + 337BOU + 84.4SN – 29.2OVV + 4.4LnBR + 0.05TUR – 21.0, F4,9 = 52.7	0.96	<0.0001
BB = 624D – 1844B3 + 387LnFL – 2903LnCSA, + 10632, F4,9 = 16.8	0.82	<0.0001

General findings

Superficially, waterholes in the Moonie and Warrego catchments appear very similar. Therefore, we would expect similar relationships between habitat variables and fish biomass. However, fish biomass, hence fish production, in the Moonie was around three times lower than that in the Warrego. This could be explained by an east–west geographic temperature gradient, with westerly regions experiencing higher temperatures and greater primary production, hence higher fish production. Fellows *et al.* (2009) reported that fish abundance and primary productivity in Cooper Creek, further west along the temperature gradient, was higher than in the Warrego, supporting the concept that temperature/latitude may be important in controlling rates of fish production in dryland rivers. Despite different levels of productivity, we would still expect patterns of fish biomass to respond similarly to physical habitat conditions and flow/seasonal variability. However, this is not what we found.

There was high spatial and temporal variation in fish biomass in both river systems, confirming the dynamic nature of dryland fish assemblages (Puckridge *et al.* 2000; Arthington *et al.* 2005). Fish species within these systems are adapted to their variable environment: they migrate among waterholes in the presence of flow, exhibit a range of breeding strategies in relation to hydrological and seasonal variability in habitat and temperature, and

Table 5.4: Standard errors and significance levels of regression coefficient for significant equations reported in Table 5.3

	Error	p
Moonie summer post-flow – BB		
Maximum depth	0.17	0.02
Constant	0.58	<0.0001
Warrego spring no-flow reduced dataset – YB		
Depth	1.7	0.02
Snags	20.2	0.003
Constant	14.4	0.007
Warrego spring no-flow reduced dataset – BB		
Volume	0.000009	0.002
Wetted perimeter	0.232	0.002
Constant	15	<0.0001
Warrego autumn post-flow reduced dataset – YB		
Depth	1.8	n.s.
Snags	18.9	0.009
Turbidity	0.008	0.001
Constant	17.9	n.s.
Warrego autumn post-flow reduced dataset – BB		
Area	0.001	n.s.
Depth	113	0.003
Cross-sectional area	413	<0.0001
Wetted perimeter	5.9	0.02
Constant		
Warrego spring no-flow full dataset – YB		
Depth	2.0	0.003
Mid-channel bars	20.0	n.s.
Eroding banks	13.7	0.03
Snags	20	0.001
Constant	15.8	0.001
Warrego spring no-flow full dataset – BB		
Depth	1.5	<0.0001
Bench 1/3–2/3	6	<0.0001
Straight-line distance	1.8	<0.0001
Cross-sectional area	4.3	<0.0001
Backwaters	12.1	<0.0001
Turbidity	0.004	<0.0001
Constant	19	<0.0001
Warrego autumn post-flow full dataset – YB		
Mid-channel bars	9.8	0.001
Boulders	53	<0.0001
Snags	8.0	<0.0001
Overhanging vegetation	4.3	<0.0001

	Error	p
Bifurcation ratio	1.4	0.01
Turbidity	0.004	<0.0001
Constant	11.8	n.s.
Warrego autumn post-flow full dataset – BB		
Depth	117	<0.0001
Bench 2/3–3/3	127	n.s.
Fetch length	127	0.01
Cross-sectional area	372	<0.0001
Constant	1632	<0.0001

See Table 5.3 for coefficient values.
n.s. = not significant at $P < 0.05$.

feed opportunistically (Arthington *et al.* 2005; Balcombe *et al.* 2006, 2007; Balcombe and Arthington 2009; Sternberg *et al.* 2008). Therefore, given the highly variable spatial and temporal patterns in fish biomass in the two rivers, it could be expected that the ecological response to physical waterhole variables would also be highly variable, or even unpredictable. This expectation was realised in the Moonie dataset for both species (excluding one weakly significant relationship between bony bream biomass and maximum depth).

The low predictability in fish production in the Moonie River challenges the relevance of the physical data used to predict fish biomass, especially the use of variables measured at a single spatial scale (whole-waterhole). Given that the major drivers of fish abundance and assemblage structure in dryland rivers are known to be nested within a hierarchy of spatial scales (Arthington *et al.* 2005), a broader dataset that incorporates both larger and finer spatial scales might be expected to reveal more predictable relationships. This seems unlikely to be the only explanation for weak relationships between physical variables and fish biomass in the Moonie, given that yellowbelly and bony bream production were highly predictable in the Warrego River using a dataset similar to that used in the Moonie River analysis.

This raises the question of why fish production was predictable from equivalent physical habitat variables in one river and not the other. Importantly, the difference in predictability between the Moonie and the Warrego cannot be explained by differences in the variability of waterhole attributes. Although waterholes were, on average, larger in the Warrego, the coefficients of variation of measured physical attributes were similar in the two systems.

An alternative explanation for the differences in predictability between the two catchments relate to the differences in their drainage networks and physical structure. The Moonie River has a comparatively simple channel with few tributaries, while the Warrego is a larger and more complex effluent system with significant tributaries and many distributaries (effluent channels). These large-scale morphological differences may be influencing the strength of relationships between fish production and waterhole morphology. Also, the Moonie catchment is more affected by agricultural development than the Warrego, with many banks and riparian zones highly eroded and in-stream barriers such as small to medium weirs dispersed along the length of the river. These barriers may impede fish movement, particularly in low- to medium-flow periods. Fish may not be able to colonise the full range of waterholes or actively select any particular waterhole, potentially forcing them to settle in suboptimal habitats. Modified flow regimes can also affect the primary and secondary productivity of dryland rivers, for example disrupting benthic algal production and dependent biota by elevating flows during normally dry spells (Bunn *et al.* 2006b).

The very high variance in fish production (70–90%), explained by using the full set of physical variables for the Warrego, demonstrates the importance of considering all relevant spatial and temporal scales. Importantly, these models were much better able to predict fish productivity than those based only on whole-waterhole scale variables. Past research in dryland rivers has shown that fish assemblages are strongly influenced by processes and habitat structure operating at more than the waterhole scale (Arthington *et al.* 2005; Balcombe *et al.* 2006, 2007).

It is interesting to note that although individual factors were consistent predictors of biomass for a given species, some were opposite in sign between the two sampling times. One explanation could be that the habitat requirements of each species or size classes of each species vary through time. Further analysis of the fish data revealed that fish sizes were significantly smaller in the autumn period, with a high proportion of new recruits of both species at that time (S. Balcombe, unpubl. data). This finding highlights the importance of considering environmental conditions, particularly antecedent flow conditions, at different time periods when developing predictive ecological models in hydrologically variable systems, where recruitment responses are likely to be tightly linked to flow history and flow events (Puckridge et *al.* 2000; Balcombe and Arthington 2009).

Conclusions

This chapter has demonstrated that it is possible to derive meaningful ecological response models for fish production in dryland rivers of the upper Murray-Darling Basin. It was not possible to transfer the models between the two catchments in this study, potentially because they are different at the landscape scale, both physically and in degree of agricultural and water resource development. While generic transferable models have obvious management applications (Arthington *et al.* 2006), our results highlight the difficulty in developing generic models that might be applied to multiple dryland catchments. It is important to understand the fundamental differences among catchments, and hence the drivers of fish production, before attempting to transfer models and any related management practices between catchments. It is possible that there are broad-scale attributes of catchments that we do not yet fully understand and that ultimately determine the predictability of fish production in dryland rivers. The search for stronger explanatory models continues.

Acknowledgments

This research was conducted under two General Queensland Fisheries permits (PRM00157K and #89212) and two Griffith University Animal Ethics Experimentation permits (AES/03/02 and AES/14/05). We would like to thank Neil Foster and Neil Saintilan for their assistance in improving our original draft. Thanks to the Dryland Refugium team members (CRC for Freshwater Ecology) for their assistance during Warrego River sampling and to Andrea Prior, Sharon Marshall and Tariq Khan for their efforts in the Moonie River sampling. We are also grateful for the access to sites provided by landowners in the Warrego and Moonie River catchments.

References

Amoros C and Bornette G (2002). Connectivity and biocomplexity in waterbodies of riverine floodplains. *Freshwater Biology* **47**, 517–39.

Arthington AH (1996). The effects of agricultural land use on tributaries of the Darling River, Australia. *GeoJournal* **40,** 115–125.

Arthington AH, Balcombe SR, Wilson GA, Thoms MC and Marshall J (2005). Spatial and temporal variation in fish assemblage structure in isolated waterholes during the 2001 dry season of an arid-zone river, Cooper Creek, Australia. *Marine and Freshwater Research* **56,** 25–35.

Arthington AH, Bunn SE, Poff NL and Naiman RJ (2006). The challenge of providing environmental flow rules to sustain river ecosystems. *Ecological Applications* **16**, 1311–1318.

Balcombe SR and Arthington AH (2009). Temporal changes in fish abundance in response to hydrological variability in a dryland floodplain river. *Marine and Freshwater Research* **60,** 146–159.

Balcombe SR, Bunn SE, McKenzie-Smith FJ and Davies PE (2005). Variability of fish diets between dry and flood periods in an arid zone floodplain river. *Journal of Fish Biology* **67,** 1552–1567.

Balcombe SR, Arthington AH, Foster ND, Thoms MC, Wilson GG and Bunn SE (2006). Fish assemblages of an Australian dryland river: abundance, assemblage structure and recruitment patterns in the Warrego River, Murray-Darling Basin. *Marine and Freshwater Research* **57,** 619–633.

Balcombe SR, Bunn SE, Arthington AH, Fawcett JH, McKenzie-Smith FJ and Wright A (2007). Fish larvae, growth and biomass relationships in an Australian arid zone river: links between floodplains and waterholes. *Freshwater Biology* **52,** 2385–2398.

Bunn SE, Balcombe SR, Davies PM, Fellows CS and McKenzie-Smith FJ (2006a). Aquatic productivity and food webs of desert river ecosystems. In *Ecology of Desert Rivers.* (Ed. RT Kingsford) pp. 76–99. Cambridge University Press, Melbourne.

Bunn SE, Thoms MC, Hamilton SK and Capon SJ (2006b). Flow variability in dryland rivers: boom, bust and the bits in between. *River Research and Applications* **22,** 179–186.

Burford MA, Cook AJ, Fellows CS, Balcombe SR and Bunn SE (2008). Sources of carbon fuelling production in an arid floodplain river. *Marine and Freshwater Research* **59,** 224–234.

Fausch KD, Torgersen CE, Baxter CV and Li HW (2002). Landscapes to riverscapes: bridging the gap between research and conservation of stream fishes. *Bioscience* **52**, 483–498.

Fellows CS, Bunn SE, Sheldon F and Beard NJ (2009). Benthic metabolism in two turbid dryland rivers. *Freshwater Biology* **54,** 236–253.

Gehrke PC, Brown P, Schiller CB, Moffatt DB and Bruce AM (1995). River regulation and fish communities in the Murray-Darling River system, Australia. *Regulated Rivers: Research and Management* **10,** 15–38.

Harris J and Gehrke PC (1997). Fish and rivers in stress: NSW rivers survey. NSW Fisheries Office of Conservation/CRC for Freshwater Ecology, Sydney.

Humphries P, King AJ and Koehn JD (1999). Fish, flows and floodplains: links between freshwater fishes and their environment in the Murray-Darling River system, Australia. *Environmental Biology of Fishes* **56**, 129–151.

King AJ, Humphries P and Lake PS (2003). Fish recruitment on floodplains: the roles of patterns of flooding and life history characteristics. *Canadian Journal of Fisheries and Aquatic Sciences* **60**, 773–786.

Kingsford R (2006). *Ecology of Desert Rivers.* Cambridge University Press, Melbourne.

Knight JT and Arthington AH (2008). Distribution and habitat associations of the endangered Oxleyan pygmy perch, *Nannoperca oxleyana* Whitley, in eastern Australia. *Aquatic Conservation Marine and Freshwater Ecosystems* **18**, 1240–1250. DOI: 10.1002/aqc.936.

Knighton AD and Nanson GC (1994). Waterholes and their significance in the anastomosing channel system of Cooper Creek, Australia. *Geomorphology* **9**, 311–324.

Labbe TR and Fausch KD (2000). Dynamics of intermittent stream habitat regulates persistence of a threatened fish at multiple scales. *Ecological Applications* **10,** 1774–1791.

Lintermans M (2007). Fishes of the Murray-Darlin Basin: an introductory guide. Murray-Darling Basin Commission, Australia. MDBC Publication No. 10/07.

Ly D (2007). Body condition of two native fish species, bony bream (*Nematolosa erebi*) and golden perch (*Macquaria ambigua*) in waterholes of the Moonie River, upper Murray-Darling Basin. Honours thesis. Australian School of Environmental Studies, Griffith University, Nathan Campus.

MDBC (2004). Native fish strategy for the Murray-Darling Basin 2003–2013. Murray-Darling Basin Commission, Canberra. Available at www.mdbc.gov.au.

Medeiros SEF and Arthington AH (2008). The importance of zooplankton in the diets of three native fish species in floodplain waterholes of a dryland river, the Macintyre River, Australia. *Hydrobiologia* **614,** 19–31.

Morton SJ, Short J and Baker RD (1995). Refugia for biological diversity in arid and semi-arid Australia. Biodiversity Series, Paper No. 4. Dept of Environment, Sport and Territories, Canberra.

Power ME (1992). Top-down and bottom-up forces in food webs: do plants have primacy? *Ecology* **73,** 733–746.

Puckridge JT, Walker KF and Costelloe JF (2000). Hydrological persistence and the ecology of dryland rivers. *Regulated Rivers: Research and Management* **16**, 385–402.

Pusey BJ, Kennard MJ and Arthington AH (2004). *Freshwater Fishes of North-eastern Australia.* CSIRO Publishing, Melbourne.

Scheurer JA, Fausch KD and Bestgen KR (2003). Multiscale processes regulate brassy minnow persistence in a Great Plains River. *Transactions of the American Fisheries Society* **132**, 840–855.

Sheldon F, Boulton AJ and Puckridge JT (2002). Conservation value of variable connectivity: aquatic invertebrate assemblages of channel and floodplain habitats of a central Australian arid-zone river, Cooper Creek. *Biological Conservation* **103**, 13–31.

Sternberg D, Balcombe S, Marshall J and Lobegeiger J (2008). Food resource variability in an Australian dryland river: evidence from the diet of two generalist native fish species. *Marine and Freshwater Research* **59,** 137–144.

Thoms MC and Sheldon F (2000). Lowland rivers: an Australian perspective. *Regulated Rivers: Research and Management* **16,** 375–383.

Thoms MC, Southwell M and McGinness HM (2005). Floodplain-river ecosystems: fragmentation and water resources development. *Geomorphology* **71,** 126–138.

Walker KF, Sheldon F and Puckridge JT (1995). An ecological perspective on dryland river ecosystems. *Regulated Rivers: Research and Management* **11**, 85–104.

Welcomme RL (2001). *Inland Fisheries Ecology and Management.* Fishing News Books, Blackwell Science, Oxford.

Welcomme RL, Bene C, Brown CA, Arthington A, Dugan P, King JM and Sugunan V (2006). Predicting the water requirements of river fisheries. In *Wetlands and Natural Resource Management.* (Eds JTA Verhoeven, B Beltman, R Bobbink and DF Whigham) pp. 123–154. Ecological Studies Vol. 190, Springer-Verlag, Berlin.

Chapter 6

Using isotopic techniques to assess trophic structure in northern Murray-Darling Basin wetlands

Jeff Kelleway, Debashish Mazumder, Glenn Wilson and Tsuyoshi Kobayashi

Introduction

Floodplain wetlands provide habitats for many species of terrestrial and aquatic biota (Junk *et al.* 1989; Junk and Wantzen 2004). Their overall productivity and biodiversity is closely linked to patterns of flooding (Kingsford and Thomas 1995; Kingsford 2000; Arthington and Pusey 2003) which mobilise resources such as carbon and nutrients from the floodplain surface and upstream watercourses. These resources are subsequently utilised by a range of organisms, from bacteria to predatory fish and waterbirds over a range of temporal and spatial scales. For example, microbial components such as bacteria and algae develop within days of inundation of the floodplain sediments (Kobayashi *et al.* 2009) and are then utilised by higher-level consumers such as zooplankton and fish (Lindholm *et al.* 2007; Burford *et al.* 2008).

Understanding the flow of nutrients and energy, as well as the trophic linkages within food webs, is essential for developing ecological models for sustainable management of aquatic ecosystems. However, food web connections are often complex and are likely to be influenced by the dynamics of physico-chemical processes. For example, the relative contribution of energy to food webs by autochthonous (derived within the watercourse, e.g. algae) and allochthonous (derived outside the watercourse, e.g. riparian leaf litter) primary sources varies between wetlands, influenced by factors such as climate, discharge regime, canopy cover, nutrient concentrations and other abiotic characteristics (Vannote *et al.* 1980; Bunn 1986; Lake *et al.* 1986; Reid *et al.* 2008). Identifying linkages between sources of primary production to lower and higher trophic-order consumers and detrital pathways (e.g. bacteria and detritivores) is fundamental to our understanding of ecosystems, and ultimately their conservation.

Carbon and nitrogen stable isotope ratios ($\delta^{13}C$ and $\delta^{15}N$) provide an important tool with which to model trophic connectivity between species and ecosystem resources. In this chapter, we review the ecological applications of stable isotope techniques to freshwater food web research and present research findings from two key wetland systems in the northern Murray-Darling Basin.

Ecological applications of stable isotope techniques

Food web research has often relied upon feeding studies and/or gut content analyses. Both these methods provide valuable information on predator–prey relationships, although they also have clear limitations. Controlled feeding studies may neither adequately replicate natural conditions nor include all food sources available to a species in the wild. In gut content analyses, it is often difficult to visually identify food items which may be in various states of digestive decomposition, and operator subjectivity may be involved. In addition, gut content analyses present only a snapshot of food consumed immediately prior to capture, and not all ingested materials are assimilated (Michener and Schell 1994). Some ingested materials are digested quickly and are therefore rarely found in the stomach (Gee 1989).

Stable isotope analyses provide a complementary and in many cases more powerful tool for tracing food sources and analysing food web structure than feeding studies and gut content analyses (DeNiro and Epstein 1978; Peterson and Fry 1987; Fry 1991; Post 2002). The technique involves preparing a dry, ground powder from material of each food web component (primary sources, consumer and/or predator species) which is then analysed through a stable isotope mass-spectrometer. The stable (non-radioactive) isotopes of carbon ($^{13}C/^{12}C$) and nitrogen ($^{15}N/^{14}N$) are most frequently used in food web research. The key to their interpretation is that the ratios or signatures of these isotopes are transferred from the food item to the consumer tissues upon consumption. During the assimilation process, there is generally a small increase in the ratio of the heavier (^{13}C and ^{15}N) isotope of the consumer relative to the source, known as trophic fractionation, which needs to be taken into account in food web interpretation.

Carbon

Carbon isotope ratios remain relatively stable as the element is cycled through food webs, with an average trophic fractionation of 0.8–1.1‰ between trophic levels (DeNiro and Epstein 1978). It has been suggested that this fractionation may be even lower for freshwater consumers (France 1996). Carbon isotopes are useful in identifying the sources of primary production supporting a food web and the sources of carbon for particular organisms.

There are several important distinctions in the carbon isotope values between key primary producers, which allow their contribution as food sources to be identified. These divisions include a difference between terrestrial C3 and C4 plants, with the latter more enriched in the heavier ^{13}C isotope due to differences in their mode of photosynthesis (Smith and Epstein 1971). Differences in carbon isotope ratios also occur between plants accessing carbon from atmospheric and aquatic sources.

Nitrogen

Nitrogen isotopes are used to indicate the trophic level of food web components. Measurements of $\delta^{15}N$ (the ratio of $^{15}N/^{14}N$ relative to atmospheric nitrogen) of a consumer's tissues are generally higher than those of its diet (Vanderklift and Ponsard 2003). One of the main reasons for this is that the consumer generally accumulates more of the heavier N, while excreting lighter N in its waste products. Using an estimated or average fractionation factor, it is possible to determine the trophic position of a sampled species relative to a sampled basal food resource. The following formula can be used to determine this relative trophic position (Post 2002):

$$\lambda + (\delta^{15}N_{\text{organism}} - \delta^{15}N_{\text{base of food web}}) / \Delta\delta^{15}N \tag{1}$$

where λ is the trophic position of the organism used to estimate $\delta^{15}N_{\text{base of food web}}$ (for a primary producer λ is 1.0).

The magnitude of the difference, or fractionation factor ($\Delta^{15}N$), in $\delta^{15}N$ from a dietary item to the consumer has generally been accepted as in the order of 3–4‰ (Minagawa and Wada 1984; Hamilton *et al.* 1992; Cabana and Rasmussen 1996; Post 2002), although the applicability of this generalisation has been debated (McCutchan *et al.* 2003).

Stable isotope analysis in Australian freshwater ecosystems

The use of stable isotope analyses in the study of Australian freshwater ecosystems has been limited, including within the Murray-Darling Basin. Of the studies that have been conducted, the most important fields of research have been the identification of sources of production sustaining freshwater ecosystems and investigations into the feeding characteristic of targeted aquatic species.

Sources of production sustaining freshwater food webs

Stable isotope analyses have been used to identify the sources of production sustaining Australian floodplain waterholes and disconnected floodplain river ecosystems. The first study (Bunn and Boon 1993), conducted in River Murray billabongs, identified one of the limitations of carbon isotope analysis – that many primary production sources (e.g. fringing, emergent or floating vegetation or benthic detritus) can have similar or overlapping $\delta^{13}C$ ratios, making the determination of their separate contributions to consumer-feeding difficult. Despite this, carbon isotope analysis was able to support some specific conclusions. For example, freshwater crayfish (*Cherax* sp.) appeared to feed primarily upon detrital material, while gastropods and some caddisfly larvae obtained a greater portion of their biomass carbon from epiphytes than from the host macrophytes.

In the Cooper Creek system west of the Murray-Darling Basin, stable isotope studies have been used to identify production dynamics in different flow conditions. During non-flood periods, filamentous benthic algae growing in a narrow band in the littoral zone were shown to be the most important source of organic carbon in floodplain waterholes, providing more than half of the carbon biomass to most of the consumer species studied (Bunn *et al.* 2003). Recent studies of in-channel food webs have suggested that benthic algae are also an important source of production in floodplain river food webs during relatively dry conditions (see case study 1). Combining stable isotope analysis with biomass estimates and measures of primary production, Burford *et al.* (2008) suggested that as much as 50% of fish carbon in a waterhole after flooding was derived from floodplain food sources, i.e. fish feeding upon the floodplain during its inundation then returning to waterholes as water levels receded.

Similar results in regard to waterhole connection have been identified in lowland streams of the Murray-Darling Basin in Victoria (Reid *et al.* 2008). During periods of stream connection (albeit slow flow, <0.05 m s^{-1}), a mixing model analysis of $\delta^{13}C$ and $\delta^{15}N$ showed allochthonous detritus represented 23–99% of assimilated material across all aquatic consumers studied. When streams were reduced to disconnected pools several months later, allochthonous detritus represented a much smaller food source for most consumers, with feeding supplemented by autochthonous sources of higher nutritional value (algae, biofilms and macrophytes).

Feeding behaviour of targeted aquatic species

Ecosystem-scale studies have been complemented by stable isotope investigations targeting a small number of consumer species. As long-lived primary consumers, surface-grazing snails are often used in stable isotope studies to assess benthic food webs (Post 2002). In Cooper Creek, carbon isotopes, C:N ratios and a feeding study were utilised to determine the main dietary sources of the freshwater gastropods *Notopala sublineata* (Viviparidae) and *Thiara balonnensis* (Thiaridae) (Sheldon and Walker 1997). Stable isotope results suggested allochthonous detritus, rather than biofilm from shallow or deep snags (coarse wood), as the most likely food source for both species. Carbon isotopes of coarse and fine particulate organic matter and dominant plant materials were also used to identify the contribution of each component to the detrital mix, and hence their value as a food source for the gastropods.

Where species are able to access a range of feeding environments (benthic, pelagic, littoral), identifying food sources may become more difficult. In the lower River Murray, differences in isotope values among potential food sources were used to suggest feeding differences between gorge (main item = littoral plants) and floodplain (main item = pelagic fine particulate organic matter) reaches for the freshwater prawn *Macrobrachium australiense* and freshwater shrimp *Paratya australiensis* (Burns and Walker 2000).

An important application of targeted stable isotope studies lies in their ability to detect dietary competition between endemic and introduced species. This approach was undertaken to identify seasonal differences in the feeding dynamics between two crayfish species in Western Australia, namely the endemic *Cherax cainii* and the introduced *Cherax destructor* (Beatty 2006). The case studies at the end of this chapter provide further isotopic comparisons between endemic and introduced species.

Areas requiring further research in Australian freshwater ecosystems

Studies including those reviewed above have enhanced our understanding of trophic dynamics in Australian freshwater ecosystems. There remain many areas of food web research, however, which require further attention. We outline some of the areas in which stable isotope techniques in particular can further our knowledge and management of Australia's river–floodplain ecosystems.

- In Australia, $\delta^{13}C$-depleted primary sources have been proposed, but have gone largely unsampled (Bunn and Boon 1993; see also case study 2). Bunn and Boon (1993) proposed that such 'missing' sources may include planktonic algae (e.g. Chlorophyta) or methanogenic bacteria. Little is known of the stable isotope signatures of such pelagic or highly depleted sources in Australian floodplains.
- Stable isotope analysis will also be useful in the detection of food web response to various flow conditions. This will include detecting the importance of large natural inundation events and, critical for wetland managers, the importance of various environmental flow scenarios to trophic structure and ecosystem functionality.
- Because of the intensive preparation and financial costs of numerous stable isotope analyses, it may be advantageous to identify appropriate indicator variables (i.e. which ecosystem components and species to monitor) for assessing food web structure and ecosystem functionality. Stable isotope and gut content methods will provide an important tool for carrying out such research and monitoring of aquatic food webs.

Northern Murray-Darling Basin floodplain wetlands: case studies

Water resources development, coupled with periods of prolonged drought, has changed the flooding regime of many floodplain wetlands in the Murray-Darling Basin, including those in the Gwydir Wetlands and the Macquarie Marshes (Fig. 6.1). Such development is suspected to have influenced trophic connectivity and ecosystem functionality. In these circumstances, environmental flow releases are used by water resource managers to specifically address the ecological needs of river–floodplain environments for inundation (Arthington and Pusey 2003; Schofield *et al.* 2003). An understanding of the trophic dynamics and ecosystem function (pathways and interactions of resources and organisms leading to food webs supporting high-order predators such as waterbirds) is of fundamental importance in determining the efficient delivery of environmental water allocations to wetland systems.

Three case studies of trophic research carried out in the Gwydir Wetlands (case studies 1 and 2) and the Macquarie Marshes (case study 3) are presented. All three studies are based on

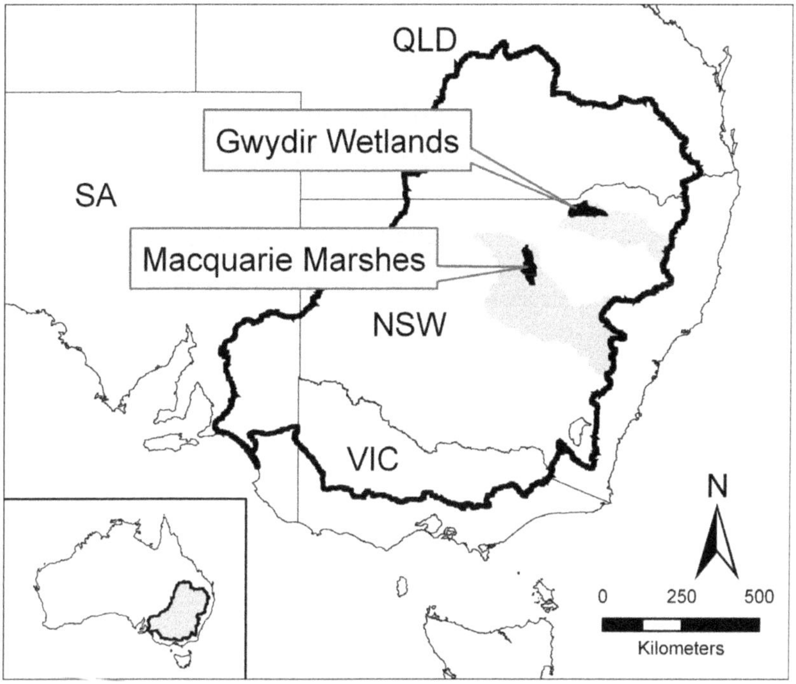

Figure 6.1: Map of the Murray-Darling Basin highlighting the Gwydir Wetlands and the Macquarie Marshes (dark shading) and their catchments (light shading).

stable isotope techniques, investigating the spatial and temporal aspects of the food web structure under varying flow conditions.

Case study 1: Using stable isotopes to identify spatial differences in food web structure across the Gwydir Wetlands

Introduction and methodology

Carbon ($\delta^{13}C$) and nitrogen ($\delta^{15}N$) isotopes were used to assess the importance of benthic sediment organic matter (SOM) to in-channel food webs and determine the trophic status of important aquatic fauna across the Gwydir Wetlands. This floodplain wetland system is serviced by three main watercourses. These are the Gingham Watercourse (low flow rates during dry periods and little riparian cover), the Lower Gwydir River (intermediate flow and immediate riparian zone largely intact) and the Mehi River (highest flow, immediate riparian zone largely intact but low floodplain connectivity).

Two benthic sediment samples of 5 cm core depth were taken from the littoral zone (<1 m water depth) of each of three sites along each watercourse. These samples were sieved in the laboratory and material <1 mm was analysed as SOM. Fyke nets were deployed at each site to collect aquatic fauna. Aquatic fauna reported here are the two decapod crustaceans, yabby (*Cherax destructor*) and prawn (*Macrobrachium australiense*), the native fish bony bream (*Nematalosa erebi*) and spangled perch (*Leiopotherapon unicolor*) and the introduced fish European carp (*Cyprinus carpio*). Five specimens of a similar body size were collected from each watercourse, with the exception of the Lower Gwydir River where only three specimens of European carp and four specimens of spangled perch were obtained. Samples of some of the dominant riparian and aquatic plant species (all are C3 plants) were also collected from various locations for analysis.

Results and implications

SOM comprises plant and animal detritus, benthic algae, bacteria and micro-crustaceans. The stable isotope values reflect mixed values of its various components and their relative contributions. Riparian sources generally dominate the carbon pools of floodplain rivers (Robertson *et al.* 1999). However, the Gwydir Wetlands SOM had $\delta^{13}C$ ratios more enriched than the sampled riparian C3 plant species (Fig. 6.2). A likely explanation is that these SOM signatures reflected their mixture of C3 plant material and more enriched primary sources. Such enriched sources may include C4 plants such as *Paspalum distichum* (–13.8‰ to –18.7‰; unpubl. data).

Consumer $\delta^{13}C$ isotope ratios highlight differences in feeding dynamics between taxa. In each watercourse, bony bream displayed similar mean $\delta^{13}C$ ratios between ~–27‰ to ~–29‰, indicating a more specialised feeding than most other taxa. These $\delta^{13}C$ values, combined with gut analyses, also suggest that while some SOM may have been consumed, more $\delta^{13}C$-depleted sources (e.g. C3 plant material, primarily as detritus, and potentially planktonic algae) provided a greater contribution to their diet. This is consistent with the classification of bony bream as a herbivore/detritivore (Harris 1995). In contrast, yabby $\delta^{13}C$ values varied between watercourses, with Mehi River yabbies significantly more enriched in $\delta^{13}C$ than yabbies from the other two watercourses (ANOVA: $df = 2,12$; $F = 12.11$; $P = 0.001$). This variation in $\delta^{13}C$ suggests opportunistic feeding by yabbies, with Mehi River specimens apparently taking advantage of the abundance of enriched source materials (e.g. the abundant C4 grass *Paspalum distichum*) observed in the one Mehi River site where yabbies were present. Together, the overall variation in carbon results show that a range of food sources are utilised in aquatic food webs and that resource use may vary both between taxa and between watercourses.

The trophic position of each species (Fig. 6.3) was calculated using Equation 1 (Post 2002). The yabby had the lowest trophic position of species sampled, with its mean position of 2.2 showing it to feed predominantly as a first-order consumer. In contrast, freshwater prawns displayed a considerably higher trophic position, despite anatomical and feeding habit similarities. The higher trophic position of the prawn is likely to reflect its role as a scavenger (*sensu* Lee and Fielder 1983) of higher trophic species. This is especially the case in the Mehi River (Mehi mean trophic position = 3.2) where prawns appear to be a particularly important food web component, recycling dead animal material into the food web.

Trophic position calculations also showed that spangled perch were primarily carnivorous in the Lower Gwydir (mean trophic position = 3.5) and Mehi rivers (3.1) and more omnivorous in the Gingham Watercourse (2.8). Gut content analyses supported this with Lower Gwydir and Mehi guts containing fish, macro-crustaceans and aquatic insects, while algae and plant material made up part of the diet in Gingham Watercourse samples. It is proposed that the trophic position of spangled perch in the Gingham may be related to low flow conditions and/or resource constraints (see case study 2).

A comparison of both the $\delta^{13}C$ and trophic position results showed competition between native and introduced fish species for food resources. Competition between European carp (*Cyprinus carpio*) and spangled perch appeared to be greatest in the Gingham Watercourse (where spangled perch is the dominant native species: Wilson *et al.* 2009), while competition with bony bream is most evident in the Lower Gwydir River (where bony bream is most abundant: Wilson *et al.* 2009). This suggests that carp may be able to adapt its feeding to exploit the dominant food resources.

Conclusions

This case study presents baseline data on the trophic dynamics of the Gwydir Wetlands under particularly dry conditions. It also identifies important implications for wetland management. First, food web structure may vary across floodplain river wetlands, both within and between

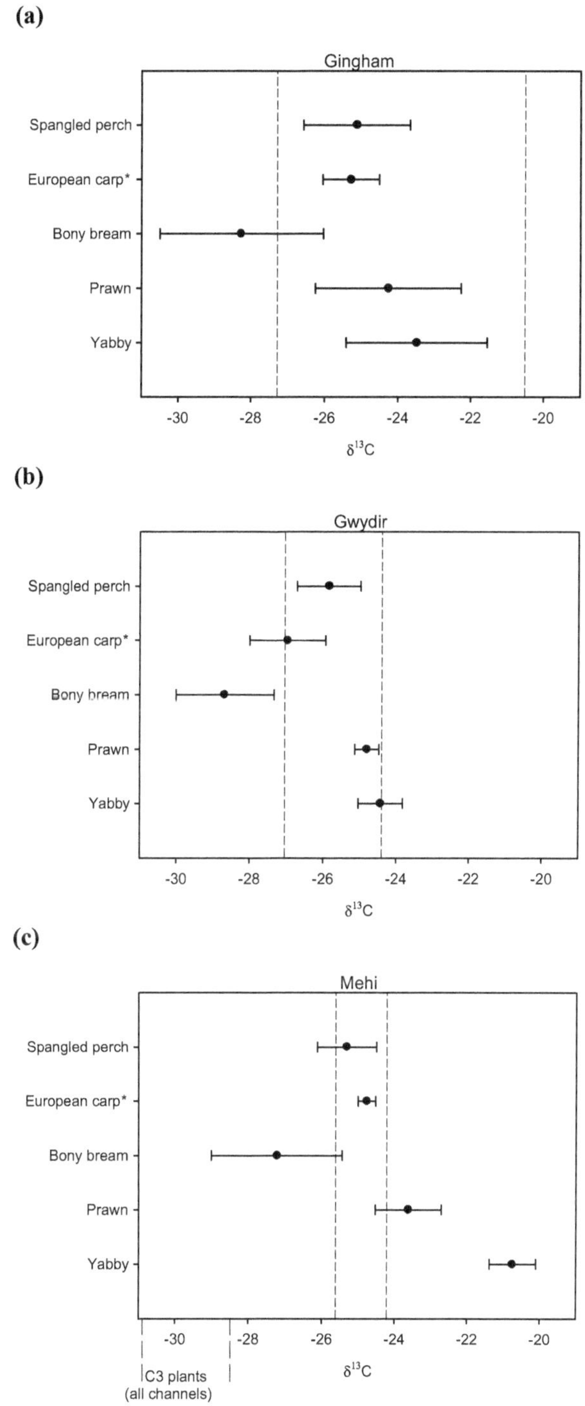

Figure 6.2: Mean carbon isotope values of aquatic species in the Gwydir Wetlands: (a) Gingham Watercourse, (b) Lower Gwydir River and (c) Mehi River. Error bars indicate one standard deviation either side of the mean. Dashed lines within graphs represent one standard deviation either side of the mean for SOM within the watercourse. Range of $\delta^{13}C$ values for all C3 plant specimens signified by dashed line below bottom graph. * = introduced species.

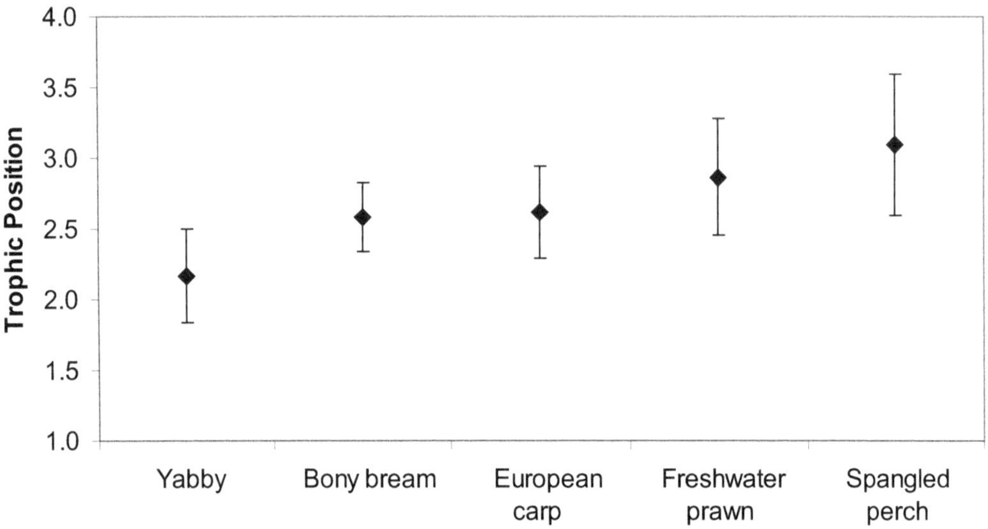

Figure 6.3: Mean trophic position of selected aquatic species in the Gwydir Wetlands. Error bars indicate one standard deviation each side of the mean. Trophic positions calculated with the formula of Post (2002) using SOM as a base resource, i.e. trophic position of 1.0. * = introduced species.

major watercourses. Second, it is apparent that a range of carbon sources are required to maintain different aquatic biota – flow conditions which maximise the input of a variety of floodplain and instream food sources are likely to benefit food web condition. Last, competition for food resources with introduced fish species is identified as a threat to several key native species under the studied (low-flow) conditions.

Case study 2: The influence of an environmental flow on food web structure, Lower Gwydir Wetlands

Introduction and methodology

The development of irrigated agriculture in the Gwydir Valley over the past 30 years (Keyte 1994; McCosker 1999) has significantly affected flow patterns into lower Gwydir Wetlands, and is suspected to have influenced food web and ecosystem functionality. A BACI (before-after-control-impact) sampling program was initiated to understand the ecological response of ecosystem components of wetlands to environmental flow releases (Wilson *et al.* 2009). In April 2007, such a release was made into the Gingham Watercourse. In our sampling design, the Gingham Watercourse was used as the impacted site subject to the environmental flow release between sampling events in March and May 2007 (average flow rate ~250 ML day^{-1}). The Lower Gwydir River was not subject to the same release between sampling periods March and May 2007 (~50 ML day^{-1}) and was therefore regarded as a control site for the purposes of this study. For consistency, the March and May collections from the Lower Gwydir were termed 'before' and 'after', respectively, in the text and figures. Similar-sized individuals of the same taxa (yabby, *Cherax destructor*; European carp, *Cyprinus carpio*; spangled perch, *Leiopotherapon unicolor*) and SOM were collected from each watercourse for the analysis of stable carbon and nitrogen isotope ratios ($\delta^{13}C$ and $\delta^{15}N$).

Results and implications

Stable isotope compositions: After the flow release, a significant change of mean SOM carbon isotope values was observed in the Gingham Watercourse, with its mean $\delta^{13}C$ value of SOM more depleted (–31.75‰) than prior to the flow (–27.11‰). Prior to the environmental flow in

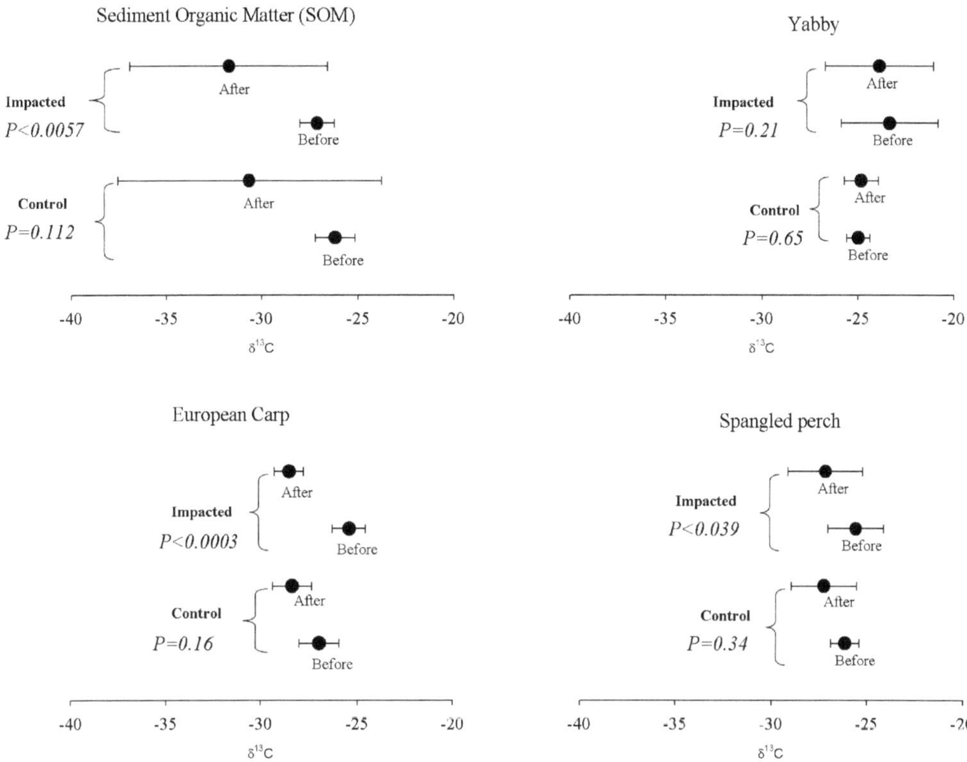

Figure 6.4: Mean (±S.D.) $\delta^{13}C$ values for SOM, yabby, carp and spangled perch collected from Gingham Watercourse (impact) and Lower Gwydir River (control) as per BACI experimental design. *P*-values presented for one-way ANOVA comparisons between control and impact locations.

this watercourse, mean $\delta^{13}C$ values for carp (–25.42‰) and spangled perch (–25.60‰) were very close, implying greater trophic competition between the species. Following the flow, mean $\delta^{13}C$ values for carp (–28.53‰) and spangled perch (–27.20‰) were significantly different. No significant difference was observed for mean $\delta^{13}C$ values of yabby between pre-flow (–23.36‰) and post-flow (–23.90‰) conditions. Across all taxa, significant trophic shifts were observed for spangled perch ($\delta^{13}C$: $P = 0.039$; $\delta^{15}N$: $P = 0.028$) and carp ($\delta^{13}C$: $P < 0.001$) after the flow release in the impacted site (Fig. 6.4). Post-flow mean $\delta^{15}N$ values for spangled perch (11.10‰) were 3.34‰ higher than carp (7.76‰), suggesting that spangled perch were most likely accessing relatively higher trophic-level food items than carp.

The lack of significant differences for $\delta^{13}C$ and $\delta^{15}N$ values of ecosystem attributes (SOM, yabby, carp and spangled perch) between March and May in the Lower Gwydir River (control site) implies that food and nutrients for these invertebrates and fish were similar between the two sampling periods (Fig. 6.4). Increased SOM variability was observed in May 2007 although, unlike in the Gingham Watercourse, differences between means were not statistically significant ($P = 0.112$).

Variations in food web length: The length of food web (difference between the lowest and highest $\delta^{15}N$ values across all sampled taxa at a given site and sampling time) increased almost three-fold after the flow release in the Gingham Watercourse. The total lengths of the food web were similar across both sampling periods in the control site (Fig. 6.5).

Stable isotope analysis showed a marked shift in the isotopic values of SOM after the environmental flow release in the Gingham Watercourse. Depletion of $\delta^{13}C$ values of SOM after the

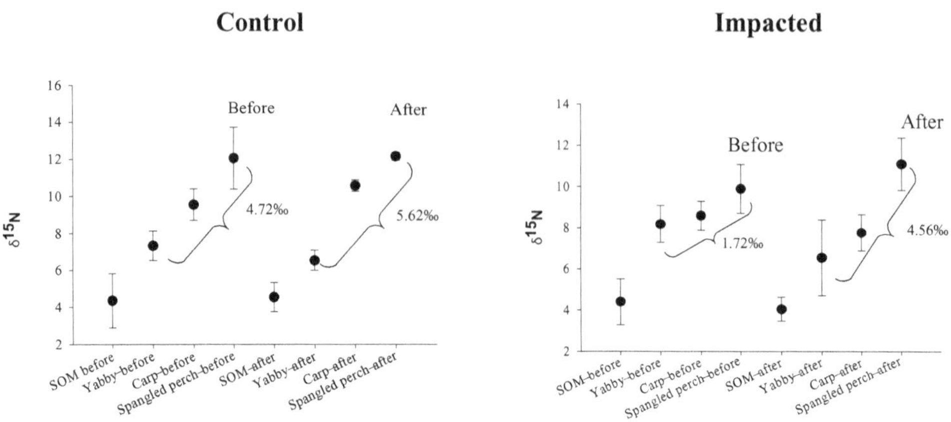

Figure 6.5: Length of the food web in the control and flow-impacted sites of the Gwydir Wetlands. The length of the food web, from invertebrate to fish, was calculated using the formula proposed by Marty *et al.* (2008): $\Delta^{15}N = \text{Max}(\delta^{15}N_{FISH}) - \text{Min}(\delta^{15}N_{INV})$.

environmental release implies some incorporation of carbon from riparian sources, which are thought to dominate the carbon pools of Australian floodplain rivers (Robertson *et al.* 1999).

Two species of fish, spangled perch (*L. unicolor*) and European carp (*C. carpio*), showed noticeable responses, appearing to shift their diets after the environmental flow release. The variation of food web length in the flow-impacted Gingham Watercourse before and after the environmental release is most likely related to hydrological regime, with lower food web length in pre-flow conditions. Theoretical (Duffy *et al.* 2005) and empirical (Pace *et al.* 1999) research has shown that food web length can have significant effects on the structure and function of ecosystems as a whole, with changes in food web length often resulting from the removal of the apical predator or a change in its degree of omnivory (Power 1995; Post and Takimoto 2007). Gut content analysis of spangled perch and carp based on the published literature suggests both species consume similar foods and display omnivorous feeding habits (Cadwallader 1979; Hall 1981; Hume *et al.* 1983; Fletcher *et al.* 1985; Pusey *et al.* 2000; Khan 2003; Medeiros 2004; Balcombe *et al.* 2005; King 2005). One benefit of our isotopic work is that it provides a more incisive tool to distinguish between feeding behaviours of different species, particularly their variable trophic responses to an environmental flow.

Food web length changes in the Gingham Watercourse after the environmental flow release were most likely due to increased availability of higher trophic-level food for spangled perch. Considering the trophic enrichment factor for $\delta^{15}N$ of 3‰ to 4‰ per trophic level (Minagawa and Wada 1984), the higher $\delta^{15}N$ value for spangled perch (11.10‰) relative to carp (7.76‰) suggests that spangled perch moved up at least one trophic level in the food web after the flow release, relative to carp. This upward movement increased the food web length, and may have reduced the trophic competition between the native spangled perch and the introduced carp. Spangled perch seem to be adapted to variability in flow, able to quickly shift from a lower to higher trophic position with altered hydrology; the carp maintained a similar trophic position while accessing food from different sources. It is suspected that constant low-flow conditions draw these species into extended periods of competition for longer than desirable. This has important implications for the conservation of spangled perch and perhaps other native fish species, as research until now has found the dietary habits of exotic species within the Murray-Darling Basin are generally quite similar to those of native species within most of the major trophic groups (Kennard *et al.* 2001).

Conclusions

Floodplain river wetlands experience highly variable conditions, with significant contractions and expansions in habitat between dry and wet periods, respectively (Bunn *et al.* 2003). These contractions and expansions would almost certainly affect the food web structure. Continued restriction of flows (consistent low-flow conditions), coupled with prolonged dry/drought conditions, is likely to increase the dietary insecurity for aquatic organisms and disrupt the healthy functionality of the ecosystem.

Evidence provided by this study of the effects of an environmental flow release on the food web suggests that spangled perch benefit from flow releases through reduced competition for resources. Stable isotopes appear to be a useful cost-effective tool in monitoring the outcomes of environmental flows, in that a small sample size of a single species may be sufficient to demonstrate a positive benefit on ecosystem structure and functionality. Long-term monitoring of the key ecosystem components is necessary to fully understand the impact of regulated flow variations and follow-on effects on ecosystem processes and functionality.

Case study 3: Longitudinal spatial pattern in the relative trophic position of yabby (Cherax destructor) in the Macquarie Marshes during a period of flow pulses

Introduction and methodology

In recent years, natural flooding of floodplain wetlands in Australia and elsewhere has become less frequent because of extensive droughts, increased levels of water extraction and highly regulated flow regimes (Kingsford and Thomas 1995; Puckridge *et al.* 1998; Kingsford 2000; Arthington and Pusey 2003). The disconnection from their associated floodplain of the rivers has become more frequent and takes place over a greater area. This has led to an increased frequency and duration of flow pulses – the period during which in-channel flows are present without inundating the floodplains (Bunn *et al.* 2006). The increased frequency and duration of the flow pulses are said to have negative impacts on aquatic biota (e.g. the disruption of the formation of productive benthic algal zones) (Bunn *et al.* 2006), although they may still provide important aquatic habitats during dry periods (Kobayashi *et al.* 2007). When the channels are hydrologically reconnected with floodplains, the aquatic biota maintained in the in-channel habitats may contribute to the initial floodplain colonisation (Jenkins and Boulton 2003).

Freshwater crayfish *Cherax destructor* (commonly termed the yabby) is found in a wide range of habitats including rivers and wetlands (Semple *et al.* 1990; Beatty 2006). Like other species of freshwater crayfish, *C. destructor* is thought to be a primarily benthic or bottom-feeder in natural habitats, consuming detritus and plant matter (or biofilms) associated with benthic sediments and riparian vegetation (Verhoef *et al.* 1998; Beatty 2006; Meakin *et al.* 2008). Based on a stable isotope study, however, Beatty (2006) reported a seasonal change in the relative trophic position of *C. destructor* in the Hutt River, Western Australia, from a primarily carnivorous level in summer (by consuming fish) to a more herbivorous trophic position in winter. Such a temporal trophic shift is thought to be a response to the seasonal change in the type of food available to *C. destructor* in the river (Beatty 2006). This temporal shift suggests that there may also be spatial change in the trophic position of *C. destructor*, reflecting possible spatial variation in food availability.

In this study, we investigated spatial variation in the relative trophic position (*RTP*) of *C. destructor* in a channel system of the Macquarie Marshes during a period of flow pulses (November 2006), using the $\delta^{15}N$ values of *C. destructor* and those of benthic SOM (<1 mm in size) as basal food resources for the species. Sampling was conducted between 7 and 10

November 2006 at a total of 12 sites along the channel system consisting of the Macquarie River downstream of the Marebone weir (location 31°41'04"S/147°50'07"E), Monkeygar Creek and Bora Channel that run through the Macquarie Marshes from the southern to northern end over an approximate channel distance of 124 km. At the time of study, the floodplain of the Marshes was dry and disconnected from the main river channels, with prime water habitats confined to the main channels.

Our study took place after the cessation of an environmental flow that had been released for the Macquarie Marshes through the Macquarie River system almost a year earlier (October and December 2005, Driver and Knight 2007). The flood pulses occurred about 10 months prior to sampling and the in-channel flow occurred without inundating the floodplain until the sampling was carried out. A sweep net and two or three yabby traps were used at each site to collect *C. destructor.* Baited yabby traps were deployed at each site for 24–48 hrs. Benthic sediment samples (~top 5 cm) were collected manually using plastic bags at three different locations within ~10 m upstream and downstream of the yabby traps to obtain one composite sediment sample of ~600 g wet weight per site. All the *C. destructor* and benthic sediment samples collected were frozen and stored at −20°C until stable isotope analyses were carried out. No chemical preservatives were added to any of the yabby and sediment samples.

Results and implications

Seventeen specimens of *C. destructor* were collected from eight out of 12 sites, with body length in the range 40–120 mm. The $\delta^{15}N$ of *C. destructor* ranged from 6.0‰ to 10.9‰ and the $\delta^{15}N$ of the SOM (<1 mm in size) ranged from 2.9‰ to 5.7‰. The *RTP* of *C. destructor* varied from 1.2 to 3.6 (median 2.7; mean 2.6) and tended to decrease downstream (Fig. 6.6). The overall relationship between the *RTP* of *C. destructor* (y) and distance downstream (x, km) of the Marebone weir (x = 0 km at the weir) was described by the least-squares linear regression equation $y = 3.82 - 0.013x$ ($r^2 = 0.52$, $n = 17$, $P = 0.0011$) using all data, and $y = 4.04 - 0.015x$ ($r^2 = 0.72$, $n = 16$, $P < 0.0001$) when one outlying data point was excluded from the analysis. The body length of *C. destructor* was not a significant predictor of *RTP* ($P = 0.173$ using all data and $P = 0.166$ when one outlying data point was excluded from the analysis). These results imply a substantial trophic shift for *C. destructor* in the Macquarie Marshes during the period of flow pulses, from a more carnivorous/omnivorous trophic position (*RTP* ~>3) to a more herbivorous trophic position (*RTP* ~<2).

Conclusions

Little is known about the ecological conditions of channel systems in inland floodplain wetlands during periods of flow pulses, including the Macquarie Marshes. Because ecological stress on inland floodplain wetlands increases with increasing regulation and water extraction (Kingsford 2000), understanding these aspects becomes important for water resource managers in modelling ecological processes and developing a comprehensive program on wetland protections.

In this study, we investigated the relative trophic position of freshwater crayfish (*C. destructor*) in the in-channel system of the Macquarie Marshes during a period of flow pulses, based on the analysis of stable nitrogen isotope ratios. The results of this and other (Beatty 2006) studies suggest that the trophic position of *C. destructor* may change temporally and spatially in hydrologically variable rivers. Such trophic flexibility may partly explain the occurrence of *C. destructor* in a wide range of habitats including rivers and wetlands.

Conclusions

Understanding the flow of energy and nutrients through ecosystems and their food webs is essential for developing ecological models which will inform the sustainable management of

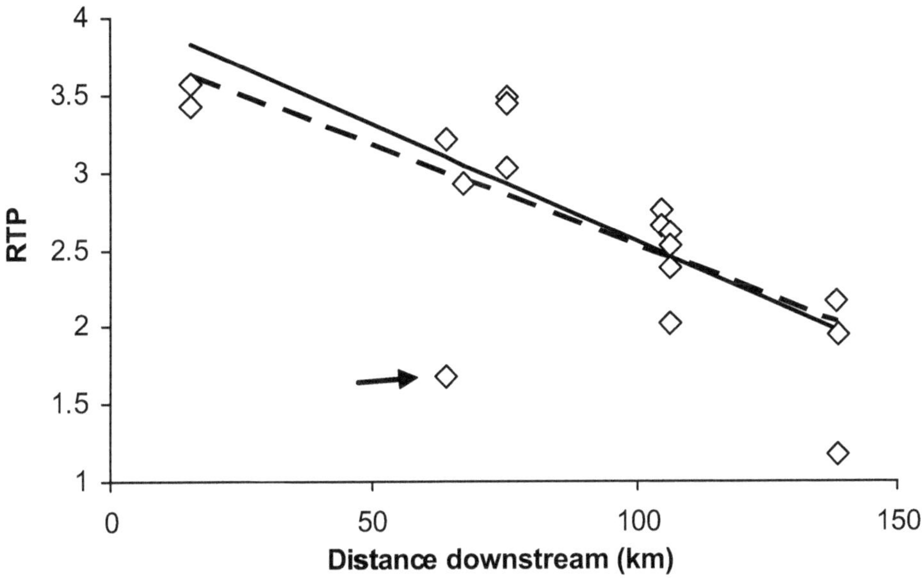

Figure 6.6: *RTP* of *C. destructor* in the in-channel system of the Macquarie Marshes in relation to distance downstream during a period of flow pulses (November 2006), using the $\delta^{15}N$ (‰) value of the benthic SOM as basal food resources at each site. The *RTP* of basal food resources was assigned to one ($\lambda = 1$) and a trophic fractionation factor ($\Delta^{15}N$‰) of 2.2 (Fry 2006) was used. The overall relationship between the *RTP* of *C. destructor* (y) and distance downstream (x, km) of the Marebone weir (where $x = 0$ km) is described by simple linear regression equation $y = 3.82 - 0.013x$ ($r^2 = 0.52$, $n = 17$, $P = 0.0011$) using all data (fitted regression line is shown by a broken line) and $y = 4.04 - 0.015x$ ($r^2 = 0.72$, $n = 16$, $P < 0.0001$) when one outlying data point (solid arrow) is excluded from the regression analysis (fitted regression line is shown by a solid line).

ecosystems. Stable isotope analyses provide a powerful tool which complements traditional techniques for tracing food sources and analysing food web structure. Studies utilising stable isotope approaches have contributed to our understanding of Australian freshwater food webs, and will continue to provide further insights into these complex systems.

Evidence from stable isotope analyses undertaken in key northern Murray-Darling Basin wetlands (Gwydir Wetlands and Macquarie Marshes) suggests that food web structure in floodplain rivers is variable through both space and time. The study of trophic structure under variable flow conditions suggests that the continued restriction of flows (nil or consistent low-flow conditions), coupled with prolonged drought conditions, is likely to increase the dietary insecurity for aquatic organisms and disrupt the healthy functionality of the ecosystem. These findings emphasise the usefulness and effectiveness of stable isotope techniques for assessing aquatic ecosystem functionality and highlight the need for further research in this field, specifically in regard to various flow scenarios.

Acknowledgments

We thank all landholders for access to their properties and for their hospitality and assistance; Bill Johnson, Renee Shepherd, Daryl Albertson, Ray Jones, Michael Maher, Andy McQuie and Tim Pritchard (DECCW) and Suzanne Hollins (ANSTO) for logistic support; Tim Ralph, Lisa Knowles, Jordan Iles and Jennifer Spencer (DECCW) for assistance with field work, sample preparation and advice; Richard Gardiner (DECCW), Julia Sisson (UNE) and Craig Thompson (ANSTO) for assistance with field work; Val Sadler (ANSTO), Laura Rayner and Leroy

Gonsalves (ACU) for assisting with sample preparation; and Rachael Thomas, Shannon Simpson, Simon Hunter and Geoff Gordon (DECCW) for advice. Funding for this research was provided by the Australian Government through the Cotton Catchment Communities CRC and the NSW Wetland Recovery Program which is jointly funded by the NSW Government and the Australian Government's Water for the Future plan, and is delivered in partnership by the NSW Department of Environment, Climate Change and Water, in association with the Central West and Border Rivers-Gwydir Catchment Management Authorities.

References

Arthington AH and Pusey BJ (2003). Flow restoration and protection in Australian rivers. *Rivers Research and Application* **19**, 377–395.

Balcombe SR, Bunn SE, McKenzie Smith FJ and Davies PM (2005). Variability of fish diets between dry and flood periods in an arid zone floodplain river. *Journal of Fish Biology* **67**, 1552–1567.

Beatty SJ (2006). The diet and trophic positions of translocated, sympatric populations of *Cherax destructor* and *Cherax cainii* in the Hutt River, Western Australia: evidence of resource overlap. *Marine and Freshwater Research* **57**, 825–835.

Bunn SE (1986). Origin and fate of organic matter in Australian upland streams. In *Limnology in Australia*. (Eds P De Deckker and WD Williams) pp. 277–297. CSIRO Publishing, Melbourne.

Bunn SE and Boon PI (1993). What sources of organic carbon drive food webs in billabongs? A study based on stable isotope analysis. *Oecologia* **96**, 85–94.

Bunn SE, Davies PM and Winning M (2003). Sources of organic carbon supporting the food web of an arid zone floodplain river. *Freshwater Biology* **48**, 619–635.

Bunn SE, Thoms MC, Hamilton SK and Capon SJ (2006). Flow variability in dry rivers: boom, bust and bits in between. *River Research and Applications* **22**, 179–186.

Burford MA, Cook AJ, Fellows CS, Balcombe SR and Bunn SE (2008). Sources of carbon fuelling production in an arid floodplain river. *Marine and Freshwater Research* **59**, 224–234.

Burns A and Walker KF (2000). Biofilms as food for decapods (Atyidae, Palaemonidae) in the River Murray, South Australia. *Hydrobiologia* **437**, 83–90.

Cabana G and Rasmussen JB (1996). Comparison of aquatic food chains using nitrogen isotopes. *Proceedings of the National Academy of Sciences USA* **93**, 10844–10847.

Cadwallader PL (1979). Distribution of native and introduced fish in the Seven Creeks River system, Victoria. *Australian Journal of Ecology* **4**, 361–385.

DeNiro MJ and Epstein S (1978). Influence of diet on the distribution of carbon isotopes in animals. *Geochimica et Cosmochimica Acta* **42**, 495–506.

Driver P and Knight C (2007). Macquarie Marshes 2005 environmental flow: responses of groundcover plants to environmental flows. Report to the Environmental Flow Reference Group. NSW Dept of Water and Energy, Sydney.

Duffy JE, Richardson JP and France KE (2005). Ecosystem consequences of diversity depend on food chain length in estuarine vegetation. *Ecology Letters* **8**, 301–309.

Fletcher AR, Morison AK and Hume DJ (1985). Effects of carp, *Cyprinus carpio* L., on communities of aquatic vegetation and turbidity of waterbodies in the lower Goulburn River basin. *Marine and Freshwater Research* **36**, 311–327.

France RL (1996). Absence or masking of metabolic fractionations of ^{13}C in a freshwater benthic food web. *Freshwater Biology* **36**, 1–6.

Fry B (1991). Stable isotope diagrams of freshwater food webs. *Ecology* **72**, 2293–2297.

Fry B (2006). *Stable Isotope Ecology*. Springer, New York.

Gee JM (1989). An ecological and economical review of meiofauna as food for fish. *Zoological Journal of the Linnean Society* **96**, 243–261.

Hall D (1981). The feeding ecology of the European carp in Lake Alexandrina and the lower River Murray, South Australia. BSc (Hons) thesis. University of Adelaide, Australia.

Hamilton SK, Lewis WM and Sippel SJ (1992). Energy sources for aquatic animals in the Orinoco River floodplain: evidence from stable isotopes. *Oecologia* **89**, 324–330.

Harris JH (1995). The use of fish in ecological assessments. *Australian Journal of Ecology* **20**, 65–80.

Hume DJ, Fletcher AR and Morison AK (1983). Interspecific hybridization between carp (*Cyprinus carpio* L.) and goldfish (*Carassius auratus* L.) from Victorian waters. *Australian Journal of Marine and Freshwater Research* **34**, 915–919.

Jenkins KM and Boulton AJ (2003). Connectivity in a dryland river: short-term aquatic microinvertebrate recruitment following floodplain inundation. *Ecology* **84**, 2708–2723.

Junk WJ and Wantzen KM (2004). The Flood Pulse concept: new aspects, approaches, and applications – an update. In *Proceedings of the 2nd International Symposium on the Management of Large Rivers for Fisheries.* (Eds RL Welcomme and T Petr) pp. 117–140. Food and Agriculture Organization/Mekong River Commission. FAO Regional Office for Asia and the Pacific, Bangkok. RAP Publication 2004/17.

Junk WJ, Bayley PB and Sparks RE (1989). The flood pulse concept in river–floodplain systems. Proceedings of the International Large River Symposium, Canadian Special Publication of Fisheries and Aquatic Sciences **106**, 110–127.

Kennard M, Pusey B and Arthington A (2001). Trophic ecology of freshwater fishes in Australia. CRC for Freshwater Ecology Scoping Study ScD6 Summary Report. CRC for Freshwater Ecology, Canberra/Centre for Catchment and In-Stream Research, Brisbane.

Keyte P (1994). Lower Gwydir Wetland Plan of Management 1994 to 1997. Report by NSW Dept of Water Resources for the Lower Gwydir Wetland Steering Committee. NSW Dept of Water Resources, Sydney.

Khan TA (2003). Dietary studies on exotic carp (*Cyprinus carpio* L.) from two lakes of western Victoria, Australia. *Aquatic Sciences* **65**, 272–286.

King AJ (2005). Ontogenetic dietary shifts of fishes in an Australian floodplain river. *Marine and Freshwater Research* **56**, 215–225.

Kingsford RT (2000). Ecological impacts of dams, water diversions and river management on floodplain wetlands in Australia. *Austral Ecology* **25**, 109–127.

Kingsford RT and Thomas RF (1995). The Macquarie Marshes in arid Australia and their waterbirds: a 50 year history of decline. *Environmental Management* **19**, 867–878.

Kobayashi T, Ryder DS, Gordon G, Shannon I, Ingleton T, Carpenter M and Jacobs SJ (2009). Short-term response of nutrients, carbon and planktonic microbial communities to floodplain wetland inundation. *Aquatic Ecology* **43**, 843–858.

Kobayashi T, Shiel RJ and Segers H (2007). First record of the rotifer *Lecane shieli* Segers & Sanoamuang, 1994 from Australia. *Australian Zoologist* **34**, 181–183.

Lake PS, Barmuta LA, Boulton AJ, Campbell IC and St Clair RM (1986). Australian streams and northern hemisphere stream ecology: comparisons and problems. *Proceedings of the Ecological Society of Australia* **14**, 61–82.

Lee CL and Fielder DR (1983). Post larval growth of the freshwater prawn *Macrobranchium australiense* Holthuis 1950 (Crustacea: Deapoda: Palaemonidae). *Aquaculture* **30**, 63–70.

Lindholm M, Hessen DO, Mosepele K and Wolski P (2007). Flood webs and energy flexes on a seasonal floodplain: the influence of flood size. *Wetlands* **27**, 775–784.

Marty J, Smokorowski K and Power M (2008). The influence of fluctuating ramping rates on the food web of boreal rivers. *River Research and Applications.* DOI:10.1002/rrq.1194.

McCosker RO (1999). Gwydir Wetlands ecological response to flooding 1998. Report prepared for NSW National Parks and Wildlife Service. Landmax Pty Ltd, Kangaroo Point, Queensland.

McCutchan JH Jr, Lewis WM Jr, Kendall C and McGrath CC (2003). Variation in trophic shift for stable isotope ratios of carbon, nitrogen, and sulfur. *Oikos* **102**, 378–390.

Meakin C, Qin J and Mair G (2008). Feeding behaviour, efficiency and food preference in yabbies *Cherax destructor*. *Hydrobiologia* **605**, 29–35.

Medeiros ESF (2004). Trophic ecology and energy sources for fish on the floodplain of a regulated dryland river: Macintyre River, Australia. PhD thesis. Griffith University, Queensland.

Michener RH and Schell DM (1994). Stable isotope ratios as tracers in marine aquatic food webs. In *Stable Isotopes in Ecology and Environmental Science*. (Eds K Lajtha and RH Michener) pp. 138–157. Blackwell, London.

Minagawa M and Wada E (1984). Stepwise enrichment of ^{15}N along food chains: further evidence and the relation between $d^{15}N$ and animal age. *Geochimica et Cosmochimica Acta* **48**, 1135–1140.

Pace ML, Cole JJ, Carpenter SR and Kitchell JF (1999). Trophic cascades revealed in diverse ecosystems. *Trends in Ecology and Evolution* **14**, 483–488.

Peterson BJ and Fry B (1987). Stable isotopes in ecosystem studies. *Annual Review of Ecology and Systematics* **18**, 293–320.

Post DM (2002). Using stable isotopes to estimate trophic position: models, methods, and assumptions. *Ecology* **83**, 703–718.

Post DM and Takimoto G (2007). Proximate structural mechanisms for variation in food-chain length. *Oikos* **116**, 775–782.

Power ME (1995). Floods, food chains and ecosystem process in rivers. In *Linking Species and Ecosystems*. (Eds CG Jones and JH Lawton) pp. 52–60. Chapman and Hall, New York.

Puckridge JT, Sheldon F, Walker KF and Boulton AJ (1998). Flow variability and the ecology of arid zone rivers. *Marine and Freshwater Research* **49**, 55–72.

Pusey BJ, Arthington AH and Read MG (2000). The dry-season diet of freshwater fishes in monsoonal tropical rivers of Cape York Peninsula, Australia. *Ecology of Freshwater Fish* **9**, 177–190.

Reid DJ, Quinn GP, Lake PS and Reich P (2008). Terrestrial detritus supports the food webs in lowland intermittent streams of south-eastern Australia: a stable isotope study. *Freshwater Biology* **53**, 2036–2050.

Robertson AI, Bunn SE, Walker KF and Boon PI (1999). Sources, sinks and transformations of organic carbon in Australian floodplain rivers. *Marine and Freshwater Research* **50**, 813–829.

Schofield N, Burt A and Connell D (2003). *Environmental Water Allocation: Principles, Policies and Practices*. Land and Water Australia, Canberra.

Semple GP, Rouse DB and Mclain KR (1990). *Cherax destructor C tenuimanas C quadricarinatus*: a comparative review of biological traits relating to aquaculture. *Freshwater Crayfish* **8**, 495–503.

Sheldon F and Walker KF (1997). Changes in biofilms induced by flow regulation could explain extinctions of aquatic snails in the lower River Murray, Australia. *Hydrobiologia* **347**, 97–108.

Smith BN and Epstein S (1971). Two categories of $^{13}C/^{12}C$ ratios for higher plants. *Plant Physiology* **47**, 380–384.

Vanderklift M and Ponsard S (2003). Sources of variation in consumer-diet $\delta^{15}N$ enrichment: a meta-analysis. *Oecologia* **136**, 169–182.

Vannote R, Minshall G, Cummins K, Sedell J and Cushing C (1980). The river continuum concept. *Canadian Journal of Fisheries and Aquatic Sciences* **37**, 130–137.

Verhoef GD, Jones PL and Austin CM (1998). A comparison of natural and artificial diets for juveniles of the Australian freshwater crayfish *Cherax destructor. Journal of the World Aquaculture Society* **29**, 243–248.

Wilson GG, Bickel TO, Berney PJ and Sisson JL (2009). Managing environmental flows in an agricultural landscape: the Lower Gwydir floodplain. Final Report to the Australian Government Department of the Environment, Water, Heritage and the Arts. University of New England/Cotton Catchment Communities CRC, Armidale.

Chapter 7

Responses of fish and waterbirds to flow variability in the Gwydir Wetlands

Glenn Wilson, Jennifer Spencer and Elizabeth Heagney

Introduction

Many Australian lowland rivers flow through low-rainfall dryland regions with extensive floodplains (Thoms and Sheldon 2000). These systems typically experience a high level of spatial and temporal flow variability which shapes the morphology of their channels and the formation, diversity and maintenance of wetlands on the surrounding floodplain (Ward 1998; Thoms *et al.* 2006; Young and Kingsford 2006). This interplay between hydrology and floodplain geomorphology influences the diversity, abundance and population dynamics of flora and fauna inhabiting these systems (Walker *et al.* 1995; Balcombe *et al.* 2006). This occurs through flows shaping the availability of structural habitat and the characteristics of flow signals that drive key processes such as spawning activity, recruitment success and species succession (Jenkins and Boulton 2003; Humphries *et al.* 2006). However, regulation of the natural flow regime through dams or weirs can affect these key ecosystem components and processes, disrupting biotic structure and function (Bunn and Arthington 2002; Young and Kingsford 2006).

In recent years there has been a growing shift towards an integrated catchment view of lowland river systems, with greater recognition of the importance of floodplain wetlands in the ecological functioning of lowland freshwater ecosystems (Bailey 1991; Kingsford 2000). However, our knowledge of how floodplain wetlands respond to particular flow events is largely in its infancy, despite the importance of these systems for the population ecology of many fauna at spatial scales beyond the boundary of the wetlands. Fish and waterbirds are a particular case in point. For lowland populations of native Murray-Darling fishes, the relative importance of high- and low-flow periods has been an active area of debate (Harris and Gehrke 1994; Humphries *et al.* 1999; Mallen-Cooper and Stuart 2003), although much of the focus has been on southern rivers dominated by late-winter flow peaks. Similarly, while waterbirds across inland Australia are known to move readily among wetland systems in response to flood events (Kingsford and Norman 2002; Roshier *et al.* 2002), detailed studies of waterbird breeding–flow relationships and subsequent recruitment are still needed for most of the northern Murray-Darling Basin.

The Gwydir River in the northern Murray-Darling Basin is a 300 km long semi-arid lowland river system with an expansive terminal floodplain wetland complex. The Gwydir Wetlands contain internationally recognised breeding areas for colonial waterbirds (Kingsford

2000) and stands of threatened wetland plants (Wilson *et al.* 2008). However, development of the floodplain for irrigated agriculture has altered flow patterns in the lower catchment (Keyte 1994; Sheldon *et al.* 2000; Growns 2008) and the core wetland area has decreased by 90% in recent decades (Crabb 1997). Flow characteristics are important for maintaining aquatic ecosystem integrity in lowland river and wetland systems such as this (Bunn and Arthington 2002; Young and Kingsford 2006). Protecting key elements of the flow regime by limiting abstraction of water during flood events or specifying certain environmental flow releases may benefit aquatic communities and ecological processes.

An effective environmental water strategy for floodplain wetlands requires a clear understanding of links between biotic structure and flow variability. Determining the ways in which assemblage structure, population replenishment and species richness are driven by flows is central to devising appropriate environmental flow rules, and will underpin the development of predictive tools for adaptive management (Ch. 8, this volume). Flow patterns can influence wetland species, assemblages and populations over a range of temporal scales (Walker *et al.* 1995; Puckridge *et al.* 1998). However, responses to longer-term flow histories (months to years) and individual flow pulses (days to months) are particularly relevant to environmental flow management.

We investigate links between flow variability and fish and colonial waterbird assemblages in the Gwydir Wetlands. We focus on flow variability, particularly in terms of the magnitude, timing and duration of flows, at the long-term and pulse scales to present insights into some of the ecological flow requirements of this ecosystem. A mix of historical and recent observations of fish and colonially nesting waterbirds is used to examine how flow has influenced population size, assemblage structure and recruitment.

Gwydir Wetlands

Channel and wetland structure and catchment hydrology

The Gwydir Wetlands cover around 24 000 ha and receive regulated flows from Copeton Dam upstream on the Gwydir River via the Lower Gwydir River and Gingham Watercourses (Fig. 7.1). Other major effluent streams with regulated flow on the lower Gwydir floodplain include the Mehi River and the Carole and Moomin creeks. Copeton Dam (1 364 000 ML) was completed in 1976 and supplies approximately 55% of total inflow to the Gwydir River (Keyte 1994). Regulatory structures at Tareelaroi, Boolooroo and Tyreel (Fig. 7.1) divert flows from the Gwydir River into the Mehi River, Carole Creek and Lower Gwydir River/Gingham Watercourse, respectively.

Comparisons of observed daily flow data with modelled pre-development conditions indicate that the Gwydir is one of the most flow-altered rivers in the region (Sheldon *et al.* 2000; Growns 2008). Water resource development has led to significant changes in the magnitude, seasonality and frequency of flows reaching the Gwydir Wetlands, including a 12% reduction in the frequency of flooding of the core wetland area (Kingsford 2000) and a 70% reduction in the frequency of large floods (Bennett and Green 1993; Sheldon *et al.* 2000).

Unregulated flows can still reach the wetlands from the Horton River and Warialda Creek (Fig. 7.1), both of which enter the Gwydir River below Copeton Dam. Mean annual rainfall in the Gwydir Wetlands varies from less than 450 mm to the west to 585 mm on the eastern fringe (BOM 2007a; Keyte 1994). Average monthly discharge from the Gwydir River peaks during summer months (December to February) (ANRA 1995).

Environmental water management

The water-sharing plan for the Gwydir Regulated River Water Source (NSW Government 2003) established an environmental water allocation (environmental contingency allowance,

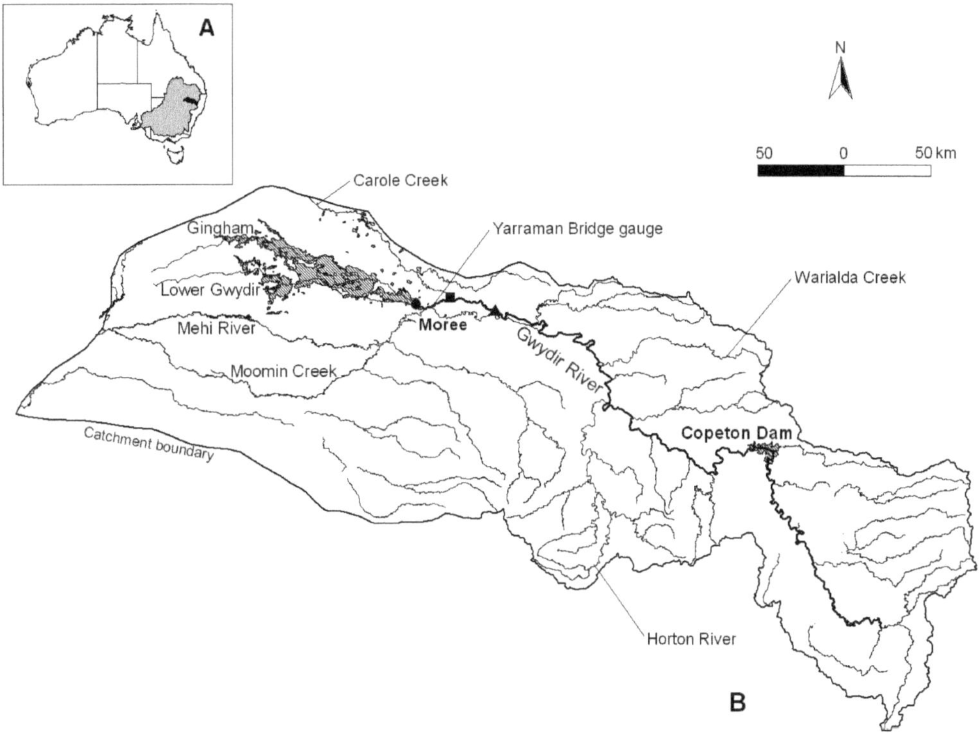

Figure 7.1: Location of (A) the Gwydir River catchment within the Murray-Darling Basin in northern NSW and (B) the Gwydir Wetlands and major channels and regulators. Shaded wetland extent in map b is based on 1990 aerial photography. Symbols refer to regulators at Tareelaroi (▲), Boolooroo (■) and Tyreel (●).

ECA) of up to 45 000 ML in Copeton Dam. This allows for environmental water (ECA) releases into the Gwydir Wetlands. A stakeholder operations advisory committee comprising industry, scientific, conservation and agency representation advises the NSW Government on the timing, rate and duration of ECA releases. These are implemented to achieve a set of nine ecological objectives including the provision of environmental flows along downstream effluent streams, the support of native fish populations and their habitats, and the support of colonially nesting waterbirds (DECCW 2008). Eight ECA releases have been delivered to the Gwydir Wetlands since 1996 (DECCW 2008), primarily to maintain core areas of wetland vegetation or to support waterbird breeding, with the assumption that other ecosystem components and management objectives would receive similar benefits. ECA discharges have typically been low (e.g. 264 ML and 278 ML per day in April and November 2007 respectively; Fig. 7.2), to maximise the efficiency of flow delivery to target wetlands while minimising the potential for losses onto the upstream floodplain.

Flow variability and fish

Assemblage structure

Several studies have described fish assemblage structure from the Gwydir Wetlands, including a total of 11 native and three exotic fish species. However, few have related their observations to hydrology and most have examined patterns at the river scale rather than separating lowland channels from upstream sites. Native species richness was relatively consistent from limited

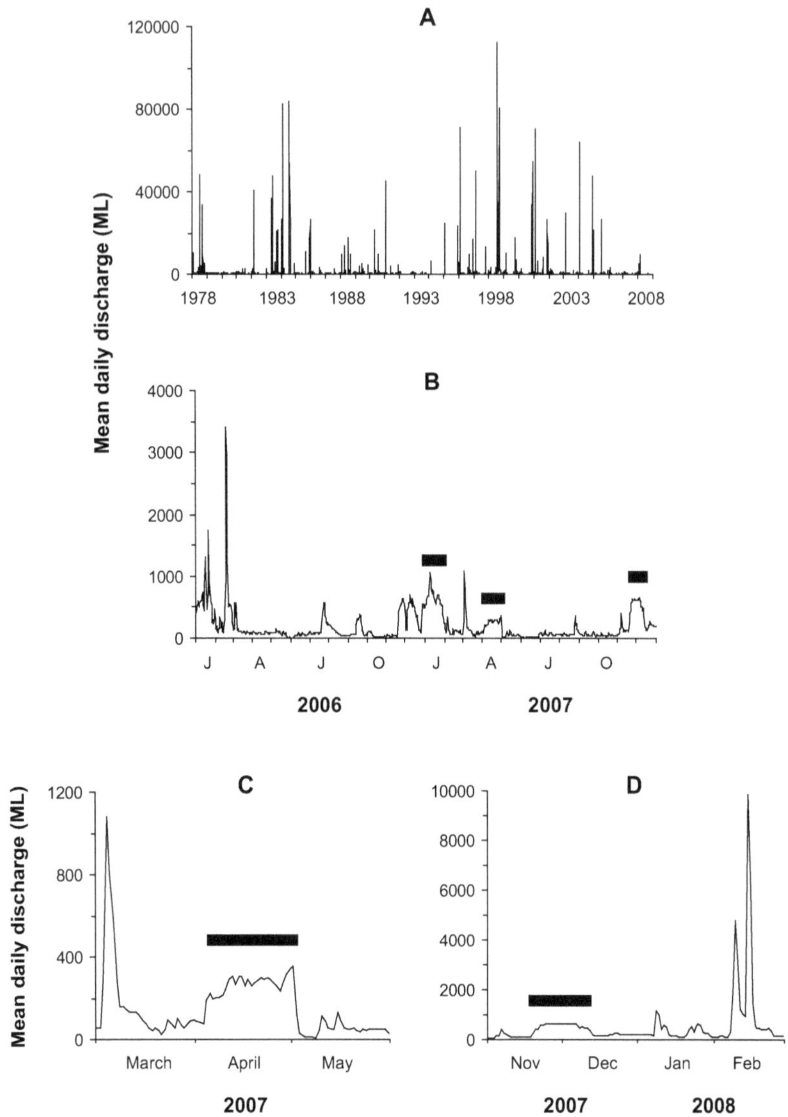

Figure 7.2: Discharge variability in the Gwydir River at the Yarraman Bridge gauge (418004) over a range of temporal scales. (a) 1978 to 2008. (b) January 2006 to December 2007, with environmental releases during this period indicated by rectangles. (c) and (d) Individual environmental flow releases commencing in April and November 2007. Note the different y-axis scales in (c) and (d).

pre-regulation surveys conducted in 1975, to post-regulation surveys conducted throughout high-rainfall periods (1994–2005) and more recent surveys conducted during a period of low rainfall and drought (2006–09). Harris and Gehrke (1997) surveyed one site in the Mehi River at Moree (Fig. 7.1) (1994–96), and detected seven native and two exotic fish species. Bony bream (*Nematolosa erebi*), Murray-Darling rainbowfish (*Melanotaenia fluviatilis*) and carp gudgeons (*Hypseleotris* spp.) were the dominant taxa, while a number of species collected more recently from the Gwydir Wetlands were absent: unspecked hardyhead (*Craterocephalus stercusmuscarum fulvus*), eel-tailed catfish (*Tandanus tandanus*) and the exotic mosquitofish

(*Gambusia holbrooki*) (Spencer *et al.* 2009; Wilson *et al.* 2009). Subsequent sampling at 10 sites in the Gwydir Wetlands by the NSW Department of Primary Industries (1999–2002; Growns and Rodgers 2002) detected the last two of those species but no studies have recorded glass perch (*Ambassis agassizii*) or silver perch (*Bidyanus bidyanus*) despite their occurrence in similar habitats elsewhere in the region (Wilson and Wright 2005).

Fish assemblage structure in lowland river systems may be affected by flows in three basic ways. First, numerous species may respond to particular flow signals in their spawning activity (discussed below), although few data are available for the lower Gwydir channels. Second, population increases may occur through local increases in recruitment or immigration. For example, European carp (*Cyprinus carpio*) are thought to have invaded the lower Gwydir system through major floods in the early 1970s (Copeland *et al.* 2003). Their establishment has undoubtedly been aided by the various low-level weirs and frequent low-flow conditions (Mallen-Cooper 2000; Sheldon *et al.* 2000) that provide an ideal environment for spawning and recruitment (Driver *et al.* 1997). Third, negative changes may occur through flow-related emigration or local extinctions. Local extinctions have been documented anecdotally in the Gwydir Wetlands after the drying of critical habitats (Copeland *et al.* 2003). Anecdotal accounts also report severe post-regulation declines in native fish populations in the catchment, including eel-tailed catfish, Murray cod (*Maccullochella peelii peelii*) and silver perch (Copeland *et al.* 2003). However, separating flow effects from those due to habitat modification or probable increases in fishing pressure on these larger iconic species is tenuous, and the mechanisms by which their abundances may have been affected by altered hydrology have not been established.

Growns (2008) examined the effects of flow regulation on fish assemblages in rivers across NSW using a three-year annual survey program. Four lower Gwydir sites were included, although the data were interpreted at the river scale. While Gwydir assemblages differed significantly from those in the other five rivers, the index of flow alteration (amended annual proportional flow deviation) by itself did not significantly explain any of the variation in fish abundance or assemblage structure.

Wilson *et al.* (2009) examined fish assemblage structure in the Lower Gwydir River, Gingham Watercourse and Mehi River through event-based monitoring from October 2006 to February 2009. Despite these channels receiving flows from a common upstream source, each had a consistent difference in species dominance: Mehi River 59% carp gudgeon (*Hypseleotris* spp), Lower Gwydir River 70% bony bream and Gingham Watercourse 39% spangled perch (*Leiopotherapon unicolor*). This made it difficult to compare responses of individual species to specific flow events across the system. Moreover, the influence of discharge on shifts in assemblage structure was inconsistent over time. For example, shifts in the Gingham Watercourse following flow events in early 2007, late 2007 and early 2008 differed considerably, despite relatively similar discharges over the preceding 30 days. Similar inconsistencies were evident in the Mehi River between February and May 2007 following natural flows and releases for irrigation and stock and domestic use. These observations highlight the difficulty of isolating hydrological influences on the ecology of regulated systems where frequent flow releases are made for a variety of purposes. They also suggest that indicators of population structure and/or processes (e.g. spawning patterns, cohort sizes or population-specific growth rates) may show more promise for detecting how flow variability influences lowland river fish species.

Early development and timing of spawning

Recruitment of juvenile fish into adult populations depends on spawning dynamics and the survival of individuals through early life-history stages. Faster-growing juveniles move more quickly through the smaller size classes that are most vulnerable to predation, and are

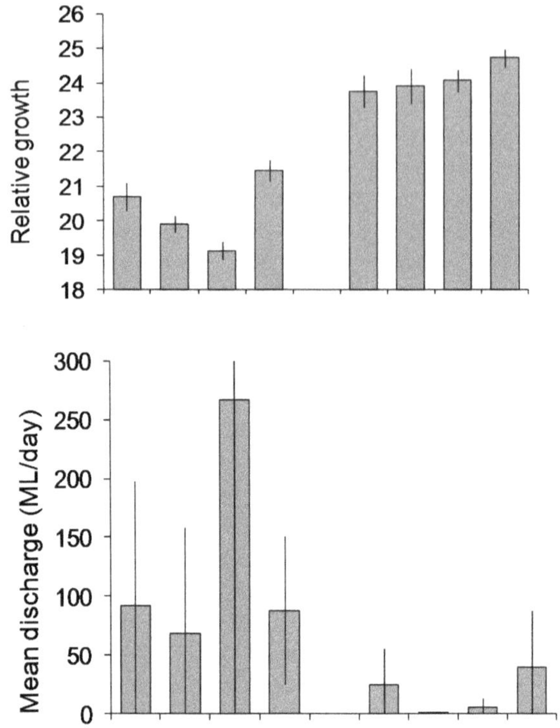

Figure 7.4: Mean relative growth in Australian smelt (*Retropinna semoni*) and mean daily discharge (± SE) over the hatch period (back-calculated from otolith increments) in the Gwydir Wetlands. Data are presented for eight sites across four channels: Carole Creek, Gwydir River, Gingham Watercourse and Mehi River. Relative growth (± SE) is represented by the mean standard length for a 77-day-old smelt from each site.

considered more likely to survive to maturity (Bailey *et al.* 1995; Meekan and Fortier 1996; Post and Parkinson 2001). Consequently, juvenile fish growth rates may be an important determinant of recruitment to adult populations. Fish growth is a plastic life-history trait that is largely determined by environmental conditions (Campana and Thorrold 2001; Boeuf and Payan 2001), and can be used as an indicator of wetland health and productivity (Tonkin *et al.* 2008b). Although no explicit link between freshwater flow variability and juvenile growth has been reported, flow may affect growth indirectly via known effects on temperature, salinity and food supply (Boeuf and Payan 2001; Tonkin *et al.* 2008a).

Heagney *et al.* (2009) used otolith-based analyses to examine the influence of hydrology on early growth rates of two native species, Australian smelt (*Retropinna semoni*) and bony bream. These species were relatively common in channels in the Gwydir Wetlands during sampling in November 2007 (Australian smelt 47% and bony bream 18% of total catch) (Spencer *et al.* 2009). The study included sites in four of the main channels feeding the Gwydir Wetlands (Carole Creek, Gwydir River, Gingham Watercourse and Mehi River). Age was determined by counting daily otolith increments (Fig. 7.3/colour plate, p. 301), and the estimates were used to back-calculate fish spawning and hatch dates. Size-at-age data were used to examine patterns in juvenile growth among channels and in relation to discharge volume and variability.

Back-calculation of hatch dates for smelt indicated that fish had spawned from late July to at least early October in spring 2007. It was unclear whether spawning was triggered by individual flow events. However, Wilson *et al.* (2009) noted that environmental flow releases initiated spawning in two other native species, bony bream and spangled perch, in lower Gwydir channels. Mean growth curves from all sample sites indicated that juvenile growth rates of Australian smelt from the Gwydir Wetlands were low to intermediate compared with values reported from other regions of the Murray-Darling Basin (Humphries *et al.* 2008; Tonkin *et al.* 2008b). However, smelt growth rates varied considerably among the eight sites sampled within the Gwydir Wetlands, ranging from 0.11–0.28 mm/day (ANCOVA, $p < 0.001$). Growth rates were negatively correlated (Spearman's, $p < 0.028$) with mean daily discharge over the hatch period (Fig. 7.4). Smelt had slower growth rates in sites in the Carole Creek and Gwydir River where discharge volumes and flow variability were high, and higher growth rates at sites in the Gingham Watercourse and Mehi River where discharge volumes and flow variability were lower (Fig. 7.4). Juvenile growth rates of bony bream among four Gwydir River and Gingham Watercourse sites also displayed a significant negative relationship with mean daily discharge throughout the spawning/hatch period (Heagney *et al.* 2009).

Colonially nesting waterbirds

At least 75 waterbird species have been recorded in the Gwydir Wetlands since the 1920s and at least 50 species have bred there (Spencer 2009) (Table 7.1). Records of the significance of the Gwydir Wetlands for colonial waterbirds date back to the early 1900s, when vast colonies of straw-necked ibis (*Threskiornis spinicollis*), intermediate egret (*Ardea intermedia*), glossy ibis (*Plegadis falcinellus*) and rufous night heron (*Nycticorax caledonicus*) were first documented (Morse 1918, 1922; Bailey 1934). Prior to river regulation, colonial waterbirds bred in seven years of every decade in these wetlands, although this declined to four years per decade following regulation (Johnson 2004). Studies of colony size and success were limited until the late 1980s when breeding events had begun to diminish; major flooding in the summers of 1995/96, 1996/97 and 1998/99 initiated more detailed surveys. Most large breeding events have occurred along the Gingham Watercourse, triggered by heavy local rainfall and flooding from upstream (McCosker 1996, 1999a, b).

The largest recorded colonial breeding event in the Gwydir Wetlands took place after major flooding (>192 000 ha) in July–September 1998 (8000–12 500 ML/day) (Fig. 7.5) when more than 150 000 breeding pairs of straw-necked ibis and large numbers of intermediate egrets (10 000 pairs), rufous night herons (15 000 pairs) and glossy ibis (15 000 pairs) were recorded (Johnson 2002). Johnson (2001, 2004) estimated that once breeding had begun, large colonies could be maintained by flows into the wetlands of 250–500 ML/day, although the volume of water required depended on antecedent conditions and water levels needed to be maintained for 10 weeks or more (Johnson 2001). Inundation of the surrounding floodplain on the Gingham and Lower Gwydir watercourses was also necessary, to provide feeding habitat to support large breeding events (McCosker 1996, 1999a, b).

Linear regression analysis was used to determine whether the size of colonial breeding events along the Gingham Watercourse between 1988 and 2007 (McCosker 1996, 1999a, b; Johnson 2002) was related to cumulative discharge from the Gwydir River preceding (April–September) or during (October–March) the breeding season (after Kingsford and Auld 2005). This analysis suggested that total number of nests along the Gingham Watercourse was related more to discharge during the breeding season ($r^2 = 0.3$, $F_{1,40} = 14.5$, $P = 0.001$) than that recorded over the preceding six months ($r^2 = 0.04$, $F_{1,40} = 1.7$, $P = 0.202$). Total flows of over 200 000 ML were necessary for straw-necked ibis to breed; egrets formed small colonies in periods of relatively low flows (approx. 55 000–145 000 ML in 1988, 1995 and 1998). A small breeding event over the 2000/01 summer (2000 pairs of intermediate egrets, 200 pairs of glossy ibis and 20 pairs of rufous night herons) did not fit this overall pattern, despite high flows

Table 7.1: Waterbird species observed in the Gwydir Wetlands historically and during field surveys in 2007–08

Family	Common name	Scientific name	Historically recorded breeding	Observed during field survey
Anatidae	Australasian shoveler	*Anas rhynchotis*	Y	Y
	Australian shelduck	*Tadorna tadornoides*	Y	
	Australian wood duck	*Chenonetta jubata*	Y	Y
	Black swan	*Cygnus atratus*	Y	Y
	Blue-billed duck	*Oxyura australis* (V)	Y	
	Chestnut teal	*Anas castanea*		
	Freckled duck	*Stictonetta naevosa* (V)	Y	
	Grey teal	*Anas gracilis*	Y	Y
	Hardhead	*Aythya australis*	Y	Y
	Musk duck	*Biziura lobata*	Y	Y
	Pacific black duck	*Anas superciliosa*	Y	Y
	Pink-eared duck	*Malocorhynchus membranaceus*	Y	Y
	Plumed whistling duck	*Dendrocygna eytoni*	Y	Y
	Wandering whistling duck	*Dendrocygna arcuata*	Y	
Anhingidae	Darter	*Anhinga melanogaster*	Y	Y
Anseranatidae	Magpie goose	*Anseranas semipalmata* (V)	Y	
Ardeidae	Australasian bittern	*Botaurus poiciloptilus* (V)		
	Black bittern	*Ixobrychus flavicollis* (V)		
	Cattle egret	*Ardea ibis*	Y	Y
	Great egret	*Ardea alba*	Y	Y
	Intermediate egret	*Ardea intermedia*	Y	Y
	Little bittern	*Ixobrychus minutus*	Y	
	Little egret	*Ardea garzetta*	Y	Y
	Rufous night heron	*Nycticorax caledonicus*	Y	Y
	White-faced heron	*Ardea novaehollandiae*	Y	Y
	White-necked heron	*Ardea pacifica*	Y	Y
Burhinidae	Bush stone curlew	*Burhinus grallarius*	Y	
Charadriidae	Banded lapwing	*Vanellus tricolor*	Y	
	Black-fronted dotterel	*Elseyornis melanops*	Y	
	Double-banded plover	*Charadrius bicinctus*		
	Inland dotterel	*Charadrius australis*		
	Masked lapwing	*Vanellus miles*	Y	Y
	Oriental plover	*Charadrius veredus*		
	Red-capped plover	*Charadrius ruficapillus*		
	Red-kneed dotterel	*Erythrogonys cinctus*	Y	Y

Family	Common name	Scientific name	Historically recorded breeding	Observed during field survey
Ciconiidae	Black-necked stork	*Ephippiorhynchus asiaticus* (E)	Y	Y
Glareolidae	Australian pratincole	*Stiltia isabella*		
Gruidae	Brolga	*Grus rubicundus* (V)	Y	Y
Jacanidae	Comb-crested jacana	*Irediparra gallinacea* (V)	Y	
Laridae	Caspian tern	*Sterna caspia*		Y
	Common tern	*Sterna hirundo*		
	Gull-billed tern	*Sterna nilotica*		Y
	Silver gull	*Larus novaehollandiae*		Y
	Whiskered tern	*Chlidonias hybridus*	Y	Y
	White-winged black tern	*Chlidonias leucopterus*		
Pelecanidae	Australian pelican	*Pelecanus conspicillatus*		Y
Phalacrocoracidae	Great cormorant	*Phalacrocorax carbo*	Y	Y
	Little black cormorant	*Phalacrocorax sulcirostris*	Y	Y
	Little pied cormorant	*Phalacrocorax melanoleucos*	Y	Y
	Pied cormorant	*Phalacrocorax varius*	Y	Y
Plataleidae	Australian white ibis	*Threskiornis molucca*	Y	Y
	Glossy ibis	*Plegadis falcinellus*	Y	Y
	Royal spoonbill	*Platalea regia*	Y	Y
	Straw-necked ibis	*Threskiornis spinicollis*	Y	Y
	Yellow-billed spoonbill	*Platalea flavipes*	Y	Y
Podicipedidae	Australasian grebe	*Tachybaptus novaehollandiae*	Y	Y
	Great-crested grebe	*Podiceps cristatus*	Y	Y
	Hoary-headed grebe	*Poliocephalus poliocephalus*	Y	Y
Rallidae	Australian crake	*Porzana fluminea*		
	Black-tailed native hen	*Gallinula ventralis*	Y	Y
	Dusky moorhen	*Gallinula tenebrosa*	Y	Y
	Eurasian coot	*Fulica atra*	Y	Y
	Purple swamphen	*Porphyrio porphyrio*	Y	Y
	Spotless crake	*Porzana tabuensis*		
Recurvirostridae	Banded stilt	*Cladorhynchus leucocephalus*		
	Black-winged stilt	*Himantopus himantopus*	Y	Y
	Red-necked avocet	*Recurvirostra novaehollandiae*		
Rostratulidae	Australian Painted snipe	*Rostratula australis* (E)	Y	
Scolopacidae	Common greenshank	*Tringa nebularia*		Y

Family	Common name	Scientific name	Historically recorded breeding	Observed during field survey
	Common sandpiper	*Actitis hypoleucos*		
	Latham's snipe	*Gallinago hardwickii*		Y
	Little curlew	*Numenius minutus*		
	Marsh sandpiper	*Tringa stagnatilis*		
	Ruff	*Philomachus pugnax*		
	Sharp-tailed sandpiper	*Calidris acuminata*		

Species listed as V (vulnerable) and E (endangered) under threatened species legislation (*Threatened Species Conservation Act 1995*, NSW) are indicated. Taxonomic nomenclature follows Marchant and Higgins (1990, 1993) and Higgins and Davies (1996).

(580 000 ML). However, that event followed three large breeding events in close succession (1995–99) which coincided with major flooding in other parts of the Murray-Darling Basin including the Macquarie Marshes (Kingsford and Auld 2005). Limited Gwydir breeding may have been due to movements of birds between wetland systems.

Overall, cumulative discharge explained only a small proportion of the total variance associated with the occurrence of breeding in colonial waterbirds. Other factors are also important for successful breeding in colonial waterbirds, including the depth of water around the nesting sites, rate of water delivery and recession, and food availability (Scott 1997). Three instances of large-scale desertions of eggs and/or chicks have been recorded in the Gwydir Wetlands:

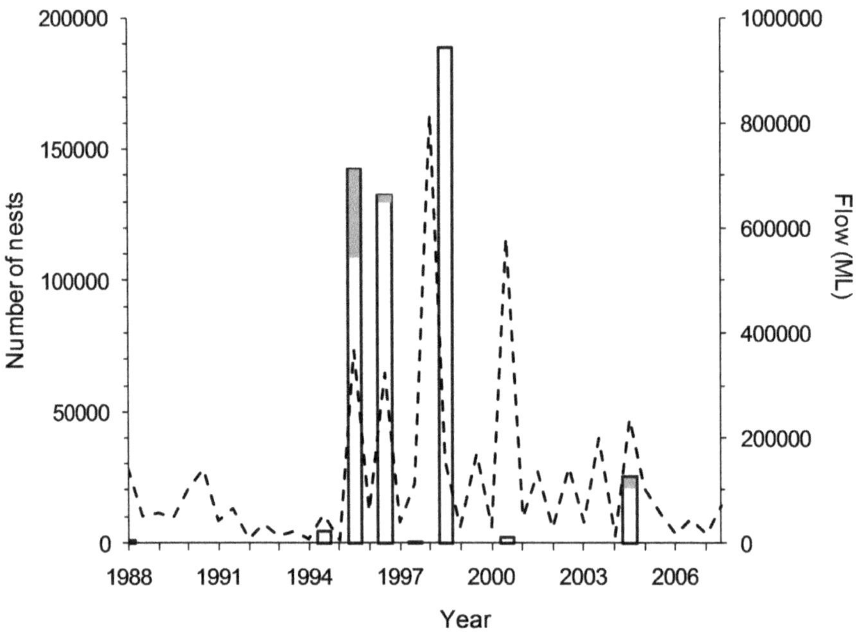

Figure 7.5: Total number of nests (bars) of colonially nesting waterbirds in the Gwydir Wetlands and six-month cumulative flow (ML) (dashed line) at Yarraman Bridge gauge (418004), Gwydir River, 1988–2008. Shaded bars indicate the estimated number of nest desertions. Estimates of colony size were taken from McCosker (1996, 1999a, b), Jones (1997) and Johnson (2002). Reproduced from Spencer (2009).

December 1995 (15 000 pairs of straw-necked ibis, 14 000 pairs of intermediate egrets), April 1997 (300 pairs of straw-necked ibis) and January 2005 (4000 pairs of straw-necked ibis) (Fig. 7.5). In straw-necked ibis, rapid recession of floodwaters around the colonies caused birds to desert their nests on all three occasions (Spencer 2009).

To determine the current status of waterbird populations in the Gwydir Wetlands, aerial and ground-based surveys were undertaken in October 2007 and April 2008 (Spencer *et al.* 2009). We concentrated on wetland habitats along the main channels and floodplains of the Lower Gwydir River, Gingham Watercourse, Mehi River and Carole Creek, and included a number of large (>2 ha) farm storages. During both survey periods, much of the floodplain was extremely dry and surface water was confined to channels, floodplain waterholes and farm storages. Forty-five waterbird species were recorded (Table 7.1), but numbers of colonially nesting waterbirds were very low (489 ± 142 SE) and no breeding activity was recorded despite some managed releases and natural inflows from December 2007 to February 2008 (Fig. 7.2). Artificial sites attracted species from the Anatidae (ducks and swans) and Phalacrocoracidae (cormorants) families, which favour deep permanent waterbodies. Other waterbird species, such as Australian white ibis (*Threskiornis molucca*), white-faced herons (*Ardea novaehollandiae*) and egrets were more common on inundated parts of the floodplain.

Conclusions and implications for management

This chapter provides a snapshot of fish and waterbird assemblages in the Gwydir Wetlands 33 years after river regulation began, following a period of protracted drought throughout much of the Murray-Darling Basin (BOM 2007b). Populations of native fish and colonial waterbirds in this ecosystem are now largely composed of generalist species with broad flow requirements for breeding and recruitment. Our synthesis of how fish and waterbirds appear to respond to flow regime in the lower Gwydir allows a more refined suite of indicators and response variables to be chosen for future monitoring of flow responses in these groups.

Attempts to characterise relationships between hydrology and fish assemblage structure in the Gwydir Wetlands did not produce a clear picture either from annual or event-based sampling. Several factors may be responsible. Patterns may be obscured by the high levels of inherent spatial variability in sampling fish populations and the difficulties in obtaining representative and time-effective assessments of fish populations in turbid and structurally complex freshwater systems (Harris and Gehrke 1997). It is unlikely that all species from these environments will be equally influenced by flow regime and its alteration. For example, lowland riverine fish possess a range of spawning modes (Humphries *et al.* 1999) with varying patterns of spawning season, duration and mode. Fish with a more protracted or flexible spawning season might be expected to cover their risk of recruitment failure under adverse flow conditions better than species with a narrower set of spawning conditions (Humphries *et al.* 2008). Flexibility in other traits, such as diet or physical habitat preference, would be expected to confer a similar advantage.

It may be more profitable to examine the effects of flow variability and alteration on lowland river fish through life-history traits such as recruitment, early growth and body condition in key species. Riverine species are influenced by flow variability in a variety of ways and through most facets of their life-history (Puckridge *et al.* 1998; Young and Kingsford 2006). Native species such as Australian smelt and bony bream display variable growth rates among rivers and regions in the Murray-Darling Basin (Heagney *et al.* 2009; Humphries *et al.* 2008; Tonkin *et al.* 2008b). Given the spatial complexity of lowland river systems, a greater understanding of how the biology of individual species varies within assemblages and among channels is necessary before we can better predict outcomes from alternative flow management scenarios.

Conceptualising the ways in which species or their environment may respond to particular points in the hydrograph or flow sequences (Gawne and Scholz 2006) will assist in identifying appropriate predictor variables for modelling ecological responses to flows. However, the difficulties in relating assemblage patterns to hydrology underscores the importance of selecting the most appropriate flow metrics and indices of deviation from natural.

Studies focused largely or solely on southern Murray-Darling Basin rivers (Humphries *et al.* 1999, 2006) have illustrated the importance of periods or areas of low-flow conditions for successful fish recruitment. The resulting low-flow recruitment hypothesis was proposed to explain how fish may gain a survivorship advantage by spawning during low-flow periods of higher prey density and more favourable temperature and hydrologic conditions (Humphries *et al.* 1999). Early growth and development in at least two native species (Australian smelt and bony bream) in this northern river system were enhanced in areas of low flow, but high flows are also important as they can initiate spawning in other native fish. In the Gwydir Wetlands, both species had higher growth rates at sites with lower discharge throughout the hatching period and larval-juvenile phase. Observations from other locations (Schlosser 1998; Humphries *et al.* 2006) have shown that low-velocity refuges may be key sites for recruitment over a broader stream scale. This has important implications for restoration efforts in the Gwydir Wetlands, where management has previously focused on maximising the efficiency of water delivery for agricultural use.

Many waterbird species in the Murray-Darling Basin have adapted to highly variable flooding regimes and thrive on the 'boom and bust' cycles that offer different breeding and feeding opportunities (Kingsford and Norman 2002; Roshier *et al.* 2002). Colonially nesting waterbirds have specific requirements for breeding, including particular vegetation for nesting and adequately timed flows of sufficient duration, depth and extent (Scott 1997). Our analysis indicated that higher discharge during the breeding season, rather than in the preceding six months, was one predictor of colonial waterbird breeding activity in the Gwydir Wetlands. While large floods were important for initiating large breeding events, these were generally infrequent. Small to medium floods were important for supporting smaller breeding events and maintaining nesting habitat between larger floods. Waterbird species that depend on ephemeral floodplain wetlands for feeding and breeding habitat, including some colonially nesting waterbirds, tend to be most affected by river regulation (Scott 1997). Species that prefer permanent areas of deep open water, such fish-eating species and some waterfowl, can persist in artificial habitats such as farm dams, irrigation channels and impoundments (Marchant and Higgins 1990), where wetland habitat is limited and breeding activity is usually absent. The frequency of large floods and the availability of suitable breeding habitat are significant limiting factors for colonial waterbirds in the Gwydir Wetlands, and both have declined since flow regulation (Kingsford 2000; Sheldon *et al.* 2000). As large areas of feeding habitat are also required for successful breeding, it is critical that the condition of the remaining semi-permanent wetlands be maintained between major floods to maximise food availability in the event of large-scale flooding. It is unlikely that ECA releases will ever be large enough to trigger colonial breeding events in the Gwydir. However, strategically timed environmental flows could be used to supplement natural flows to maximise fledging success. Restoring and protecting remaining colonial waterbird breeding habitat (particularly *Typha* sp., lignum *Muehlenbeckia florulenta* and floodplain woodland species such as belah *Casuarina cristata* and cooba *Acacia salicina*) should also be a priority for future use of environmental water in the Gwydir Wetlands.

Flow variability including season, flooding frequency and duration, and periods of low flow, are needed to maintain healthy fish and waterbird populations in the Gwydir Wetlands. Management that focuses on restoring and maintaining natural flow variability is also likely to enhance overall wetland productivity, as flow regimes that include regular wetting and drying cycles are known to maximise productivity across a range of trophic levels, including macrophytes, aquatic invertebrates, fish and waterbirds.

Acknowledgments

This work was jointly funded by the Australian Government and the NSW Department of Environment and Climate Change through the NSW Wetland Recovery Program and Rivers Environmental Restoration Program. Additional support was provided by the Cotton Catchment Communities CRC and University of New England. John Porter (UNSW), Ash Fowler (UTS), Richard Allman (UNSW), Julia Sisson (UNE), Tobias Bickel (UNE) and numerous volunteers provided invaluable assistance with field surveys and laboratory analyses. We gratefully acknowledge the various Gwydir landholders for freely allowing access to their property. Ivor Growns, Heather Malan and John Porter are thanked for their comments on earlier versions of the manuscript.

References

ANRA (1995). Australian Natural Resources Atlas. Water Resources – Availability – Surface Water Measurement Station: Gwydir River D/S Copeton Dam DWR (ID: 418026). Available at www.anra.gov.au/topics/water/availability/swms/swms-418026.html. Accessed 26 May 2009.

Bailey RF (1934). New nesting records of glossy ibis. *Emu* **33**, 279–291.

Bailey PB (1991). The flood-pulse advantage and the restoration of river-floodplain systems. *Regulated Rivers: Research and Management* **6**, 75–86.

Bailey KM, Canino MF, Napp JM, Spring SM and Brown AL (1995). Contrasting years of prey levels, feeding conditions and mortality of larval walleye pollock *Theragra chalcogranmia* in the western Gulf of Alaska. *Marine Ecology Progress Series* **119,** 11–23.

Balcombe SR, Arthington AH, Foster ND, Thoms MC, Wilson GG and Bunn SE (2006). Fish assemblages of an Australian dryland river: abundance, assemblage structure and recruitment patterns in the Warrego River, Murray-Darling Basin. *Marine and Freshwater Research* **57**, 619–633.

Bennett M and Green J (1993). Preliminary assessment of Gwydir Wetlands water needs. NSW Dept of Water Resources Technical Services Division Report. NSW Dept of Water Resources, Sydney.

Boeuf G and Payan P (2001). How should salinity influence fish growth? *Comparative Biochemistry and Physiology Part C* **130**, 411–423.

BOM (2007a). Climate records for Moree weather stations. Commonwealth of Australia, Bureau of Meteorology.

BOM (2007b). Special Climate Statement 14. Six years of widespread drought in southern and eastern Australia, November 2001–October 2007. National Climate Centre, Bureau of Meteorology, Melbourne.

Bunn SE and Arthington AH (2002). Basic principles and ecological consequences of altered flow regimes for aquatic biodiversity. *Environmental Management* **30**, 492–507.

Campana SE and Thorrold SR (2001). Otoliths, increments and elements: keys to a comprehensive understanding of fish populations? *Canadian Journal of Fisheries and Aquatic Science* **58**, 30–38.

Copeland C, Schooneveldt-Reid E and Neller S (2003). Fish Everywhere: An Oral History of Fish and their Habitats in the Gwydir River. NSW Fisheries, Ballina.

Crabb P (1997). *Murray-Darling Basin Resources.* Murray-Darling Basin Commission, Canberra.

DECCW (2008). Gwydir environmental contingency allowance annual flow plan 2008–2009. NSW Department of Environment, Climate Change and Water, Armidale.

Driver PD, Harris JH, Norris RH and Closs GP (1997). The role of natural environment and human impacts in determining biomass densities of common carp in New South Wales rivers. In *Fish and Rivers in Stress: The NSW Rivers Survey.* (Eds JH Harris and PC Gehrke) pp. 225–250. NSW Fisheries Office of Conservation/CRC for Freshwater Ecology, Sydney.

Fowler AM (2008). Determination of suitable otolith preparation techniques for *Retropinna semoni* (Australian smelt) and *Nematalosa erebi* (bony bream): a technical report. Final report to the Rivers and Wetlands Unit, NSW Dept of Environment and Climate Change. University of Technology, Sydney.

Gawne B and Scholz O (2006). Synthesis of a conceptual model to facilitate the management of ephemeral deflation basin lakes. *Lakes and Reservoirs: Research and Management* **11**, 177–188.

Growns I (2008). The influence of changes to river hydrology on freshwater fish in regulated rivers of the Murray-Darling Basin. *Hydrobiologia* **596**, 203–211.

Growns I and Rodgers M (2002). Integrated monitoring of environmental flows. 5th Progress Report. NSW Fisheries Office of Conservation, Port Stephens.

Harris JH and Gehrke PC (1994). Modelling the relationship between streamflow and population recruitment to manage freshwater fisheries. *Fisheries Resources Modelling and Assessment* **6**, 28–30.

Harris JH and Gehrke PC (1997). Fish and Rivers in Stress: The NSW Rivers Survey. NSW Fisheries Office of Conservation/CRC for Freshwater Ecology, Sydney.

Heagney EC, Fowler AM, Allman RP and Spencer JA (2009). Growth and condition of juvenile Australian smelt and bony bream in the Gwydir wetlands: an otolith-based analysis. Final report to the NSW Dept of Environment and Climate Change. University of New South Wales, Sydney.

Higgins PJ and Davies SJJF (1996). *Handbook of Australian, New Zealand and Antarctic Birds.* Vol. 3: Snipe to Pigeons. Oxford University Press, Melbourne.

Humphries P, King AJ and Koehn JD (1999). Fish, flows and flood plains: links between freshwater fishes and their environment in the Murray-Darling river system, Australia. *Environmental Biology of Fishes* **56**, 129–151.

Humphries P, Cook RA, Richardson AJ and Serafini LG (2006). Creating disturbance: manipulating slackwaters in a lowland river. *River Research and Applications* **22**, 525–542.

Humphries P, Wilson G, Richardson A and Ellison T (2008). Impacts of managed flows on fish spawning and recruitment. Report to the Murray-Darling Basin Commission for project R5002. Murray-Darling Freshwater Research Centre, Wodonga.

Jenkins KM and Boulton AJ (2003). Ecological connectivity in a dryland river: short-term aquatic macroinvertebrate recruitment following floodplain inundation. *Ecology* **84**, 2708–2723.

Johnson B (2001). Water for the environment in the Gwydir River: providing water for the environment from Copeton Dam. Unpublished report to the NSW National Parks and Wildlife Service. NSW National Parks and Wildlife Service, Dubbo.

Johnson W (2002). Notes on waterbird breeding in the Gwydir Wetlands. Unpublished report to the NSW National Parks and Wildlife Service. NSW National Parks and Wildlife Service, Dubbo.

Johnson W (2004). Breeding of colonially nesting waterbirds. Unpublished report to the NSW National Parks and Wildlife Service. NSW National Parks and Wildlife Service, Dubbo.

Jones R (1997). Gwydir Wetlands: colonially nesting waterbird breeding: February/May 1997. NSW National Parks and Wildlife Service, Dubbo.

Keyte PA (1994). Lower Gwydir River Wetland Plan of Management 1994–1997. Report by NSW Dept of Water Resources for the Lower Gwydir Wetland Steering Committee. NSW Dept of Water Resources, Sydney.

Kingsford RT (2000). Ecological impacts of dams, water diversions and river management on floodplain wetlands in Australia. *Austral Ecology* **25**, 109–127.

Kingsford RT and Auld KM (2005). Waterbird breeding and environmental flow management in the Macquarie Marshes, arid Australia. *River Research and Applications* **21**, 187–200.

Kingsford RT and Norman FI (2002). Australian waterbirds: products of the continent's ecology. *Emu* **102**, 47–69.

Mallen-Cooper M (2000). Review of barriers to fish passage in the Lower Gwydir River floodplain. Report to Wetland Care Australia. Fishway Consulting Services, Sydney.

Mallen-Cooper M and Stuart I (2003). Age, growth and non-flood recruitment of two potamodromous fishes in a large semi-arid temperate river system. *River Research and Applications* **19**, 697–719.

Marchant S and Higgins PJ (1990). *Handbook of Australian, New Zealand and Antarctic Birds.* Vol. 1: Ratites to Ducks. Oxford University Press, Melbourne.

Marchant S and Higgins PJ (1993). *Handbook of Australian, New Zealand and Antarctic Birds.* Vol. 2: Raptors to Lapwings. Oxford University Press, Melbourne.

McCosker RO (1996). Gwydir Wetlands Ecological Response to Flooding 1995-96. Report prepared for the Gwydir Environmental Flows Committee. LANDMAX Pty Ltd, Natural Resource Management Services, Armidale.

McCosker RO (1999a). Gwydir wetlands: ecological response to flooding 1997. Report prepared for the NSW National Parks and Wildlife Service. LANDMAX Pty Ltd, Kangaroo Point.

McCosker RO (1999b). Gwydir wetlands: ecological response to flooding 1998. Report prepared for the NSW National Parks and Wildlife Service. LANDMAX Pty Ltd, Kangaroo Point.

Meekan MG and Fortier L (1996). Selection for fast growth during the planktonic life of Atlantic cod *Gadus morhua* on the Scotian Shelf. *Marine Ecology Progress Series* **137,** 25–37.

Merritt W, Powell S, Pollino C and Jakeman T (2010). IBIS: a decision support system for managers of environmental flows into wetlands. In *Ecosystem Response Modelling in the Murray-Darling Basin.* (Eds N Saintilan and I Overton) pp. 119–136. CSIRO Publishing, Melbourne.

Morse FC (1918). Nesting notes from Moree. *Emu* **28**, 65–71.

Morse FC (1922). Egret and glossy ibis rookeries. *Emu* **22**, 36–39.

NSW Government (2003). Water Sharing Plan for the Gwydir Regulated River Water Source 2003 Order. NSW Government, Sydney.

Post JR and Parkinson EA (2001). Energy allocation strategy in young fish: allometry and survival. *Ecology* **82**, 1040–1051.

Puckridge JT, Sheldon F, Walker KF and Boulton AJ (1998). Flow variability and the ecology of large rivers. *Marine and Freshwater Research* **49**, 55–72.

Roshier DA, Robertson AI and Kingsford RT (2002). Responses of waterbirds to flooding in an arid region of Australia and implications for conservation. *Biological Conservation* **106**, 399–411.

Schlosser IJ (1998). Fish recruitment, dispersal, and trophic interactions in a heterogeneous lotic environment. *Oecologia* **113**, 260–268.

Scott A (1997). Relationships between waterbird ecology and river flows in the Murray-Darling Basin. Technical Report No. 5/97. CSIRO Land and Water, Canberra.

Sheldon F, Thoms MC, Berry O and Puckridge J (2000). Using disaster to prevent catastrophe: referencing the impacts of flow changes in large dryland rivers. *Regulated Rivers: Research and Management* **16**, 403–420.

Spencer JA (2009). Historical records of waterbirds and fish populations in the Gwydir wetlands. NSW Wetland Recovery Program. Rivers and Wetlands Unit, NSW Dept of Environment and Climate Change, Sydney.

Spencer JA, Heagney EC and Porter JL (2009). Final report on the Gwydir waterbird and fish habitat study. NSW Wetland Recovery Program. Rivers and Wetlands Unit, NSW Dept of Environment and Climate Change, Sydney.

Thoms MC and Sheldon F (2000). Lowland rivers: an Australian introduction. *Regulated Rivers: Research and Management* **16**, 375–383.

Thoms MC, Beyer PJ and Rogers KH (2006). Variability, complexity and diversity: the geomorphology of river ecosystems in dryland regions. In *Ecology of Desert Rivers*. (Ed. RT Kingsford) pp. 47–75. Cambridge University Press, Melbourne.

Tonkin Z, King AJ and Ramsey DSL (2008a). Otolith increment width responses of juvenile Australian smelt *Retropinna semoni* to sudden changes in food levels: the importance of feeding history. *Journal of Fish Biology* **73**, 853–860.

Tonkin ZD, Ramsey DSL and King AJ (2008b). Larval and juvenile Australian smelt *Retropinna semoni* somatic and otolith growth parameters: implications for growth interpretation of wild fish. *Ecology of Freshwater Fish* **17**, 489–494.

Walker KF, Sheldon F and Puckridge JT (1995). A perspective on dryland river ecosystems. *Regulated Rivers: Research and Management* **11**, 85–104.

Ward JV (1998). Riverine landscapes: biodiversity patterns, disturbance regimes and aquatic conservation. *Biological Conservation* **83**, 267–278.

Wilson G and Wright A (2005). Fish recruitment in the Border Rivers and Moonie catchments. CRC for Freshwater Ecology/Murray-Darling Freshwater Research Centre, Goondiwindi.

Wilson GG, Berney PJ, Ryder DS and Price JN (2008). Stage 2: Grazing/landuse in the Macquarie Marshes and Gwydir Wetlands. Final report to the NSW Dept of Environment and Climate Change. University of New England, Armidale.

Wilson GG, Bickel TO, Berney PJ and Sisson JL (2009). Managing environmental flows in an agricultural landscape: the Lower Gwydir floodplain. Final report to the Australian Government Department of the Environment, Water, Heritage and the Arts. University of New England/Cotton Catchment Communities CRC, Armidale.

Young WJ and Kingsford RT (2006). Flow variability in large unregulated dryland rivers. In *Ecology of Desert Rivers*. (Ed. RT Kingsford) pp. 11–46. Cambridge University Press, Melbourne.

Chapter 8

IBIS: a decision support system for managers of environmental flows into wetlands

Wendy Merritt, Sue Powell, Carmel Pollino and Tony Jakeman

Introduction

Predicting the outcomes of environmental flows in wetland systems is a complex interdisciplinary task requiring an understanding of the processes governing the climate, hydrology, hydraulics and ecology of a system and the interaction between these system components. Adding to this complexity are the spatial (continental, basin, catchment and wetland) and temporal (event, long-term, lag) scales at which these processes operate. Decision support systems (DSS) can assist resource managers by integrating across processes (using models and other forms of knowledge) and scales, and by providing a systematic approach to decision-making by exploring management options (e.g. current and alternate environmental flow scenarios).

The IBIS DSS (IBIS is not an acronym) is built on an adaptive system of decision support. It acts as a warehouse of information, containing maps, publication materials, facts sheets, models and model documentation. The DSS captures the best available scientific knowledge in a way that is relevant to decision-makers, and that is highly interactive and easy to use. Being adaptive, the materials and the models housed in the DSS can be updated as new knowledge is obtained. The DSS setup can be applied to any inland wetland system. At present, there are two environmental flow wetland applications of IBIS – the Gwydir Wetlands and Narran Lakes. In this chapter, we outline the development of the IBIS DSS to assist the management of environmental flows in wetlands, using as a case study the application developed for the Gwydir Wetlands in north-west NSW. For more information on the Narran case study, see Merritt *et al.* (2009).

The IBIS application for the Gwydir Wetlands consists of a daily water balance model which aims to represent the complex hydrologic and hydraulic behaviour of the wetland system. Using the water balance model predictions of, for example, event timing, duration, depth and frequency of flooding, likely ecological responses are modelled using discrete Bayesian network (BN) models. The DSS interface interacts with the integrated water balance model and the ecological response model, displaying the characteristics and ecological outcomes of flows for historic events, future climatic conditions and alternative water-sharing plan rules. The ecological response models for the Gwydir case study consider native fish and carp spawning and recruitment. The intended model users are wetland managers, who can use the DSS to inform and support on-ground management decisions, and model developers and researchers, who can input new data, rules, and test hypotheses (e.g. altering the form of, or thresholds in, flow to ecology relationships).

Managing environmental flows into the Gwydir Wetlands

Gwydir Wetlands

The Gwydir Wetlands, located in the Gwydir catchment of north-west NSW, are among the largest and most significant terminal wetland systems in the region (Keyte 1994). Two distinct wetland areas exist – the Gingham Wetlands and the Lower Gwydir Wetlands (Fig. 8.1). The wetlands support a diverse community of rare, vulnerable and endangered species and are important for supporting migratory bird populations and diverse frog populations. Aquatic vegetation is dominated by water couch, rushes and sedges (NSW Government 2007). The wetlands provide habitat for native and exotic fish species, and are considered a breeding hotspot for the introduced common carp, *Cyprinus carpio* (http://www.mdbc.gov.au/subs/fish-info/alien/carp.html, accessed 25 May 2009).

The hydrologic regime of the Gwydir River and associated wetlands has greatly changed since the completion of the Copeton Dam in 1976 and the rapid development of the local irrigation industry in the 1980s. The natural variability of the system has been lessened by river regulation, with extractions for irrigation having reduced the volume of water moving down the Gwydir and Gingham channels and the nearby Mehi River. These changes, exacerbated by the construction of channels and levees within the wetlands, have reduced the extent and duration of flooding of the wetlands (Kingsford 2000). As a result of the changed flooding regime, vegetation patterns in the Lower Gwydir and Gingham watercourses have altered, with degradation of some aquatic vegetation and terrestrial vegetation moving onto areas previously colonised by aquatic vegetation (Keyte 1994; Kingsford 2000). Although there is less

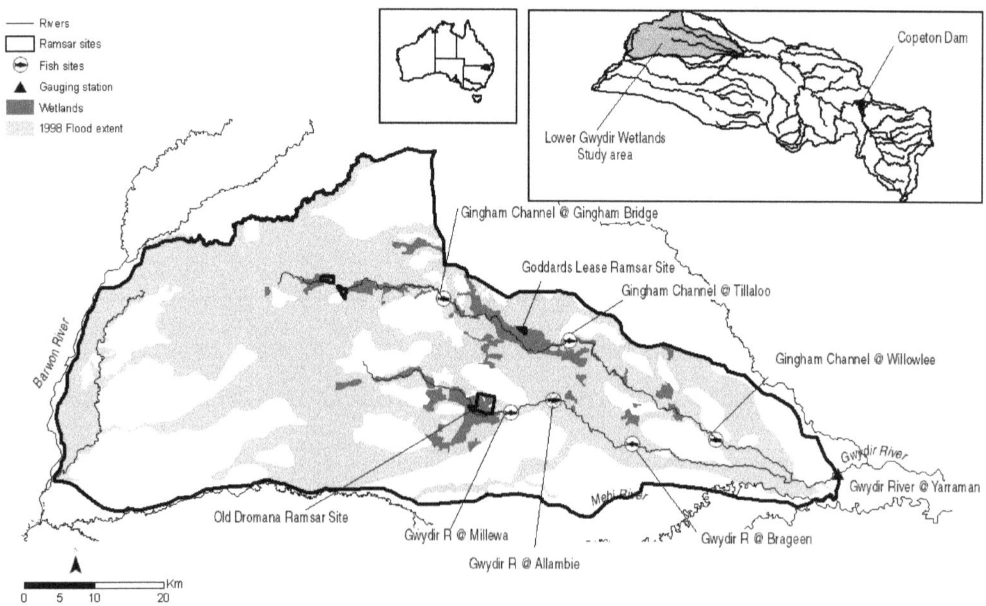

Figure 8.1: The Gwydir IBIS application models the hydrology of the Gingham and Gwydir channels downstream of the Gwydir River flow gauge at Yarraman. Fish models have been developed for six sites. Vegetation models of the Old Dromana Ramsar site and the Goddards Lease Ramsar site are under construction.

evidence for change in wetland and aquatic fauna due to regulation and extraction (Kingsford 2000), this may be due to lack of formal monitoring programs. Contributors to an oral history of fish and their habitats in the Gwydir River generally viewed native fish populations as being depleted or of more uncertain abundance since the commencement of regulation and irrigation (Copeland *et al.* 2003).

Decision-making context: environmental flows

To improve the management of the Gwydir Wetlands and protect and recover the stressed wetland ecology, it is essential to determine the environmental flows requirements of the Gwydir River, wetlands and floodplains. This information would need to guide decision-making at national, state and local levels. According to decision-makers, specific decision-making needs are:

- knowing which ecological assets and functions require protection;
- knowing how much water is required to obtain a desired environmental outcome;
- knowing how to prioritise allocations of water sources to achieve outcomes.

Sound knowledge of how water behaves within the system (both natural flows and regulated components) and the ecological response to flows (e.g. volume, timing and extent of inundation) is required.

Since 1995, the Gwydir system has been subject to the application of environmental flow rules that aim to maintain or improve wetland vegetation, breeding and migration of native fish and opportunities for waterbird breeding. The water-sharing plan for the Gwydir Regulated River Water Source (2003) outlines provisions on the use of water within the catchment, which are designed to reserve water above the extraction limit for the environment, protect low flows, establish entitlements for off-allocation water (water available above the licence limit) and provide water for environmental contingencies (45 000 ML), which is available to enhance river and wetland health (DIPNR 2004). The environmental flow rules were established in a climate of uncertainty, and the need to monitor and review the rules has been recognised.

The complexity of the natural system, together with incomplete and imperfect knowledge about the hydrological regime and ecological response to the regime, complicates the decision-making process. A DSS has been constructed to explore how the flow rules and system hydrology affect ecological outcomes.

Decision support systems

A DSS can assist decision-making processes by providing a means by which decisions are:

- consistent and repeatable;
- transparent (it documents existing knowledge and gaps, data and/or assumptions used in the decision-making process);
- scientifically rigorous (uses best available science or knowledge);
- updated or revised as new knowledge is gained.

A DSS can be at the centre of the planning process, linking decision-making and science in a participatory environment.

Development

Generally, a DSS is a computer-based system designed to support decision-making processes by organising and integrating existing knowledge and/or models into a tool that resource managers and policy-makers can use to compare different strategies (McIntosh *et al.* 2008). They have been promoted as one way of assisting decision-makers in managing a complex system, particularly if there are multiple values (environmental, cultural etc.) and/or competing uses of the system's resources (irrigation, environmental flows etc.). They also help integrate science with management, exploring causality through linking management decisions with resource outcomes and dealing with the associated uncertainties across time and space.

Underlying a DSS there is generally a model that describes system behaviour and interactions. Jakeman *et al.* (2006) proposed a 10-step iterative process for the development and evaluation of environmental models and stressed the importance of close partnerships between model developers, clients and other interested parties. As with models, good practice in the development and evaluation of a DSS is essential. 'Good practice' in DSS development includes whether the DSS is a robust product that addresses the needs of decision-makers and can be readily used by stakeholders. Both the DSS and underlying models need to be scoped and developed in close collaboration with decision-makers to ensure that the tools meet decision-maker needs and to increase their confidence in the DSS and its outputs. The DSS should support and enhance management (Rizzoli and Young 1997), be flexible enough to adapt to new situations (e.g. a different wetland system, flow rules or changed climatic conditions) and present models and information in a form that is rigorous and scientifically defensible but suitable for extension to resource managers, policy-makers and other stakeholders.

DSSs are increasingly used to support management of wetland systems and to identify economic, social and environmental implications and trade-offs associated with the regulation and allocation of water. An example of a DSS that has been developed for the Gwydir catchment is the Water Allocation Decision Support System (WAdss – Letcher 2005). WAdss is a software tool developed to consider socio-economic trade-offs associated with changes to access and allocation across unregulated and regulated surface water systems and groundwater systems. An environmental flow DSS, linking hydrology and ecology, had not been previously constructed for the Gwydir system.

DECCW Wetlands DSS project

Under the NSW Rivers Environmental Restoration Program, funded by the NSW Department of Environment, Climate Change and Water and the Commonwealth of Australia, prototype DSSs have been developed for the Gwydir Wetlands and the Narran Lakes. These DSSs are intended to provide a framework to support decision-making, particularly with respect to the use of environmental flows to maintain and improve ecological assets.

In consultation with DECCW program managers, DECCW wetland managers and wetland researchers, it was decided that the Gwydir DSS should have the following functionality:

- simulate the depth, duration and expected ecological outcomes of a given volume of water delivered to a part of the Gwydir system;
- model the volume of water needed to support known ecological assets given their associated water requirements;
- provide information on how to intervene in or supplement natural events in order to maintain or improve opportunities for natural events, such as bird breeding.

The DSS should be able to be used to explore such outcomes in a dynamic and catchment context (e.g. volume and timing of dam releases, natural floods and droughts). It was important that the DSS links the hydrological and ecological system components.

IBIS DSS

The model structure and interface of the IBIS DSS allows for the development of stand-alone applications for different wetland systems. Prototype applications have been developed for the Narran Lakes (Merritt *et al.* 2009) and Gwydir Wetlands.

The Gwydir application focuses on exploring outcomes of environmental flow releases into the Gwydir and Gingham channels below Yarraman gauge, which is the entry point to the wetlands from the Gwydir River (Fig. 8.1) and is used for assessing environmental flow delivery to the wetlands, as detailed in the water-sharing plan. The application couples a water balance model of the Gwydir Wetland (Powell *et al.* 2008) to ecological models of vegetation and fish response. Hydrological and ecological outcomes can be represented at different temporal and spatial scales (Fig. 8.2/colour plate, p. 302). The interface enables rapid data input and scenario testing and provides access to supporting information and model documentation as well as reporting and data export functionalities.

DSS structure

Model structure

The underlying model of the Gwydir application comprises a daily water balance model linked to event-based fish response models. The water balance model simulates wetland behaviour on the delivery of flow to the Yarraman gauge and outputs are daily. The ecological models are Bayesian networks (BNs). The BN models use summaries of the inundation time series from the water balance model, together with other data (e.g. spawning adult abundance within the wetland), to describe the likely ecological response to the hydrology regime.

Software platform

The water balance and fish models in the prototype Gwydir application have been coded using the Integrated Component Modelling System. This is a model-building and delivery environment developed by CSIRO Land and Water (Reed *et al.* 2000) in which objects and links are used to develop the integrated model.

Interface

The IBIS DSS interface overlays the integrated model project and allows users to navigate through the tool, run the model and view (and export) model outputs and supporting documentation. Support documentation includes maps, a description of the DSS and Gwydir Wetlands and reporting functions. Any flow sequence delivered to the Yarraman gauge (generated externally) can be input to the DSS via the interface. The hydrological and ecological outcomes of this flow sequence across temporal and spatial scales are displayed within the interface.

IBIS has two types of user groups: expert users, who can modify, define and set hydrological and ecological rules and thresholds; and default users who cannot alter model settings. The default model settings are for non-technical users, expert users are technical users who may want to alter parameter settings within the underlying models.

Water balance model

A semi-distributed water balance model of the Gwydir system, developed by Powell *et al.* (2008), was adapted for use in the Gwydir application of the IBIS DSS to simulate flood dynamics of the floodplain wetland. The model consists of a number of submodels – localised floodplains, channels, junctions, waterholes and time delays across the wetland system. These are linked to represent the Gwydir and Gingham channels and their associated localised wetland environments. Channels of a defined length and flow capacity flow through a floodplain. If flow in the channels exceeds the channel capacity, water overflows into the floodplain flow paths, which are defined as a set surface area based on volume of water. Outflow from a flow path is assumed to be proportional to the volume of free surface water. Outflows from core wetlands or flow paths can return to the next channel component or be routed directly to other wetlands or flow paths. A water balance is calculated for each channel, flow path or core wetland considering inflows, outflows, antecedent conditions (i.e. soil moisture), precipitation, infiltration and evapotranspiration.

As with all models, the water balance model is a simplification of the Gwydir system. It nonetheless provides a reasonable representation of historically available flooding data in the eastern areas of the wetland. This area includes the core Ramsar listed wetland areas, the critical areas for assessing flow-related outcomes. It does not represent flooding from very small flows to targeted locations and it tends to overestimate the volume of water towards the western end of the system. For further detail on the water balance model, see Powell *et al.* (2008). Future versions of the Gwydir application will incorporate a hydrodynamic model that is being developed.

Ecology

As stated above, the ecological models in the Gwydir application are Bayesian networks. These were developed in consultation with wetland researchers. The BNs are coupled to the water balance model. The water balance model outputs are inputs to the BNs; they include timing, duration, extent and depth of flooding to the ecological component BNs. The relationships in a BN are causal (linked variables are used to describe cause and effect) and the relationships between variables are described probabilistically. The distributions used to construct the BNs are based on generalised knowledge of species and groups of species responding to flow (published) and local observational data. A limitation of the BN approach is its imperfect representation of time, space and feedbacks.

Conceptualisation

A review of literature was conducted to identify major hydrological variables that drive the following components of, or factors influencing, ecological response in floodplain wetlands: water quality, soil properties, habitat connectivity, vegetation community structure, invasive plant species, invertebrates and colonial nesting birds and fish (Table 8.1). Although the Gwydir Wetlands are adapted to a highly variable flooding regime, the adaptability of fauna and plant species in the wetlands is limited to the flood regime under which they developed (Bunn and Arthington 2002).

The time scales over which the DSS can be run include:

- event-based – days to weeks or months;
- long-term, considering changing management actions and variability in climate.

The outputs of the water balance model were daily, representing a more dynamic system behaviour. However, the ecological BNs are more lumped and are probabilistic. Consequently, the data from the water balance model needed to be aggregated before being passed to the ecological model. Such outputs include depth, duration and frequency of inundation. The thresholds of ecological response for hydrologic events are based on information from experts, observational data and published materials. These thresholds may vary for different components of the ecology (e.g. fish versus wetland vegetation) or different species (*Paspalum distichum* versus *Phyla cannescens*), depending on hydrological requirements.

Common carp and native fish models

The relationship between the composition of the fish community and flow has been the subject of a study at the University of New England (Wilson *et al.* 2008a; Ch. 7, this volume), which analysed changes in fish communities at several sites of the Gingham and Gwydir channels before and after environmental flow releases. Preliminary analysis of the data suggests there is no clear relationship between flow and fish, and further investigation is required. Overall, the

Table 8.1: Significant hydrological factors influencing wetland ecology

	Flood frequency	Water depth	Area inundated	Flood duration	Drying cycles	Inflow peak and volume	Rate of water level change	Timing and tempera-ture	Existing vegetation structure	Water quality of inflows
Abiotic – water quality	✓	✓	✓		✓[1]					✓[2]
Abiotic – soil properties		✓[3,4,5]	✓[5]	✓[3,4,5]	✓[1,5,6]				✓	
Habitat connectivity		✓[7]	✓[4,7]							
Vegetation community structure	✓[8,9,10,11,12]	✓[8,11,12,13,14]		✓[8,10,11,12,14]	✓[9,10,15]			✓[12,16]	✓[2,8,9,12,14]	✓[17,18]
Invasive plant species		✓[9]		✓[9]	✓[9]					
Invertebrates				✓[19,20]	✓[19,20]			✓[21]	✓[21]	
Colonial nesting birds	✓[22]	✓[23]		✓[23]	✓[22]	✓[23]	✓[23]	✓[16]	✓[16]	

1 Nielsen and Chick (1997); 2 Reid and Brooks (2000); 3 Pezeshki (2001); 4 Thoms (2003); 5 Ponnampeeruma (1984); 6 Roberts and Marston (2000); 7 Overton (2005); 8 Casanova and Brock (2000); 9 Mawhinney (2003); 10 Bennett and Green (1993); 11 Brock and Casanova (1997); 12 Warwick and Brock (2003); 13 Roberts *et al.* (2000); 14 Webb *et al.* (2006); 15 Briggs (1998); 16 Kingsford and Johnson (1998); 17 Hart *et al.* (1990); 18 Boulton and Brock (1999); 19 Boulton and Lloyd (1991); 20 Driver *et al.* (2003); 21 Balla and Davis (1995); 22 Young *et al.* (1999); 23 Kingsford and Auld (2005).

fish population in the Gwydir Wetlands can be considered fairly depauperate, being largely composed of generalist species which have a broad range of habitat (including flow) requirements for spawning and recruitment (Ch. 7, this volume).

There are two fish models in the IBIS prototype for the Gwydir which focus on native fish (spangled perch and bony bream) (Fig. 8.3), and a pest species, the common carp (not shown). Bony bream and spangled perch have been observed to initiate spawning in response to environmental flow releases in the lower Gwydir channels (Wilson *et al.* 2008a), and common carp has flourished in the system since its invasion which is thought to have occurred in the 1970s (Ch. 7, this volume).

The native fish and carp models were developed by the authors in collaboration with researchers at the University of New England and contain estimates of the relationship between flow, spawning and recruitment derived using monitoring data, literature values and expert opinion. The model has been run for six sites in the wetland system: downstream of Brageen Crossing (Gwydir channel), downstream of Allambie Bridge (Gwydir channel), Birrah (Gwydir channel), Willowlee (Gingham channel), Westholme (Gingham channel) and Boyanga Waterhole (Gingham channel) (Fig. 8.1).

The models examine the likely outcomes in recruitment (to the juvenile stage) of fish in response to a flow event. The flow event is a trigger for spawning and establishes suitable conditions for habitat and food supplies for reaching. Predator densities and starting abundances of spawning adults are included in the models. The duration of wetting of eggs is included in the carp model. A limitation of the models is the poor knowledge of life-history requirements of fish in the wetlands and the lack of knowledge relating fish responses to flow events. Future development of the fish models will focus on analysing existing datasets for the Gwydir and similar wetland systems to improve the representativeness and rigour of the BN models of common carp and native fish.

Vegetation models

A conceptual model of vegetation response to hydrological regime has been developed by the authors based on literature from the Gwydir Wetlands and the wider Murray-Darling Basin (Fig. 8.4). This conceptual model will be used as a basis for the development of BN models of vegetation response in future development of the Gwydir application of the IBIS DSS.

Wetland plant response occurs in response to inundation (soil saturation plus some depth of surface water). The degree of response depends on the timing, depth and duration of inundation. Antecedent conditions such as time since the last event, duration of the dry spell and soil moisture are also important. Conditions must be suitable first for either germination (if from seed) or growth from perennial root stock (e.g. rhizomes), then for growth to maturity and finally for seed production and development of below-ground biomass. Other factors that have been shown in the literature to influence vegetation response to flooding include temperature and/or day length (Warwick and Brock 2003) and grazing pressure.

The amount and quality of information describing vegetation response to flooding regimes is variable. Some location-specific information has been developed for the Gwydir (Mawhinney 2003), but the majority of available information relates to other parts of the Murray-Darling Basin (Blanch *et al.* 1999; Roberts and Marston 2000; Young *et al.* 2003). Several authors have demonstrated that seedbanks respond to a range of flood regimes and seasonality (Brock and Casanova 1997; Casanova and Brock 2000; Warwick and Brock 2003). However, the link between hydrological regime and vegetative response (as opposed to seedbank germination) is not well understood, although it is known that many wetland species are able to resprout from basal or below-ground plant parts.

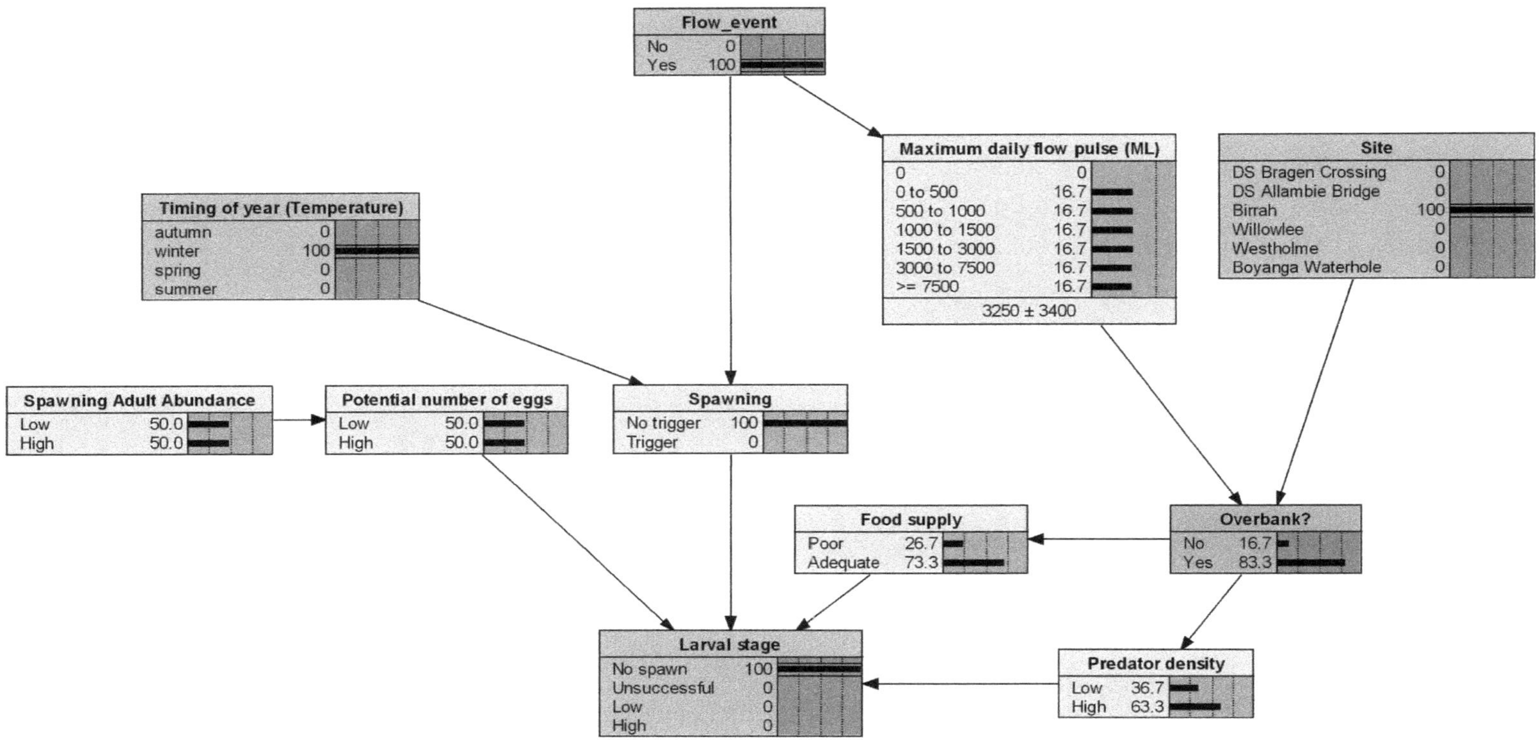

Figure 8.3: Bayesian network model for spangled perch and bony bream recruitment. This model interacts with the water balance models, as shown in Figure 8.2.

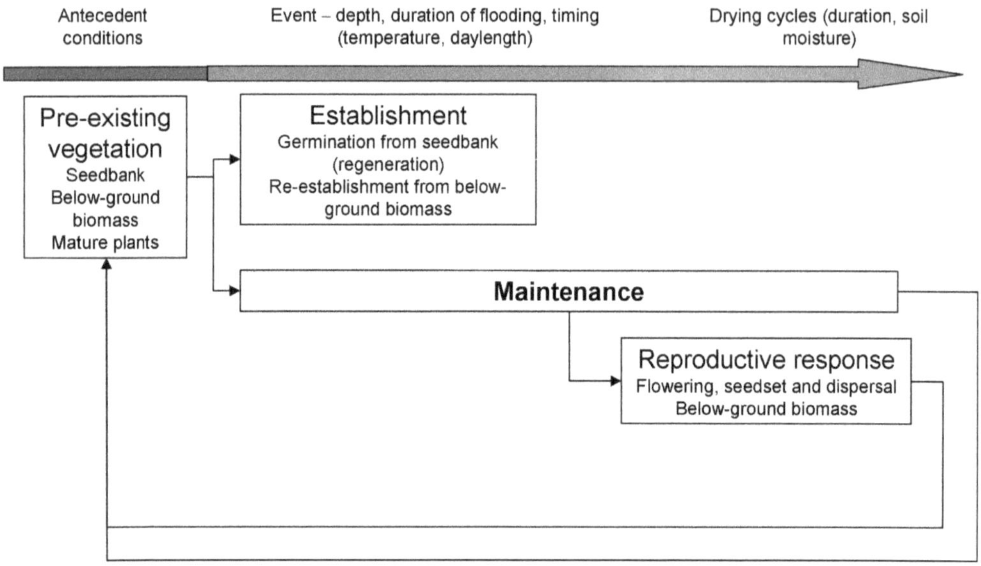

Figure 8.4: Conceptual model of vegetation response to flow regimes.

A simplified model framework for vegetation response is given in Figure 8.5, that shows two species common in the Gwydir system – *Paspalum distichum* (water couch) and *Phyla cannescens* (lippia). The response of each species to a defined flood is determined by the timing and duration of the event and its suitability for the species, antecedent conditions (e.g. duration since last event), the distribution of the species and the relationship between grazing density and species response. The likely community structure is determined by the response of each species and the pre-exisiting community structure. For example, if lippia is present in the system, if hydrologic conditions during a flow event (or series of events) favour lippia over water couch and if there is a high level of grazing (where animals preferentially graze water couch), we could expect the community structure to move towards being lippia-dominated.

Using the DSS

The IBIS DSS is designed to meet short- (event and annual) and long-term (up to 100 years) planning horizons. The Gwydir application of the IBIS DSS has three major uses:

- to characterise flow events into the wetland based on climate variability and water-sharing plan arrangements and predict the likely ecological responses of these within the Gwydir system;
- to alter and test key flow or ecological parameters in the model;
- to develop, test and compare environmental flow scenarios.

Ecological response to flows

The final Gwydir application will be used to link scenarios of climate variability and water-sharing plan arrangements to flooding extent, duration and depth in the wetland. These flow events are linked to ecological responses. These response models are probabilistic, allowing the uncertainty associated with a predicted response to be incorporated into the outcome. Within the DSS application the characteristics of each event (a flow output time series) from the water balance model can be plotted in the interface along with the likely response of all variables in the ecological response models.

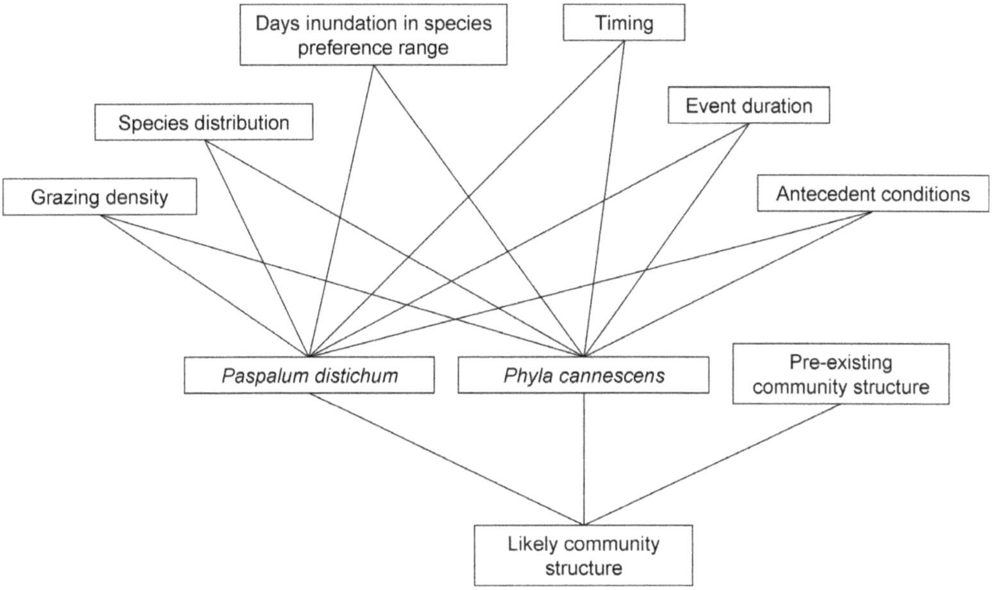

Figure 8.5: Simplified model framework for vegetation response.

Testing and setting model parameters

Relationships between the hydrology and some aspects of ecological response are not well established, and during the development of the Narran and Gwydir prototype applications of the IBIS DSS it became apparent that flexibility was required in model structure and model parameters.

An important component of the water balance model is describing flow events. We need to know what constitutes the start and end of an event, and the important characteristics of an event. The nature of what constitutes an event varies according to an ecological response. Fish, waterbirds, floodplain and wetland vegetation groups all have different triggers or water requirements, some of which may be well understood (e.g. water couch – Bennett and Green 1993; Blanch *et al.* 1999; Mawhinney 2003) and others which are not. In IBIS, these triggers are implemented as settings that can be altered through the interface to account for hypothesis testing, acquisition of new knowledge, development of a new application (i.e. a different focus wetland) or extension of an existing application to model, for example, an additional species. These settings are used to describe individual relationships between a variable in the BN (e.g. 'Spawning' in Fig. 8.3, p. 127) and each input variable (e.g. 'Flow_event' in Fig. 8.3).

In addition to knowing how an event is characterised, we need to know the strength of the influence of different input variables and whether there are tipping-points in characterising ecological responses. Figure 8.3 (p. 127) shows that the success of the larval stage for spangled perch and bony bream is influenced by four variables: the potential number of eggs, whether or not spawning occurred, the adequacy of the food supply and the density of predators. Assuming the individual relationships between each input variable and a successful larval stage are established, how do these individual impacts combine? IBIS allows users to investigate various combinations for describing how different aspects of the system affect outcomes. Alternative combinations include:

- weighted average – users identify and apply weights to each link and use them to weight probabilities;
- override (partial or complete) – in some conditions, a particular variable may override all others. For example, if the duration of inundation is insufficient for waterbirds then

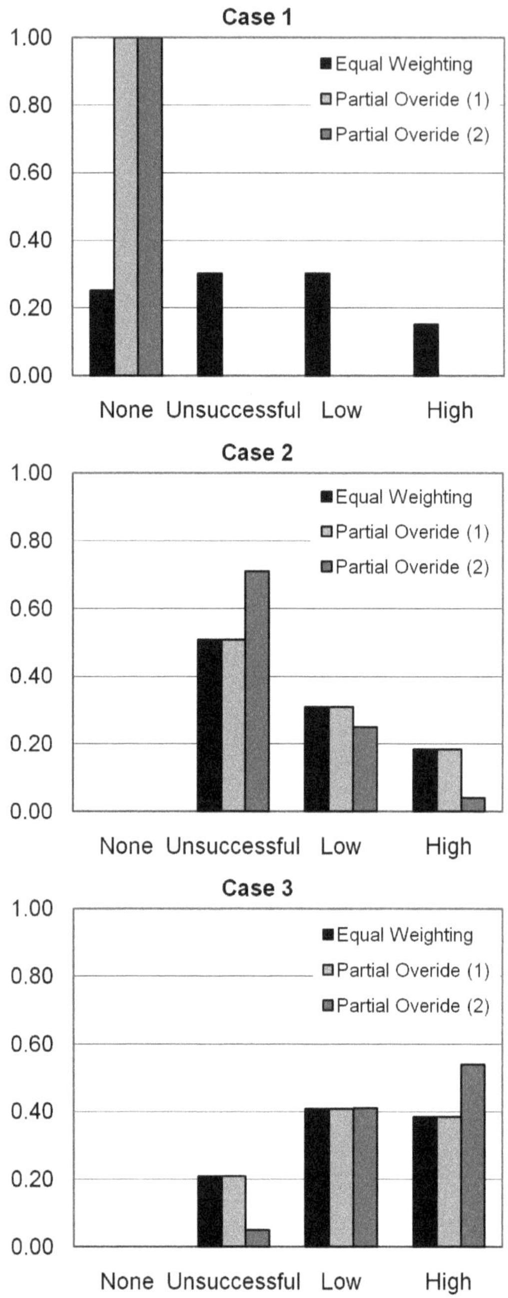

Figure 8.6: Impact of BN definition rules on the likely recruitment of spangled perch and bony bream to larval stage.

abandonment may be highly likely regardless of the depth under nests, the change in water surface elevation or temperature. Users can select a complete override, where an input variable controls the outcome under all its possible states;

- worst case – this option selects the worst case from each individual input when generating the probability distributions;

Table 8.2: Impact of BN definition rules on likely recruitment of spangled perch and bony bream to larval stage

	State		
Input variable	**Case 1**	**Case 2**	**Case 3**
Egg numbers	Low	High	Low
Predator density	Low	High	Low
Food supply	Poor	Poor	Adequate
Spawning	No trigger	Trigger	Trigger

- best case – this option selects the best case from each individual input when generating the probability distributions.

An example of how these rules affect outcomes for waterbirds in the Narran Lakes is explored in Merritt *et al.* (2009).

The effect of some of these rules on the likely recruitment of spangled perch and bony bream is shown in Figure 8.6 for the three cases summarised in Table 8.2. The relationship between each input variable (potential number of eggs, spawning, food supply and predator density) and the larval stage variable is listed in Table 8.3. In each panel of Figure 8.6, the black columns represent the probability distribution when each input variable is assumed to have equal weighting. Case 1 has only 25% chance of no spawning, despite there being no spawning trigger, which is highly unrealistic. The light grey columns show the probability distribution when a partial override is implemented: when there is no spawning trigger the outcome of the larval stage variable is determined solely by the spawning variable. Otherwise, an equal weighting is applied. This results in no spawning for case 1 ('None'). For cases 2 and 3, when spawning is triggered there is no difference in the probability distribution calculated for the 'equal weighting' and 'partial override (1)' rules. The dark grey columns show the probability

Table 8.3: Impact of input variable states on likely recruitment of spangled perch and bony bream to larval stage

	Larval stage states			
Individual impact	**None**	**Unsuccessful**	**Low**	**High**
Potential number of eggs				
– Low	0.00	0.50	0.50	0.00
– High	0.00	0.20	0.40	0.40
Predator numbers				
– Low	0.00	0.00	0.40	0.60
– High	0.00	0.80	0.20	0.00
Food supply				
– Poor	0.00	0.70	0.30	0.00
– Adequate	0.00	0.00	0.40	0.60
Spawning trigger				
– No trigger	1.00	0.00	0.00	0.00
– Trigger	0.00	0.33	0.33	0.33

distribution for recruitment success when a partial override is implemented if there is no spawning trigger and, when spawning does occur, a weighted average is applied where the predator density (weight = 0.6) variable exerts a stronger influence on larval stage than food supply (weight = 0.3) or the potential number of eggs (weight = 0.1).

Environmental flow scenarios

The DSS can be used to develop, test and compare environmental flow scenarios, including flows in future climates or testing current or alternative flow rules. The DSS can be used to plan over a range of time horizons, from event and annual to long-term (50–100 years) predictions. Future DSS developments include linking the DSS to hydology models, such as IQQM, and the development of a ecological response models for vegetation and waterbirds.

Conclusions

The Gwydir application of the IBIS DSS is a prototype developed for the NSW DECCW to demonstrate the features and capabilities of a DSS that will support the environmental flows decision-making process in the Gwydir Wetlands. IBIS:

- is an adaptable tool that can be built on as new knowledge and data are obtained and as decisions are made (an updatable knowledge base);
- can be used to interrogate knowledge, specifically to explore flow scenarios (strategic and responsive), assess trade-offs for prioritisation and target outcomes;
- can be used to communicate science to a range of stakeholders.

The DSS promotes transparent use and analysis of model outputs by documenting the strengths and weaknesses of current knowledge, data quality, model structure and assumptions (risk-based decisions). It allows for iterative updating and development, so has a continuing role in adaptive management. It assists scientists and decision-makers to focus on integration needs, particularly in linking climate, hydrological-hydraulic and ecological information. When finalised, the Gwydir application of the IBIS DSS will enable transparent analysis of decisions, support consistency in decision-making and promote decisions based on best available science. The strengths and weaknesses of current knowledge, data quality, model structure and assumptions are documented in the tool.

Acknowledgments

The development of the Gwydir application of the IBIS DSS is funded by the Australian Government and the NSW Department of Environment and Climate Change under the Rivers Environmental Restoration Program. The fish BN models were developed in collaboration with Dr Glenn Wilson. Comments from two anonymous reviewers helped improve this chapter.

References

Balla SA and Davis JA (1995). Seasonal variation in the macroinvertebrate fauna of wetlands of differing water regime and nutrient status on the Swan Coastal Plain, Western Australia. *Hydrobiologia* **299**, 147–161.

Bennett M and Green AJ (1993). Preliminary assessment of the Gwydir wetland water needs. TS 93.085. NSW Dept of Water Resources, Sydney.

Blanch S, Ganf G and Walker K (1999). Tolerance of riverine plants to flooding and exposure indicated by water regime. *Regulated Rivers: Research and Management* **15**, 43–62.

Boulton AJ and Brock MA (1999). *Australian Freshwater Ecology: Processes and Management.* Gleneagles Publishing, Adelaide.

Boulton AJ and Lloyd LN (1991). Macroinvertebrate assemblages in floodplain habitats of the lower River Murray, South Australia. *Regulated Rivers: Research and Management* **6**, 183–201.

Briggs SV (1998). Guidelines for management of inland wetlands in southern New South Wales. *Wetlands Australia* **8**, 1–2.

Brock MA and Casanova MT (1997). Plant life at the edges of wetlands; ecological responses to wetting and drying patterns. In *Frontiers in Ecology: Building the Links.* (Eds N Klomp and I Lunt) pp. 181–192. Elsevier Science, Oxford.

Bunn SE and Arthington AH (2002). Basic principles and ecological consequences of altered flow regimes for aquatic biodiversity. *Environmental Management* **30**, 492–507.

Casanova MT and Brock MA (2000). How do depth, duration and frequency of flooding influence the establishment of wetland plant communities? *Plant Ecology* **147**, 237–250.

Copeland C, Schooneveldt-Reid E and Neller S (2003). Fish Everywhere: An Oral History of Fish and their Habitats in the Gwydir River. NSW Fisheries: Ballina.

Driver P, Hardwick L, Hohnberg D and Chessman B (2003). Composition of billabong macroinvertebrate communities in relation to long-term river flow regimes and succession in three lowland rivers of New South Wales, Australia, NISORS Conference, Albury.

Hart BT, Bailey P, Edwards R, Hortle K, James K, McMahon A, Meredith C and Swadling K (1990). Effects of salinity on river, stream and wetland ecosystems in Victoria, Australia. *Water Resources* **24(9)**, 1103–1117.

Jakeman AJ, Letcher RA and Norton JP (2006). Ten iterative steps in development and evaluation of environmental models. *Environmental Modelling and Software* **21**, 602–614.

Keyte P (1994). Lower Gwydir Wetland Plan of Management 1994 to 1997. Report prepared by NSW Dept of Water Resources for the Lower Gwydir Wetland Steering Committee. NSW Dept of Water Resources, Sydney.

Kingsford RT (1990). Ecological impacts of dams, water diversions and river management on floodplain wetlands in Australia. *Austral Ecology* **25**, 91–104.

Kingsford RT (2000). Ecological impacts of dams, water diversions and river management on floodplain wetlands in Australia. *Austral Ecology* **25**, 109–127.

Kingsford RT and Auld KM (2005). Waterbird breeding and environmental flow management in the Macquarie Marshes, arid Australia. *Rivers Research and Applications* **21(2–3)**, 187–200.

Kingsford RT and Johnson W (1998). Impact of water diversions on colonially nesting waterbirds in the Macquarie Marshes of arid Australia. *Colonial Waterbirds* **21(2)**, 159–170.

Letcher RA (2005). Implementation of a water allocation decision support system in the Namoi and Gwydir valleys. In *MODSIM05: International Congress on Modelling and Simulation.* December 2005. pp. 1456–1552. Modelling and Simulation Society of Australia and New Zealand.

Mawhinney W (2003). Restoring biodiversity in the Gwydir wetlands through environmental flows. *Water Science and Technology* **48**, 73–81.

McIntosh BS, Giupponi C, Voinov VV, Smith C, Matthews KB, Monticino M, Kolkman MJ, Crossman N, van Ittersum M, Haase D, Haase A, Mysiak J, Groot JCJ, Siber S, Verweih P, Quinn N, Waeger P, Gaber N, Hepting D, Scholten H, Sulis A, van Delden H, Gaddis E and Assaf H (2008). Bridging the gaps between design and use: developing tools to support environmental management and policy. In *Environmental Modelling, Software and*

Decision Support: State of the Art and New Perspectives. (Eds AJ Jakeman, AA Voinov, AE Rizzoli and SH Chen) pp. 33–48. Elsevier, Amsterdam.

Merritt WS, Pollino C, Powell S and Jakeman AJ (2009). Integrating hydrology and ecology models into flexible and adaptive decision support tools: the IBIS DSS. In *MODSIM 2009: International Congress on Modelling and Simulation*. July 2009. Modelling and Simulation Society of Australia and New Zealand.

Nielsen DL and Chick AJ (1997). Flood-mediated changes in aquatic macrophyte community structure. *Marine and Freshwater Research* **48**, 153–157.

NSW Government (2007). Gwydir Wetlands. Available at http://www.naturalresources.nsw.gov.au/water/ wetlands_area_gwydir.shtml. Accessed 17 November 2008.

Overton IC (2005). Modelling floodplain inundation on a regulated river: integrating GIS, remote sensing and hydrological models. *River Research and Applications* **21**, 991–1001.

Pezeshki SR (2001). Wetland plant responses to soil flooding. *Environmental and Experimental Botany* **46(3)**, 299–312.

Ponnampeeruma FN (1984). Effects of flooding on soils. In *Flooding and Plant Growth*. (Ed. TT Kozlowski) pp. 10–46. Academic Press, New York.

Powell SJ, Letcher RA and Croke BFW (2008). Modelling floodplain inundation for environmental flows: Gwydir wetlands, Australia. *Ecological Modelling* **211**, 350–362.

Reed M, Cuddy SM and Rizzoli AE (2000). A framework for modelling multiple resource management issues: an open modelling approach. *Environmental Modelling and Software* **14**, 503–509.

Refsgaard JC, Henriksen HJ, Harrar WG, Scholten H and Kassahun A (2005). Quality assurance in model based water management: review of existing practice and outline of new approaches. *Environmental Modelling and Software* **20**, 1201–1215.

Reid MA and Brooks JJ (2000). Detecting effects of environmental water allocations in wetlands of the Murray-Darling Basin, Australia. *Regulated Rivers: Research and Management* **16**, 479–496.

Rizzoli AE and Young WJ (1997). Delivering environmental decision support systems: software tools and techniques. *Environmental Modelling and Software* **12**, 237–250.

Roberts J and Marston F (2000). Water Regime of Wetland and Floodplain Plants in the Murray-Darling Basin: A Source Book of Ecological Knowledge. CSIRO Land and Water, Canberra.

Roberts J, Young B and Marston F (2000). Estimating the water requirements for plants of floodplain wetlands: a guide. Occasional Paper 04/00. Land and Water Resources Research and Development Corporation, Canberra.

Thoms M (2003). Floodplain-river ecosystems: lateral connections and the implications of human interference. *Geomorphology* **56**, 335–349.

Ticehurst JL, Letcher RA and Rissik D (2008). Integration modelling and decision support: a case study of the Coastal Lake Assessment and Management (CLAM) tool. *Mathematics and Computers in Simulation* **78**, 435–449.

Warwick NWM and Brock MA (2003). Plant reproduction in temporary wetlands: the effects of seasonal timing, depth, and duration of flooding. *Aquatic Botany* **77**, 153–167.

Webb M, Reid M, Capon S, Thoms M, Rayburg S and James C (2006). Are flood plain wet-land plant communities determined by seed bank composition or inundation periods? In *Proceedings of the Symposium on Sediment Dynamics and the Hydromorphology of Fluvial Systems*. pp. 241–248. IAHS Publications, Dundee, UK.

Wilson GG, Bickel TO, Berney PJ and Sisson JL (2008a). Managing environmental flows in an agricultural landscape: the lower Gwydir floodplain. University of New England, Third

Year Report to the Commonwealth Dept of the Environment, Water, Heritage and the Arts, Armidale.

Wilson G, Berney PJ, Ryder DS and Price JN (2008b). Stage 2: Grazing/landuse in the Macquarie Marshes and Gwydir Wetlands. Final report to the NSW Dept of Environment and Climate Change. University of New England, Armidale.

Wilson GG, Spencer JA and Heagney EC (2010). Responses of fish and waterbirds to flow variability in the Gwydir wetlands. In *Ecosystem Response Modelling in the Murray-Darling Basin.* (Eds N Saintilan and I Overton) pp. 103–118. CSIRO Publishing, Melbourne.

Young WJ, Cuddy SM, Marston FM and Farley TFN (1999). Environmental flows decision support system: a prototype. Vol. 1: Systems description. Technical Report 11/99. CSIRO Land and Water, Canberra.

Young WJ, Scott AC, Cuddy SM and Rennie BA (2003). Murray flow assessment tool: a technical description. Client Report. CSIRO Land and Water, Canberra.

Chapter 9

Inundation response of vegetation communities of the Macquarie Marshes in semi-arid Australia

Rachael Thomas, Sharon Bowen, Shannon Simpson, Steve Cox, Neil Sims, Simon Hunter and Yi Lu

Introduction

Floodplain wetlands in semi-arid regions of the Murray-Darling Basin rely on highly variable flow regimes of rivers for flooding (Frazier and Page 2006; Kingsford 2000; Young and Kingsford 2006) and are particularly vulnerable to the effects of water resource development (Bunn and Arthington 2002; George *et al.* 2005; Kingsford 2000). Inundation is one of the most influential factors governing the ecological responses of floodplain wetland ecosystems (Bunn and Arthington 2002; Roberts *et al.* 2000). It sustains the life-cycles of vegetation communities (Brock and Casanova 1997) and fauna populations (Jenkins and Boulton 2003) and dictates their distribution, abundance and composition at varying spatial and temporal scales. There is good evidence that flows in regulated rivers have been altered (Arthington and Pusey 2003; Frazier and Page 2006; Kingsford 2000; Maheshwari *et al.* 1995) to the detriment of the ecological integrity at varying spatial and temporal scales (Arthington and Pusey 2003; Bunn *et al.* 2006; Jenkins and Boulton 2007; Kingsford 2000; Kingsford *et al.* 2006). To mitigate these impacts, there has been a strong emphasis on developing management strategies that restore environmental flows with the aim of sustaining or rehabilitating ecosystem structure and function (Arthington and Pusey 2003; Jenkins 2006).

Governments in Australia have committed to buying back water-use entitlements under the Water for the Future plan (Australian Government 2008) with the aim of restoring more environmental flows to the rivers of the Murray-Darling Basin. Such an investment requires a quantitative understanding of how flows can be used effectively, as water is scarce in semi-arid regions and climate change scenarios predict further reductions in rainfall (CSIRO 2008; Herron *et al.* 2002). For wetlands dependent on highly variable water regimes, the challenge is to measure water regime components (e.g. inundation) and ecosystem responses at the appropriate temporal and spatial scales. Ecosystem responses are intrinsically scale-dependent both spatially and temporally, and for wetland vegetation communities are relative to their structural growth form, longevity and functional characteristics (Brock and Casanova 1997; Roberts and Dyer 2007; Roberts *et al.* 2000). Floodplain wetlands are large and diverse (Ward *et al.* 2002) and often require a landscape-scale approach to measuring ecosystem response. Long-term data are also necessary, to gain insight into ecosystem responses over the range and diversity of water regime scales.

The Landsat satellite offers a cost-effective data source for monitoring highly variable landscapes that are large, remote and often inaccessible. Its unique long-term historical archive (Wulder *et al.* 2008) and moderate spectral, spatial and temporal resolutions make it ideal for mapping inundation (Frazier and Page 2000; Hudson and Colditz 2003; Kingsford 2000; Lunetta and Balogh 1999; Shaikh *et al.* 2001), a key component of hydrology. Mapping inundation patterns is necessary for assessing the ecological outcomes of environmental flows over varying time scales (Kingsford 2000; Roberts *et al.* 2000; Young *et al.* 2006). Delineating wetland vegetation from aerial photography is a powerful way of tracking changes in community distributions over different time periods (years, decades), especially when linked to changes in inundation regimes.

In this chapter we describe the inundation response of broad vegetation communities over a long-term water regime scale (1979–2006) in the Macquarie Marshes Nature Reserve. This information will contribute to management options for the strategic delivery of environmental flows and assessment of their ecological benefit.

Macquarie Marshes

The Macquarie Marshes occupy a large area of about 200 000 ha on the Macquarie River floodplain (Kingsford and Thomas 1995; Fig. 9.1). Regulated flows from two major reservoirs in the upper Macquarie River catchment – Burrendong Dam (operational in 1967) and Windamere Dam (operational in 1984) – control 70% of flows into the Macquarie Marshes (Kingsford and Auld 2005). Inundation of the Macquarie Marshes is mostly dependent on these regulated releases, but unregulated tributary flows downstream of the major dams are also important (Kingsford and Auld 2005). Localised rainfall can sometimes provide significant flooding in the Macquarie Marshes (B. Johnson, *pers. comm.* 2008). Inundation is influenced by the landform mosaic (Paijmans 1981) and the complex network of ephemeral creeks that traverse the floodplain (Jenkins 2006; Shelly 2005).

The Macquarie Marshes Nature Reserve (Fig. 9.1) was gazetted in 1971 in three sections for the conservation of the wetland vegetation communities that support more than 60 species of waterbirds (Kingsford and Thomas 1995). Part of this nature reserve, the northern and southern sections, was internationally recognised as a significant wetland under the Ramsar Convention in 1986 primarily to protect the vegetation communities that support large populations of colonially nesting waterbirds (Kingsford and Auld 2005; Kingsford and Johnson 1998). Despite its global conservation importance, the Nature Reserve's ecological integrity continues to be compromised by reductions in river flows and flooding (Jenkins *et al.* 2005; Kingsford 2000; Kingsford and Auld 2005).

The vegetation communities characteristic of the Northern Nature Reserve, that are frequently flooded, include river red gum (*Eucalyptus camaldulensis* Denh.) forests and woodlands, lignum *(Muehlenbeckia florulenta* Meisn.) and river cooba (*Acacia stenophylla,* A.Cunn. ex Benth) as well as common reed (*Phragmites australis* (Cav.) Trin ex Streud.), cumbungi (*Typha domingensis* Pers.) and water couch (*Paspalum distichum* L.) grasslands interspersed with open water lagoons (Jenkins *et al.* 2005; Paijmans 1981; Shelly 2005). Where there is less frequent flooding, vegetation communities are characterised by coolibah (*E. coolabah* Blakely & Jacobs) and black box (*E. largiflorens* F.Muell.) woodlands, and myall (*Acacia pendula* A.Cunn. & G.Don), and grasslands distributed at high elevations on the floodplain (Bowen and Simpson 2009; Kingsford and Auld 2005; Paijmans 1981). Comparatively, in the southern nature reserve there are large areas of grasslands on gilgaied floodplain (Paijmans 1981). Historically, an extensive common reed bed and a chain of open water lagoons occurred in the southern nature reserve (Brander 1987; Paijmans 1981). Collectively, these vegetation communities provide significant foraging and nesting habitat for waterbirds (Kingsford and Johnson 1998) and other fauna species (Shelly 2005).

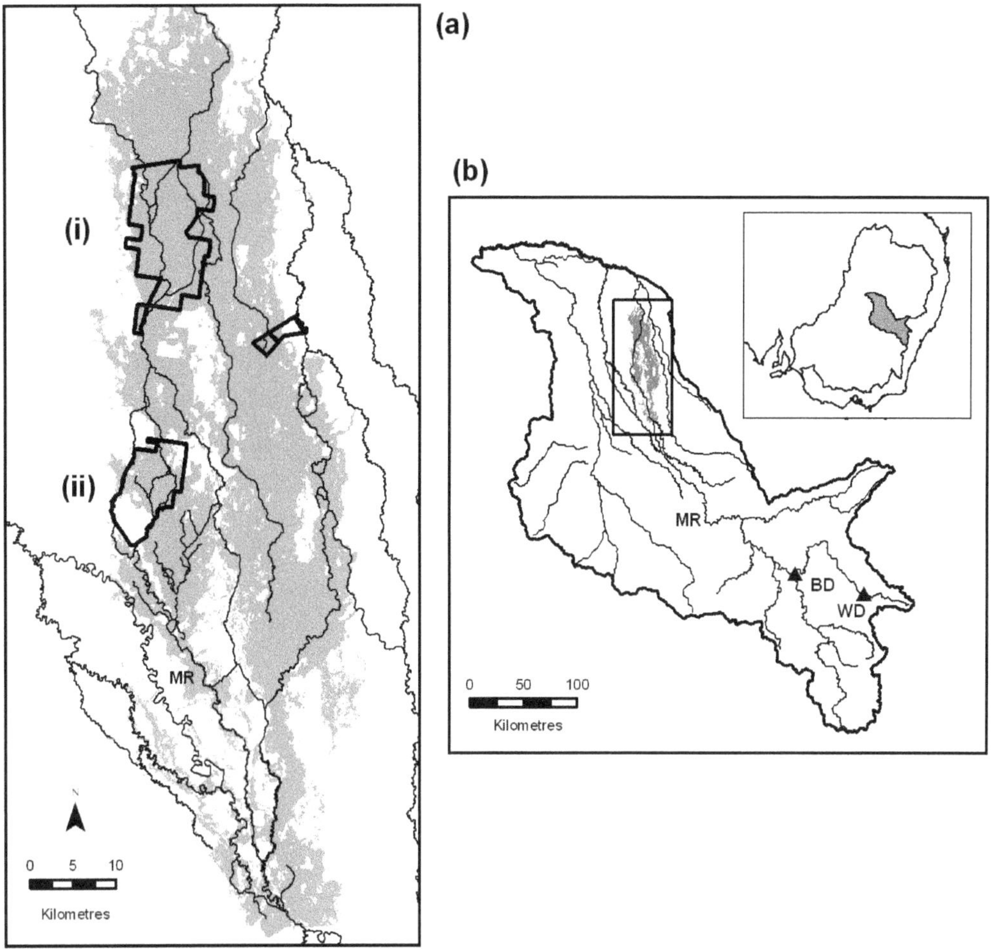

Figure 9.1: (a) Location of the Macquarie Marshes (grey), main rivers and creeks (black lines) and the Nature Reserve sections (thick black lines) – Northern (i) and Southern (ii). (b) The Macquarie River (MR) catchment of the Murray-Darling Basin. Locations of the main river flow regulators – Burrendong dam (BD) and Windamere dam (WD).

The vegetation communities of the northern and southern nature reserves have been mapped from aerial photography interpretation in the last 17 years on two occasions – 1991 (Wilson 1992) and 2008 (Bowen and Simpson 2009). Wilson (1992) used black and white aerial photography to delineate the 1991 vegetation communities (Fig. 9.3a/colour plate, p. 304). Bowen and Simpson (2009) delineated the 2008 vegetation communities by updating the vegetation boundaries and attributes of Wilson (1992) using digital colour aerial photography acquired in 2008 and systematic botanical surveys. These two vegetation maps define the time periods over which inundation and associated changes are examined in this chapter.

Inundation patterns

Few studies have examined floodplain inundation patterns and their change over time in the Macquarie Marshes. Kingsford and Thomas (1995) tracked inundation patterns in the Macquarie Marshes by linking a time series (1983–93) of floods digitised from hard copy photographic Landsat imagery to historical river flows and used this to explain the waterbird population response over the same period of time. DWLC (2000b) used 20 hard copy photographic Landsat images between 1989 and 1994 to map flooded areas by visual interpretation. The purpose was to create flood zones correlated to different flow magnitudes, in order to predict flood frequencies under modelled river flow scenarios.

Thomas *et al.* (2010) derived annual inundation from 28 near-spring digital Landsat images between 1979 and 2006. Inundated area was highly correlated to the magnitude of antecedent flow up to six months prior to image acquisition (Thomas *et al.* 2010). We used the annual inundation maps to construct an index of inundation frequency for two periods prior to each of the 1991 and 2008 vegetation maps for the northern and southern sections of the nature reserve. The first frequency index used 13 years of annual inundation mapping from 1979 to 1991 (Fig. 9.2a/colour plate, p. 303) and the second index used 15 maps between 1992 and 2006 (Fig. 9.2b, p. 303). Each frequency index counted the number of times a pixel was inundated for each time period. Inundation counts were then standardised to account for the different periods of time over which inundation frequency was calculated (Fig. 9.2a, b). High-inundation frequency zones received 10 or more floods in the 13 years (0.77–1.0) of the first period (Fig. 9.2a) and 11 or more in the 15 years (0.73–1.0) of the second period (Fig. 9.2b). Moderate-frequency zones received five to nine floods in the first period and six to 10 in the second period. Inundation change was mapped by subtracting the first period frequency index from the second period index. The subsequent change gradient between –1.0 and 1.0 was divided into inundation frequency change zones at 0.07 intervals (Fig. 9.2c, p. 303).

Vegetation change

Previous studies have tracked the extent and distributions of vegetation types in the Macquarie Marshes prior to 1991 (Brander 1987; DLWC 2000a) but there was a limited attempt to link these changes as responses to environmental parameters, except anecdotally. Bray (1994) used a vegetation cover index to measure expansions of cumbungi and common reed between 1983 and 1994 in three transects representative of the entire Macquarie Marshes but did not quantitatively associate these with changes in inundation. DWLC (2000b) inferred potential vegetation change under different water management scenarios. Bowen and Simpson (2009) determined the recent (1991–2008) changes to vegetation community extent and condition in the Macquarie Marshes Nature Reserve by comparing the 1991 (Wilson 1992) and 2008 (Bowen and Simpson 2009) vegetation maps.

In this chapter we used dominant canopy species as the primary indicator of response to inundation frequency differences between the two periods. To simplify the analysis of vegetation change, we grouped vegetation communities by the dominant canopy species mapped in 1991 according to their broad inundation frequency requirements – high-frequency (wet), low-frequency (dry) and no inundation (terrestrial) – and according to their structural growth form (non-woody or woody) (Fig. 9.3a/colour plate, p. 304). This was also done for the vegetation communities mapped in 2008, but for that period we included the presence of the terrestrial perennial chenopod shrub species black rolypoly (*Sclerolaena muricata* (Moq.) Domin) and buck bush (*Salsola kali* L. var. *kali*). They are often colonising species after disturbance (Jacobs 1990) and their distribution has been noted to have changed substantially between 1991 and 2008 (Bowen and Simpson 2009). Their relatively short life-cycles and intolerance to flooding make them a more suitable indicator of response than the long-lived woody vegetation species, whose responses to inundation changes may not be evident for decades (Bacon *et al.* 1993; Capon 2005; George *et al.* 2005). Vegetation change classes were created by comparing the broad groupings of 1991 with those of 2008 in the same location (Table 9.1 and Fig. 9.3b/colour plate, p. 304). Each change class represents a transition between these periods from one vegetation community to another. These classes were the primary indicators of response to changes in inundation frequency, and included a no-change class (Table 9.1 and Fig. 9.3b). Responses that occurred in small areas (<20 ha) of the terrestrial non-woody class (Table 9.1) are not discussed in this chapter.

Inundation response

Vegetation responses to changes in inundation frequency were assessed by overlaying the inundation frequency change map with the vegetation change map (Fig. 9.4/colour plate, p. 305). For the northern and southern nature reserves we tabulated the percentage area of each vegetation change class within each inundation frequency change zone (Fig. 9.5/colour plate, p. 306).

The vegetation responses detected in the wet vegetation communities (non-woody and woody) were mostly (more than 80% of the area) distributed in locations with decreased inundation frequency between the two time periods (Fig. 9.5). The chenopod replacement of 2430 ha of wet non-woody communities – common reed (*Phragmites australis*), water couch (*Paspalum distichum*) and cumbungi (*Typha domingensis*) – was strongly associated with reductions in inundation frequency in both the northern and southern nature reserves (Figs 9.4 and 9.5). In the southern nature reserve the majority (83%) of the 1380 ha was distributed in areas that experienced a decrease in inundation frequency equivalent to a loss of four or more near-spring floods in 13 years (<–0.28, Fig. 9.5) by the next time period. This chenopod response included the replacement of almost all of the 1991 non-woody communities: 91% of cumbungi, 96% of water couch and 95% of the extensive (1151 ha) common reed mapped in 1991 (Table 9.1 and Fig. 9.3a) (Bowen and Simpson 2009). Evidence of this trend was reported by Driver and Knight (2007), who found a decline in common reed in their surveys since 1999. In early 2006, Driver and Knight (2007) observed a complete absence of common reed (including in the southern nature reserve) and the dominance of terrestrial vegetation due to insufficient inundation to recharge soil moisture. Soil moisture may offer protection to common reed rhizomes during prolonged dry periods (Hocking *et al.* 1983; Roberts 2000), but further investigations would be required to confirm this. Historical evidence suggests that the common reed beds were declining prior to this study in the southern nature reserve due to continued incision and widening of the channels since the late 1960s (Brander 1987; Brereton 1994), decreasing the likelihood of widespread inundation in this area because of the need for greater volumes of water (Ralph 2008). In the northern nature reserve, 1050 ha of wet non-woody communities were replaced by chenopods. About half of the 1991 water couch community and about a quarter of the common reed community were replaced by chenopods in the northern nature reserve (Table 9.1 and Fig. 9.4). This response was associated with a decrease in inundation frequency. About 60% of the area was distributed in locations that lost four or more spring floods by the next time period (<–0.28, Fig. 9.5) and about 30% was distributed where one to three floods were lost (–0.14 to –0.28, Fig. 9.5).

The chenopod colonisation of 308 ha of wet non-woody vegetation communities in the northern nature reserve (Table 9.1) was also associated with a decrease in inundation frequency. About half the area was distributed in locations where floods were reduced by one to three floods (–0.14 to –0.28, Fig. 9.5) and about 33% of the area was distributed in areas where inundation frequency decreased by four or more floods (Figs 9.4 and 9.5). Almost a third of the cumbungi community was affected (Table 9.1, Fig. 9.4) by chenopod colonisation in the northern nature reserve. In the southern nature reserve, the remaining area of the 1991 common reed community was colonised by chenopods, in zones of decreasing inundation frequency (Figs 9.4 and 9.5). The shift from open water to dry depressions favoured chenopod colonisation, particularly in the southern nature reserve where all lagoons were colonised. This was entirely distributed in zones of reduced inundation frequency (Table 9.1, Fig. 9.5).

Table 9.1: Response descriptions and area (ha) of 12 vegetation change classes and area of no vegetation change between 1991 and 2008 for the total reserve, the northern nature reserve (12260 ha) and the southern nature reserve (6292 ha)

Change class	Response in 2008	Total ha	NNR ha (%)[a]	%[b]	SNR ha (%[a])	%[b]
No change (no Δ)	Vegetation communities mapped in 1991 remained the same	2945	2945 (24)		0 (0)	
	Common reed		1463	68	-	
	River red gum		1388	23	-	
	Open water		43	76	-	
	Other		51	-	-	
Change among wet non-woody (Δ WnW)	Colonisation/replacement among the 1991 wet-non woody communities	379	377 (2)		2 (<1)	
Open water to chenopod (OW→C)	Chenopod[c] colonisation of 1991 open water lagoons	39	13 (<1)	24	26 (<1)	100
Wet non-woody to chenopod (WnW→C)	Chenopod[c] replacement of 1991 wet non-woody species	2430	1050 (9)		1380 (22)	
	Common reed		503	23	1151	95
	Water couch		502	55	210	96
	Cumbungi		35	13	18	91
Wet non-woody to wet non-woody-chenopod (WnW→WnW/C)	Chenopod[c] colonisation of 1991 wet non-woody communties	370	308 (3)		62 (1)	
	Common reed		78	4	55	5
	Water couch		149	16	7	3
	Cumbungi		80	31	-	-
Wet non-woody to wet woody (WnW→WW)	Recruitment of river red gum into 1991 wet non-woody communities	46	43 (<1)		3 (6) (<1)	
	Common reed		19	1	-	-
	Water couch		24	3	3	1
Wet non-woody to wet woody-chenopod (WnW→WW/C)	Recruitment of river red gum with chenopod[c] understorey into 1991 wet non-woody communities	82	82 (1)		0 (0)	

Change class	Response in 2008	Total ha	NNR ha (%)[a]	%[b]	SNR ha (%[a])	%[b]
	Water couch		41	5	-	-
	Cumbungi		42	15	-	-
Wet woody to wet woody-chenopod (WW→WW/C)	Chenopod[c] colonisation of 1991 understorey of wet woody communities	5012	4842 (39)		171 (3)	
	River red gum		4677	77	171	100
	River cooba		7	100	-	-
	Lignum		6	39	-	-
Dry woody to dry woody-chenopod (DW→DW/C)	Chenopod[c] colonisation of 1991 understorey of dry woody communities	549	411 (3)		137 (2)	
	Black box		110	100	51	100
	Coolibah		457	94	43	100
Terrestrial non-woody to terrestrial non-woody-chenopod (TnW→TnW/C)	Chenopod[c] colonisation of 1991 grasslands	6365	1863 (15)	98	4507 (72)	100
Terrestrial non-woody to wet non-woody-chenopod (TnW→WnW/C)	Mixed marsh/chenopod[c] replacement of 1991 grassland/cleared	19	19 (<1)	1	0 (0)	-
Terrestrial non-woody to wet woody-chenopod (TnW→WW/C)	Recruitment of river red gum with a chenopod[c] understorey into 1991 grassland/cleared land	19	19 (<1)	1	0 (0)	-
Terrestrial woody to terrestrial woody-chenopod (TW→TW/C)	Chenopod[c] colonisation of 1991 understorey of myall communities	3	3 (<1)		0 (0)	

a: % of each nature reserve section.

b: % of the 1991 vegetation communities. See Bowen and Simpson (2009) for detailed vegetation community descriptions.

c: Terrestrial shrubs black roly-poly (*Sclerolaena muricata* (Moq.) Domin) and buck bush (*Salsola kali* L. var. *kali*)

The presence of the invasive chenopod shrubs in river red gum (*E. camaldulensis*) understorey (Table 9.1) was spatially associated with a drying trend between the two time periods. In the northern nature reserve, 4677 ha of the river red gum understorey was colonised by chenopods, affecting just over three-quarters of the 1991 river red gum extent. Almost half of the colonised extent was located where inundation frequency reduction was large (four or more floods) (<–0.28, Fig. 9.5). The colonisation occurred in the entire 1991 river red gum community understorey of the southern nature reserve (171 ha) but its distribution across the inundation change gradient was more varied – where flooding decreased by one to three floods, no change, or where flooding increased – although areas were relatively small. Short-lived chenopod shrubs such as buck bush and black rolypoly have historically been opportunistic colonisers of river red gum woodland understorey in drought years (Paijmans 1981). With a continual drying trend they can persist in the understorey of river red gums and compete for resources. This drying trend of over 10 years will ultimately have a negative impact on the long-lived river red gum communities that rely on flooding, soil moisture and groundwater sources (Bond *et al.* 2008; Cunningham *et al.* 2007; George *et al.* 2005).

The response of chenopod shrubs moving into the understorey of dry woody communities (coolibah and black box) was not spatially associated with a decrease in inundation frequency in the northern nature reserve in this study. Most (65%) of the colonised area (Fig. 9.5) was distributed in locations where inundation frequency remained at the same low frequency (Figs 9.2 and 9.4). In the southern nature reserve, about 50% of the colonised area was confined to zones of moderate inundation frequency decreases (two to three floods) (Fig. 9.5) and about 40% of the area was distributed in zones that did not flood in either time period (Fig. 9.4). The lack of substantial spring flooding after 2000 (Thomas *et al.* 2010) may have been enough recent drying for terrestrial chenopods to persist in the landscape. The chenopod colonisation of the large terrestrial grassland that dominated the southern reserve in 1991 (4503 ha, Table 9.1, Fig. 9.3) occurred in zones that did not receive inundation in either time period (Fig. 9.4) and so the response could not be attributed to a change in inundation in this study. However, there is evidence that these areas in the Southern Nature Reserve were dominated by wetland vegetation prior to our study period having had an earlier history of flooding (Brander 1987; Paijmans 1981). This may suggest that these areas were 'terrestrialised' before 1979, and that there is a general persistence of chenopods in the area as the landscape has dried out. In the northern reserve, the chenopod colonisation was also located where there was no inundation in both time periods (Figs 9.2 and 9.4). It was also spatially associated with areas of no change in low-inundation frequency, and those that experienced some drying (two or three floods) (Fig. 9.5).

River red gum recruitment (wet non-woody to wet woody) in the northern reserve was relatively small in extent (43 ha; Table 9.1) and distributed over the gradient of inundation frequency: 35% was in no-change zones, 32% in moderately decreasing and 27% in moderately increasing frequency zones (Figs 9.4 and 9.5). At some point after 1991 localised conditions were suitable for the establishment of river red gum seedlings. Ground surveys suggest that this recruitment was recent (Bowen and Simpson 2009), so the inundation history may not have adequately reflected the time scale of this response (Table 9.1, Figs 9.4 and 9.5). However, additional areas (82 ha) of river red gum recruitment did not receive the inundation they required and chenopods colonised those areas.

Changes among the wet non-woody communities (common reed, water couch, cumbungi) were almost exclusive to the northern nature reserve. They were mostly distributed in areas that were drying but were also in areas that received slightly more frequent flooding (two to three floods). Their inundation frequency requirements to maintain vigour are interannual or near-annual (one to two years) (Roberts and Marston 2000). A detailed analysis of these changes in relation to inundation was beyond the scope of this case study.

In the northern nature reserve, zones of high inundation frequency were maintained between the two time periods (Fig. 9.2). The vegetation communities spatially associated with this zone, mostly common reed and river red gum did not change (Figs 9.3 and 9.4). About 68% of the 1991 common reed extent (1463 ha) remained in the northern nature reserve (Table 9.1, Fig. 9.5), a stark contrast to the southern reserve where only 5% of the 1991 common reed extent was retained and colonised by chenopods. About 23% of the 1991 river red gum community extent in the northern reserve (6065 ha) remained unchanged, with intact wet understorey located along drainage channels (Bowen and Simpson 2009) where high-inundation frequency was maintained. There is, however, emerging evidence that the river red gum trees in the northern reserve are suffering water stress and lack regenerative potential (Nairn 2008) as described for other river red gum communities in the Murray-Darling Basin (Chesterfield 1986; Cunningham *et al.* 2009; George *et al.* 2005; MDBC 2005).

Environmental flows and monitoring

Understanding the temporal and spatial variability of inundation patterns is critical for environmental flow management and monitoring of vegetation response. By measuring the spatial extent of inundation, the amount of vegetation affected, how often (frequency) it is affected and for how long (duration) can be estimated. Inundation responses of vegetation communities are inherently scale-dependent. The dynamic nature of vegetation communities can mask inundation responses at different scales because they respond seasonally, as they grow and as environmental parameters change (Roberts *et al.* 2000). This is particularly so for those with high intrinsic variability such as river red gums (Chesterfield 1986), which occur widely across the inundation frequency gradient (Bacon *et al.* 1993; Bren 1992; Roberts and Marston 2000) and respond differently to flooding according to their age (George *et al.* 2005). They respond physiologically (Bacon *et al.* 1993; Cunningham *et al.* 2007) and spatially to an inundation event or to a historical surface water regime, and to groundwater availability, but with substantial lag-times (Bond *et al.* 2008; George *et al.* 2005). Responses that occur over the short or long term and at small or large spatial scales are equally important when assessing the effectiveness of environmental flows. The type and magnitude of vegetation response to changes in flooding is also relative to the structural growth form (trees, shrubs or grasslands) and functional characteristics of vegetation (*sensu* Brock and Casanova 1997). Components of these characteristics that have the strongest influence on vegetation response to inundation relative to spatial setting will most likely be detectable, measurable (Roberts and Dyer 2007) and predictable for the management of environmental flows.

This chapter has demonstrated the response of flood-dependent vegetation communities to the gradual drying of the floodplain in terms of reduced spring inundation frequency over nearly three decades. Our study suggests that the loss of four or more spring floods in a 13-year period cannot maintain the water requirements of semi-permanent wetland vegetation, including the understorey of longer-lived woody communities such as river red gums. Non-woody semi-permanent vegetation communities that require regular flooding (once or twice a year) either disappeared or were dominated by vegetation adapted to relatively dry conditions. Comparatively, we did not show a decrease in the extent of river red gums in response to drying. Rather, there was a colonisation of terrestrial species in the understorey, as found in other surveys of the Macquarie Marshes (Driver and Knight 2007). These findings mirror the impacts of river regulation reported elsewhere, which include reduced flow variability and flow volume and inundation extent (Arthington and Pusey 2003; Kingsford 2000), and reduced frequency and altered timing of floods (DLWC 2000b).

Evidence of river red gum recruitment was in small pockets, which is typical in variable environments (George *et al.* 2005). In our study it was difficult to relate this to changes in inundation frequency at this spatial and temporal scale. Chenopod colonisation into such areas suggests that inundation was adequate to trigger establishment of river red gums at a point in time (Bowen and Simpson 2009), but the chenopod colonisation indicates that the subsequent flooding may be insufficient to maintain growth to maturity (George *et al.* 2005). Monitoring of recruitment in response to environmental flows should be considered, but it can be difficult to measure (Kingsford 2000) without estimates of population structure and viability (George *et al.* 2005). Also, the observed spring inundation frequency may be insufficient to detect some increased inundation frequency, due to the seasonal shift from spring to summer inundation (DLWC 2000b; Roberts *et al.* 2000). Such seasonal shifts have altered understorey species composition in river red gum woodlands in other locations of the Murray-Darling Basin (Robertson *et al.* 2001).

Monitoring the ecological effectiveness of environmental flows therefore requires a multi-scale approach within an adaptive framework for environmental flow management (Ch. 2, this volume). Robust and rigorous techniques for the analysis of inundation patterns and vegetation distribution at the landscape scale and at ecologically meaningful time scales are necessary (Kingsford 2000). This case study has demonstrated the long-term inundation responses of floodplain wetland vegetation using a Landsat-derived inundation frequency index. With the advent of digital high-resolution photography, the opportunity for archiving the distribution of floodplain wetland vegetation over time is greatly enhanced. The Landsat satellite provides the opportunity to monitor inundation patterns over environmental flow event scales, due to the regularity of acquisition over the same location (every 16 days) suitable for inundation flow modelling. Quantitative response measures and condition variables measured on-ground, linked to remotely sensed variables, can be used to estimate responses over large areas as a direct consequence of environmental flows.

Relating inundation to river flows allows modelling of long-term impacts of river regulation on inundated areas (Frazier and Page 2006; Ren *et al.* in press) to guide the management of environmental flows via decision support systems (Overton 2005; Powell *et al.* 2008). Such models can also be used to infer long-term vegetation changes (Roberts *et al.* 2000) over large areas. If linked to ecological responses (Shaikh *et al.* 1998; Sims and Thoms 2002) or measures of condition (Cunningham *et al.* 2007), they can be used to assess the ecological success of environmental flows.

Conclusions

Environmental flows are crucial for the maintenance and rehabilitation of the structure and functioning of flood-dependent ecosystems throughout the Murray-Darling Basin. To maximise the ecological benefit of environmental flows, it is important to monitor the flows and the ecological response to them within an adaptive management framework. This chapter has demonstrated one approach to monitoring vegetation response to changes in inundation patterns over the long term and at the landscape scale.

This chapter indicates that, while the distribution of iconic vegetation communities such as river red gum may not have changed spatially in the Macquarie Marshes Nature Reserve over the period of study, the composition of the community has changed and is indicative of drying. In particular, terrestrial chenopod colonisers are more prevalent in the understorey of river red gum communities and have replaced some of the semi-permanent wetland vegetation communities. The ability to predict ecological response to water regime scenarios is an important part of the adaptive management of environmental flows in the Macquarie Marshes. This chapter has demonstrated that a loss of four or more spring floods in a 13-year period will reduce the spatial extent of frequently inundated areas; in response to this contraction semi-permanent wetland vegetation will disappear and terrestrial colonisers will dominate the landscape. Spring inundation frequency is a major component of the inundation regime of the Macquarie Marshes and its strong influence on vegetation distribution, as shown in this chapter, can therefore be used as a basis for ecosystem response modelling for predicting the effects of long-term water management scenarios.

Our case study has demonstrated the importance of using long-term data (nearly 30 years) to gain insight into the spatial response of flood-dependent vegetation communities that persist over varying time scales (years to decades). The use of the long-term Landsat archival data, coupled with high-resolution photography, for delineating vegetation communities and quantitative coincident ground survey (Ch. 2, this volume) provides a robust technique for monitoring the extent and condition of flood-dependent vegetation communities across these large, highly variable and valuable ecosystems.

Acknowledgments

We are grateful to the landholders of the Macquarie Marshes who assisted with the background research projects and gave access to their land. Ray Jones, Miranda Kerr, Darren Shelly and Terry Mazzer of the NSW Department of Environment, Climate Change and Water provided field assistance. Renee Shepherd (DECCW) provided much appreciated program support. We thank Bill Johnson (DECCW) and Professor Richard Kingsford (UNSW) for sharing their knowledge and insight. Jennifer Spencer (DECCW) commented on an earlier draft. Two reviewers improved this chapter. This work was undertaken as part of the NSW Wetland Recovery Program and NSW Rivers Environmental Restoration Program, jointly funded by the Australian Government's Water for the Future plan and the NSW DECCW.

References

Arthington AH and Pusey BJ (2003). Flow restoration and protection in Australian rivers. *River Research and Applications* **19**, 377–395.

Australian Government (2008). The Australian Government's plans for water. Available at http://www.environment.gov.au/water/publications/action/pubs/water-future.pdf. Accessed 10 June 2009.

Bacon PE, Stone C, Binns DL, Leslie DJ and Edwards DW (1993). Relationships between water availability and *Eucalyptus camaldulensis* growth in a riparian forest. *Journal of Hydrology* **150**, 541–561.

Bond N, Lake P and Arthington A (2008). The impacts of drought on freshwater ecosystems: an Australian perspective. *Hydrobiologia* **600**, 3–16.

Bowen S and Simpson S (2009). Changes in extent and condition of the vegetation communities of the Macquarie Marshes Floodplain 1991–2008. Report for the NSW Wetland Recovery Program, Australian Government Water for the Future Program. NSW Dept of Environment and Climate Change, Sydney.

Brander D (1987). Environmental changes in the southern Macquarie Marshes: 1934 to 1987. Honours thesis. University of New South Wales, Sydney.

Bray N (1994). Vegetation changes at selected sites in the Macquarie Marshes: 1983–1994. Natural Resource Management Strategy.

Bren LJ (1992). Tree invasion of an intermittent wetland in relation to changes in the flooding frequency of the River Murray, Australia. *Australian Journal of Ecology* **17**, 395–408.

Brereton G (1994). Macquarie Marshes Management Strategy. Stage 1: Biophysical Investigations. Summary Report. Natural Resources Management Strategy.

Brock M and Casanova M (1997). Plant life at the edge of wetlands: ecological responses to wetting and drying patterns. In *Frontiers in Ecology: Building the Links.* (Eds N Klomp and I Lunt) pp. 181–192. Elsiever Science, Oxford.

Bunn SE and Arthington AH (2002). Basic principles and ecological consequences of altered flow regimes for aquatic biodiversity. *Environmental Management* **30**, 492–507.

Bunn SE, Balcombe PM, Davies PM, Fellows CS and McKenzie-Smith FJ (2006). Aquatic productivity and food webs of desert river ecosystems. In *Ecology of Desert Rivers.* (Ed. RT Kingsford) pp. 76–99. Cambridge University Press, Melbourne).

Capon S (2005). Flood variability and spatial variation in plant community composition and structure on a large arid floodplain. *Journal of Arid Environments* **60**, 283–302.

Chesterfield EA (1986). Changes in the vegetation of the river red gum forest at Barmah, Victoria. *Australian Forestry* **49**, 4–15.

CSIRO (2008). Water availability in the Macquarie-Castlereagh. Report to the Australian Government from the CSIRO Murray-Darling Basin Sustainable Yields Project. CSIRO, Australia.

Cunningham SC, MacNally R, Read J, Baker P, White M, Thomson J and Griffioen P (2009). A robust technique for mapping vegetation condition across a major river system. *Ecosystems* **12**, 207–219.

Cunningham SC, Read J, Baker PJ and MacNally R (2007). Quantitative assessment of stand condition and its relationship to physiological stress in stands of *Eucalyptus camaldulensis* (Myrtaceae). *Australian Journal of Botany* **55**, 692–699.

DLWC (2000a). *Historical Vegetation Mapping in the Macquarie Marshes 1949–1991.* NSW Dept of Land and Water Conservation.

DLWC (2000b). *A Review of Recent Studies investigating Biological and Physical Processes in the Macquarie Marshes.* NSW Dept of Land and Water Conservation.

Driver PD and Knight C (2007). Macquarie Marshes 2005/06 Environmental Flows. Response of groundcover plants to environmental flows. Report to the Environmental Flows Reference Group. NSW Dept of Natural Resources.

Frazier PS and Page KJ (2000). Water body detection and delineation with Landsat TM data. *Photogrammetric Engineering and Remote Sensing* **66**, 1461–1467.

Frazier PS and Page K (2006). The effect of river regulation on floodplain wetland inundation, Murrumbidgee River, Australia. *Marine and Freshwater Research* **57**, 133–141.

George AK, Walker KF and Lewis MM (2005). Population status of eucalypt trees on the River Murray floodplain, South Australia. *River Research and Applications* **21**, 271–282.

Herron N, Davis R and Jones R (2002). The effects of large-scale afforestation and climate change on water allocation in the Macquarie River catchment, NSW, Australia. *Journal of Environmental Management* **65**, 369–381.

Hocking PJ, Finlayson CM and Chick AJ (1983). The biology of Australian weeds. 12. *Phragmites australis* (Cav.) Trin. ex Steud. *Journal of the Australian Institute of Agricultural Science* **49,** 123–132.

Hudson PF and Colditz RR (2003). Flood delineation in a large and complex alluvial valley, lower Panuco basin, Mexico. *Journal of Hydrology* **280**, 229–245.

Jacobs SWL (1990). Chenopodiaceae. In *Flora of New South Wales Vol. 1.* (Ed. GJ Harden). NSW Press, Royal Botanic Gardens, Sydney.

Jenkins K (2006). Links between flow, aquatic productivity and diversity: setting rehabilitation targets for regulated floodplain wetlands. Final report to the State Wetland Advisory Committee on Ecosystem Management. University of New England, Armidale.
Jenkins KM and Boulton AJ (2003). Connectivity in a dryland river: short-term aquatic microinvertebrate recruitment following floodplain inundation. *Ecology* **84**, 2708–2723.
Jenkins KM and Boulton AJ (2007). Detecting impacts and setting restoration targets in arid-zone rivers: aquatic micro-invertebrate responses to reduced floodplain inundation. *Journal of Applied Ecology* **44**, 823–832.
Jenkins K, Boulton A and Ryder D (2005). A common parched future? Research and management of Australian arid-zone floodplain wetlands. *Hydrobiologia* **552**, 57–73.
Kingsford R (2000). Ecological impacts of dams, water diversions and river management on floodplain wetlands in Australia. *Austral Ecology* **25**, 109–127.
Kingsford RT and Auld KM (2005). Waterbird breeding and environmental flow management in the Macquarie Marshes, arid Australia. *River Research and Applications* **21**, 187–200.
Kingsford RT and Johnson WJ (1998). The impact of water diversions on colonially nesting waterbirds in the Macquarie Marshes in arid Australia. *Colonial Waterbirds* **21**, 159–170.
Kingsford R and Thomas R (1995). The Macquarie Marshes in arid Australia and their waterbirds: a 50-year history of decline. *Environmental Management* **19**, 867–878.
Kingsford RT, Lemley AD and Thompson JR (2006). Impacts of dams, river management and diversions on desert rivers. In *Ecology of Desert Rivers*. (Ed. RT Kingsford) pp. 203–247. Cambridge University Press, Melbourne.
Kingsford R, Brandis K, Jenkins KM, Nairn L and Rayner T (2010). Measuring ecosystem responses to flow across temporal and spatial scales. In *Ecosystem Response Modelling in the Murray-Darling Basin*. (Eds N Saintilan and I Overton) pp. 15–36. CSIRO Publishing, Melbourne.
Lunetta RS and Balogh ME (1999). Application of multi-temporal Landsat 5 TM imagery for wetland identification. *Photogrammetric Engineering and Remote Sensing* **65**, 1303–1310.
Maheshwari BL, Walker KF and McMahon T (1995). Effects of regulation on the flow regime of the River Murray, Australia. *Regulated Rivers: Research and Management* **10**, 15–38.
MDBC (2005). Survey of river red gum and black box health along the River Murray in NSW, Victoria and South Australia, 2004. Murray-Darling Basin Commission, Canberra.
Nairn L (2008). Redgum health in the Macquarie Marshes Nature Reserve. Unpublished report to the NSW Dept of Environment and Climate Change. University of New South Wales, Sydney.
Overton IC (2005). Modelling floodplain inundation on a regulated river: integrating GIS, remote sensing and hydrological models. *River Research and Applications* **21**, 991–1001.
Paijmans K (1981). The Macquarie Marshes of inland northern New South Wales, Australia. CSIRO Australia, Division of Land Use Research. Technical Paper No. 41.
Powell SJ, Letcher RA and Croke BFW (2008). Modelling floodplain inundation for environmental flows: Gwydir wetlands, Australia. *Ecological Modelling* **211**, 350–362.
Ralph TJ (2008). Channel breakdown and floodplain wetland morphodynamics in the Macquarie Marshes. PhD thesis. Macquarie University, Sydney.
Ren S, Kingsford RT and Thomas RF (In press). Modelling flow to and inundation of the Macquarie Marshes in arid Australia. *Environmetrics*.
Roberts J (2000). Changes in *Phragmites australis* in south-eastern Australia: a habitat assessment. *Folia Geobotanica* **35**, 353–362.
Roberts J and Dyer F (2007). Magnitude of response: selecting sites for monitoring vegetation response to environmental flows. In *Proceedings of the 5th Australian Stream Management Conference. Australian Rivers: Making a Difference*. (Eds AL Wilson, RL Dehaan, RJ Watts,

KJ Page, KH Bowmer and A Curtis) pp. 330–335. Charles Sturt University, Thurgoona, NSW.

Roberts J and Marston F (2000). *Water Regime of Wetland and Floodplain Plants in the Murray-Darling Basin: A Source Book of Ecological Knowledge*. CSIRO Land and Water, Canberra.

Roberts J, Young B and Marston F (2000). Estimating the water requirements for plants of floodplain wetlands: a guide.Occasional Paper 04/00. Land and Water Resources Research and Development Corporation, Canberra.

Robertson AI, Bacon P and Heagney G (2001). The responses of floodplain primary production to flood frequency and timing. *Journal of Applied Ecology* **38**, 126–136.

Shaikh M, Brady AT and Sharma P (1998). Applications of remote sensing to assess wetland inundation and vegetation response in relation to hydrology in the Great Cumbung Swamp, Lachlan Valley, NSW, Australia. In *Wetlands for the Future*. (Eds AJ McComb and JA Davis) pp. 1–7. Gleneagles Publishing, Adelaide.

Shaikh M, Green D and Cross H (2001). A remote sensing approach to determine environmental flows for wetlands of the Lower Darling River, New South Wales, Australia. *International Journal of Remote Sensing* **22**, 1737–1751.

Shelly D (2005). Flora and fauna of the Macquarie Marshes Region. NSW Dept of Infrastructure, Planning and Natural Resources, Dubbo.

Sims NC and Thoms MC (2002). What happens when flood plains wet themselves: vegetation response to inundation on the lower Balonne flood plain. In *Proceedings of an International Symposium. The Structure, Function and Management Implications of Fluvial Sedimentary Systems*. Alice Springs, Australia. pp. 195–202. IAHS Publication no. 276.

Thomas RF, Kingsford RT, Lu Y and Hunter SJ (2010). Mapping annual inundation (1979–2006) of the Macquarie Marshes in semi-arid Australia, using the Landsat archive. Final Report to the NSW Wetland Recovery Program. NSW Dept of Environment, Climate Change and Water.

Ward JV, Tockner K, Arscott DB and Claret C (2002). Riverine landscape diversity. *Freshwater Biology* **47**, 517–539.

Wilson R (1992). Vegetation map of the Macquarie Marshes. NSW National Parks and Wildlife Service.

Wulder MA, White JC, Goward SN, Masek JG, Irons JR, Herold M, Cohen WB, Loveland TR and Woodcock CE (2008). Landsat continuity: issues and opportunities for land cover monitoring. *Remote Sensing of Environment* **112**, 955–969.

Young WJ and Kingsford RT (2006). Flow variability in large unregulated dryland rivers. In *Ecology of Desert Rivers*. (Ed. RT Kingsford) pp. 11–46. Cambridge University Press, Melbourne.

Young W, Brandis K and Kingsford RT (2006). Modelling monthly streamflows in two Australian dryland rivers: matching model complexity to spatial scale and data availability. *Journal of Hydrology* **331**, 242–256.

Chapter 10

Water requirements of biota, geomorphology and climate change in the Macquarie Marshes

Kerrylee Rogers, Tim Ralph and Sarah Imgraben

Introduction

The Macquarie Marshes, on the lower reaches of the Macquarie River, are the largest and one of the most ecologically significant floodplain wetland systems in the Murray-Darling Basin of south-east Australia. All wetlands, such as the Macquarie Marshes, rely on the supply of water for their long-term survival (Mitsch and Gosselink 2000) and distinct flood events provide inundation which maintains their ecological character. In accordance with the flood pulse concept (Junk *et al.* 1989), flooding is the 'principal driving force responsible for the existence, productivity and interactions of the major biota' (Junk *et al.* 1989, p. 110) in most floodplain wetlands. Although the flood pulse concept theoretically does not apply to river systems with highly variable flow regimes, Walker *et al.* (1995) proposed that biota in variable environments and habitats develop strategies such as opportunism and flexibility as adaptations to flow variability. They suggest that the flood pulse concept may apply to floodplains with variable flow regimes, such as the Macquarie Marshes. To adequately account for the ecological role of flooding on floodplain biota, Junk *et al.* (1989, p. 112) claimed that 'the resulting physico-chemical environment causes the biota to respond by morphological, anatomical, physiological, phenological, and/or ethological adaptations, and produce characteristic community structures' in floodplain wetlands. Thus, ecosystems such as those in the Macquarie Marshes are likely to be both shaped and sustained by floodwaters from upstream, since these flows influence the wetland environments within the system.

There are numerous pressures on water availability and water delivery to the Macquarie Marshes, such as regulation, abstraction and diversion. Climate change, which encompasses both natural climate variability and anthropogenic climate change due to enhanced carbon dioxide (CO_2) emissions and global warming, is also projected to have significant implications for water availability and water security within the Murray-Darling Basin. It threatens the delicate ecological balance between hydrology and biota in the Macquarie Marshes (CSIRO 2008; Hassall and Associates 1998; Jones *et al.* 2002; Pollino *et al.* 2007; Tighe *et al.* 2007a). According to the future climate and current development scenario best estimate in the CSIRO water availability in the Macquarie-Castlereagh report (CSIRO 2008), water availability in the Macquarie-Castlereagh system would decrease overall by 8%, with a 9% decrease in end-of-system flows and a 15% decrease in outflows from the Macquarie Marshes at Carinda. The impact on the Macquarie Marshes, according to the best estimate, by 2030 would be a 10% increase in the period between important inundation events, less frequent flood events would be 5% smaller and the total higher flood level volume would decrease by 16% (CSIRO 2008, p. iii).

Despite the strong intrinsic relationship between floodplain and aquatic biota and hydrology, there is relatively little scientific literature describing the response of flora and fauna to flood perturbations in the Macquarie Marshes. The absence of scientific literature on the relationship between hydrology and biota in the Macquarie Marshes is surprising given the ecological significance of the system, the ecosystem services provided by the wetlands and the fact that the purpose of river flow objectives under the NSW government water reform program is to improve survival of ecosystems and aquatic biodiversity, improve the health of in-stream and riparian vegetation and promote healthier wetlands (DNR 2007). Furthermore, since precipitation in central-western NSW is projected to decrease according to current climate change scenarios (CSIRO and BOM 2007), it is timely to consider the water requirements of biota within floodplain wetlands, particularly those that occur in semi-arid regions. By coupling an understanding of the water requirements of biota with predictions of water availability, we can identify individual species, functional groups of species and/or wetland areas vulnerable to loss or decline. Management efforts may then be directed to areas where the predicted water availability meets the water requirements of the established biota, or to threatened species. Understanding the ecohydrology and water requirements of biota is the first step towards managing floodplain aquatic ecosystems in the face of climate change and declining water resources.

This chapter presents the methods and key findings from a synthesis of scientific literature regarding the water requirements of biota in the Macquarie Marshes. Species-specific results will be available in future publications (Rogers and Ralph, in prep). Consideration is also given to species' ability to adapt to climate change, particularly to a changing floodplain wetland inundation regime. While climate change will have direct effects such as increased atmospheric CO_2 concentrations, increased temperature and associated evapotranspiration and increased solar radiation (Bernstein *et al.* 2007), the indirect effects through alterations to the hydrology of the Macquarie Marshes are likely to have the greatest impact on species distribution and abundance in the floodplain wetlands (Fig. 10.1/colour plate, p. 307). Therefore, this chapter is limited to describing the possible impacts of climate change-induced alterations to hydrology on the biota and geomorphology of the Macquarie Marshes.

Water requirements of biota in the Macquarie Marshes

Study site

The Macquarie River in central-western NSW is a major tributary of the Murray-Darling catchment (Fig. 10.2/colour plate, p. 308). The Macquarie River drains inland from its steep subhumid headwaters (mean annual rainfall 800–1000 mm, mean annual potential evapotranspiration 1400–1600 mm; BOM 2005) and passes through a partly confined bedrock valley in its middle reaches before it debouches onto an unconfined alluvial plain. The lower third of the Macquarie River exists entirely on this semi-arid, low-relief plain which receives, on average, 400–500 mm of rainfall per year and has 2000–2400 mm of mean potential evapotranspiration per year (BOM 2005). In this lowland–dryland setting, the meandering channel of the Macquarie River breaks down into a series of distributary channels, floodouts and floodplain wetlands within the Macquarie Marshes (Ralph and Hesse in press).

The Macquarie Marshes rely on flow from the upper catchment and the river has a downstream decline in tributary inputs and total discharge as it enters the region of channel breakdown on alluvial plain (Ralph and Hesse in press). Semi-permanent wetlands cover an area of around 400 km^2 in the system and large floods can cover an area of around 2500 km^2. The perennial and ephemeral wetlands are fed by overbank flooding and floodout from numerous creeks and distributary channels. Two core areas of wetlands exist in the reserve system – the southern Macquarie Marshes and the northern Macquarie Marshes – which are each made up of reed beds, mixed marshlands and riparian woodlands (Fig. 10.2). The main rivers and creeks in the system are the Macquarie River (numerous segments), Monkeygar Creek, The Breakaway, Bulgeraga Creek, Monkey Creek, Oxley Break, Buckiinguy Break, Bora Creek, Gum Cowal and Terrigal Creek.

The hydrology and geomorphology of the Macquarie system is similar to several other inland floodplain wetlands in the Murray-Darling Basin (Ralph 2008), although all the systems have been altered by floodplain development and/or river regulation and diversions. Water flow along the Macquarie River has a long history of natural variability, despite a reduction in flow variation since large-scale regulation began with the construction of Burrendong Dam in 1967 (Kingsford and Thomas 1995). In particular, intraseasonal, interannual and interdecadal climatic and hydrological variability in Australia are known to be strongly influenced by Indo-Pacific fluctuations in sea surface temperature, air pressure and rainfall, related to the Madden-Julian Oscillation (MJO; Donald *et al.* 2006; Hendon and Liebmann 1990), El niño-Southern Oscillation (ENSO; Chiew *et al.* 1998; Walker *et al.* 1995), Interdecadal Pacific Oscillation (IPO; Power *et al.* 1999; Verdon and Franks 2005; Verdon *et al.* 2006) and Pacific Decadal Oscillation (PDO; Power *et al.* 1999; Verdon *et al.* 2004), and consequently influence biota within Australia (Norman and Nicholls 1991; Thoms *et al.* 2007). Nevertheless, the maintenance of biota and ecological communities within the Macquarie Marshes relies on the continuance of adequate water availability and a variable flow regime in the Macquarie River.

Literature review

The supply of water to the Macquarie Marshes is essential for its long-term survival and maintenance and the flood pulse is the primary driving force controlling biota within floodplain wetlands (Junk *et al.* 1989). Each flood pulse is unique and the response of biota will vary, depending on characteristics of the flood pulse. The water regime may be defined as the prevailing pattern of flood pulses over a period of time. We primarily aimed to compile and synthesise the scientific literature regarding the water requirements of key species of flora and fauna with specific reference to aspects of the water regime, including flood frequency, duration, depth, timing, interflood dry period and the rates of flood rise and fall. Biotic factors that relate to the water regime were also considered to be important, especially for waterbirds and fish. These included germination timing, reproduction stimuli, season and duration and waterbird breeding lag time, which refers to the period of time taken for waterbirds to build up fat reserves, establish mating pairs and construct nests before commencing breeding.

Both peer-reviewed and non-peer-reviewed scientific literature were used to compile information on the water requirements of relevant aquatic biota. Information on the water requirements of a species rarely conflicted in the different sources of literature. When conflict did occur, consideration was given to the reliability of the data (peer-reviewed literature was prioritised over non-peer-reviewed work) and the geographic distribution of the data (literature based on studies at sites with environmental characteristics similar to the Macquarie Marshes) was prioritised. Water requirements were compiled for 35 plant species, 52 waterbirds and 11 fish (Table 10.1).

Climate change projections

General circulation models (GCMs) are regarded as the best tools for generating large-scale scenarios that reflect the projected anthropogenic climate change (Bates *et al.* 1998; Oelschlägel 1995). In fact, Randall *et al.* (2007) state:

Table 10.1: Floodplain and wetland plants, waterbirds and fish included in the literature review. Species were selected due to their occurrence at the Macquarie Marshes

Type	Functional group	Scientific name	Common name
Plants[a]	Trees	*Eucalyptus camaldulensis*	River red gum
		Eucalyptus largiflorens	Black box
		Eucalyptus coolabah	Coolibah
		Acacia pendula	Weeping myall
		Acacia stenophylla	River cooba
	Shrubs	*Muehlenbeckia florulenta*	Lignum
	Grasses	*Paspalum distichum*	Water couch
		Phragmites australis	Common reed
	Sedges and rushes	*Typha orientalis*	Broadleaf cumbungi
		Typha domingensis	Narrow-leaved cumbungi
		Juncus flavidus	Gold rush
		Juncus usitatus	Billabong rush
		Juncus aridicola	Tussock rush
		Bolboschoenus caldwellii	Marsh club rush
		Cyperus bifax	Downs nutgrass
		Cyperus concinnus	Trim flat-sedge
		Cyperus difformis	Rice sedge
		Cyperus rigidellus	Curly flat-sedge
		Eleocharis acuta	Common spike-rush
		Eleocharis pallens	Pale spike-rush
		Eleocharis plana	Flat spike-rush
		Eleocharis pusilla	Small spike-rush
		Eleocharis sphacelata	Tall spike-rush
	Aquatic macrophytes	*Vallisneria species*[b]	Vallisneria
	Herbs and forbs	*Isotoma auxillaris*	Rock isotome
		Isotoma fluviatilis	Swamp isotome
		Ludwigia peploides	Water primrose
		Marsilea costulifera	Narrow-leaf nardoo
		Marsilea drummondii	Common nardoo
		Nymphoides crenata	Wavy marshwort
		Pratia concolor	Poison pratia
		Pratia purpurascens	Whiteroot
		Ranunculus inundatus	River buttercup
		Ranunculus pumilio	Ferny buttercup
		Ranunculus undosus	Swamp buttercup

Type	Functional group	Scientific name	Common name
Waterbirds[c]	Fish-eaters	*Podiceps cristatus*	Great crested grebe
		Poliocephalus poliocephalus	Hoary-headed grebe
		Tachybaptus novaehollandiae	Australasian grebe
		Pelecanus conspicillatus	Australian pelican
		Anhinga melanogaster	Darter
		Phalacrocorax carbo	Great cormorant
		Phalacrocorax melanoleucos	Little pied cormorant
		Phalacrocorax sulcirostris	Little black cormorant
		Phalacrocorax varius	Pied cormorant
		Ardea alba	Great egret
		Ardea intermedia	Intermediate egret
		Egretta garzetta	Little egret
		Ardea pacifica	Pacific heron
		Egretta novaehollandiae	White-faced heron
		Nycticorax caledonicus	Rufous night heron
		Larus novaehollandiae	Silver gull
		Chlidonias hybridus	Whiskered tern
		Sterna caspia	Caspian tern
		Sterna nilotica	Gull-billed tern
	Deep-water foragers	*Cygnus atratus*	Black swan
		Aythya australis	Hardhead
		Biziura lobata	Musk duck
		Oxyura australis	Blue-billed duck
		Fulica atra	Eurasian coot
	Dabbling ducks	*Anas castanea*	Chestnut teal
		Anas rhynchotis	Australasian shoveler
		Anas superciliosa	Pacific black duck
		Malacorhynchus membranaceus	Pink-eared duck
		Stictonetta naevosa	Freckled duck
	Grazing waterfowl	*Chenonetta jubata*	Maned duck
		Dendrocygna eytoni	Plumed whistling-duck
		Tadorna tadornoides	Australian shelduck
	Shoreline foragers	*Gallinula ventralis*	Black-tailed native hen
		Porphyrio porphyrio	Purple swamphen

Type	Functional group	Scientific name	Common name
		Vanellus miles	Masked lapwing
		Vanellus tricolor	Banded lapwing
	Large waders	*Platalea flavipes*	Yellow-billed spoonbill
		Platalea regia	Royal spoonbill
		Plegadis falcinellus	Glossy ibis
		Threskiornis molucca	Australian white ibis
		Threskiornis spinicollis	Straw-necked ibis
		Grus rubicunda	Brolga
	Small waders	*Himantopus himantopus*	Black-winged stilt
		Recurvirostra novaehollandiae	Red-necked avocet
		Calidris acuminata	Sharp-tailed sandpiper
		Charadrius ruficapillus	Red-capped plover
		Elseyornis melanops	Black-fronted dotterel
		Erythrogonys cinctus	Red-kneed dotterel
		Limosa limosa	Black-tailed godwit
		Tringa nebularia	Common greenshank
		Tringa stagnatilis	Marsh sandpiper
		Stiltia isabella	Australian pratincole
Fish	Fish	*Craterocephalus stercusmuscarum*	Un-specked or fly-specked hardyhead
		Nematalosa erebi	Bony herring
		Hypseleotris species	Carp gudgeons
		Philypnodon grandiceps	Flathead gudgeon
		Melanotaenia fluviatilis	Murray-Darling rainbowfish
		Maccullochella peelii	Murray cod
		Macquaria ambigua	Golden perch
		Tandanus tandanus	Freshwater catfish
		Retropinna semoni	Australian smelt
		Bidyanus bidyanus	Silver perch
		Leiopotherapon unicolor	Spangled perch

a: Plant functional classification based on that used by Roberts and Marston (2000).

b: The taxonomy of Vallisneria species in NSW is problematic. Specimens have previously been referred to as *V. gigantea*, *V. spiralis* or *V. americana* (Jacobs and Frank 1997).

c: Waterbird functional classification based on that used by Roshier *et al.* (2002).

There is considerable confidence that Atmosphere–Ocean General Circulation Models provide credible quantitative estimates of future climate change, particularly at continental scales and above.

Despite a greater consensus on the reliability of GCMs at a global scale, climate models still show significant errors. The most significant is related to the incorporation of regional and small-scale processes within GCMs. These limitations may be attributed to limitations in computing power and scientific understanding of some physical processes (Randall *et al.* 2007). Other techniques, such as down-scaling, are employed to analyse climate change at regional and local scales. Considerable research effort is being directed towards down-scaling GCMs so that they are suitable for climate change impact and adaptation studies at a regional or local scale (Thatcher and McGregor 2008; Timbal 2004; Timbal *et al.* 2009; Timbal and Jones 2008). The Intergovernmental Panel on Climate Change (IPCC) Fourth Assessment Report indicated that the application of down-scaled models is the current direction of climate change impacts research (Carter *et al.* 2007).

Application of both statistical and dynamical down-scaled models for hydrological applications should be undertaken cautiously due to mismatches between climate model outputs and the spatial scale of hydrological models (Wood *et al.* 2004). In particular, climate models are typically grid-based, while hydrological models reflect drainage basins and are commonly linear-shaped. Both statistical and dynamical down-scaling scenarios are not available for application within the Macquarie Marshes region. It is therefore inappropriate to directly apply current climate models to ecological impact and adaptation research in the Macquarie Marshes.

Despite the lack of down-scaled projections for application into impact and adaptation studies in the Macquarie Marshes, the consistency of models at global, national and regional scales with regard to temperature indicate that warming is likely to occur at the Macquarie Marshes (CSIRO and BOM 2007; Hennessy *et al.* 2004; Meehl *et al.* 2007). While there may be minor decreases in temperature at regional scales, due to the confounding impacts of increased precipitation on temperature, it is more likely that most regions in Australia will experience some degree of warming (CSIRO and BOM 2007). Rainfall trends are not as consistent as temperature, with some models projecting increases in rainfall and others significant decreases, but at a national level there is a 60% probability that rainfall will decrease by 2070. Droughts are likely to increase in frequency (CSIRO and BOM 2007; Meehl *et al.* 2007). Even at a regional level (NSW) there is a greater tendency for rainfall to decrease by 2070 with 'business as usual' projections ranging from –40% to +20% (Hennessy *et al.* 2004). Approximately 75% of the GCMs used by Hennessy *et al.* (2004) indicate that rainfall will decrease in central-western NSW, compared to approximately 60% of models indicating that rainfall will decrease in coastal and western NSW.

Previous studies in the region of the Macquarie Marshes also present consistent projected climate change impacts (CSIRO 2008; Herron *et al.* 2002; Jones *et al.* 2001, 2002; Jones and Page 2001; O'Neill *et al.* 2004; Pollino *et al.* 2007; Tighe *et al.* 2007a). The most pertinent change is the projected reduced water availability for the Macquarie Marshes. This reduced water availability will be reflected in changes to the water regime such as reduced flood frequency and variability (CSIRO 2008; Hassall and Associates 1998; Pollino *et al.* 2007; Tighe *et al.* 2007a). In particular, it is likely that base flows to the Macquarie Marshes may increase due to current and/or future development of water resources, while large floods will occur less frequently (CSIRO 2008; Hassall and Associates 1998; Pollino *et al.* 2007; Tighe *et al.* 2007a).

Detailed analysis of the impacts of climate change on aquatic biota in the Macquarie Marshes did not form part of this study, for two reasons:

- down-scaled (both dynamical and statistical) climate models at a scale suitable for application at the Macquarie Marshes are not yet available;
- detailed elevation models of floodplain topography in the system and models of the likely lateral distribution of flooding across the Macquarie Marshes are not yet available.

Nevertheless, based on the water requirements of biota it is possible to provide preliminary predictions of the vulnerability and adaptation of aquatic biota to climate change. By applying the consistent trends of a decrease in all aspects of the water regime, particularly flood frequency and flood variability, consideration can be given to the adaptive capacity of biota to climate change.

Climate change adaptations in the Macquarie Marshes

Geomorphology

Low-sinuosity distributary channels and shallow marsh channels convey water to perennial and ephemeral wetlands in the Macquarie Marshes, while extensive unchannelised areas occur on the floodplain. Surface drainage lines, low levees, floodbasins, depressions and gilgai (mounds and hollows generated by the shrink–swell behaviour of clays in episodically wet locations) are important geomorphic units (Ralph 2008). These features are related to processes of channel and floodplain sedimentation and erosion, and affect the amount of overbank flooding and the distribution of floodwaters in the system, particularly during small–moderate events (Ralph 2008). Many channels break down completely in the wetlands where sediment is trapped by vegetation, while some feed water into small lagoons. Floodwaters also tend to reconverge through small channels at the downstream limits of the wetlands. In other cases channels do not reform and the floodwaters dissipate on land and/or pond in shallow depressions, where they evaporate.

The impacts of climate change on the geomorphology of the Macquarie Marshes have not previously been studied, although a few conditions may still be discussed based on our understanding of the contemporary system and recent historical changes related to the water regime. In particular, the character and behaviour of marsh channels and antecedent floodplain conditions are likely to change in a climate with less rainfall and greater evapotranspiration and a hydrological regime with reduced large flood frequency and flood variability. A reduction in overbank flooding due to a probable reduction in moderate or large floods would result in greater drying of the floodplain and extensive deep cracking of gilgai on the peripheries of the wetlands and in unchannelised areas. The extent of semi-permanent wetlands would probably also decrease. An increase in the proportion of low-level flows in channels due to the ongoing effects of river regulation may encourage in-channel erosion where bank sediments are no longer armoured by vegetation and can be undercut and excavated. This would lead to the addition of sediment into the system, which would accumulate at the ends of the channels in the core wetlands. The frequency and extent of overbank flooding may decrease further due to excessive channelisation, while channel incision may lead to a faster throughput of floodwaters in the system and less ponding on the floodplain. When a flood could inundate the broader floodplain, significantly more water would be required to achieve the same extent and duration of flooding under drier antecedent conditions.

Vegetation

Brock and Casanova (1997) described the impact of increasing and decreasing aspects of the water regime on plants, such as a decrease in flood frequency or an increase in interflood drying. By applying these descriptions to preliminary estimates of the impact of climate change on the water regime in the Macquarie Marshes, it is evident that there will be an increase in the area of floodplain wetlands that will be permanently dry. This will encourage the encroachment of dryland species. Indeed, encroachment of dryland species has already occurred in response to regulation of the Macquarie River (Bowen and Simpson 2008; Kidson *et al.* 2000). Between 1946, when Burrendong Dam was commissioned on the Macquarie River, and 1997 there was a 42% decrease in the extent of semi-permanent wetland and a 40% decrease in ephemeral wetland (Hassall and Associates 1998). It is likely that reductions in the extent of floodplain wetland vegetation will continue in the future.

A climate change-induced decrease in flood frequency may significantly alter the distribution of plant species and associations in the Macquarie Marshes. Trees, particularly river red gum (*Eucalyptus camaldulensis*), will likely exhibit signs of stress and a decline in reproductive ability. Since river red gum does not maintain a soil seedbank, water stress resulting from reduced flood frequency severely limits its ability to regenerate and maintain viable populations. Black box (*Eucalytpus largiflorens*) may be able to store seed within its canopy and release seed when conditions are suitable, but water stress may reduce the number and viability of seeds and limit its adaptation to the altered water regime. Both species may have some capacity to migrate to areas with ideal flood frequencies, but their limited reproductive capacity would significantly decrease the rate of migration and the ability of trees to contract to core floodplain wetland areas. Even though river red gum and black box may survive under reduced flood frequencies, maintenance of populations will be severely limited and will result in the death of trees as they age. Conditions will be more suitable for dryland species such as bimble box (*Eucalyptus populnea*), belah (*Casuarina cristata*) and wilga (*Geijera parviflora*). A decrease in flood variability is likely to change the flood regime from variable to more static. This will favour species that are adapted to more permanently wet or more permanently dry habitats and is likely to significantly reduce the frequency of large floods. The response of trees to a less variable water regime is largely unknown, and varies between species. If trees can adapt to reduced flood frequency and relocate to core habitat areas where their water requirements are met, populations may be maintained. However, it is more likely that reduced flood variability will limit the number of large floods which inundate floodplain tree species, resulting in conditions more suitable for dryland trees.

Unlike floodplain trees, lignum (*Muehlenbeckia florulenta*) exhibits smaller, sparser cover in the absence of frequent inundation, yet responds opportunistically to flooding with lush vegetative growth and prolific seed production (Chong and Walker 2005; Thoms *et al.* 2007). Provided that flooding occurs at least every 10 years (Craig *et al.* 1991), populations may be maintained in the short term. However, reliance on one flood every 10 years to maintain lignum populations is hazardous, particularly if seed germination is limited due to other factors such as frost or low temperatures. Since seeds may be dispersed by floodwater, there is some capacity for contraction of lignum to core wetland areas. However, as lignum is typically located at a mid-floodplain elevation, it may be severely affected by a reduction in flood variability. A reduced frequency of flooding of lignum will limit opportunities for regeneration. A reduction in flood variability may also increase the frequency of smaller flood events. Individuals located at lower elevations may receive flooding at higher than ideal frequencies and will be exposed to increased competition from species adapted to higher flood frequencies. In fact, reduced flood variability as a consequence of river regulation may be the cause of observed reductions in lignum extent in the Macquarie Marshes (Bowen and Simpson 2008).

Grasses, sedges and rushes, and herbs and forbs occur at the ecotone between frequently wet and frequently dry environments and are regarded as amphibious species (Brock and Casanova 1997). Upper areas of the ecotone will be flooded less frequently and may no longer be regarded as part of the ecotone, shrinking the size of this zone. While these ecotone habitats will be maintained, contraction in size will increase competition between plants for space and resources. Species that are able to expand vegetatively, such as common reed (*Phragmites australis*), may migrate to remain within the ecotone, independent of its requirements for regeneration by seed. In accordance with the environmental sieve model (van der Valk 1981) of succession, species that maintain a soil seedbank, such as nardoo (*Marsilea drummondii*), can respond opportunistically to flood events despite increased competition for space and resources. These species may appear absent and will regenerate when water requirements for germination are suitable. Dispersal-dependent species may be severely restricted in their ability to adapt to the smaller ecotone. Flood frequency is essential for the long-term maintenance of these populations, as they are reliant on flooding for regeneration.

Likewise, reduced flood variability will favour species adapted to more static water levels. For example, while common reed prefers fluctuating water levels, it commonly grows in static waters at depths of 45 ± 20 cm (Coops *et al.* 1996) or 20 cm above the waterline to 50 cm below (Haslam 1970). Reduced flood variability also favours species that require flooding for greater periods. Some species, such as marshwart (*Nymphoides crenata*) and cumbungi (*Typha* species), will be maintained within more static conditions.

Provided that river flows continue within watercourses, the impact of reduced flood frequency on submerged aquatic macrophytes is limited. Populations will be sustained, but opportunities to expand into new habitats may be limited. Similarly, submerged aquatic macrophytes will be favoured when flood variability is reduced, particular if base water flows increase.

Waterbirds

The adaptive capacity of waterbirds to flow variability is particularly evident when comparing Australian waterbirds to those in the northern hemisphere. Australian waterbirds tend not to have clearly predictable patterns of movement, feeding ecology, reproductive ecology, moult and habitat use (Kingsford and Norman 2002). Rather, they exhibit many aspects of opportunism. In particular, many waterbirds exhibit great dispersal capabilities, and are able to search for ideal wetland patches within the fluctuating mosaic of wetlands in arid and semi-arid Australia (Roshier *et al.* 2001). Strong relationships have been established between waterbird abundance and climatic variability (Thoms *et al.* 2007). In fact, the response of waterbirds to flooding and their innate ability to track floodwaters has long fascinated bird watchers in Australia (e.g. the account of the silver gull, Hobbs 1961).

As waterbirds commonly rely on flooding to stimulate reproduction (Briggs 1990), climate change-induced alterations to the water regime, particularly a reduction in flood frequency and variability, may have significant implications for the reproduction of waterbirds and maintenance of population numbers, particularly at a local scale. Due to the relationship between large flood events and floods of long duration, reduced flood frequency is likely to negatively affect waterbirds that require long lag times (≥ six months) to enhance their physical condition or that have long breeding durations (≥ four months). Reduced frequency may severely affect waterbirds that require long flood durations due to extremely long lag times, breeding durations or both (nine to12 months). This group of waterbirds includes the colonial nesting waterbirds known to breed in the Macquarie Marshes. Waterbirds with relatively short life expectancies, such as ducks, and resident breeding waterbirds need to maximise breeding opportunities to maintain population numbers. A decrease in flood frequency will reduce the number of breeding opportunities for these waterbirds and may severely affect the long-term stability of their populations.

A reduction in flood variability and associated increase in low-flow events that are likely to have shorter flood durations will favour waterbirds that require short or no lag times. Waterbirds such as the whiskered tern, gull-billed tern, black swan, grey teal, Pacific black duck, Australian shelduck and brolga respond rapidly to flooding, with breeding commencing within one or two months of flooding. Similarly, waterbirds that exhibit short breeding durations of one to two months, such as the great crested grebe, hoary-headed grebe, whiskered tern, Eurasian coot and glossy ibis, are likely to adapt to a reduction in flood variability. Waterbirds that exhibit short or no lag times and have short breeding durations are well placed for successful breeding should the water regime be altered according to climate change projections. Alternatively, reduced flood variability will result in less interflood drying in some parts of the Macquarie Marshes; this may affect the breeding success of waterbirds that rely to some extent on interflood drying (Crome 1986, 1988), including some colonial nesting waterbirds, ducks and terns.

Many waterbirds, including most of the colonial nesting waterbirds and some ducks, require flooding near or around their nests to commence breeding (Higgins and Davies 1996; Marchant and Higgins 1990, 1993). Reduced flood frequency and variability is likely to decrease the occurrence of suitable conditions for breeding and limit the breeding opportunities for these waterbirds.

Since waterbirds are heterotrophs and are heavily reliant on other biota for their survival, climate change is also likely to affect waterbirds indirectly due to alterations in the quality and quantity of habitats and consumed items. For example, floodplain trees such as river red gum and black box have limited capacity for adapting to reduced flood frequency and variability. Some waterbirds, such as the little pied cormorant, little black cormorant, cattle egret, little egret, intermediate egret, great egret, white-necked heron, rufous night heron, yellow-billed spoonbill, royal spoonbill and Australian white ibis, prefer to breed in live river red gum. Others, such as the darter, great cormorant, pied cormorant and Pacific heron, prefer to breed in dead river red gum (Briggs *et al.* 1997; Leslie 2001; Marchant and Higgins 1990). As reduced flood frequency may result in the loss of vigour and the death of river red gum, available breeding habitat for waterbirds that breed in dead river red gum may temporarily increase. There will be a corresponding decline in breeding habitat for waterbirds that utilise live river red gum. Similarly, loss of vigour of floodplain trees may result in the generation of more hollows within floodplain trees, thereby increasing available breeding habitat for many ducks that nest within hollows (Marchant and Higgins 1990). However, due to the continued decomposition of dead floodplain trees, the generation of these habitats will only be of short-term benefit; the long-term result will be a net decline in available habitat for waterbirds using both live and dead floodplain trees.

Lignum is used as a nesting site for some waterbirds, such as the hoary-headed grebe, Australian pelican, whiskered tern, hardhead, musk duck, freckled duck, black-tailed native hen, glossy ibis, straw-necked ibis and black-winged stilt (Higgins and Davies 1996; Marchant and Higgins 1990). As lignum responds rapidly to flooding, as long as flooding occurs at least every 10 years, it is possible that lignum will be sustained and there will be only slight changes to this breeding habitat.

Apart from the dispersal-dependent species, many wetland and aquatic plants exhibit some adaptive capacity, such as vegetative expansion or seedbanks that may enable migration of these habitats to core wetland areas where flood frequency requirements are maintained. Should these plants adapt to climate change by migrating to core wetland areas, it is probable that foraging and/or breeding habitat for waterbirds such as the great crested grebe, hoary-headed grebe, Australasian grebe, great cormorant, intermediate egret, rufous night heron, whiskered tern, black swan, hardhead, musk duck, blue-billed duck, Eurasian coot, chestnut teal, grey teal, purple swamphen, yellow-billed spoonbill, royal spoonbill, straw-necked ibis, brolga and black-winged stilt may be maintained, albeit at a diminished level (Higgins and Davies 1996; Marchant and Higgins 1990, 1993).

As reduced flood variability will favour plant species adapted to more static water conditions, such as some grasses, sedges and rushes, breeding of waterbirds that utilise these habitats may be enhanced. Waterbirds that prefer to breed on more static water levels, such as the great crested grebe, hoary-headed grebe, Australasian grebe, great cormorant, silver gull and black swan, may exhibit little change in breeding success (Higgins and Davies 1996; Marchant and Higgins 1990). Waterbirds that prefer to forage in areas with more static water conditions may also adapt to changes in the water regime. In particular, the fish-eating functional group of birds described by Roshier *et al.* (2002), may adapt, as the more permanent water conditions established with more persistent base flows are conducive to the establishment and maintenance of fish populations. Similarly, deep-water foragers may exhibit some resilience to climate change as their foraging habitats may increase due to the increased base flows.

Fish

Reduced flood frequency is more likely to negatively affect the more ephemeral habitats which are utilised opportunistically by native fish, such as the floodplain and to some extent the wetlands and billabongs, and to influence fish that prefer these habitats, such as the fly-specked hardyhead, bony herring, carp gudgeon, golden perch, freshwater catfish, Australian smelt, silver perch and spangled perch. However, as all reviewed fish species use the main channel during their life-cycles as part of their natural adaptation to a variable flood regime, they exhibit some capacity to adapt to reduced flood frequency in the Macquarie Marshes. The main channel is likely to act as a refuge for fish during periods of low flow, while localised ponding and backflooding of channels and riparian areas near weir pools may provide alternative habitats for spawning.

The main channel may act as a refuge and ensure the survival of fish during periods of low flow, but maintenance of fish populations depends on the availability of suitable spawning habitats and flow conditions which provide critical feeding opportunities and a larval dispersal mechanism. Fish species that are more characteristic of larger waterbodies, such as the bony herring, Murray cod, golden perch, silver perch and spangled perch, are reliant on increasing flows to stimulate rigorous and widespread spawning. These species are likely to be vulnerable to flow regime changes, specifically a reduction in large floods. The bony herring and spangled perch are particularly vulnerable, as they have short life-spans of approximately three years and therefore have limited opportunities to spawn during prolonged droughts of periods without flow.

Reduced flood variability, which is likely to be characterised by increased base flows and a reduced frequency of large floods, may have varying impacts on the fish assemblages in the Macquarie Marshes. A reduction in flood variability is likely to promote the survival of fish within the main channel, particularly those that spawn there and do not require an increase in flows as a stimulus. These conditions are particularly suitable for introduced species such as carp (*Cyprinus carpio*), gambusia (*Gambusia holbrooki*) and goldfish (*Carassius auratus*). However, the reduced frequency of large floods will limit opportunities for large-bodied native species to spawn and use the floodplain and wetlands opportunistically.

Food webs and trophic linkages

As the indirect impacts of climate change are mediated through hydrology, climate change will have a flow-on effect throughout all components of the Macquarie Marshes, particularly the interactions between biota and hydrology (Fig. 10.1). In fact, the impact of climate change on lower trophic-level species may uncouple the trophic linkages between biota and cause significant changes to high-level species and the ecology of an ecosystem (Winder and Schindler 2004). In particular, a reduction in floodplain inundation frequency will affect nutrient inputs, which are the building-blocks of productive food webs. For example, the boom of waterbird and fish populations on the River Murray has been attributed to the pulse of biota at lower trophic levels, such as macro-invertebrates (Jenkins *et al.* 2005). While this chapter focuses on floodplain and wetland plants, waterbirds and fish, it is essential for flooding to maintain

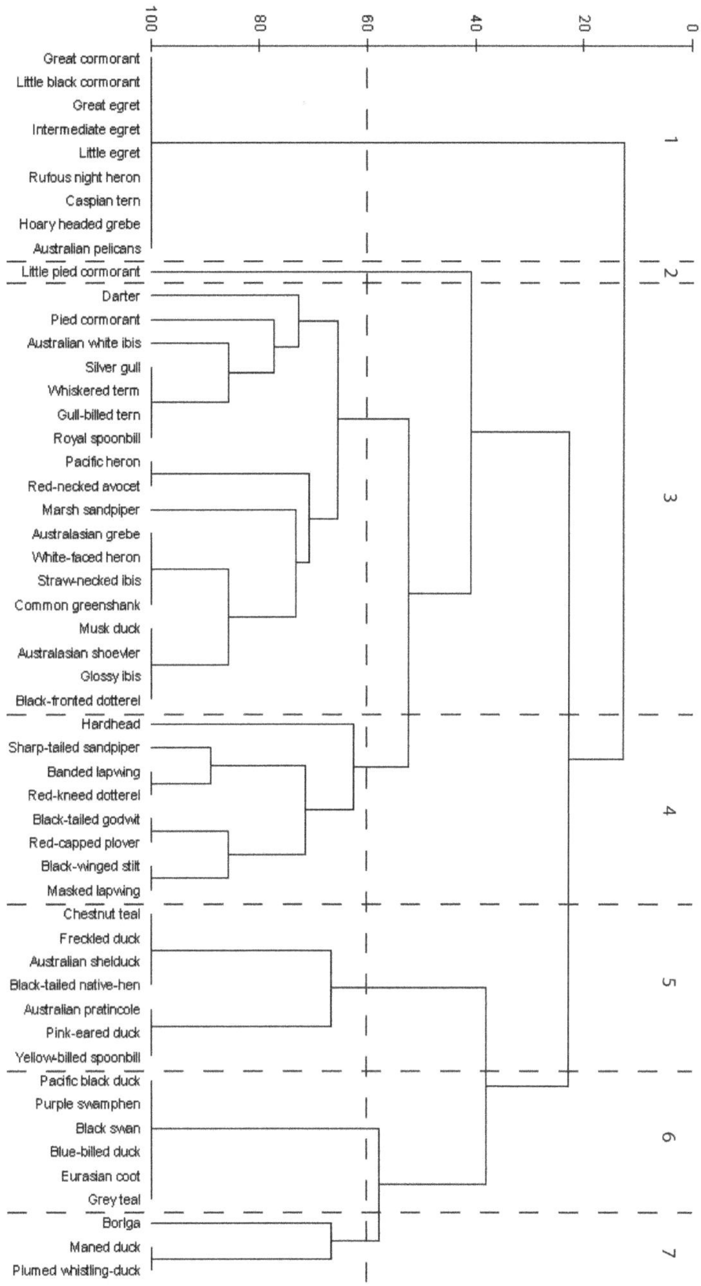

Figure 10.3: Cluster analysis, using the Bray-Curtis similarity index, of preferred dietary items of waterbirds. At the 60% similarity level, seven groups were evident. 1 = fish-eaters, waterbirds whose diet primarily consists of fish. 2 = crustacean-eaters, waterbirds whose diet primarily consists of crustaceans. 3 = indiscriminate carnivores, waterbirds whose diet is primarily carnivorous but includes a range of fauna items, including fish, insects, crustaceans, molluscs and worms. 4 = indiscriminate omnivores, waterbirds whose diet includes a range of fauna items and aquatic plants. Interestingly, waterbirds did not exhibit a preference for both fish and aquatic plants, therefore no waterbirds in this group consumed fish as a preferred food item. 5 = insect-eaters, waterbirds whose diet primarily consists of insects but may include some aquatic plants. 6 = aquatic plant-eaters, waterbirds whose diet primarily consists of aquatic plants. 7 = indiscriminate grazers, waterbirds whose diet consists primarily of aquatic and terrestrial plants.

sources of primary production and population abundance and diversity at lower trophic levels if biota are to adapt to climate change. Jenkins *et al.* (2005) indicated that an absence of flooding beyond 20 years will markedly reduce ecosystem productivity and diversity on the River Murray.

Cluster analysis, using a Bray-Curtis similarity index, of the dietary requirements of waterbirds (Fig. 10.3) indicates that waterbirds with more indiscriminate dietary requirements, such as the Australasian grebe, straw-necked ibis, marsh sandpiper, silver gull, hardhead and red-kneed dotterel, are likely to be more resilient to ecosystem changes in the Macquarie Marshes. Those with more specific diets are more vulnerable to ecosystem changes in the Macquarie Marshes. For example, should changes to the water regime in the Macquarie Marshes result in a collapse of the fish populations, it is probable that there will be a corresponding decline in the population of fish-eaters. Similarly, a decline in the extent and quality of aquatic plants may reduce the feeding capacity of aquatic plant-eaters.

Biophysical linkages and climate change

The climate and hydrology of the Macquarie River and its catchment and the resultant flow regime in the semi-arid lowland reaches drive geomorphic changes in the Macquarie Marshes, including channel and floodplain development. Potential changes to the frequency, magnitude and duration of overbank flooding and the shape and size of marsh channels in future climate change scenarios would probably lead to changes in the distribution and extent of floodwaters in the Macquarie Marshes. Flood-reliant species and ecological communities would also be affected by habitat changes related to these geomorphic adjustments.

Floodplain wetland biota display distinct species responses to flooding and exhibit unique water requirements for survival, maintenance, reproduction and regeneration. Since flooding is the primary factor controlling the distribution of species within floodplain wetlands, it is evident that the response of plants to flooding markedly influences the distribution of plants within a floodwater gradient and promotes zonation of plant species and plant associations, and consequently influences the distribution of fauna along a floodwater gradient. Floodplain wetland biota exhibit population booms (Kingsford *et al.* 1999) in response to increased water availability. Flooding or increasing flow is recognised as a primary stimulus for waterbird breeding and fish spawning, with reproductive success dependent on characteristics of the flood and the associated regional climate.

Conceptual models are a means of evaluating, synthesising and organising existing understanding of the key ecological components, processes and linkages within an ecosystem (Ogden 2005). Conceptual models have been developed using the CSIRO and BOM (2007) projections of reduced water availability by 2070 (Figs 10.4–10.7/colour plates, pp. 309–312) to illustrate the ecological risk to biota and geomorphology from climate change, and the adaptive capacity of biota and geomorphology to respond to climate change. Ecological risk is particularly prevalent by 2070 and it is likely that the core wetland area will contract during the wet phase from its size in the current conditions. It is also probable that there will be considerable loss of floodplain vegetation and fewer seed and egg banks in the soil, due to poor conditions and reduced opportunities for breeding. Reductions in nutrient cycling and nutrient release during the wet phase may also occur due to a reduction in flooding, which will cause a decrease in the size and frequency of invertebrate population booms. There may be a decline in the use of the floodplain by some waterbirds during the wet phase. During the dry phase, use of the floodplain by terrestrial animals may decrease due to smaller and fewer watering points. There may be greater opportunity for the expansion of dryland plants and woody weeds during the dry phase. While the models show that there is significant risk to the Macquarie Marshes, there is also likely to be some adaptation by the biota to the hydrological changes induced by climate change. This is primarily illustrated through the movement of aquatic vegetation to core wetland areas. Despite the reduction in use of the floodplain by some waterbirds, other waterbirds may find the altered conditions suitable for foraging and breeding and consequently increase their use of the wetlands. Floodplain vegetation may undergo some loss or decline in health, but some vegetation species may exhibit resilience due to their capacity to store seeds in the seedbank or within their canopies.

Climate change impacts modelling

A decision support system, known as Exploring CLimAte Impacts on Management (EXCLAIM) (Tighe *et al.* 2007b; Fu *et al.* 2009) has been developed for the Central West Catchment Management Authority to help assess the likelihood of meeting natural resource condition targets under a range of Intergovernmental Panel on Climate Change projected climate change scenarios. EXCLAIM relies on a Bayesian network modelling approach, which presents findings as probabilities and uses a range of information, including data (numerical values) or expert opinion. This type of model can cope well with the uncertainty inherent in natural ecosystems (Tighe *et al.* 2007b).

The Bayesian network that underpins EXCLAIM is a conceptual model of the impacts of climate change on the Macquarie River and Macquarie Marshes (Tighe *et al.* 2007b). Published and unpublished data relating to climate, hydrology and ecological factors were incorporated into the model, although much of the data were incomplete and of varying quality (Tighe *et al.* 2007b). Efforts were made to account for knowledge gaps and the model was designed to be reassessed and updated as new information became available. Version 1 of EXCLAIM provides ecological outputs for four key indicators of wetland health:

1 waterbird breeding event (probability of colonial nesting waterbird breeding occurring, according to a selected IPCC scenario);
2 lippia weed response (probability of lippia growth being sustained);
3 aquatic/terrestrial emergent plant response (probability of growth being sustained);
4 river red gum average growth rate (probability of river red gum growth rate exhibiting a decrease, slight increase, medium increase or large increase).

These four ecological indicators and a general water quality indicator are used to predict the likely health of the Macquarie Marshes under different climate change scenarios (Tighe *et al.* 2007b).

Conclusions

This chapter has highlighted the unique water requirements of individual species in the Macquarie Marshes. It is evident that ideal water conditions for one species may be less than ideal for another species. For example, supplying water to the Macquarie Marshes to maximise river red gum growth may not account for the differing water requirements of lignum and may therefore have a detrimental impact on lignum health over extended periods. Similarly, supplying water to meet the water needs of aquatic/terrestrial emergent plants may not supply the conditions necessary for flow-dependent fish to spawn. The literature review undertaken for this study is being used to increase EXCLAIM's capacity to model the response of a range of biota to changing water regimes related to climate change in the Macquarie Marshes. Version 2 of EXCLAIM (Fu *et al.* 2009) has been commissioned by the NSW Department of Environment, Climate Change and Water, and the Central West Catchment Management Authority. It incorporates ecological indicators/outputs based on probabilities developed from this literature review (Fu and Pollino 2009).

Further climate change impacts modelling and monitoring of key components and processes of floodplain wetlands is essential to increase our capacity to predict the response and adaptations of floodplain wetland biota to climate change. This modelling may be applied to habitat distributions, based on current mapping, to forecast the spatial response of biota to climate change. Changes to the trophic linkages and food webs within the wetland ecosystem will be harder to gauge due to their complexity and the lack of data explaining trophic dynamics. Nevertheless, this will enable management actions that promote the protection of significant ecological habitats in the Macquarie Marshes.

Acknowledgments

The authors wish to thank Sue Cuddy, Carmel Pollino and Baihua Fu from the Integrated Catchment Assessment and Management centre, Australian National University, for their assistance in developing this manuscript. Their willingness to help is greatly appreciated.

References

Bates BC, Charles SP and Hughes JP (1998). Stochastic downscaling of numerical climate model simulations. *Environmental Modelling and Software* **13**, 325–331.

Bernstein L, Bosch P, Canziani O, Chen Z, Christ R, Davidson O, Hare W, Huq S, Karoly D, Kattsov V, Kundzewicz Z, Liu J, Lohmann U, Manning M, Matsuno T, Menne B, Metz B, Mirza M, Nicholls N, Nurse L, Pachauri R, Palutikof J, Parry M, Qin D, Ravindranath N, Reisinger A, Ren J, Riahi K, Rosenzweig C, Rusticucci M, Schneider S, Sokona Y, Solomon S, Stott P, Stouffer R, Sugiyama T, Swart R, Tirpak D, Vogel C and Yohe G (2007). Climate change 2007: synthesis report. Summary for policy makers. p. 22. Intergovernmental Panel on Climate Change, Spain, 2007.

BOM (2005). Australian climate monthly summary data. Commonwealth Bureau of Meteorology.

Bowen S and Simpson S (2008). Changes in the extent and condition of the vegetation communities of the Macquarie Marshes floodplain 1991–2008: Draft. NSW Dept of Environment and Climate Change, Sydney.

Briggs SV (1990). Waterbirds. In *The Murray.* (Eds N MacKay and D Eastburn). Murray-Darling Basin Commission, Canberra.

Briggs SV, Thornton SA and Lawler WG (1997). Relationships between hydrological control of river red gum wetlands and waterbird breeding. *Emu* **97**, 31–42.

Brock MA and Casanova MT (1997). Plant life at the edge of wetlands: ecological responses to wetting and drying patterns. In *Frontiers in Ecology: Building the Links.* (Eds NK Klomp and I Lunt) pp. 181–192. Elsevier Science, Oxford.

Carter TR, Jones RN, Lu X, Bhadwal S, Conde C, Mearns LO, O'Neill BC, Rounsevell MDA and Zurek MB (2007). New assessment methods and the characterisation of future conditions. In *Climate Change 2007: Impacts, Adaptation and Vulnerability. Contribution of Working Group II to the 4th Assessment Report of the Intergovernmental Panel on Climate Change.* (Eds ML Parry, OF Canziani, JP Palutikof, PJ van der Linden and CE Hanson) pp. 133–171. Cambridge University Press, Cambridge.

Chiew FHS, Piechota TC, Dracup JA and McMahon TA (1998). El Nino/Southern Oscillation and Australian rainfall, streamflow and drought: links and potential for forecasting. *Journal of Hydrology* **204**, 138–149.

Chong C and Walker KF (2005). Does lignum rely on a soil seed bank? Germination and reproductive phenology of *Muehlenbeckia florulenta* (Polygonaceae). *Australian Journal of Botany* **53**, 407–415.

Coops H, van den Brink FWB and van der Velde G (1996). Growth and morphological responses of four helophyte species in an experimental water-depth gradient. *Aquatic Botany* **54**, 11–24.

Craig AE, Walker KF and Boulton AJ (1991). Effects of edaphic factors and flood frequency on the abundance of lignum (*Muehlenbeckia florulenta* Meissner) (Polygonaceae) on the River Murray floodplain, South Australia. *Australian Journal of Botany* **39**, 431–443.

Crome FHJ (1986). Australian waterfowl do not necessarily breed on a rising water level. *Wildlife Research* **13**, 461–480.

Crome FHJ (1988). To drain or not to drain? Intermittent swamp drainage and waterbird breeding. *Emu* **88**, 243–248.

CSIRO (2008). Water availability in the Macquarie-Castlereagh. Report to the Australian Government from the CSIRO Murray-Darling Basin Sustainable Yields Project. CSIRO, Australia.

CSIRO and BOM (2007). Climate change in Australia. CSIRO and Bureau of Meteorology, Canberra.

DNR (2007). How do the New South Wales water reforms relate to wetlands and their management in NSW? NSW Dept of Natural Resources, Sydney.

Donald A, Meike H, Power B, Maia d HN, Wheeler M, White N, Stone RC and Ribbe J (2006). Near-global impact of the Madden-Julian Oscillation on rainfall. *Geophysical Research Letters* **33**, L09704. doi: 10.10292005GL025155.

Fu B and Pollino C (2009). EXCLAIM2 (EXploring CLimAte Impacts on Management). Working document on water quality and ecological health modelling. iCAM report 2009/02. Integrated Catchment Assessment and Management Centre, Fenner School of Environment and Society, Australian National University, Canberra.

Fu B, Pollino C, Andrews A and Cuddy S (2009). EXCLAIM2 (EXploring CLimAte Impacts on Management). Product Requirements and Prototype Specifications. iCAM report 2009/3. Integrated Catchment Assessment and Management Centre, Fenner School of Environment and Society, Australian National University, Canberra.

Haslam SM (1970). The performance of phragmites communis Trin. in relation to water-supply. *Annals of Botany* **34**, 867–877.

Hassall and Associates (1998). Climate change scenarios and managing the scarce water resources of the Macquarie River. Prepared for the Australian Greenhouse Office.

Hendon HH and Liebmann B (1990). The intraseasonal (30-50 day) oscillationof the Australiam summer monsoon. *Journal of Atmospheric Sciences* **47**, 2909–2923.

Hennessy K, Page CM, McInnes KL, Jones R, Bathols J, Collins D and Jones D (2004). Climate change in New South Wales. I. Past climate variability and projected changes in average climate. CSIRO Division of Atmospheric Research, Melbourne.

Herron N, Davis R and Jones R (2002). The effects of large-scale afforestation and climate change on water allocation in the Macquarie River catchment, NSW, Australia. *Journal of Environmental Management* **65**, 369–381.

Higgins PJ and Davies SJJF (1996). *Handbook of Australian, New Zealand and Antarctic Birds. Vol. 3: Snipe to Pigeons.* Oxford University Press, Melbourne.

Hobbs JN (1961). The birds of south-west New South Wales. *Emu* **61**, 21–55.

Jacobs SWL and Frank KA (1997). Notes on *Vallisneria* (Hydrocharitaceae) in Australia, with descriptions of two new species. *Telopea* **7**, 111–118.

Jenkins KM, Boulton AJ and Ryder DS (2005). A common parched future? Research and management of Australian arid-zone floodplain wetlands. *Hydrobiologia* **552**, 57–73.

Jones RN and Page CM (2001). Assessing the risk of climate change on the water resources of the Macquarie River catchment. In *Modsim 2001: International Congress on Modelling and Simulation* (Eds F Ghassemi, PH Whetton, R Little and M Littleboy) pp. 673–678. Canberra, Australia.

Jones R, Page CM, Herron N, Davis R, O'Neill R, Bennett D, McClintock D and Associates HA (2001). Climate change and the risk to long-term water supply in the Murray-Darling Basin.

Jones R, Whetton P, Walsh K and Page CM (2002). Future impacts of climate variability, climate change and land use change on water resources in the Murray-Darling Basin. CSIRO, Melbourne.

Junk WJ and Welcomme RL (1990). Floodplains. In *Wetlands and Shallow Continental Water Bodies* (Ed. BC Patten) pp. 491–524. SPB Academic Publishers, The Hague.

Junk WJ, Bayley PB and Sparks RE (1989). The flood pulse concept in river-floodplain system. *Canadian Special Publication of Fisheries and Aquatic Sciences* **106**, 110–127.

Kidson R, Witts T and Martin W (2000). Historical vegetation mapping in the Macquarie Marshes 1949–1991. (Ed. CWD Riverine Environment Unit, NSW Dept of Land and Water Conservation). NSW Dept of Land and Water Conservation, Dubbo, NSW.

Kingsford RT and Norman FI (2002). Australian waterbirds: products of the continent's ecology. *Emu* **102**, 47–69.

Kingsford R and Thomas R (1995). The Macquarie Marshes in arid Australia and their waterbirds: a 50-year history of decline. *Environmental Management* **19**, 867–878.

Kingsford RT, Curtin AL and Porter J (1999). Water flows on Cooper Creek in arid Australia determine 'boom' and 'bust' periods for waterbirds. *Biological Conservation* **88**, 231–248.

Leslie DJ (2001). Effect of river management on colonially nesting waterbirds in the Barmah-Millewa forest, south-eastern Australia. *Regulated Rivers: Research and Management* **17**, 21–36.

Marchant S and Higgins PJ (1990). *Handbook of Australian, New Zealand and Antarctic Birds. Vo. 1: Ratites to Ducks.* Oxford University Press, Melbourne.

Marchant S and Higgins PJ (1993). *Handbook of Australian, New Zealand and Antarctic Birds. Vol. 2: Raptors to Lapwings.* Oxford University Press, Melbourne.

Meehl GA, Stocker TF, Collins WD, Friedlingstein P, Gaye AT, Gregory JM, Kitoh A, Knutti R, Murphy JM, Noda A, Raper SCB, Watterson IG, Weaver AJ and Zhao Z-C (2007). Global climate projections. In *Climate Change 2007: The Physical Science Basis. Contribution of Working Group I to the 4th Assessment Report of the Intergovernmental Panel on Climate Change.* (Eds S Solomon, D Qin, M Manning, Z Chen, M Marquis, KB Averyt, M Tignor and HL Miller). Cambridge University Press, Cambridge.

Mitsch WJ and Gosselink JG (2000). *Wetlands.* John Wiley and Sons, New York.

Norman FI and Nicholls N (1991). The Southern Oscillation and variations in waterfowl abundance in southeastern Australia. *Austral Ecology* **16**, 485–490.

Oelschlägel B (1995). A method for downscaling global climate model calculations by a statistical weather generator. *Ecological Modelling* **82**, 199–204.

Ogden JC (2005). Everglades ridge and slough conceptual ecological model. *Wetlands* **25**, 810–820.

O'Neill R, Rawson D, Jones R and Page CM (2004). Possible impacts of climate change in the Macquarie Valley. NSW Dept of Infrastructure, Planning and Natural Resources/CSIRO Division of Atmospheric Research, Sydney.

Pollino CA, Tighe M, Cuddy SM and Whitfield S (2007). Alternative system views of climate change in the central west of New South Wales (Australia). In *MODSIM 2007: International*

Congress on Modelling and Simulation. December 2007. (Eds L Oxley and D Kulisarir) pp. 643–649. Modelling and Simulation Society of Australia and New Zealand.

Power S, Casey T, Folland C, Colman A and Mehta V (1999). Inter-decadal modulation of the impact of ENSO on Australia. *Climate Dynamics* **15**, 319–324.

Ralph TJ (2008). Channel breakdown and floodplain wetland morphodynamics in the Macquarie Marshes, south-eastern Australia. PhD thesis, Dept of Physical Geography, Macquarie University.

Ralph TJ and Hesse PP (in press). Downstream hydro-geomorphic changes along the Macquarie River, south-eastern Australia, leading to channel breakdown and floodplain wetlands. *Geomorphology*. doi: 10.1016/j.geomorph.2009.12.007

Randall DA, Wood RA, Bony S, Colman R, Fichefet T, Fyfe J, Kattsov V, Pitman A, Shukla J, Srinivasan J, Stouffer RJ, Sumi A and Taylor KE (2007). Climate models and their evaluation. In *Climate Change 2007: The Physical Science Basis. Contribution of Working Group I to the 4th Assessment Report of the Intergovernmental Panel on Climate Change*. (Eds S Solomon, D Qin, M Manning, Z Chen, M Marquis, KB Averyt, M Tignor and HL Miller). Cambridge University Press, Cambridge.

Roberts J and Marston F (2000). Water regime of wetland and floodplain plants in the Murray-Darling Basin. CSIRO Land and Water, Canberra.

Rogers K and Ralph T (in prep). Floodplain Wetland Biota in the Murray-Darling Basin: Water and Habitat Requirements. CSIRO Publishing, Melbourne.

Roshier DA, Robertson AI, Kingsford RT and Green DG (2001). Continental-scale interactions with temporary resources may explain the paradox of large populations of desert waterbirds in Australia. *Landscape Ecology* **16**, 547–556.

Roshier DA, Robertson AI and Kingsford RT (2002). Responses of waterbirds to flooding in an arid region of Australia and implications for conservation. *Biological Conservation* **106**, 399–411.

Thatcher M and McGregor J (2008). Dynamically downscaled Mk3.0 climate projection for the Marine and Tropical Sciences Research Facility (MTSRF) from 2060–2090. Report prepared for the Marine and Tropical Sciences Research Facility. CSIRO, Melbourne.

Thoms M, Capon S, James C, Padgham M and Rayburg S (2007). The Narran Ecosystem Project: the response of a terminal wetland system to variable wetting and drying. Final report to the Murray-Darling Basin Commission. Murray-Darling Basin Commission, Canberra.

Tighe M, Pollino CA, Cuddy SM and Whitfield S (2007a). A Bayesian approach to assessing regional climate change pressure on natural resource conditions in the central west of NSW, Australia. In *MODSIM 2007: International Congress on Modelling and Simulation*. December 2007. (Eds L Oxley and D Kulisari). pp. 233–239. Modelling and Simulation Society of Australia and New Zealand.

Tighe M, Bennett N, Pollino CA and Cuddy SM (2007b). Characterisation of the Macquarie catchment for use with the EXCLAIM DSS. iCAM report 2007/26. Integrated Catchment Assessment and Management Centre, Fenner School of Environment and Society, Australian National University, Canberra.

Timbal B (2004). Southwest Australia past and future rainfall trends. *Climate Research* **26**, 233–249.

Timbal B and Jones D (2008). Future projections of winter rainfall in southeast Australia using a statistical downscaling technique. *Climatic Change* **86**, 165–187.

Timbal B, Fernandez E and Li Z (2009). Generalization of a statistical downscaling model to provide local climate change projections for Australia. *Environmental Modelling and Software* **24**, 341–358.

van der Valk AG (1981). Succession in wetlands: a Gleasonian approach. *Ecology* **62**, 688–696.

Verdon DC and Franks SW (2005). Indian Ocean sea surface temperature variability and winter rainfall: eastern Australia. *Water Resources Research* **41**, 1–10.

Verdon DC, Wyatt AM, Kiem AS and Franks SW (2004). Multidecadal variability of rainfall and streamflow: eastern Australia. *Water Resources Research* **40**, 1–8.

Verdon DC, Franks SW and Hameed T (2006). Climate variability impacts on water resources in the Lachlan River Valley. In *30th Hydrology and Water Resources Symposium*, Launceston, Tas.

Walker KF, Sheldon F and Puckridge JT (1995). An ecological perspective on dryland river ecosystems. *Regulated Rivers: Research and Management* **11**, 85–104.

Winder M and Schindler DE (2004). Climate change uncouples trophic interactions in an aquatic ecosystem. *Ecology* **85**, 2100–2106.

Wood AW, Leung LR, Sridhar V and Lettenmaier DP (2004). Hydrologic implications of dynamical and statistical approaches to downscaling climate model outputs. *Climatic Change* **62**, 189–216.

PART III

ECOSYSTEM RESPONSE MODELLING IN THE SOUTHERN BASIN

Introduction: Murray Basin

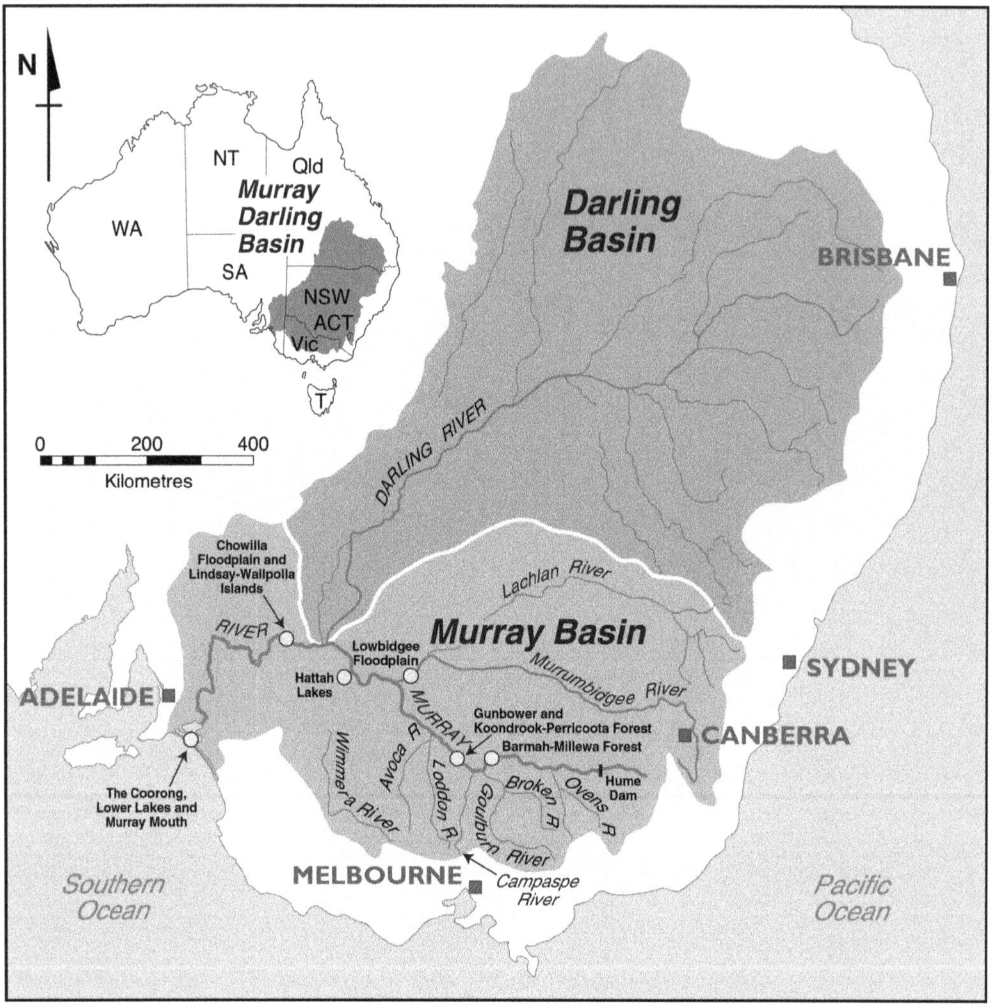

The southern part of the Murray-Darling Basin covers the drainage area of the River Murray and includes the river basin below the confluence of the Darling River from the northern basin. The basin is defined by the River Murray, which begins in the Alps, ends at the Southern Ocean and has an annual average discharge of approximately 24 000 GL. The river is 2375 km long and drains a basin of approximately 430 000 square kilometres.

The southern basin includes the Murrumbidgee, Murray and Goulburn catchments. These have much less variability in run-off than other parts of the basin, with annual flow variability ranging from 30% to 300% of the long-term mean. Flows are highest in spring, resulting from snow melt, and lowest in mid-autumn. The south-west catchments, including the Campaspe, Loddon, Avoca and Wimmera, have a much lower variability in their winter-dominated rainfall than other parts of the basin. Variability of run-off is very high throughout the year, with many rivers and terminal lakes dry during summer and autumn. The small catchments of the eastern Mount Lofty Ranges in South Australia, including the Angas River and the Finniss River, are considered part of the Murray-Darling Basin because they flow into Lake Alexandrina. These rivers are seasonal, being usually dry in the summer, but their winter rainfall and stream flow is more reliable than those of the south-west rivers. The Ovens River is the most unregulated of the major rivers in the southern basin. Future climate change scenarios suggest a reduction of 15% in available water, resulting in less flooding and more River Murray Mouth closures.

The basin is home to many wetlands of international and national significance, including the Ramsar-listed Barmah-Millewa Forest, the Chowilla Floodplain and Lindsay and Wallpolla Islands, the Banrock wetlands and the Coorong, Lower Lakes and Murray Mouth. The Murray-Darling Basin Living Murray Program also recognises the Gunbower-Perricoota Floodplain, the Hattah Lakes wetlands and the River Murray Channel. In addition, other important wetlands include the Lower Murrumbidgee and the Great Cumbung Swamp.

Modelling flows and ecological responses in this part of the basin is challenging, due to the high degree of river regulation in the lower part of the system, the large degree of hydrological alteration from pre-regulation conditions, and the large amount of surface water and groundwater interactions from the shallow aquifers.

Chapter 11

Ecosystem response modelling needs of the Living Murray Initiative

Lex Cogle, Stuart Little, Jong Lee, Jody Swirepik, David Hohnberg and Michelle Akeroyd

Introduction

The declining health of waterways around the world has prompted high-level government strategies and community action (International River Foundation 2008). In Australia, the Murray-Darling Basin Ministerial Council announced its First Step Decision for The Living Murray in 2003, the largest river restoration program in Australia. More recently, these strategies have led to new arrangements and planning for managing water resources in the Murray-Darling Basin, captured in the *Water Act 2007*. Essentially, there is acknowledgment that water, commonly called environmental water, is needed to restore and rehabilitate stressed ecosystems, in addition to its allocation to irrigated agriculture and urban supplies. Specific environmental management sections have been written in new legislation and will draw on information and expertise gathered from implementation of the Living Murray Initiative (TLM) and other environmental watering activities.

This chapter summarises ecological response knowledge needs for TLM and shows how models have supported the development and implementation of TLM. It emphasises the importance of monitoring to validate ecological responses and assist in adaptive management.

Living Murray Initiative

The Living Murray was established in 2002 in response to concerns about the environmental and economic health of the River Murray system. It is the largest river restoration program in Australia and has a vision of 'a healthy River Murray system, sustaining communities and preserving unique values'. In November 2003, the Murray-Darling Basin Ministerial Council announced its First Step Decision to recover 500 GL of water for environmental use. Commencing in 2004–05, the program was intended to spend $700 million over five years to recover water for the environment, and approximately $270 million over eight years on an environmental works and measures program.

The First Step Decision focuses on achieving environmental objectives at six significant ecological assets or icon sites identified for their high conservation, recreation, cultural, heritage and economic value. The icon sites are the Barmah-Millewa Forest, the Gunbower-Koondrook-Perricoota forests, the Hattah Lakes, the Chowilla Floodplain and Lindsay-Wallpolla Islands,

Table 11.1: First Step Decision icon site ecological objectives

Icon site and ecological objectives
Lower Lakes, Coorong and Murray Mouth • An open Murray Mouth at all times • More frequent estuarine fish spawning • Enhanced migratory wader bird habitat in the lakes
Chowilla Floodplain, Lindsay-Wallpolla Islands • Retain current area of river red gum and maintain at least 20% of the original area of black box • Maintain high-value wetlands
Hattah Lakes • Successful breeding events of colonial waterbirds in at least two years out of 10 • Increase population size and breeding events of hardyheads, Australian smelt, gudgeons and other wetland fish • Restoration of the aquatic vegetation zone in at least 50% of the lakes to increase native fish and bird breeding and survival
Gunbower-Koondrook-Perricoota Forests • 30% of river red gum forest in healthy condition • Successful breeding of thousands of colonial waterbirds in at least three years out of 10 • Healthy populations of resident native fish in wetlands • 80% of permanent and semi-permanent wetlands in healthy condition
Barmah-Millewa Forest • Successful breeding of thousands of colonial waterbirds in at least three years out of 10 • Healthy vegetation in at least 55% of the area of the forest (including virtually all the giant rush, moira grass and river red gum forest)
Murray River Channel • Overcoming barriers to migration of native fish species between the sea and Hume Dam • Increased frequency of higher, ecologically significant flows in spring, and maintaining channel stability

the Murray Mouth, Coorong and Lower Lakes and the River Murray Channel (Fig. 11.1/colour plate, p. 313). Specific ecological objectives were identified for each icon site, covering fish, birds and vegetation (Table 11.1).

To meet the ecological objectives, the TLM water recovery program was implemented to return, on average, 500 GL/yr (long-term cap equivalent) to the River Murray by a target date of 30 June 2009. The long-term cap equivalent is defined as the volume registered for a particular recovery work or measure, calculated using the accepted best practice computer model for that system as long-term average contribution to the long-term cap equivalent or as potential contribution to the long-term average flows in the relevant river valley. The long-term cap equivalent is defined by the Ministerial Council as 'the volume of water that would have been diverted under 1993/94 levels of development'. The recovered water includes contributions of up to 249 GL (NSW), 214 GL (Vic.), 35 GL (SA) and 2 GL (ACT) and totals approximately 6% of diversions in the southern connected basin.

There is a range of activities and/or mechanisms that can be implemented to address the ecological objectives of TLM. These include explicit water delivery approaches (pumping, weir manipulation, flow enhancement and retaining water on floodplains) and other important approaches not directly related to water delivery, such as resnagging and fishways. All these activities and mechanisms require some form of infrastructure to support their delivery.

The environmental works and measures program was initiated to provide structural works for the effective and efficient use of recovered water to address the icon site ecological objectives. The works include regulators, weirs, pumping infrastructure and fishways at the icon sites. The target date for implementation is 2011, but final construction timetables will be

Table 11.2: EWG decision framework

Primary objective	**Environmental benefit, in terms of stated icon site environmental objective**
Ranking criteria	Significance of outcome Amount of benefit for the volume of water (including opportunity to take advantage of other events) Risk of not watering Certainty/likelihood of benefit

determined by available building resources and capacity at icon sites. A key feature of the proposed works is their ability to allow managed watering of ecologically significant floodplain areas and, importantly, enable their inundation, independent of floods, at regulated flow levels (Burns *et al.* 2008). Examples of works include:

- the use of regulators and levees to control flooding within the Gunbower-Koondrook-Perricoota forests following the construction of channels from the Torrumbary weir pool and the use of fishways to provide fish passage in and out of the forests;
- the construction of a pumping station to deliver 1000 ML/day from the River Murray to the Hattah Lakes system and the use of regulators and levees to control flooding.

Delivery of environmental water to the icon sites to meet the ecological objectives necessitates a decision by water managers to allocate a volume of water to the intervention. To date, these decisions are made by a multi-jurisdictional working group called the Environmental Watering Group. That group has used various decision frameworks to support their decisions incorporating objectives and criteria, as shown in Table 11.2.

Scientific information that supports decisions based on that process includes expert opinion, data from past and present research and monitoring and the Environmental Watering Group's own knowledge of the system. Decisions are made within the broader policy environment that includes diverse community interests. An ability to synthesise all information in a rapid and logical format would therefore be of value.

Ecosystem response modelling

Ecosystem response models range from simple static representation of factors such as cause and effect relationships between specific ecosystem components (e.g. vegetation responds to watering), to complex dynamic models that attempt to link the interaction and processes of ecosystem components (e.g. the impact of watering on habitat). The latter requires significantly greater information and is therefore a difficult undertaking. For example, the dynamics of the fish population are related to a range of environmental drivers such as flow (some fish are flow-dependent spawners), water quality (dissolved oxygen is required for fish survival, but concentrations vary both spatially and temporally and in response to the dynamics of other biological populations), habitat availability (watercourses with shallow-flowing habitat may be required for spawning, but deep waterholes may provide protection) and predation (there is a range of potential fish predators including other fish species and birds). Complex models, due to their data requirements and limited understanding of interactions, are generally difficult to run and hence apply to management decisions.

Frameworks that transparently illustrate decision processes based on available scientific knowledge are often captured within the broad categorisation of modelling. These are generally called decision support systems.

Prior to the First Step Decision, the development of TLM required a comprehensive assessment of the costs and benefits to the environment, industries and communities of delivering environmental water to the River Murray. The Ministerial Council chose three environmental flow reference points as the basis for this analysis (350 GL/yr, 750 GL/yr and 1500 GL/yr) and noted that structural and operational modifications should be assessed (Young *et al.* 2003). A scientific reference panel was assembled to address this requirement. It examined five environmental flow assessment tools for their value in advising on the Ministerial Council requirements. The environmental flows decision support system (Young *et al.* 1999) was chosen as the most appropriate tool, as it was:

- structured as a decision support system with the ability to estimate ecological responses to specified flow regimes for riverine environments at a range of scales in a consistent repeatable manner;
- applicable to the River Murray system or readily adaptable to meet specified time frames;
- effective using existing data and able to use structured expert opinion;
- applicable to local, regional and system-wide assessments.

Following the incorporation of several modifications to make the decision support system more applicable, the scientific reference panel identified the new decision support system or tool as the Murray Flow Assessment Tool (Young *et al.* 2003). The Murray Flow Assessment Tool was used with regional evaluation groups under the guidance of the scientific reference panel to deliver advice to the Ministerial Council. In other words, ecosystem modelling has been a very important component of TLM for many years and delivered important underlying information of water requirements prior to its commencement.

At the beginning of TLM, it was unclear how the environmental objectives could be achieved. It was recognised that water recovery alone would be insufficient because floodplain inundation is required to achieve many objectives. It was also realised that 500 GL/year (long-term cap equivalent) is insufficient to achieve that aim in many locations under natural flooding processes. Scoping of potential water delivery options began through the Murray-Darling Basin Commission's Environmental Works and Measures program. Digital elevation models, hydraulic models and feasibility studies led to the development of five major landscape-scale floodplain inundation works proposals that can deliver water to the icon sites without the need for the volume of water and flow rates associated with natural flood events. While the major floodplain inundation works were being developed, the Commission's Water Recovery program was also scoped, developed and implemented. In 2007, the details of both programs became clear and the need to ensure they were compatible was apparent. A second major phase of hydrologic modelling was instigated, with the aim of testing the feasibility of supplying and delivering the volumes of water required by the major works proposals. The first stages of this modelling have been completed; they indicate that 500 GL will be sufficient to deliver the required water in most years.

Following this improved understanding of water availability and delivery options, there is a renewed emphasis on understanding ecosystem responses to environmental water. This will ensure that decisions by water managers on environmental water delivery across the six icon sites will be based on best available ecosystem knowledge to address the respective ecological objectives. An important method to communicate this understanding is via the use of ecosystem response models.

The earlier Murray Flow Assessment Tool experience identified the decision support system approach as valuable. This approach permits the use of many possible inputs including unquantified expert opinion, experimental data sets and process model outputs. It is probable that the complexity of the ecosystem understanding required for TLM necessitates a continuation of this approach to service its ecosystem modelling requirements. The approach also allows a transparent representation of the state of knowledge as water managers and policy-makers address increasingly tight time frames for decisions.

The broad requirements for the ecosystem response models in TLM would be:

- the ability to predict or assess the ecological impact of differing volumes, timing of watering and flows. This would provide information on when and how much water to deliver to an ecosystem, via the newly constructed infrastructure. For example, whether 10 GL in March is as useful as the same volume in September or whether two 5 GL volumes should be delivered, and whether there are certain flow rates that would create optimum environmental conditions (e.g. for spawning of native fish);
- the ability to predict or assess the spatial impact of the watering. This could provide information on the area of a floodplain that requires water and the water depth required to make an effective environmental outcome (e.g. for a waterbird breeding event);
- to identify a trigger point that will initiate or possibly ensure the commencement of watering. Trigger points could be based on water balance models and identify the permanent wilting points of various vegetation communities, or could identify water volumes, temperatures and water quality parameters required in waterholes for fish survival;
- to support the identification of a sustainable diversion limit that would allow the achievement of all objectives. A sustainable diversion limit is a goal of all water management plans, and advice from ecosystem models will inform the trade-offs central to its determination. In TLM, it is useful to think of this as applying to the cumulative amount across all icon sites, but it must also be seen in the context of the entire Murray-Darling Basin;
- to assist in the identification of a baseline trajectory for TLM ecological objectives. Models have an ability to make predictions, based on calibration using known inputs and results. TLM would benefit from trajectories of ecological response into the past so that a baseline which recognises both man-made and natural impacts (e.g. river regulation, drought) can be identified;
- the ability to utilise and communicate new and current information, knowledge and monitoring in an adaptive sense to a range of audiences.

The Murray-Darling Basin Authority has partnered with the eWater CRC to address some of the needs for ecosystem modelling, recognising that the Murray Flow Assessment Tool was a useful contribution to the development of TLM. The CRC's Ecological Modeller program is central to this partnership, the development of which will run until June 2010 with the following main activities:

- implementing and updating MFAT relationships for all icon sites into Ecological Modeller;
- developing new ecological models for key indicators, particularly building on any parallel activities underway in the project or the CRC. The indicators will be informed by the ecological objectives for fish, birds and vegetation and informed by available scientific research collated as part of the project;
- developing software that will allow all types of ecological response models to exist in the framework and development of interpretation tools.

Monitoring

The development of ecosystem models requires information on ecosystem components and the underlying relationships and processes linking those components. The information will, where possible, integrate landscape-scale impacts on the ecosystem, including climate and intersite attributes such as connectivity. Monitoring is essential to obtain this information, with outputs allowing calibration and validation of hypothesised relationships and processes in ecosystem models. Implementation of monitoring programs also allows the further refinement of ecosystem models as new data and information become available. Ideally, this continued development is undertaken with all interested parties, which may improve ownership of a management regime that is adaptively responding to new knowledge. Indeed, for optimum ownership of model outputs, a work environment where model developers, model users, data providers and water managers work together ensures that model improvements address all needs.

The TLM business plan requires a consistent approach to monitoring across the six icon sites and a monitoring framework was developed by TLM working groups. Monitoring advises on the attainment of icon site ecological objectives as defined by the Ministerial Council's First Step Decision. It also informs the environmental management plans developed for the icon sites. The Murray-Darling Basin Authority has worked with icon site managers to deliver these monitoring requirements, based on three broad categories:

- system monitoring, which aims to determine if the health of the River Murray system has improved following implementation of TLM. This is implemented across all icon sites by the Murray-Darling Basin Authority;
- condition monitoring, which aims to determine if the condition of the icon sites improves following implementation of TLM. Implementation of condition monitoring is via condition monitoring plans developed by icon site managers;
- intervention monitoring, which assesses the ecological response to a range of interventions or environmental management actions based on hypothesis testing. These interventions relate to the delivery of water to support an ecological objective and are quite targeted in their conceptual application. In the context of ecosystem response models, it is more likely that targeted studies of a cause and effect will provide the information to refine model.

Due to their sampling frequency, there are similarities between system and condition monitoring. However, intervention monitoring may include higher-frequency monitoring so that process understanding is obtained. A further type of monitoring, which can advise the success of environmental watering, is undertaken during specific environmental watering events and is also of relatively high frequency. This has been called event monitoring and has some complementarities to intervention monitoring. The synthesis of monitoring data and information and its use for improving ecosystem response models, so that their predictions are more certain, is an essential feature of their continuous improvement.

Conclusions

This chapter has provided an overview of how TLM has utilised ecosystem models in the past. Clearly, TLM has benefited from these models and supports their future application and development. The models will play a large role in providing key information to identify trade-offs on environmental watering decisions. However, if ecosystem models are to be used in the future, they must readily communicate, to a range of interested parties, the confidence of their predictions and must engender ownership. The models must also be able to communicate the impacts and associated uncertainty of environmental watering options implemented as part of TLM. Finally, we need to recognise that while ecosystem models are valuable, they are only one of the tools and criteria that water managers use when making environmental watering decisions.

References

Burns I, Howell G, Everingham P and Swirepik J (2008). Blueprints for the future of the Murray's iconic floodplains and wetlands. Available at http://www.riversymposium.com/index.php?page=2008. Accessed 3 February 2009.

International River Foundation (2008). River journeys. International River Foundation, Brisbane.

Young WJ, Cuddy SM, Marston FM and Farley TFN (1999). Environmental flows decision support system: a pPrototype. Vol. 1: System description. Vol. 2: Technical supplement. Technical Report 11/99. CSIRO Land and Water, Canberra.

Young WJ, Scott AC, Cuddy SM and Rennie BA (2003). Murray Flow Assessment Tool: a technical description. Client Report, 2003. CSIRO Land and Water, Canberra.

Chapter 12

Ecosystem response models for lower Calare (Lachlan River) floodplain wetlands: managing wetland biota and climate change modelling

Patrick Driver, Shahadat Chowdhury, Tahir Hameed, Martin O'Rourke and Mustak Shaikh

Introduction

Rivers and floodplain wetlands have had their water security and natural flow regimes substantially undermined by flow regulation (Ward and Stanford 1995; Junk 2005), including within Australia (Walker *et al.* 1995; Leslie 2001; Kingsford 2000). The impact on wetlands and wetland fauna is being exacerbated by climate change (Herron *et al.* 2002; Chambers *et al.* 2005). Models that focus on the response of ecosystems to river flows are required to quantify the impacts, and the interaction of flow regulation, climate and climate change. Conceptual and numeric ecosystem response models are critical communication tools, especially when natural resource managers are involved in their development (Jakeman *et al.* 2006). Such models can provide some level of prediction of the effects of management decisions (e.g. extent of flooding resulting from implementation of an environmental flow rule), accurate to the extent that better management decisions are made and more useful if natural resource managers are aware of model limitations (Young *et al.* 2002; Jakeman *et al.* 2006). To this end, this chapter describes ecosystem response models for the Lachlan River floodplain wetlands, some of which have been used in the development of a water-sharing plan. They can be used to mitigate undesirable impacts of flow regulation, climate and climate change.

The Lachlan River, in the Wiradjuri language the Calare (Sturt 1833), and its wetlands is in the southern Murray-Darling Basin in NSW. The area is subject to a high level of flow regulation relative to the rest of the Murray-Darling Basin, with effects on aquatic biota such as reduced wetland availability and reduced species diversity (Kingsford 2000; Roshier *et al.* 2001; Closs *et al.* 2006). The Lower Lachlan commences where floodplains develop at the confluence of the Lachlan and Abercrombie rivers, where the major regulatory structure of Wyangala Dam is situated (van der Beek and Bishop 2003; DPI 2005). The Calare catchment (90 649 km^2) contains 4710 km^2 of wetlands and extends for 1450 km from the Great Dividing Range through to the Great Cumbung Swamp (Fig. 12.1; O'Brien and Burne 1994; Kingsford *et al.* 2004). The Calare supports numerous nationally significant wetlands and species listed under international migratory bird agreements (Wettin *et al.* 1993; DEWHA 2008a) with the notable exceptions of Lakes Cowal and Brewster (Wettin *et al.* 1993; Driver *et al.* 2002). Many of these wetlands are associated with the Lachlan alluvial fan (Fig. 12.1), from Willandra Creek in the

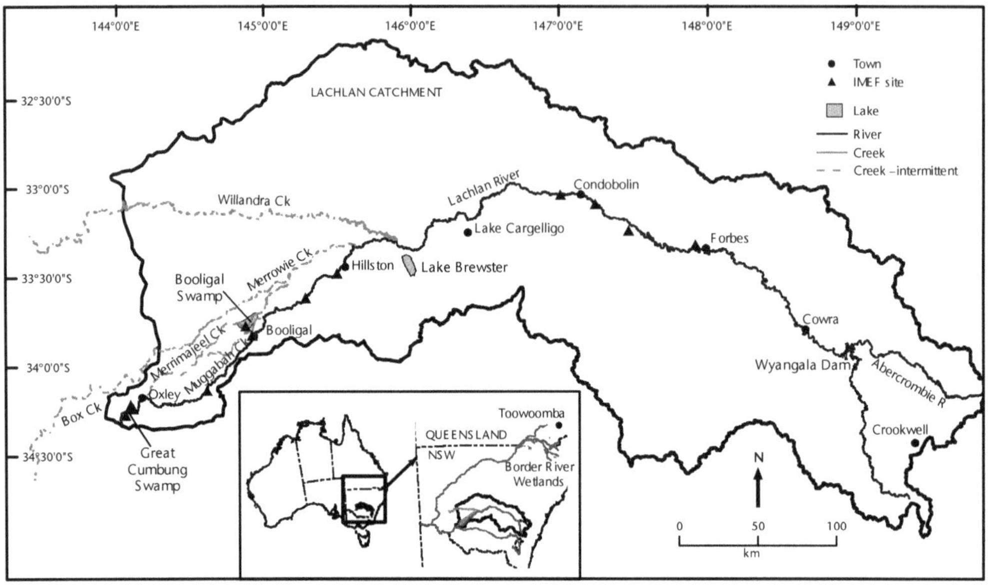

Figure 12.1: Lachlan Valley catchment, NSW.

north to the present river course that is a few thousand years old (Adamson *et al.* 1987). The fan includes the Merrowie-Box Creek (25 km^2) and the Merrimajeel-Muggebah Creek complex (50 km^2). The present river course includes and dissects Lachlan Swamp (66 km^2) and the Great Cumbung Swamp (160 km^2), which corresponds with observations in the 1830s of large areas of wetland vegetation referred to as the Lachlan Marshes (Sturt 1833). In addition to the management problems associated with highly variable river flows (compared to other rivers world-wide and other Murray-Darling Basin rivers) (Thoms and Sheldon 2000), the Lachlan catchment has a small wetland area-surface water availability ratio relative to most other Murray-Darling Basin wetlands (2.4 compared to 3.6, 13.3, 15.4 and 30.0 ML/ha in the Macquarie-Castlereagh, Gwydir, Murrumbidgee and Murray Rivers respectively; derived from Kingsford *et al.* 2004 and CSIRO 2008b). Natural wetland inundation regimes are typically intermittent, with some seasonality driven by winter/spring rainfall. Many Lachlan wetlands, particularly those in the lower watercourse to the west, are subject to a semi-arid climate with an average rainfall of ~250 mm and an average daily discharge of, for example, 753 ML/day at Booligal (Chessman *et al.* 2003; DECCW flow data). The shallow Coonambidgil aquifer underlies this region, is recharged by large floods and supports wetland plants. Water management to maintain this aquifer is critical (Hatton and Evans 1998; Driver *et al.* 2004).

Ecosystem response modelling and aims

Ecosystem response models for the Lachlan catchment have existed at least since Frith (1959) described ducks breeding in response to floodwater, and there have been many subsequent ecological studies. We describe models from 1998 onwards, to focus on wetland models that informed the development of environmental flow rules. Although management of environmental water allocations has occurred since 1992, environmental flow rules have been operating from 1998 and were formalised in the 2004 Lachlan water-sharing plan (LWSP, LRMC 2002; DIPNR 2004; Wettin *et al.* 2007). These rules include translucent environmental flows, which provide the most water for lower Calare wetlands with up to 350 000 ML/yr (~23% of average total river system inflows) between May and November every year. A translucent flow is the partial release of an inflow into the main river storage at the time of the inflow occurrence. Additionally, every year a 20 000 ML volume of water, called an environmental contingency allowance, is banked for environmental purposes. Since 2006, purchased adaptive environmental water has also been used. As such water is an absent or negligible component of the ecosystem response models described in this chapter, however, it is not discussed further.

From 1998, ecosystem response models supported the development of environmental flow rules and a water-sharing plan (Shaikh *et al.* 1998; Driver *et al.* 2002, 2004). Such models were initially developed for reporting on the success of managed intra-annual environmental flow events and predicting the likely future success of such events as part of the Integrated Monitoring of Environmental Flows program. That program was based on before–after intervention monitoring (Chessman *et al.* 2003) and played a key role in informing the river management committees who developed water-sharing plans (Hillman *et al.* 2003). The objectives of integrated monitoring of environmental flows included testing hypotheses about how altered wetland inundation regimes (hydrological change) might affect wetland function (ecological change) across the floodplains of water-sharing plan areas. Critically, the monitoring regimes broadened the spatial scope of models for water management beyond a few iconic assets to a representation of an entire floodplain, in accordance with the jurisdictional areas of water-sharing plans. The success of an event was typically measured in relation to its similarity to an estimate of natural river flow and wetland inundation patterns. For example, a translucency flow in flow year 1999/2000 was assessed to determine the extent to which tributary flows passed through the major storage unimpeded then travelled along the river as a pulse without being extracted into an off-river storage (Driver *et al.* 2000). For assessment of long-term (~100 years) environmental water management regimes, simulations of the current management regime were compared against undeveloped simulations (where the effects of river structures were removed) using a river hydrology model, the integrated quality and quantity model (Hameed and Podger 2001; Driver *et al.* 2002, 2005a, b; LRMC 2002).

Our objective is to illustrate the utility, constraints and potential use of wetland ecosystem response models not only for the Lachlan catchment, but for wider application throughout the Murray-Darling Basin and elsewhere. We provide an overview of three broad types of Lachlan wetland models (Driver *et al.* 2002, 2005a, b) that are based on critical hydro-ecological thresholds (*sensu* Arthington *et al.* 2006). Consistent with their intended primary use, all three model types have been linked to the integrated quality and quantity model to compare long-term water-sharing plan scenarios. The models have also been used in reviews of wetland flow requirements (DEWHA 2008b), for estimating water savings options (Hameed *et al.* 2005) and for predicting the impacts of forecast changes in climate (CSIRO 2008a). To address this new use of Lachlan ecosystem response models in climate change modelling, we also discuss Booligal Swamp and Great Cumbung Swamp models in detail, including models within CSIRO (2008a) that use Lachlan models to predict the effects of flow regulation and climate change impacts.

Billabong ecosystem response models

The billabong model was developed for billabongs (oxbow lakes) and back swamps (scale 1–10s km^2, up to 10s km in length, Driver *et al.* 2005a, b). Eight billabongs and four swamps from the Lower Lachlan River floodplain, selected as representative of nationally significant swamps (Wettin *et al.* 1993) and Lower Lachlan billabongs affected by environmental flows, were used for data collection. Field studies (Chessman *et al.* 2003; Driver *et al.* 2005a) were used to determine:

- the Commence-To-Fill (CTF, ML/day, i.e. the discharge when the wetland starts to fill);
- the estimated water depth at and above which aquatic plant growth occurs (W);
- the time a wetland takes to drain from rising above W to falling below W (Tw).

These measurements were used in the model to make predictions about the impacts of planned environmental flows. The model indicated that when most environmental flows occur naturally and delivered (June–November), wetlands downstream of Lake Brewster and, to a lesser extent, between Forbes and Condobolin have benefited from the implementation of the environmental flow rules (Driver *et al.* 2005a).

Distributary creek wetland ecosystem response models

The distributary creek wetland model for creeks associated with the alluvial fan (locally referred to as effluent creeks) are based on field measurements of specific fill events during 1998–2000, landholder knowledge of the relationship between Lachlan River and distributary creek discharge (ML/day) and corresponding hydrology records (Hameed *et al.* 2005; scale 10–100s km^2, 100s km in length). For creeks not used for irrigation, the CTF was determined for the creek's offtake from the Lachlan River. Creeks used for irrigation are already modelled in detail within the integrated quality and quantity model. Long-term (~100 year) simulations of the regulated Lachlan River system against undeveloped conditions have been used, for example, to show that irrigation-affected creeks are receiving a greater net annual flow under flow regulation (e.g. ~177% increase in Willandra Creek's annual discharge). As a result, downstream wetlands such as Booligal Swamp receive reduced flows (Driver *et al.* 2002).

Great Cumbung Swamp ecosystem response model

The Great Cumbung Swamp water balance model (Driver *et al.* 2005b; scale 100–1000s km^2) is underpinned by a coupled water and solute mass-balance model (Smith and Barr 2002) that was developed to predict how much of the swamp would be inundated in relation to river and meteorological conditions. The model was constructed using surface water data and meteorological data in government databases, as well as surface water and aquifer data from Brady *et al.* (1998). The relationship between river discharge at Oxley (just upstream of the swamp) versus river flows upstream, inundation area, flood (landscape) relief and salinity were based on the literature and NSW government expertise. Based on these sources, surface water inflow for 1971–98 was estimated using assumptions for groundwater outflow (infiltration rate = 25 cm/month), evapotranspiration (pan coefficient = 1.0) and river solute concentration (300 mg/L). The Great Cumbung Swamp was represented as an idealised, low-terrain (bed slope = 0.01°) inverted cone. The spatial extent was limited by levee containment areas to avoid modelling events >1500 ML/day at Oxley, which are largely outside the influence of water managers. To simulate the minimum flooding requirements for the fringing river red gum forest, Driver *et al.* (2005b) used published estimates (Bacon *et al.* 1993; Roberts and Marston 2000) to suggest a minimum duration (t) of 90 days (at 430 ML/day) to flood 4000 ha, the area of reed bed (*Phragmites australis*). A flood this size just extends into the river red gum forest. This model, when linked to the integrated quality and quantity model, showed there had been a 28% reduction in the flood frequency required for river red gum and *P. australis* maintenance during a relatively wet period. It must be noted, however, that a flood of 4000 ha was assumed to recharge the Coonambidgil aquifer and thereby sustain river red gums over a larger area. This assumption is not tested, but is consistent with the conceptual model for surface water–groundwater– plant root interactions developed for the Great Cumbung Swamp (Driver *et al.* 2004). This has practical relevance for environmental flow delivery, given that the flows required to inundate the river red gum forest (~1500 ML/day) are much larger. Flows of ~400 ML/day are much more achievable, especially during drought.

Case study 1: Booligal Swamp model and its use in climate change forecasting

Booligal Swamp, on Merrimajeel Creek (Fig. 12.1), supports some of the largest recorded breeding populations of glossy ibis (*Plegadis falcinellus*) in Australia (Magrath 1992).

Straw-necked ibis (*Threskiornis spinicollis*), sacred ibis (*T. molucca*) and glossy ibis breed within high-density lignum (*Muehlenbeckia florulenta*) between June to December during flood years. Glossy ibis move out of coastal refuges to the north later than straw-necked ibis during the southern summer (Griffioen and Clarke 2002; Dingle 2004), and consequently breed later (October–January, Marchant and Higgins 1990; Magrath 1992; Fig. 12.2). Natural colonial nesting events are artificially maintained using water from the environmental contingency allowance, allowing fledglings to develop to independence. Colonial waterbird events since 1984 have been triggered by flows ≥2500 ML/day at Booligal Weir associated with flooding at Booligal Swamp (≥50, 30 and 29 km^2 of floodwater from the main regulator for this system, the Torriganny Weir to the rookery in 1990, 1998 and 2000 respectively; unpublished Landsat TM image analysis). This discharge value is used in the Booligal Swamp model to indicate the relative frequency of different periods of continuous inundation at Booligal Swamp (a specialised distributary creek wetland model; Driver *et al.* 2005a). For this model the relationship between the number of nests and the duration of natural inundation was calculated using a generalised linear model (McCullagh and Nelder 1989). Subsequently, the frequency of different flood durations under undeveloped versus current water-sharing plan scenarios was calculated for Booligal Swamp using the integrated quality and quantity model. This showed that as a result of the environmental flow rules there is now a greater frequency of colonial nesting waterbird events lasting more than 150 days, in association with ≥37 400 ibis nests. This finding, although based on only a few events, indicates that as a result of targeted use of the environmental contingency allowance and at least in relation to colonial nesting events at Booligal Swamp, the objective of the Lachlan water-sharing plan – to mimic natural flow conditions so as to provide for the abundance of native waterbirds – has been met.

Future water availability in Booligal Swamp is modelled using the climate change model of CSIRO (2008a). Floods corresponding with ≥2500 ML/day at Booligal Weir and inundation for at least two months (Table 12.1, based on Fig. 2 in Driver *et al.* 2005a) were used to represent the minimum hydrological conditions that would support significant waterbird breeding. Their results indicate that water resource development has made the average period between winter–spring floods 34% longer. This, as CSIRO (2008a, p. 112) says, is similar to the 38% reduction in inundation days per year calculated for the Lower Lachlan swamps (including the Great Cumbung and Booligal swamps; Driver *et al.* 2005a). However, the CSIRO (2008a) model could measure the effects of flow regulation on water availability in the Booligal Swamp more accurately if it used a lower value for the minimum natural flood duration for colonial nesting events (the Lachlan water-sharing plan rules state that once a colonial nesting event has commenced it will be maintained). The CSIRO (2008a) model used 60 days of natural inundation, which corresponds with ≥15 000 nests and excludes 25% of the events since 1984 (Driver *et al.* 2005a). If, however, all colonial nesting events were included in this CSIRO (2008a) model by using a 19-day flood duration cut-off, which is the minimum duration of recorded nesting events, the simulated effects of flow regulation would be greater because previous modelling showed that colonial nesting events of ≤40 000, and particularly ≤20 000, nests are rarer under flow regulation (Driver *et al.* 2005a; Chowdhury and Driver 2008).

We tested our assertions regarding the effects of flow regulation on Booligal Swamp using CSIRO scenarios (2008a), which allowed us to explore climate change effects. By applying the same rules to the data we gained similar results for the average period between floods (years) in Booligal Swamp (≥2500 ML/day floods for ≥60 days, equating to ≥15 000 nests; Table 12.1). Our results were on average ~20% higher for the length of period between events compared to those of CSIRO (2008a). The inconsistencies could be explained by differences in rounding of decimal points between our and CSIRO algorithms. As predicted, we found the effects of flow regulation were relatively greater for the 19- versus 60-day scenario (94% versus 23% increase

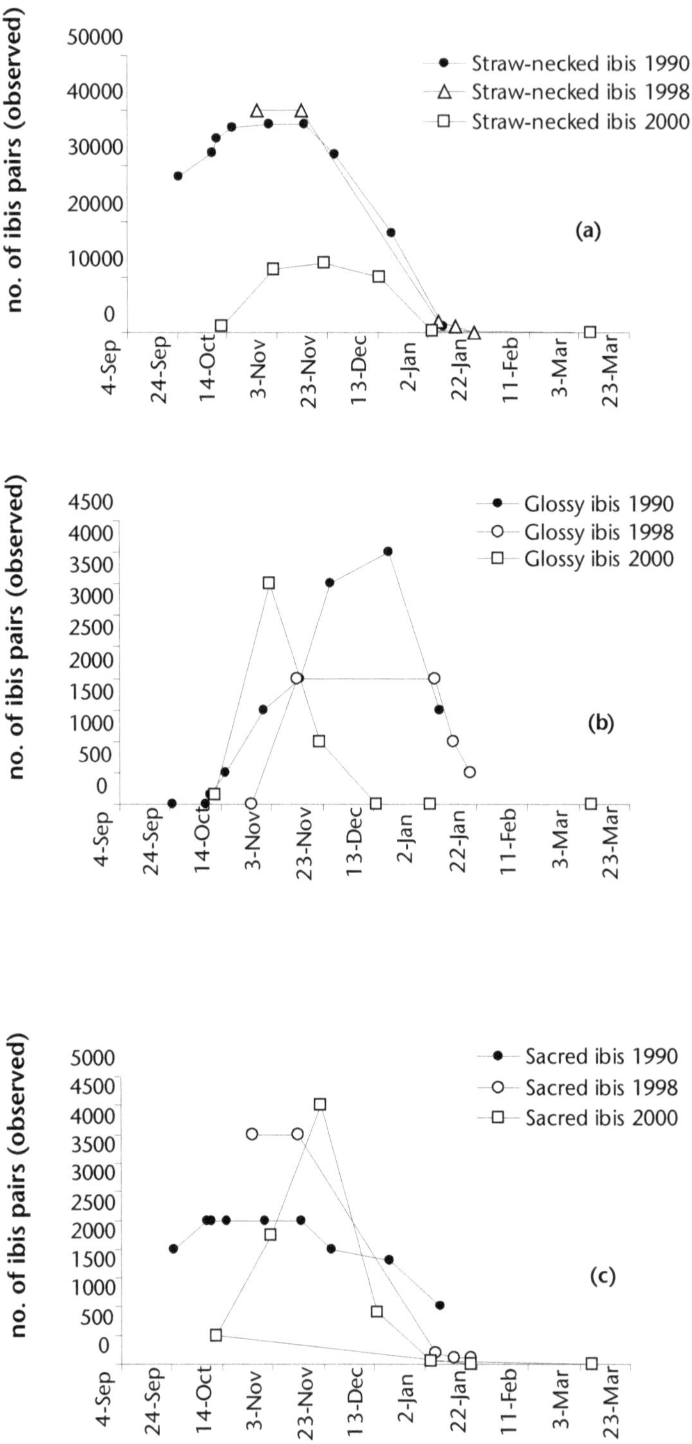

Figure 12.2: Number of breeding pairs of (a) straw-necked ibis, (b) glossy ibis and (c) sacred ibis during 1990, 1998 and 2000 in Booligal Swamp.

in the average period between floods). However, climate change effects were generally smaller for the less prolonged events (e.g. 21% versus 45% increase in the average period between floods for 19- and 60-day C_{mid} scenarios respectively; Table 12.1).

The effects of flow regulation for Booligal Swamp described above are relevant for the current scenario where local flow manipulation, aided by the use of the environmental contingency allowance, extends flood duration. In this scenario it is reasonable to assume that events of 19 days or more have ecological value because flood duration will be extended. However, without such local flow manipulation and for nearby wetlands in the Merrimajeel-Muggebah Creek complex (e.g. Upper and Lower Gum Swamps; Magrath 1992; Chessman *et al.* 2003), the events which lead to ibis recruitment, to the development of aquatic plant, macro-invertebrate and bird communities and to soil water recharge for locally common overstorey species lignum and river red gum will flood for ~>90 days (Bacon *et al.* 1993; Roberts and Marston 2000; Driver *et al.* 2005b). In this situation, the effect of flow regulation is ~50% reduction in flood frequency, and a further ~20% reduction in flood frequency because of climate by 2030 (Table 12.1).

Climate scenario models for Lachlan wetlands could include the most recent models, which that show ibis colony size is positively correlated with inundation period and show the effects of climate in other areas within the Murray-Darling Basin, and preceding drought. Chowdhury and Driver (2008) used a mixed regression model to predict the occurrence of a colonial nesting event and, if an event occurred, to predict the ibis colony size (using a Poisson distribution with excess dispersion at zero; Bohning *et al.* 1999; Table 12.2). Flood

Table 12.1: Some predicted effects of climate change on water availability in Booligal Swamp and the Great Cumbung Swamp within the operational window for the translucency rule, 15 May–15 November

	Without development	Current surface (1999–2000) and groundwater development			
	P	A	C_{dry}	C_{mid}	C_{wet}
Booligal Swamp	Years		% change from scenario A		
Average period between ≥90 day, ≥2500 ML/day floods	12	19 (50%)	50	20	–34
Average period between ≥60 day, ≥2500 ML/day floods	7	9 (23%)	118	45	–19
Average period between ≥19 day, ≥2500 ML/day floods	1.7	3.3 (94%)	109	21	–3
Great Cumbung Swamp	Years		% change from scenario A		
Average period between ≥1 day, ≥3000 ML/day floods	1.3	2.6 (100%)	138	27	–8
Average period between ≥60 day, ≥3000 ML/day floods	9	14 (49%)	60	33	–12
Average period between ≥90 day, ≥3000 ML/day floods	22	22 (0%)	400	25	0
Average period between ≥60 day, ≥800 ML/day floods	3	5 (56%)	102	38	6
Average period between ≥90 day, ≥800 ML/day floods	5	9 (102%)	71	9	–15

A = 1895–2006 climate and current development, P = 1895–2006 climate and no development, C = future (2030) climate and current development (with mid-range estimates predicting 10% reduction in mean annual run-off by ~2030). A 'dry', 'mid' (best estimate of median) and 'wet' variant are presented for scenario C.

Table 12.2: Model for predicting the size of colonial nesting ibis events at Booligal Swamp, including the % deviation explained by variables

Variable	Intercept (β)	Residual deviance	df	Deviation explained	Pseudo-R^2 with addition of each variable
Intercept (β)/null model (deviance)	6.90	838 239	17		
Number of days August–December discharge >2500 ML/day at Booligal (days)	0.078	200 094	16	638 145 (76%)	0.76
Months since last inundation at Booligal (gap)	0.02	149 470	15	50 624 (6%)	0.82
Polynomial number of days August–December discharge >2500 ML/day at Booligal ($durS^2$)	-0.0004	102 154	14	47 316 (6%)	0.88
Annual Queensland inflow into the border rivers (QldF)	-0.0011	69 939	13	32 215 (4%)	0.92

Values are based on nest counts in thousands (Chowdhury and Driver 2008).

duration in August–December explained 82% of the deviance (~non-parametric variance) of colony size. The incorporation of two other factors – the duration of preceding drought and the availability of wetlands in the northern Murray-Darling Basin – increased the deviance explained to 92%. Such modelling is open to risks of over-parameterisation, but is justified where variable inclusion is based on hypotheses and strong field evidence and when it helps further understand ecosystem function and future management decisions (Jakeman *et al.* 2006). Modelling with these justifications, Chowdhury and Driver (2008) found that when a wetland floods sufficiently to support ibis breeding, the size of the colony is positively correlated to the length of the preceding drought. This is supported by the finding that inland Murray-Darling Basin wetlands produce more invertebrate prey for waterbirds when flooding follows a period of complete drying (e.g. in Merrimajeel Creek; Maher and Carpenter 1984). In the ibis family (Ciconiiformes), larger breeding events have been shown to correspond with longer periods between breeding events (Florida Everglades; Frederick and Ogden 2001).

Models could also consider other flooded wetlands throughout the Murray-Darling Basin which could compete with Booligal Swamp in attracting colonial nesting birds. Ibis colony size was found to be smaller when there is greater nesting opportunity in the Border Rivers area (Table 12.2; Chowdhury and Driver 2008). This was only a small effect in the model, but it illustrates how wetland complexes operate at a landscape level (*sensu* Roshier *et al.* 2001). Straw-necked ibis typically occupy coastal refuges in north and eastern Australia, New Guinea and Wallacea during June–August, except when inland Australia exhibits wet conditions (McKilligan 1975; Gosper *et al.* 1983; Griffioen and Clarke 2002; Dingle 2004). Similarly, straw-necked ibis, sacred ibis and glossy ibis are less common in south-east Queensland/north-east NSW coastal refuges when rainfall is locally higher, possibly indicating wetland availability nearby (Gosper *et al.* 1983; Woodall 1985). When ibis leave these refuges they are likely to be seeking feeding or recruitment opportunities inland. This could include use of Booligal wetlands as well as closer wetland locations in the northern Murray-Darling Basin. The Border Rivers wetlands, which also support large colonial nesting events

(1088 km^2; Ley 1998; Kingsford 1999; Kingsford *et al.* 2004), were therefore chosen as a competing location for nesting.

Case study 2: Great Cumbung Swamp models and their use in climate change forecasting

A model of the impacts of climate change on the ecological responses of the Great Cumbung Swamp should include estimates of volumes required to inundate the swamp, and how long the flood should be maintained. The effect of hydrological changes in the swamp, induced by climate change, was modelled by CSIRO (2008a) on flows >3000 ML/day at Booligal Weir (~1500 ML/day at Oxley), based on estimates from Brady *et al.* (1998) of the flows required to flood the Great Cumbung Swamp. This model does not include the duration of flow required to flood the swamp in spite of published estimates of required flood duration (one month, LRMC 2002; 90 days, Driver *et al.* 2005b). Using wetland flow requirements that are more specific to particular ecosystems should improve predictions and would be more meaningful than reporting a high inundation frequency of 'floods' with flows of ~3000 ML/day at Booligal Weir and which last one or more days (Table 12.1). Extended medium-size floods that inundate most of the fringing river red gum forest would occur much less often, and even less frequently under flow regulation (Sims 1996).

As with the Booligal Swamp, we tested our assertions regarding the effects of flow regulation and climate change on the Great Cumbung Swamp using CSIRO scenarios (2008a). We gained similar results to CSIRO (2008a) for the average period between floods (years) for the Great Cumbung Swamp for ≥1-day duration, ≥3000 ML/day flood scenarios (Table 12.1). Our results showed on average a ~7% more prolonged non-flood period compared to CSIRO (2008a) scenarios. If durations of 60 or 90 days were used, the relative impacts on the time between floods were surprisingly small (respectively, P = 9 versus A = 14 days, or P = 22 versus A = 22), suggesting that prolonged large events at 3000 ML/day are largely outside the effects of flow regulation. However, climate change impacts are potentially profound (111, 28 and 22 years between ≥90-day duration, ≥3000 ML/day flood events for scenarios C_{dry}, C_{mid} and C_{wet} respectively). Our analysis then focused on the flows used to sustain floods in the Great Cumbung Swamp operationally (~400 ML/day at Oxley or ~800 ML/day at Booligal Weir). When sustained, these tend to push flows up into the edge of the river red gum forest and into lakes at the periphery (Driver *et al.* 2004, 2005b). These scenarios indicated that flow regulation and, to a lesser extent, climate change affect sustained floods (≥60 and ≥90 days, Table 12.1). The results also indicate that prolonged flood events required for river red gum maintenance (≥90 days) will be approximately half as frequent under the combined effects of flow regulation and climate change (scenario C_{mid}).

Including flows from the Murrumbidgee River, the major tributary for the Great Cumbung Swamp during very large floods, could improve the accuracy of future models. Murrumbidgee flows upstream of the Great Cumbung Swamp have negligible influence on the swamp's surface water or shallow groundwater under most flow conditions, so should not be included in the models with these surface water or shallow groundwater interactions (Driver *et al.* 2004; CSIRO 2008a). However, Shaikh *et al.* (1998) legitimately predicted area inundated using both Lachlan and Murrumbidgee River flows because they included the 1990 flood, a one-in-seven-year event where the two rivers join (O'Brien and Burne 1994).

Conclusions

Ecosystem response modelling for Murray-Darling Basin wetlands requires continuous improvement. Our models, a result of collaboration among various scientific disciplines, management agencies and stakeholders, in accordance with recommendations by Richter *et al.* (2006) and Jakeman *et al.* (2006), have helped quantify the environmental impacts on wetlands under the Lachlan water-sharing plan. For example, Lachlan ecosystem response models highlight the severe water shortage for the Great Cumbung Swamp under flow regulation, which could become profoundly worse under climate change. There may be many decades between the floods required for river red gum health. The models also show the benefits of environmental flows, such as for colonial nesting birds in Booligal Swamp. Notwithstanding the importance of these achievements, Lachlan models need to be subject to ongoing revision and augmentation. Studies that contribute to this process include investigations based on more direct measures of river red gum health in association with changing water availability in the Booligal Wetlands (Armstrong 2008). There are also efforts to integrate information from existing billabong models and distributary creek wetland models within landscape-scale wetland models, based on satellite imagery (Barma *et al.* 2009). Continued ecosystem response model revision is required, to maintain a dialogue between water managers and scientists (Jakeman *et al.* 2006), because of inevitable new climatic conditions that are outside model limitations and because of new modelling technologies, insights and water management arrangements. To this end, we have described how Lachlan wetland models can help climate change models progress to more ecosystem-specific models. For instance, they can incorporate the effects of preceding drought on waterbird responses in the Booligal Swamp model, or describe the Great Cumbung Swamp's need for sustained inflows of ≥400 ML/day for ≥90 days to sustain river red gum health. Such additional ecological insights can help to increase the efficacy of ecosystem response model predictions and inform wetland management if they form part of a dialogue between water managers and scientists.

Acknowledgments

We thank Drs Alex Meehan, Neil Saintilan (DECCW) and Naomi Rea (recently of Charles Darwin University), as well as Mr Paul Wettin (consultant), and Allan Raine (DECCW) for their constructive comments. Lachlan River ecological response models could only be progressively developed with the long-term support of DECCW colleagues such as Dr Michael Maher, Mr Peter Lloyd-Jones and Mr Ken Harris and State Water Corporation officers such as Mr Barry Orr. We are also grateful to Tim O'Connor and Vicki Martin (DECCW) for creating various versions of maps for this chapter. The views expressed in this chapter are those of the authors.

References

Adamson D, Williams MAJ and Baxter J (1987). Complex late Quaternary alluvial history in the Nile, Murray-Darling and Ganges Basins: three river systems presently controlled by the Southern Oscillation. In *International Geomorphology 1986. Proceedings of the 1st International Conference on Geomorphology.* (Ed. V Gardiner) pp. 875–887. John Wiley and Sons, Chichester.

Armstrong J (2008). The effect of river regulation on the Booligal Wetlands river red gum (*Eucalyptus camaldulensis*) swamps. Honours thesis. University of New South Wales.

Arthington AH, Bunn SE, Poff NL and Naiman RJ (2006). The challenge of providing environmental flow rules to sustain river ecosystems. *Ecological Applications* **16**, 1311–1318.

Bacon PE, Stone C, Binns DL, Leslie DJ and Edwards DW (1993). Relationships between water availability and *Eucalyptus camaldulensis* growth in a riparian forest. *Journal of Hydrology* **150**, 541–561.

Barma D, Wettin P, Powell S, Hughes D, Hameed T, Driver P, Harris K and Terrill P (2009). Lower Lachlan wetlands scoping study. Draft Report (Stage 1). May 2009. Barma Water Resource Consulting Pty Ltd for the Dept of Water and Energy.

Bohning D, Dietz E, Schlattman P, Mendonca L and Kirchner U (1999). The zero-inflated Poisson model and the decayed, missing and filled teeth index in dental epidemiology. *Journal of the Royal Statistical Society, Series A* **162**, 195–209.

Brady A, Shaikh M, King A, Ross J and Sharma P (1998). The Great Cumbung Swamp: assessment of water requirements. NRMS Project R-5048. Centre for Natural Resources, Ecosystem Management Branch, NSW Department of Land and Water Conservation.

Chambers LE, Hughes L and Weston MA (2005). Climate change and its impact on Australia's avifauna. *Emu* **105**, 1–20.

Chessman *et al.* (2003). Integrated monitoring of environmental flows. State summary report 1998–2000. NSW Dept of Infrastructure, Planning and Natural Resources.

Chowdhury S and Driver P (2008). An ecohydrological model of waterbird nesting events to altered floodplain hydrology. In *MODSIM 2007: International Congress on Modelling and Simulation*. December 2007. (Eds. L Oxley and D Kulasiri) pp. 2896–2902. Modelling and Simulation Society of Australia and New Zealand.

Closs GP, Balcombe S, Driver P, McNeil D and Shirley M (2006). The importance of floodplain wetlands to Murray-Darling fish: What's there? What do we know? What do we need to know? Native fish and wetlands in the Murray-Darling Basin. Action plan, knowledge gaps and supporting papers (Ed. B Phillips) pp. 14–28. Murray-Darling Basin Commission, Canberra.

Cowling JH and Lowe KW (1981). Studies of ibises in Victoria. I. Records of breeding since 1955. *Emu* **8**, 33–39.

CSIRO (2008a). Water availability in the Lachlan. Report to the Australian Government from the CSIRO Murray-Darling Basin Sustainable Yields Project. CSIRO, Canberra.

CSIRO (2008b). Water availability in the Murray-Darling Basin. Report to the Australian Government from the CSIRO Murray-Darling Basin Sustainable Yields Project. CSIRO, Canberra.

DEWHA (2008a). Department of Environment, Water, Heritage and the Arts. Australian heritage database. Available at http://www.environment.gov.au/heritage/ahdb/. Accessed 3 November 2008.

DEWHA (2008b). Water requirements of important wetlands in the Murray-Darling Basin: initial overview assessment. Final report. April 2008. Dept of the Environment, Water, Heritage and the Arts, Fluvial Systems and Lloyd Environmental. Environmental Water Branch, Canberra.

Dingle H (2004). The Australo-Papuan bird migration system: another consequence of Wallace's Line. *Emu* **104**, 95–108.

DIPNR (2004). Water-sharing plan for the Lachlan regulated river water source. As amended 1 July 2004. NSW Dept of Infrastructure, Planning and Natural Resources.

DPI (2005). Final recommendation aquatic ecological community in the natural drainage system of the lowland catchment of the Lachlan River. Fisheries Scientific Committee, NSW Dept of Primary Industries. File No. FSC 03/05, Ref No. FR 25.

Driver P, Raisin G, Wettin P and Sritharan S (2000). Lachlan River 1999/2000 environmental flow rules. Operation and performance report. Dept of Land and Water Conservation, Central West Region, July 2000.

Driver P, Higgins C, Lloyd-Jones P, Mackenzie-McHarg A, Raisin G, Unthank S and Wettin P (2002). Lachlan floodplain wetlands: adaptive water management framework. Final report. NSW Dept of Land and Water Conservation, Natural Heritage Trust and Lachlan River Management Committee.

Driver P, O'Rourke M, Robinson M, Jones J, Raisin G and Wettin P (2004). Great Cumbung Swamp water balance assessment. Final report. Resource Analysis Unit, NSW Dept of Infrastructure, Planning and Natural Resources.

Driver P, Chowdhury S, Wettin P and Jones H (2005a). Models to predict the effects of environmental flow releases on wetland inundation and the success of colonial bird breeding in the Lachlan River, NSW. In *Proceedings of the 4th Annual Stream Management Conference: Linking Rivers to Landscapes.* 19–22 October 2004. Launceston, Tasmania. (Eds ID Rutherfurd, I Wiszniewski, MJ Askey-Doran and R Glazik) pp. 192–198. Tasmanian Dept of Primary Industries, Water and Environment.

Driver P, Hameed T, Smith T, Wettin P and Raisin G (2005b). Rehabilitation of wetland inundation regimes in the Lachlan River, NSW, Australia: achievements and ongoing challenges. In *Proceedings of the 8th International River Symposium.* 6–9 September 2005. Brisbane, Queensland. Available at http://www.riversymposium.com/index.php?element=17. Accessed 3 November 2008.

Frederick PC and Ogden JC (2001). Pulsed breeding of long-legged wading birds and the importance of infrequent severe drought conditions in the Florida Everglades. *Wetlands* **21**, 484–491.

Frith HJ (1959). The ecology of wild ducks in inland New South Wales. *Wildlife Research* **4**, 156–181.

Gosper DG, Briggs SV and Carpenter SM (1983). Waterbird dynamics in the Richmond Valley, New South Wales, 1974–77. *Australian Wildlife Research* **10**, 319–327.

Griffioen PA and Clarke MF (2002). Large-scale bird-movement patterns evident in eastern Australian atlas data. *Emu* **102**, 99–125.

Hameed T and Podger G (2001). Use of IQQM simulation model for planning and management of a regulated river system. In *Integrated Water Resources Management.* (Eds MA Mariño and SP Simonovic) pp. 83–89. International Association of Hydrological Sciences. IAHS Publication No. 272.

Hameed T, Ribbons C, Driver P, Wettin P and Orr B (2005). Analyses of Lower Lachlan effluent flows for undeveloped *versus* developed conditions. Report for the Lachlan Customer Service Committee. July 2005. NSW Dept of Infrastructure, Planning and Natural Resources.

Hatton T and Evans R (1998). Dependence of ecosystems on groundwater and its significance to Australia. Land and Water Resources Research and Development Corporation (LWRRDC). Occasional Paper No. 12/98.

Herron N, Davis R and Jones R (2002). The effect of large-scale afforestation and climate change on water allocation in the Macquarie River Catchment, NSW, Australia. *Journal of Environmental Management* **65**, 369–381.

Hillman M, Aplin G and Brierley G (2003). The importance of process in ecosystem management: lessons from the Lachlan Catchment, New South Wales, Australia. *Journal of Environmental Planning and Management* **46**, 219–237.

Jakeman AJ, Letcher RA and Norton JP (2006). Ten iterative steps in development and evaluation of environmental models. *Environmental Modelling and Software* **21**, 602–614.

Junk WJ (2005). Flood pulsing and the linkages between terrestrial, aquatic, and wetland systems. *Verhandlungen Internationale Vereinigung für theoretische und angewandte Limnologie* **29**, 11–38.

Kingsford RT (1999). The Border Rivers region of arid Australia: government, irrigation and the environment. *Wetlands Ecology and Management* **7**, 25–35.

Kingsford R (2000). Ecological impacts of dams, water diversions and river management on floodplain wetlands in Australia. *Austral Ecology* **25**, 109–127.

Kingsford RT, Brandis K, Thomas RF, Crighton P, Knowles E and Gale E (2004). Classifying landform at broad spatial scales: the distribution and conservation of wetlands in New South Wales, Australia. *Marine and Freshwater Research* **55**, 17–31.

Leslie DJ (2001). Effect of river management on colonial nesting waterbirds in the Barmah-Millewa Forest, south-eastern Australia. *Regulated Rivers: Research and Management* **17**, 21–36.

Ley A (1998). The response of waterbirds to the 1997 flood in the Narran Lake Nature Reserve, New South Wales. *Australian Bird Watcher* **17**, 337–341.

LRMC (2002). Draft water-sharing plan for the Lachlan regulated river water source. Lachlan River Management Committee, NSW Dept of Land and Water Conservation.

Magrath MJL (1992). Waterbird study of the lower Lachlan and Murrumbidgee Valley wetlands in 1990/91. NSW Dept of Water Resources.

Maher M and Carpenter S (1984). Benthic studies of water fowl breeding habitat in New South Wales, chronomoid populations. *Australian Journal of Marine and Freshwater Research* **35**, 97–110.

Marchant S and Higgins PJ (1990). *Handbook of Australian, New Zealand and Antarctic Birds. Vol. 1.* Melbourne University Press, Melbourne.

McCullagh P and Nelder JA (1989). *Generalized Linear Models.* Chapman and Hall, London.

McKilligan NG (1975). Breeding and movements of the straw-necked ibis in Australia. *Emu* **75**, 199–212.

O'Brien PE and Burne RV (1994). The Great Cumbung Swamp. Terminus of the low gradient Lachlan River, eastern Australia. *AGSO Journal of Australian Geology and Geophysics* **15**, 223–233.

Richter BD, Warner AT, Meyer JL and Lutz K (2006). A collaborative and adaptive process for developing environmental flow recommendations. *River Research and Applications* **22**, 297–318.

Roberts J and Marston F (2000). Water regime of wetland and floodplain plants in the Murray-Darling Basin: a source book of ecological knowledge. Technical Report 30-00. October 2000. CSIRO Land and Water, Canberra.

Roshier DA, Whetton PH, Allan RJ and Robertson AI (2001). Distribution and persistence of temporary wetland habitats in arid Australia in relation to climate. *Austral Ecology* **26**, 371–384.

Shaikh M, Brady AT and Sharma P (1998). Applications of remote sensing to assess wetland inundation and vegetation response in relation to hydrology in the Great Cumbung Swamp, Lachlan Valley, NSW, Australia. In *Wetlands for the Future.* (Eds AJ McComb and JA Davis) pp. 595–606. Gleneagles Publishing, Adelaide.

Sims NC (1996). The impact of land and water development on the hydrology of the Great Cumbung Swamp. BSc (Hons) thesis. University of Canberra.

Smith T and Barr T (2002). Great Cumbung Swamp water balance assessment. Client report to the Dept of Land and Water Conservation, Central West Region, NSW. December 2002. CSIRO Land and Water, Canberra.

Sturt C (1833). *Two Expeditions into the Interior of Southern Australia during the Years 1828, 1829, 1830, 1831 with Observations on the Soil, Climate and General Resources of the Colony of New South Wales.* Smith, Elder and Co., London.

Thoms M and Sheldon F (2000). Lowland rivers: an Australian introduction. *Regulated Rivers: Research and Management* **16**, 375–383.

Van der Beek P and Bishop P (2003). Cenozoic river profile development in the Upper Lachlan (SE Australia) as a test of quantitative fluvial incision. *Journal of Geophysical Research* **108,** B6, 2309. doi:10.1029/2002JB002125.

Walker KF, Sheldon F and Puckridge JT (1995). A perspective on dryland river ecosystems. *Regulated Rivers: Research and Management* **11**, 85–104.

Ward JV and Stanford JA (1995). The serial discontinuity concept: extending the model to floodplain rivers. *Regulated Rivers: Research and Management* **10**, 159–168.

Wettin PG, Winning G and Murray M (1993). New South Wales. In *A Directory of Important Wetlands in Australia.* (Eds S Usback and R James) pp. 4/1–4/56. Australian Nature Conservation Agency, Canberra.

Wettin P, Driver P, Maher M and Thurtell L (2007). Environmental flows on the Lachlan River: 20 years of policy, science and adaptive management. In *Proceedings of the 10th International River Symposium and Environmental Flows Conference.* 3–6 September 2007. Brisbane. Available at http://www.riversymposium.com/2007_Presentations/B2A_Wettin.pdf. Accessed 16 January 2009.

Woodall PF (1985). Waterbird populations in the Brisbane region, 1972–83, and correlates with rainfall and water heights. *Australian Wildlife Research* **12**, 495–506.

Young W, Lam D, Ressel V and Wong M (2002). Development of an environmental flows decision support system. *Environment Modelling and Software* **15**, 257–265.

Chapter 13

Ecosystem response modelling in the Murrumbidgee River

Robyn Watts

Introduction

The Murrumbidgee River is one of the larger rivers of the Murray-Darling Basin in south-east Australia, having a catchment of approximately 84 000 km^2. It rises in the Snowy Mountains and flows generally westward to its confluence with the River Murray, 1600 km from its source (DLWC 1999; Fig. 13.1). The Murrumbidgee catchment contains 14 dams and eight large weirs, and has downstream irrigation areas containing over 10 000 km of irrigation canals (DSNR

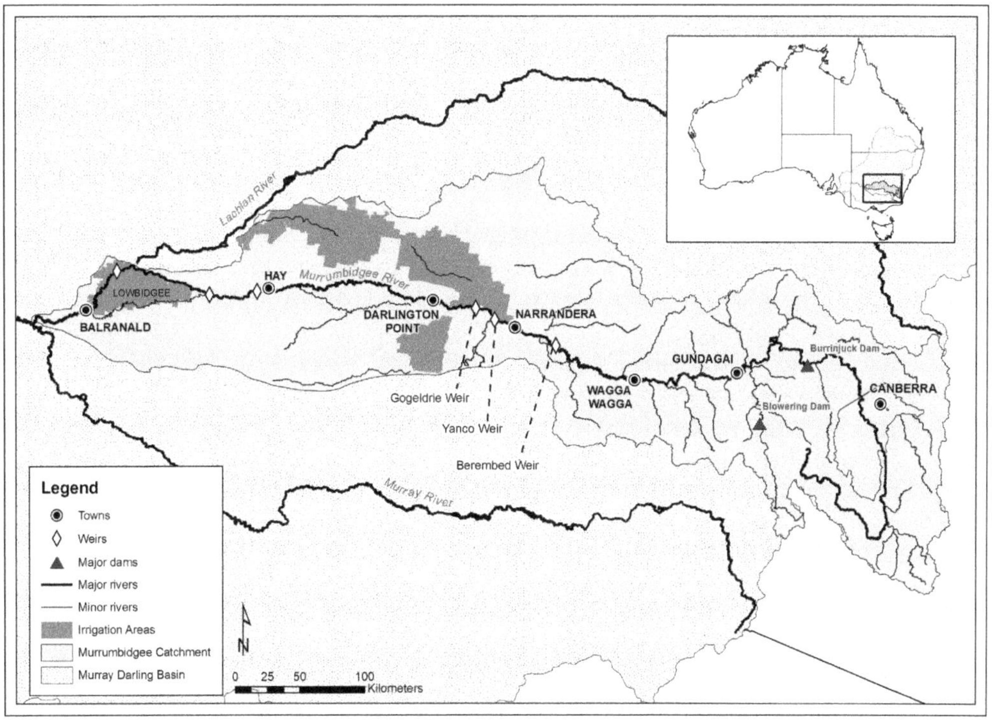

Figure 13.1: Murrumbidgee catchment showing major towns, dams, weirs and irrigation areas.

2003). The river is regulated by two major reservoirs – Burrinjuck Dam on the Murrumbidgee River and Blowering Dam on the Tumut River (MCMB 1999; Fig. 13.1). Most of the major structures regulating the flows of the Murrumbidgee River were constructed between 1912 and 1968 (Wen 2008). Downstream of Burrinjuck Dam the Murrumbidgee River flows through a confined gorge and then through a wide valley with some large meander scars and anabranches (Page and Nanson 1996). From Narrandera to Carrathool there are many large paleaeochannel meander cut-offs, whereas from Carrathool to Hay the river has a very narrow floodplain with few wetlands (Page and Nanson 1996). Downstream of Hay, the Murrumbidgee reaches the Lower Murrumbidgee floodplain, which is the most extensive wetland area in the catchment.

The natural flow regime of the Murrumbidgee River was characterised by low average flows in summer and autumn, and higher average flows in winter and spring (Page *et al.* 2005). Flows in the Murrumbidgee would have naturally been quite variable, with rainfall events in the upper and mid catchment creating flow pulses that would have travelled down the river, possibly as far as the most downstream reaches.

River regulation has considerably altered the flow regime of the Murrumbidgee River in terms of total volume, seasonal patterns of discharge, magnitude and frequency of floods, and frequency and duration of floodplain inundation. The effects of river regulation are different upstream and downstream of the main irrigation off-takes. A comparison of historical flows with modelled natural flows using integrated quality and quantity modelling shows that, between Burrinjuck Dam and the first major irrigation off-take at Berembed Weir (Fig. 13.1), average flows between 1970 and 1998 increased by diversions of approximately 550 000 ML per year from the Snowy Mountain Scheme (DLWC 1995). Upstream of the major irrigation off-takes there has been a seasonal redistribution of flows, with higher average flows in summer and autumn and lower flows in winter and spring (Page *et al.* 2005; Wen 2008). Downstream of the main irrigation areas, the seasonal pattern of flows is similar to natural, but flows have been reduced in all seasons. In addition, the variability of discharge has decreased during summer irrigation flows and in autumn and winter when minimum base flows are released during the filling phase of dam operations. River regulation has significantly reduced the magnitude of relatively frequent floods in the Murrumbidgee River. Between Gundagai and Balranald, regulation has reduced discharges at flood return periods from 1.25 to five years by 25–40% (Page *et al.* 2005; Frazier *et al.* 2005). Peaks-over-threshold recurrence interval data (Frazier *et al.* 2005) show the average time interval between bankfull flows has been doubled by flow regulation. River regulation has also affected a range of other parameters, such as water temperature, which at times is lower than natural for up to 300 km downstream of Blowering Dam (Keenan 2000).

The Murrumbidgee River and its floodplain wetlands contain significant environmental assets including nationally listed threatened species and ecological communities (Environment Australia 2001; *Threatened Species Conservation Act 2002* NSW), migratory shorebirds protected under the Japan–Australia and China–Australia Migratory Bird Agreements (JAMBA, CAMBA), Ramsar sites, wetlands of national importance and a national park. The aquatic ecological community of the Murrumbidgee River and tributaries downstream of Burrinjuck and Blowering Dams is listed as an endangered ecological community under the *Fisheries Management Act 1994*. The wetlands between Narrandera and Carrathool, collectively called the mid Murrumbidgee wetlands, are listed in the Directory of Important Wetlands in Australia (Environment Australia 2001). There are substantial forests of river red gum on the Murrumbidgee floodplain downstream of Burrinjuck Dam. At least 20 species of frogs, including the endangered southern bell frog, and 24 species of native fish including the endangered and iconic species Murray cod, trout cod and silver perch, and the iconic Murray crayfish, have been recorded downstream of Burrinjuck Dam (Tyler 1994; Lugg 2000; Environment Australia 2001; Jansen and Healey 2003).

The Murrumbidgee River and floodplain ecosystem has experienced a decline in ecological health. The fish and plant communities contain a high proportion of exotic species, many native species have low recruitment and regeneration, and many species have become locally extinct or experienced reduced distribution within the system (Watts *et al.* 2003a). The main threats to the ecology of the Murrumbidgee River include alterations to the natural timing of flows, reduced connections between the river and floodplain, cold water pollution, in-stream barriers and introduced species (Watts *et al.* 2003a). In addition, the riparian zone has been significantly altered by reduced flooding, clearing, logging, grazing by domestic livestock and the introduction of exotic plants (Watts *et al.* 2003a).

This chapter reviews the ecosystem response modelling undertaken in the Murrumbidgee catchment as part of the First Step of the Living Murray Initiative of the Murray-Darling Basin Commission, and provides an overview of other studies that have examined ecosystem responses to flow in the Murrumbidgee. It discusses the need for an integrated ecosystem response model for the Murrumbidgee River, and argues that there are sufficient data to populate such a model. This model would provide a tool to facilitate scenario testing for effective use and delivery of environmental and consumptive water in the Murrumbidgee catchment.

Ecosystem response modelling in the Murrumbidgee River

The best example of an ecosystem response model that has been applied on the Murrumbidgee River system is the Murray Flow Assessment Tool (MFAT). MFAT is a decision support system for predicting the ecological benefits and impacts of different flow scenarios along the River Murray system (Young *et al.* 2003). The Murrumbidgee River was one of 10 zones in the southern Murray-Darling Basin where the MFAT was applied as part of the First Step of the Living Murray Initiative of the (then) Murray-Darling Basin Commission. A scientific reference panel undertook the assessment, informed by regional evaluation groups that assessed the 10 zones. The MFAT was used to assess the ecological impacts of different flow scenarios at various localities on the condition of physical habitat for native fish, waterbirds, floodplain vegetation communities and wetland vegetation communities, and to assess the growth of algal blooms (Young *et al.* 2003; Fig. 13.2). The MFAT used daily river flows as the primary input data to a number of flow–ecology models that were based on scientific studies or expert opinion, and generated habitat condition indices for each of the 108 years of modelled flows. Threshold analysis was used to group these indices into four categories:

- worse (% of years each scenario is more than 10% below the reference scenario);
- no significant change (% of years each scenario is within a band of 10% below to 10% above the reference scenario);
- some improvement (% of years each scenario is within a band of 10–50% above the reference scenario);
- substantial improvement (% of years each scenario is more than 50% above the reference scenario) (SRP 2003).

For the First Step of the Living Murray Initiative the MFAT was run for several flow scenarios for each river zone, including a 'natural' scenario (flow conditions without flow regulation or diversion), 'current' scenario (flow conditions under the level of development, land use, regulation and diversions in 2003) and 'reference' scenario (flow condition as at the 93/94 basin cap, which represents the limit placed on the volume of water that can be diverted from the basin's rivers for consumption). The flow scenarios assessed for the Murrumbidgee River

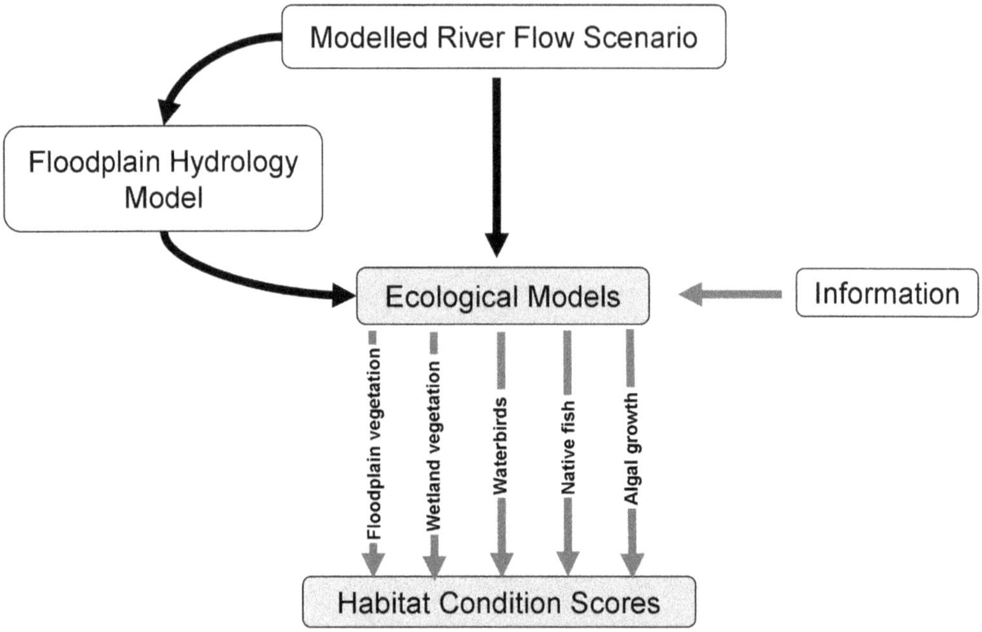

Figure 13.2: Murray Flow Assessment Tool framework (MDBC).

were 'natural', 'reference', 'current' and 'Year 6 water-sharing plan', and three scenarios based on reducing current allocations by a total of approximately 350 GL, '10%' (approx. 750 GL) and '20%' (1500 GL) (Watts *et al.* 2003b). After setting up the model and geographic weightings, the results were analysed to provide integrated assessments across river zones and the entire river system (Fig. 13.2) (SRP 2003).

The scientific reference panel concluded that a significant allocation of water would be required to secure a healthy working River Murray system. Annual allocations of water at the lower end of the range assessed (350 GL) may provide local benefits for some parts of the river, but substantial whole-of-river benefits only start to appear at the 750 GL and more generally at the 1500 GL scenario (SRP 2003). In the Murrumbidgee catchment, most of the individual habitat condition indices assessed by the MFAT showed an improvement at least once every 20 years, however, few indices showed an improvement at least once every five years (Watts *et al.* 2003b). The combined habitat condition index generally increased with higher environmental flow allocations (Fig. 13.3). The index was 'much better' than reference for 12% of years under the WSP6 flow scenario and 'much better' than reference for 18% of the years under the 20% flow scenario (Fig. 13.3) (SRP 2003).

There are several general limitations of the MFAT and some specific limitations in relation to its use for the assessment of flow scenarios for the Murrumbidgee River. When evaluating the assessment process, the scientific reference panel and external reviewers considered the MFAT to be an improvement on the expert panel approach previously used to predict responses to environmental flows, because the MFAT provided transparent documentation of expert opinion (Davies and Acreman 2003; SRP 2003). Use of the CSIRO Environmental Flow Decision Support System as the basis for developing MFAT as a decision support tool was considered by Davies and Acreman (2003) to have a number of significant advantages over other approaches:

- it had already been trialled within the basin at a catchment scale;
- it was based on existing expert knowledge of ecology–flow relationships;

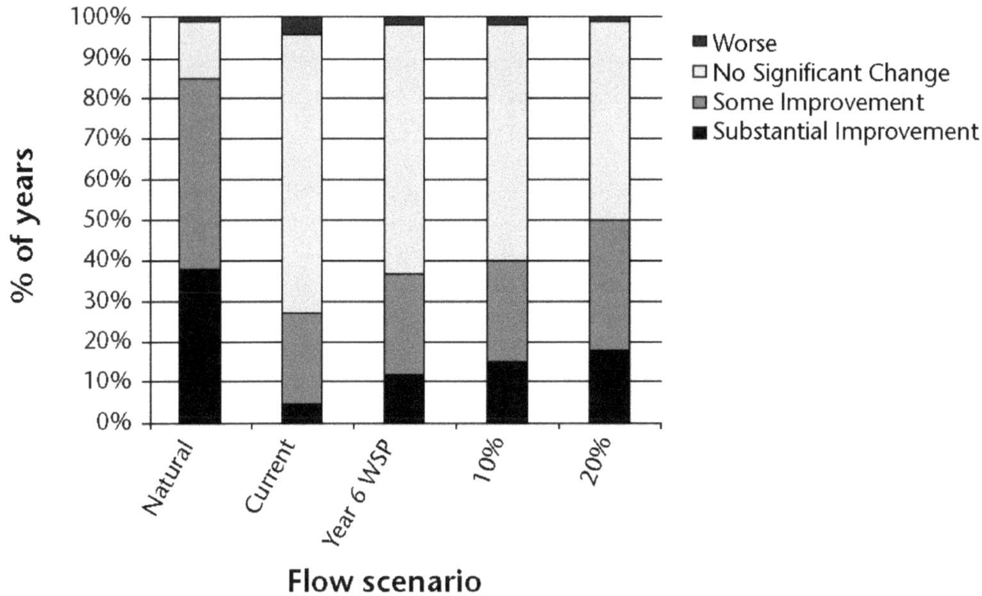

Figure 13.3: The improvement in combined habitat condition in zone J (Murrumbidgee River) relative to the reference scenario, defined as the number of years (n = 108) that the scenario/ reference statistic falls into the improvement categories. Worse =% of years each scenario is more than 10% below the reference scenario; no significant change =% of years each scenario is within a band of 10% below and 10% above the reference scenario; some improvement =% of years each scenario is within a band of 10–50% above the reference scenario; substantial improvement =% of years each scenario is more than 50% above the reference scenario (SRP 2003).

- it was based on a number of accepted scientific principles;
- it could be readily updated;
- it was already known by many of the scientists who were to be engaged in the panel process;
- the original developers were still available and could supply further developmental support.

Although the MFAT could provide estimates of responses to flow management scenarios, it was limited by the available knowledge and aspects of the decision support system. One of the main limitations of the MFAT is that it is a decision support tool and not an ecological model. Davies and Acreman (2003) stressed that MFAT outputs should not be interpreted as ecological outcomes, but as highly qualified indicators of relative potential based on flow-related habitat only. The MFAT provides an assessment of change in river habitat condition (SRP 2003), but habitat is not the sole factor affecting population distribution, abundance, recruitment and persistence. As a consequence, it was not possible to use the MFAT to predict the overall ecological response to environmental flows. More detailed hydrology–ecology models are required to predict the outcomes of a flow regime on species or communities. Furthermore, the MFAT used only river hydrological discharge data and was unable to include any hydraulic data (e.g. water depth, water velocity), although these are what most organisms essentially respond to (SRP 2003).

Another limitation is that the MFAT assessment was based on the modelling of only five biological indicator groups – native fish, waterbirds, wetland vegetation, floodplain vegetation and algae. These indicators were chosen because of the availability of scientific knowledge of these groups, which could support development and population of the ecological models (SRP 2003). However, these groups do not represent all components of the river ecosystem. In particular, they do not represent ecological processes and function. While including more assessments would not necessarily improve the accuracy of the assessment, it is unfortunate that some of these other indicators were not included in the model.

MFAT was designed to predict the outcomes of environmental flows for the whole river based on an aggregation of assessments at the river zone scale (SRP 2003). A limitation of this approach is that the whole-of-river assessment can be influenced and biased by the selection of sites. Sites for the Living Murray Assessment were selected to be representative of the range of conditions along a river zone or because they were key or iconic sites. In some cases, site selection was influenced by factors such as proximity to gauging stations and availability of ecological and hydrological information. For example, in the assessment of the Murrumbidgee River it was difficult to obtain hydrological information for some areas of the Lowbidgee, so these areas could not be included in the assessment. Only eight localities from the Murrumbidgee could be included in the MFAT assessment despite this being the largest of the 10 river zones. This was considered to be an inadequate representation of sites from that system, and a major constraint on the assessment (Watts *et al.* 2003b).

One of the main limitations of the Murrumbidgee River assessment was that the flow scenarios did not include full modelling of the environmental flow options. Only one flow delivery strategy was modelled for each of the flow scenarios. Furthermore, under the 10% and 20% flow scenarios, water was delivered to the environment according to the current flow rules. Therefore, the flow scenarios that were assessed in this report were not likely to produce the most beneficial ecological outcomes for the volume of water.

Ecosystem responses to flow

In addition to the ecosystem modelling of the Murrumbidgee River undertaken using the MFAT, there have been many individual studies of ecosystem responses to flow and flooding in this river system. Ecosystem responses to flows have been assessed using field-based studies during environmental flow releases, or field-based studies comparing reaches with different flow characteristics. In addition, ecosystem responses to flows have been examined using field and laboratory experimental simulation of different flow regimes, which can be a powerful way to compare the ecosystem responses to flow regimes within a controlled setting. Each approach has contributed to our broader understanding of ecosystem responses to flows in the Murrumbidgee River.

Responses of the in-stream ecosystem to flow

Upstream of the main irrigation off-takes in the Murrumbidgee River, the in-stream habitats experience higher water velocities during summer and less variable discharge compared to the natural flow regime, whereas in-stream habitats downstream of the main irrigation off-takes receive reduced discharge year-round (Page *et al.* 2005; Wen 2008). Several studies have examined the responses of the in-stream ecosystem in the Murrumbidgee River to aspects of the flow and flooding regime.

Ryder and Vink (2007) compared the chemical character and metabolic response to a natural flow event, an environmental flow release from Burrinjuck Dam and a piggy-backed flow (environmental water released to top up a natural flow from a tributary) in the Murrumbigdee River. They found that catchment run-off events trigger a different response compared with environmental flow releases from impoundments. Ryder and Vink (2007) found that using environmental flows as artificial floods can mobilise and transport nutrients and salts stored within the channel during low flows or, conversely, may dilute carbon concentrations and nutrient-rich natural floods with nutrient-poor and/or cold water from dams. Topping up rainfall-driven events with environmental flow releases that do not connect with the floodplain and entrain organic material may also dilute ecologically beneficial nutrients and prevent them from reaching downstream river reaches and aquatic food webs (Ryder and Vink 2007).

The biofilm community in the Murrumbidgee River downstream of Burrinjuck is strongly influenced by water level variability and water velocity. An experimental study examined the effects of in-channel water level variability on the biomass and productivity of epixylic biofilms grown on red gum blocks immersed 15 cm under water for up to 75 days in the Murrumbidgee River (Ryder 2004). The red gum blocks were used to experimentally simulate three regimes of water level variability common to the Murrumbidgee River: low inundation/emersion frequency (permanent inundation); intermediate inundation/emersion frequency (two cycles of 11 days inundated, 21 days emersed); and high inundation/emersion frequency (five cycles of five days inundated, nine days emersed).

All treatments were run over a 75-day period. The two variable flow regimes simulated the 1998 and 1999 environmental flow releases in the Murrumbidgee River respectively (Foreman 1999a, b), where the 1999 environmental flows were delivered as several smaller and shorter-duration pulses than the 1998 flows. Ryder (2004) showed that permanent inundation resulted in increased biofilm dry mass over time due to entrainment of particulate matter from the water column. In contrast, under permanent inundation biofilm net primary productivity and chlorophyll *a* increased until day 29 then declined as the biofilm became congested with sediment (Ryder 2004). The regime that had a wetting duration of 11 days separated by 21 days of drying between pulses had the highest net primary productivity and resulted in the biofilm with the highest biodiversity (Ryder 2004). This study demonstrated that extended periods of low discharge variability during irrigation releases can affect the productivity of biofilms, and therefore would have an influence on the aquatic food web. Varying the frequency and duration of inundation and desiccation of environmental flow releases can influence biomass and productivity of epixylic biofilms in regulated lowland rivers.

Water velocity can also affect the biomass and taxonomic composition of biofilms. Ryder *et al.* (2006) examined the effects of flow velocity on biofilms in an experimental flume. Blocks of river red gum wood were immersed in a regulated reach of the Murrumbidgee River for 30 or 70 days to enable biofilm to colonise the blocks, then either permanently immersed (wet) or dried for 20 days. Blocks from each of the four pre-treatments, i.e. each combination of colonisation period (30 or 70 days) and condition (wet or dried), were individually exposed to no velocity (control) or velocities of 0.3 m sec^{-1} (low), 0.55 m sec^{-1} (intermediate) or 1 m sec^{-1} (high) in a flume. Biofilms exposed to the experimental velocities had significantly lower dry mass and chlorophyll *a* than the controls (Ryder *et al.* 2006). All velocities resulted in a substantial reduction in taxonomic richness among all treatments, and filamentous Chlorophytes were completely removed by velocities of 0.55 m sec^{-1}. These results indicate the potential to delineate thresholds for the response of biofilms to water velocity based on knowledge of antecedent conditions that regulate biofilm development and condition (Ryder *et al.* 2006). The study suggested there is the potential to model and predict ecological responses to environmental flow releases in low-gradient floodplain rivers.

The response of the in-stream aquatic invertebrate community to flow regime has been examined in the Murrumbidgee River. Watts *et al.* (2001) found a shift in the timing of peak abundance of mayfly nymphs in the river reaches downstream of Blowering and Burrinjuck Dams, compared to two nearby unregulated rivers. The abundance of mayfly nymphs peaked in the regulated Murrumbidgee and Tumut rivers during the winter months when discharge was low, whereas the peak abundance in the unregulated Goodradigbee and Goobarragandra rivers occurred during the summer months when there is low discharge in those systems (Watts *et al.* 2001). The implication is that the timing of food availability for fish and other predators has been significantly altered in the upper reaches of the Murrumbidgee River.

Altered water velocities and water temperature due to river regulation also affect the in-stream communities in the Murrumbidgee River. Field surveys showed higher abundances of the shrimp *Paratya australiensis* and the crayfish *Euastacus armatus* in reaches downstream of irrigation off-takes than in reaches upstream of irrigation off-takes (Asmus 1999; Watts *et al.* 2001). *P. australiensis* breeds during spring and summer (Williams 1977; Hancock and Bunn 1999) and *E. armatus* carries eggs until October (Geddes 1990). Thus, the most vulnerable early larval stages of these species are present during the summer irrigation season when there is usually high discharge, high velocities and colder water temperatures than under natural flow conditions. Studies elsewhere in Australia have shown that *P. australiensis* aggregate in areas where current is slower (Williams 1977), larvae and berried females are poor swimmers (Hancock and Bunn 1999) and velocity can influence larval development and recruitment (Hancock *et al.* 1998). Thus, the timing and magnitude of irrigation flows in the Murrumbidgee River may limit recruitment and reduce the abundance of these species upstream of the irrigation off-takes in the Murrumbidgee River. Buik (2008) examined the habitat occupancy of different life-stages of two species of shrimps (*P. australiensis* and *Macrobrachium australiense*) in habitats such as coarse woody debris, backwater, bare bank and channel that differed in flow velocities. Buik (2008) found that larvae or berried females were more likely to occupy habitats with slower water velocities. Non-berried adults and juveniles of both species occupied a wider range of habitats.

There may be similar negative effects of regulated flows on the larvae of native fish species that inhabit reaches upstream of the irrigation off-takes, because the early larval stages of many native fish are present during the summer irrigation season. A study of the swimming capabilities of larval fish under different water velocities showed that early larval stages of Murray cod, trout cod, golden perch and silver perch are poor swimmers and that the maximum water velocity at which these larvae can swim is very low (Kopf, unpubl. data). This suggests that larval fish would be unable to utilise or move between a large proportion of in-stream habitats in the Murrumbidgee River during the irrigation season.

Responses of riverbank ecosystems to patterns of inundation

The riverbanks upstream of the main irrigation off-takes in the Murrumbidgee River are inundated for long periods and have relatively stable water levels, particularly during the summer irrigation period, compared with the natural inundation patterns (Lowe 2002). Studies of riverbank sediments from the Murrumbidgee River upstream of the irrigation off-takes have shown that they can contribute significant quantities of carbon during high flow events (Burns and Ryder 2001). The carbon supplied by flow pulses can stimulate food webs and so sustain fish populations (Watts *et al.* 2009). Thus, altered patterns of riverbank inundation have the potential to vastly alter in-stream river food webs in this river system.

Riverbanks upstream of irrigation off-takes in the Murrumbidgee River that are inundated for long periods and have relatively stable water levels in summer have lower plant species richness, lower vegetative cover and different plant community composition from riverbanks in reaches downstream of the irrigation off-takes (Lowe 2002). Lowe *et al.* (in press) implemented an outdoor laboratory experiment to examine the separate and interactive effects of inundation duration and sediment depth on survival, growth and vegetative reproduction of riverbank plants. Two herbaceous clonal riverbank plant species, *Carex bichenoviana* and *Calotis scapigera*, were grown under three levels of sediment addition and two levels of inundation duration (10 days and 30 days) and were harvested after 60 days. Both species survived up to 30 days inundation in the absence of sediment, responding to inundation by increasing shoot biomass relative to root biomass (Lowe *et al.* in press). In contrast, sediment deposition during periods of inundation had a negative impact on the survival of both species (Lowe *et al.* in press). The results indicated that the combined effects of inundation and sediment deposition during irrigation flows can be detrimental for riverbank plants in regulated rivers (Lowe *et al.* in press).

Lowe (2002) also examined the effects of different patterns of inundation on the survival, growth and reproduction of two riverbank plant species. Individuals of *C. scapigera* and *C. bichenoviana* were transplanted into riverbanks in three river reaches with differing flow regimes, and plant survival and growth was monitored at regular intervals from late winter through to the following autumn. The different patterns of inundation within a single year resulted in different survival, growth and reproduction responses of perennial riverbank plants (Lowe 2002). Both the upright perennial sedge, *C. bichenoviana*, and the low-growing clonal daisy, *C. scapigera*, were unable to survive long periods of inundation (>30 days) during the summer months. However, both species were capable of surviving multiple short periods of inundation (less than five days), regardless of season. Growth and reproduction of both species were significantly less when exposed to long periods of inundation over the spring and summer months or in the summer months only, compared with multiple short periods of inundation (Lowe 2002).

Ecosystem responses to wetland hydrology

The frequency, duration and timing of wetland inundation both upstream and downstream of the main irrigation off-takes in the Murrumbidgee River have been altered by river regulation. Some wetlands have been become permanently inundated by weir pools (Frazier *et al.* 2005; Frazier and Page 2006) and some lowlying wetlands occurring upstream of the irrigation off-takes are more regularly inundated by summer irrigation releases. In contrast, wetlands with river connections higher than the level of irrigation flows have reduced flooding frequency (Frazier and Page 2006). Downstream of the main irrigation off-takes, the frequency and duration of connection between wetlands and the river is significantly reduced due to both river regulation and floodplain development. Kingsford and Thomas (2004) estimated that 26% of the total wetland area in the Lowbidgee was lost before 1975, and an additional 32% was lost between 1975 and 1998. Several studies have examined ecological responses to these altered flooding regimes.

Gell and Little (2006) used diatom-based palaeolimnological techniques to trace the water quality history of eight wetlands along the Murrumbidgee River floodplain between Wagga Wagga and Balranald, from the origin of each wetland to the present. Most of the wetlands changed post-European settlement from shallow clear wetlands supporting a diversity of macrophytes to wetlands that are eutrophic, affected by salinity and with an accumulation of sediments (Gell and Little 2006).

The altered flooding regime of Murrumbidgee wetlands has consequences for the health of trees on the floodplain. Increased inundation due to regulation of flows in the river has killed some river red gums and detrimentally affected a large proportion of open water wetlands in the mid Murrumbidgee (Thornton and Briggs 1994). Roberts (1994) showed that a combination of water regime and environmental factors affected vegetation characteristics of riparian and aquatic vegetation of wetlands of the Murrumbidgee. Increased periods between flooding, combined with the effects of grazing, has been shown to promote ant–seed predator activity (Meeson *et al.* 2002), and hence may affect regeneration of any species with seeds harvested by ants, notably river red gum.

Drying is a normal part of the ecology of the Murrumbidgee wetlands and many wetland plant and zooplankton species persist in these environments because they have dormant seed and eggs. The length of the dry period between flooding events may affect species resilience, as the viability of the seed and egg bank decreases with increasing elapsed time. Podnar (2008) examined the effects of long-term drying and permanent inundation on the emergence of plants and zooplankton from sediments of 20 Murrumbidgee wetlands with varying flooding histories. The study was undertaken at an outdoor experimental facility where soil samples from wetlands were exposed to either damp or flooded conditions to induce plant germination, or were submerged within a net to induce zooplankton eggs to hatch. The results showed the egg and seedbanks of the wetlands were long-lived, and persisted through dry periods of up to 11 years. However, plant and zooplankton taxa richness was much lower in wetlands last flooded 30 years ago than in wetlands last flooded two, seven and 11 years ago. There was lower taxonomic richness of plants but higher zooplankton taxa richness emerging from sediments of the permanently flooded wetlands, compared with other treatments. The finding that extended periods of inundation reduce the viability of the seedbank, result in less diverse plant communities and prevent the growth of terrestrial plants is consistent with findings from elsewhere (Nielsen and Chick 1997). The higher diversity of zooplankton in permanent wetlands suggests that plants are more susceptible to prolonged inundation than zooplankton. However, this finding was not consistent with studies that have shown that permanent flooding removes cues necessary to trigger hatching and breaking dormancy of some zooplankton species (Nielsen *et al.* 2000).

Healy (2007) examined the relationship between water regime and macro-invertebrates in eight wetland sites on the floodplain of the middle Murrumbidgee River. Macro-invertebrate community composition and water quality were compared between permanent and temporary wetland sites in three habitats – coarse woody debris, open water and littoral. Permanent wetlands were deeper, contained fewer macrophytes and had lower invertebrate abundance, lower species richness and different community composition than macro-invertebrate communities in temporary wetlands (Healy 2007). The findings of this study indicated that macro-invertebrate communities differ in response to wetland water regime and are driven by complex interactions of water regime including water depth, water permanence and the presence of macrophytes.

The fish communities in Murrumbidgee wetlands are also influenced by wetland hydrology. Budd (2005) surveyed four permanent and four temporary wetlands on the floodplain of the mid Murrumbidgee River to determine the distribution and abundance of fish species. Of 23 species (including five exotic species) which have historically or recently been recorded from Murrumbidgee River wetlands (Lugg 2000), only eight species (including three exotic) were sampled during this study. Fish were more abundant in permanent billabongs and these communities were dominated by carp gudgeon species. Juvenile common carp were the most common fish in temporary billabongs (Budd 2005). The decline of fish communities in Murrumbidgee wetlands is most likely linked to the isolation of some wetlands from the river and unseasonal flooding of others, pollution from urban and agricultural run-off, and the invasion of exotic fish species. Watts and Budd (2006) suggested that these processes need to be reversed or mitigated on landscape scales before natural recolonisation or restocking of fishes will be successful.

In Chapter 14 of this book, Wassens discusses how wetlands with differing hydroperiods support different frog communities. Non-burrowing frogs usually use permanent waterbodies for refuge sites but abandon these as soon as surrounding areas are flooded, moving to breed in the newly flooded areas even if the permanent wetlands remain in good condition. Adults and new recruits may later move long distances to return to the permanent wetlands

(Schwarzkopf and Alford 2002; Wassens *et al.* 2007, 2008a). As many frogs have relatively inflexible activity and breeding periods, some species may fail to respond if wetlands flood at times outside their normal breeding periods (Ch. 14, this volume). Reduced duration of flooding has an impact on species with longer development times and limits the number of breeding opportunities over time, thus playing a significant role in population persistence. For example, the altered flooding regime of the Lowbidgee has had a significant effect on the vulnerable southern bell frog, which was the dominant frog species in river red gum wetlands and black box/lignum wetlands in the Lowbidgee in 2001 and 2002 (Wassens 2005). Since 2001 the volume of water entering the Lowbidgee floodplain has declined substantially. This has affected populations of southern bell frog in two ways: direct loss of habitat resulting in reduced recruitment and mortality, and loss of connectivity and hence gene flow between populations (Ch. 14).

Wassens *et al.* (2008b) examined the response of frogs to watering of wetlands in the Lowbidgee in December 2007 and January 2008. Six wetlands were targeted for flooding in a bid to maintain extant *Litoria raniformis* populations, and the breeding response and recruitment success were monitored until April 2008. The response of *L. raniformis* to flooding was mixed. Recruitment was observed at two of the target wetlands and a third incidental site, but no recruitment was observed in two other sites that contained adult *L. raniformis* (Wassens *et al.* 2008b). The probability of occupancy by *L. raniformis* increased with increasing cover of emergent and submerged vegetation and increasing turbidity. Wetlands that had previously experienced annual flooding were more likely to support *L. raniformis* than those flooded less frequently. High densities of common carp may have contributed to poor recruitment at some of the wetlands. Clearly, *L. raniformis* is more sensitive to changing flooding frequency and to drought periods than was previously thought.

Wetland flooding regimes also have a major influence on the abundance and spatial and temporal distribution of waterbirds in Murrumbidgee wetlands. The considerable reduction of wetland area in the Lowbidgee resulted in a significant decline in wetland-dependent birds of more than 80%, from about 117 000 in 1983–86 to about 18 500 in 1997–2000 (Kingsford and Thomas 2004).

Conclusions

A good understanding of ecosystem responses to flow and flooding has emerged through these studies undertaken in the Murrumbidgee catchment. The breadth of information available on ecosystem responses to flow is considerably greater than that used by the MFAT ecosystem modelling undertaken for the Living Murray Assessment (Watts *et al.* 2003b). In addition to information on fish, birds, plants and algae there is knowledge on ecosystem processes, biofilms, invertebrates and frogs, and there is better coverage of studies in all the geomorphic zones downstream of Burrinjuck Dam. However, many of the studies have focused on a single species, a group of related organisms or selected river reaches of the Murrumbidgee catchment. Some of the studies have shown there are different ecosystem responses in different reaches of the Murrumbidgee River and, in some cases, that ecosystem components responded to the same flow or flooding event in different ways.

Other than the use of the MFAT to assess the outcome of flow scenarios for the Living Murray Initiative (Watts *et al.* 2003b), there has been no tool or mechanism to integrate the findings from the individual studies. Our understanding of ecosystem responses to flow in the Murrumbidgee catchment has now developed to a stage where there is a need to integrate the findings of the different studies and there are sufficient data to populate an integrated ecosystem model.

There is a need for an ecosystem model to be developed specifically for the Murrumbidgee River. It should incorporate the data from the studies discussed here, plus other transferable results from studies in similar catchments. There are sufficient data to populate an ecosystem model for the Murrumbidgee. The Murrumbidgee Wetland Resource Book (Murray 2008) and a wetland database being developed by the Murrumbidgee Catchment Management Authority provide essential information that could contribute to such a model. A wetland inundation model similar to RiMFIM developed for the River Murray (Overton *et al.* 2006) is being developed for Lowbidgee, however, there is a need for this type of model to be developed for the whole regulated river system of the Murrumbidgee River. Ideally, the ecosystem model would be able to incorporate hydrology and flow hydraulics into in-channel assessments and be able to link in-stream flow to habitat and groundwater conditions. It should be able to incorporate information on key species and communities as well as key ecosystem processes. It should provide regional- and reach-scale information on ecological outcomes of flow management scenarios and possibly be coupled with management of the other factors influencing river health in the catchment (e.g. water quality, water temperature, exotic species).

In the current water policy environment in Australia, there is a need for a much better ecological understanding of ecosystem, population and species responses to flow regime, to support environmental flow allocations and other water management decisions. An ecosystem model such as that proposed for the Murrumbidgee River would provide a tool to facilitate improved modelling and scenario testing for effective use and delivery of environmental and consumptive water in the Murrumbidgee River.

Acknowledgments

The author was supported by a Charles Sturt University Research Centre Fellowship while writing this manuscript. Thanks to Deanna Duffy, CSU Spatial Analysis Unit, for production of the catchment map. Thanks to two reviewers for their comments on this chapter.

References

Asmus MW (1999). Effects of recreational fishing on populations of Murray crayfish *Euastacus armatus* in the Murrumbidgee River. Honours thesis. Charles Sturt University, Wagga Wagga.

Budd C (2005). The distribution and abundance of fishes in middle Murrumbidgee River billabongs, New South Wales. Honours thesis. Charles Sturt University, Wagga Wagga.

Buik S (2008). Habitat occupancy of freshwater shrimps in an Australian lowland river system. Honours thesis. Charles Sturt University, Wagga Wagga.

Burns A and Ryder DS (2001). Responses of bacterial extracellular enzymes to inundation of floodplain sediments. *Freshwater Biology* **46**, 1299–1307.

Davies P and Acreman M (2003). Review of the interim report of the SRP, the Murray Flow Assessment Tool and the SRP-REG scientific process. Report to the Murray-Darling Basin Commission, Canberra. Available at http://xlivingmurray.mdbc.gov.au/__data/page/255/peer_rev_sept_03.pdf. Accessed 12 May 2009.

DLWC (1995). State of the Rivers Report Murrumbidgee Catchment 1994–1995. Vol. 1. NSW Dept of Land and Water Conservation.

DLWC (1999). Stressed Rivers Assessment Report, Murrumbidgee Catchment. NSW Dept of Land and Water Conservation.

DSNR (2003). A guide to the water-sharing plan for the Murrumbidgee regulated river water source. NSW Dept of Sustainable Natural Resources.

Environment Australia (2001). *A Directory of Important Wetlands in Australia.* Environment Australia, Canberra.

Foreman M (1999a). Hydrological examination of the 1998 environmental flows on the Murrumbidgee River. NSW Dept of Land and Water Conservation.

Foreman M (1999b). The 1999 Murrumbidgee River environmental flows. NSW Dept of Land and Water Conservation.

Frazier P, Page KJ and Read A (2005). Effects of flow regulation on flow regime in the Murrumbidgee River, south eastern Australia: an assessment using a daily estimation hydrological model. *Australian Geographer* **36(3)**, 301–314.

Frazier P and Page KJ (2006). The effect of river regulation on floodplain wetland inundation, Murrumbidgee River, Australia. *Marine and Freshwater Research* **57(2)**, 133–141.

Geddes MC (1990). Crayfish. In *The Murray.* (Eds N Mackay and D Eastburn) pp. 302–307. Murray-Darling Basin Commission, Canberra.

Gell P and Little F (2006). Long-term water quality changes in Murrumbidgee floodplain wetlands revealed by diatom assemblages. In *Wetlands of the Murrumbidgee River Catchment: Practical Management in an Altered Environment.* (Eds I Taylor, P Murray and S Taylor) pp. 43–65. Fivebough and Tuckerbil Trust, Leeton.

Hancock MA and Bunn SE (1999). Swimming response to water current in *Paratya australiensis* Kemp, 1917 (Decapoda, Atyidae) under laboratory conditions. *Crustaceana* **72**, 313–323.

Hancock MA, Hughes JM and Bunn SE (1998). Influence of genetic and environmental factors on egg and clutch sizes among populations of *Paratya australiensis* Kemp (Decapoda: Atyidae) in upland rainforest streams, south-east Queensland. *Oecologia* **115**, 483–491.

Healy S (2007). The impact of altered water permanency on the aquatic macroinvertebrate communities of floodplain wetlands of the middle Murrumbidgee River, New South Wales. Honours thesis. Charles Sturt University, Wagga Wagga.

Jansen A and Healey M (2003). Frog communities and wetland condition: relationships with grazing by domestic livestock along an Australian floodplain river. *Biological Conservation* **109**, 207–219.

Keenan H (2000). Issues paper No. 4: thermals. Report for the Murrumbidgee Regulated River Management Committee. NSW Dept of Land and Water Conservation.

Kingsford RT and Thomas RF (2004). Destruction of wetlands and waterbird populations by dams and irrigation on the Murrumbidgee River in arid Australia. *Environmental Management* **34**, 383–396.

Lowe BJ (2002). Impacts of altered flow regime on riverbank vegetation in a lowland floodplain river. PhD thesis. Charles Sturt University, Wagga Wagga.

Lowe BJ, Watts RJ, Roberts J and Robertson A (in press). The effect of experimental inundation and sediment deposition on the survival and growth of two herbaceous riverbank plant species. *Plant Ecology.*

Lugg A (2000). The Bidgee: bringing back the fish. The status of fish populations in the Murrumbidgee River system and how they might be restored. Information paper prepared for the Murrumbidgee River Management Committee and Murrumbidgee Unregulated Streams Management Committee. NSW Fisheries, Nowra.

MCMB (1999). River management. Murrumbidgee Catchment Management Board. Available at http://murrumbidgee-catchment.org.au/frame/river.htm. Accessed 12 June 2003.

MCMB (2003). Murrumbidgee Catchment blueprint. NSW Dept of Land and Water Conservation.

MCMC (1998). Murrumbidgee Catchment action plan. Murrumbidgee Catchment Management Committee, Wagga Wagga.

Meeson N, Robertson AI and Jansen A (2002). The effects of flooding and livestock on post-dispersal seed predation in river red gum habitats. *Journal of Applied Ecology* **39**, 247–258.

Murray PA (2008). Murrumbidgee Wetlands resource book. Murrumbidgee Catchment Management Authority, Wagga Wagga.

Nielsen DL and Chick AJ (1997). Flood-mediated changes in aquatic macrophyte community structure. *Marine and Freshwater Research* **48(2),** 153–157.

Nielsen DL, Smith FJ, Hillman TJ and Shiel RJ (2000). Impact of water regime and fish predation on zooplankton resting egg production and emergence. *Journal of Plankton Research* **22(3),** 433–446.

Overton IC, McEwan K and Sherrah JR (2006). The Murray River floodplain inundation model: Hume Dam to Lower Lakes. Technical Report 2006. CSIRO Water for a Healthy Country. CSIRO, Canberra.

Page KJ and Nanson GC (1996). Chronology of Murrumbidgee River, palaeochannels on the riverine plain, southeastern Australia. *Journal of Quaternary Science* **11(4)**, 311–326.

Page KJ, Read A, Frazier P and Mount N (2005). The effect of altered flow regime on the frequency and duration of bankfull discharge: Murrumbidgee River, Australia. *River Research and Applications* **21(5),** 567–578.

Podnar K (2008). Does time since flooding affect emergence of micro-invertebrates and plants from wetland sediments? Honours thesis. Charles Sturt University, Wagga Wagga.

Roberts J (1994). Riverbanks, plants and water management. In *Murrumbidgee River: Past and Present.* (Eds J Roberts and R Oliver) pp. 70–82. CSIRO, Griffith.

Ryder DS (2004). Response of epixylic biofilm metabolism to water level variability in a regulated floodplain river. *Journal of the North American Benthological Society* **23**, 214–223.

Ryder D and Vink S (2007). Managing regulated flows and contaminant cycles in floodplain rivers. In *Salt, Nutrient, Sediment and Interactions: Findings from the National River Contaminants Program.* (Ed. A Mortlock) pp. 109–126. Land and Water Australia, Canberra.

Ryder DS, Watts RJ, Nye E and Burns A (2006). Can flow velocity regulate epixylic biofilm structure in a regulated floodplain river? *Marine and Freshwater Research* **57(1),** 29–36.

Schwarzkopf L and Alford RA (2002). Nomadic movements in tropical toads. *OIKOS* **96**, 492–506.

Scientific Reference Panel (2003). Ecological assessment of environmental flow reference points for the River Murray system. Interim report for the Murray-Darling Basin Commission Living Murray Initiative. CRC Freshwater Ecology, Canberra.

Thornton SA and Briggs SV (1994). A survey of hydrological changes to wetlands of the Murrumbidgee River. *Wetlands* **13**, 1–13.

Tyler MJ (1994). Frogs of eastern New South Wales. In *Future of the Fauna of Western New South Wales.* (Eds D Lunney, S Hand, P Reed and D Butcher) pp. 155–160. Royal Society of NSW, Sydney.

Wassens S (2005). The use of space by the endangered southern bell frog (*Litoria raniformis*) in the semi-arid region of New South Wales, Australia. PhD thesis. Charles Sturt University, Wagga Wagga.

Wassens S (2010). Flooding regimes for frogs in lowland rivers of the Murray-Darling Basin. In *Ecosystem Response Modelling in the Murray-Darling Basin.* (Eds N Saintilan and I Overton) pp. 213–228. CSIRO Publishing, Melbourne.

Wassens S, Roshier DA, Watts RJ and Robertson AI (2007). Spatial patterns of a southern bell frog (*Litoria raniformis*) population in an agricultural landscape. *Pacific Conservation Biology* **13**, 104–110.

Wassens S, Watts RJ, Jansen A and Roshier D (2008a). Movement patterns of southern bell frogs (*Litoria raniformis*) in response to flooding. *Journal of Wildlife Research* **35**, 50–58.

Wassens S, Arnaiz O, Healy S, Watts RJ and Maguire J (2008b). Identification of hydrological and habitat requirements to maintain viable southern bell frog (*Littoria raniformis*) populations on the Lowbidgee floodplain: phase 1. Final report. DECCW, Canberra.

Watts RJ and Budd C (2006). Fish communities in floodplain wetlands of the mid-Murrumbidgee River. In *Wetlands of the Murrumbidgee River Catchment: Practical Management in an Altered Environment*. (Eds IR Taylor, PA Murray and SG Taylor). Fivebough and Tuckerbil Wetlands Trust, Leeton.

Watts RJ, Ryder D, Chisholm L and Lowe B (2001). Assessment of environmental flows for the Murrumbidgee River: selecting indicators: developing ecological indicators for assessing river flow management. Report to the NSW Water Management Fund, Charles Sturt University, Wagga Wagga.

Watts RJ, Jansen A, Thompson L, Page K and Mullins B (2003a). Review of the ecological health of the Murrumbidgee River and its floodplain downstream of Burrinjuck Dam. Report to MDBC. CSU Johnstone Centre Report No. 181, Wagga Wagga. Available at http://www.csu.edu.au/research/jcentre/reports/jcec_report181.pdf. Accessed 10 January 2009.

Watts R, Read A, Page K, Crook D, Frazier P, Hardwick L, Jansen A, Lowe B, Lugg A, Roshier D and Ryder D (2003b). Ecological assessment of flow scenarios in the Murrumbidgee River (Zone J) for the Living Murray Initiative. Report to Murray-Darling Basin Commission, Canberra. Available at http://thelivingmurray.mdbc.gov.au/__data/page/256/InterimFinalOct03ZoneJ.pdf. Accessed 10 January 2009.

Watts RJ, Allan C, Bowmer KH, Page KJ, Ryder DS and Wilson AL (2009). Pulsed flows: a review of environmental costs and benefits and best practice. Waterlines report. National Water Commission, Canberra. http://www.nwc.gov.au/resources/documents/waterlines_report_No_16_-_Pulsed_flows_full_version.pdf. Accessed 10 January 2009.

Wen L (2008). Reconstruction natural flow in a regulated system, the Murrumbidgee River, Australia, using time series analysis. *Journal of Hydrology* **364**, 216–226.

Williams WD (1977). Some aspects of the ecology of *Paratya australiensis* (Crustacea: Decapoda: Atyidae). *Australian Journal of Marine Freshwater Research* **28**, 403–415.

Young WJ, Scott AC, Cuddy SM and Rennie BA (2003). Murray Flow Assessment Tool: a technical description. Client Report, 2003. CSIRO Land and Water, Canberra.

Chapter 14

Flooding regimes for frogs in lowland rivers of the Murray-Darling Basin

Skye Wassens

Introduction

The Murray-Darling Basin supports 31 frog species, three of which (*Litoria raniformis, Litoria booroolongensis* and *Crinia sloanei*) are listed nationally as vulnerable or endangered under the *Environment Protection Biodiversity Conservation Act 1999* (EPBC Act). Frogs occur in nearly all freshwater habitats within the basin, from ephemeral rain-fed pools in semi-arid regions to permanent lakes and streams across the Great Dividing Range. Frogs are biphasic (have a distinct tadpole and adult frog stage) and the species present in the Murray-Darling Basin all require standing water for reproduction and tadpole development. These species exhibit a range of behavioural and physiological adaptations which allow them to persist in landscapes where water resources are often variable over space and time. About half the Murray-Darling Basin frogs are classed as burrowing species. The ability to burrow, coupled with an ability to slow their metabolism to minimise energy and water use, allows these species to survive underground for extended periods (Withers and Thompson 2000). Burrowing species are common in wetlands within the drier areas of the Murray-Darling Basin, such as the Darling River system, where water resources are most variable. Non-burrowing species are common in areas with more persistent waterbodies and reliable flooding. Although they do not have specific adaptations for burrowing, this group can survive for short periods by sheltering in moist locations such as logs, mats of vegetation and deep clay cracks. Non-burrowing species may abandon wetlands as they dry in favour of nearby permanent waterbodies, as is the case for *Litoria raniformis* (Wassens *et al.* 2008b). This chapter will focus on non-burrowing species, as this group is most likely to be affected by flow regulation and environmental watering.

The response of frogs to flooding is rarely considered when planning environmental flow strategies. This may be due to the misconceptions that frogs are poorly understood and difficult to predict or that their requirements are catered for when targeting other taxonomic groups. This chapter outlines the key parameters that drive the response of frog communities following wetland flooding, and presents a case study of flooding management to maintain populations of a threatened frog species. The chapter describes how broad-scale parameters, such as the spatial distribution of waterbodies and connectivity, influence frog community structure and their response to flooding. Following this, the key factors that influence frog occupancy patterns and recruitment success at individual wetlands are described.

Landscapes and mosaics

Many frog species can move reasonable distances overland to access resources (Harrison and Taylor 1997; Marsh *et al.* 1999, 2000; Marsh and Trenham 2001; Trenham *et al.* 2001; Smith and Green 2005). This movement between waterbodies, whether a one-off dispersal movement or a regular movement in response to changes in the distribution of waterbodies, plays an important role in determining the persistence of frog populations and how they respond to flooding.

Movement between waterbodies is important at two different spatial and temporal scales. At large scales the proximity and degree of connectivity between waterbodies determines the potential for recolonisation of waterbodies following local extinctions (Smith and Green 2005). This can be an important consideration when predicting and monitoring the outcomes of environmental watering at wetlands that have been dry for an extended period (more than three years). If the area targeted for watering is isolated and potential movement corridors, such as flood runners or canals, are limited, the ability of frogs to recolonise the wetland will also be limited. As a result the response of frogs (in terms of increases in diversity and abundance) may be limited compared to the response of more mobile species such as waterbirds.

At smaller scales, floodplain wetlands typically occur within a mosaic of waterbodies with differing hydrological regimes, for example, a permanent river or stream which is associated with seasonally flooded oxbow lagoons and temporary rain-fed depressions. Individuals of many frog species regularly move around the wetland mosaic in response to wetting and drying patterns (Marsh *et al.* 2000; Wassens *et al.* 2007b, 2008b). For example, *L. raniformis* occupies permanent waterbodies during dry periods but will immediately abandon these sites in favour of newly flooded areas even though the habitat characteristics of permanent waterbodies remain unchanged (Wassens *et al.* 2008b). Breeding is more likely to occur in newly flooded areas than in permanent waterbodies, and adults along with the new generation can move quite long distances to return to permanent waterbodies during dry periods (Schwarzkopf and Alford 2002; Wassens *et al.* 2008b). A very simplified example of this processes is shown in Figure 14.1. Here, frogs occupy permanent waterbodies during dry periods, move into temporary waterbodies to breed then retreat back to permanent waterbodies with the new recruits. One of the most important consequences of this opportunistic habitat use is that past flooding history across the mosaic can play the same or a greater role in driving abundance and persistence than habitat features at a particular time, especially during dry periods. That is, frogs congregate within permanent waterbodies during dry periods and can appear to be very abundant, but this is often due to past flooding of surrounding areas. The second important consideration when planning environmental flooding is that the proximity and condition of permanent waterbodies in the area will influence how frogs respond to flooding. From a manager's perspective, simply asking 'Where is the nearest permanent waterbody?' and 'Does it look OK?' when targeting environmental water can give a good indication of which frog species will be present and how they will respond to flooding.

Local habitat features

In addition to the mosaic structure, a number of wetland parameters can be important in predicting which frog species will be present and how they will respond to a specific flooding event. Wetland hydroperiod, flooding frequency, the timing of inundation, vegetation structure, predator abundance and inter- and intraspecific competition can also influence recruitment outcomes and community composition (Babbitt and Tanner 2000; Semlitsch 2000; Snodgrass *et al.* 2000; Hazell 2003).

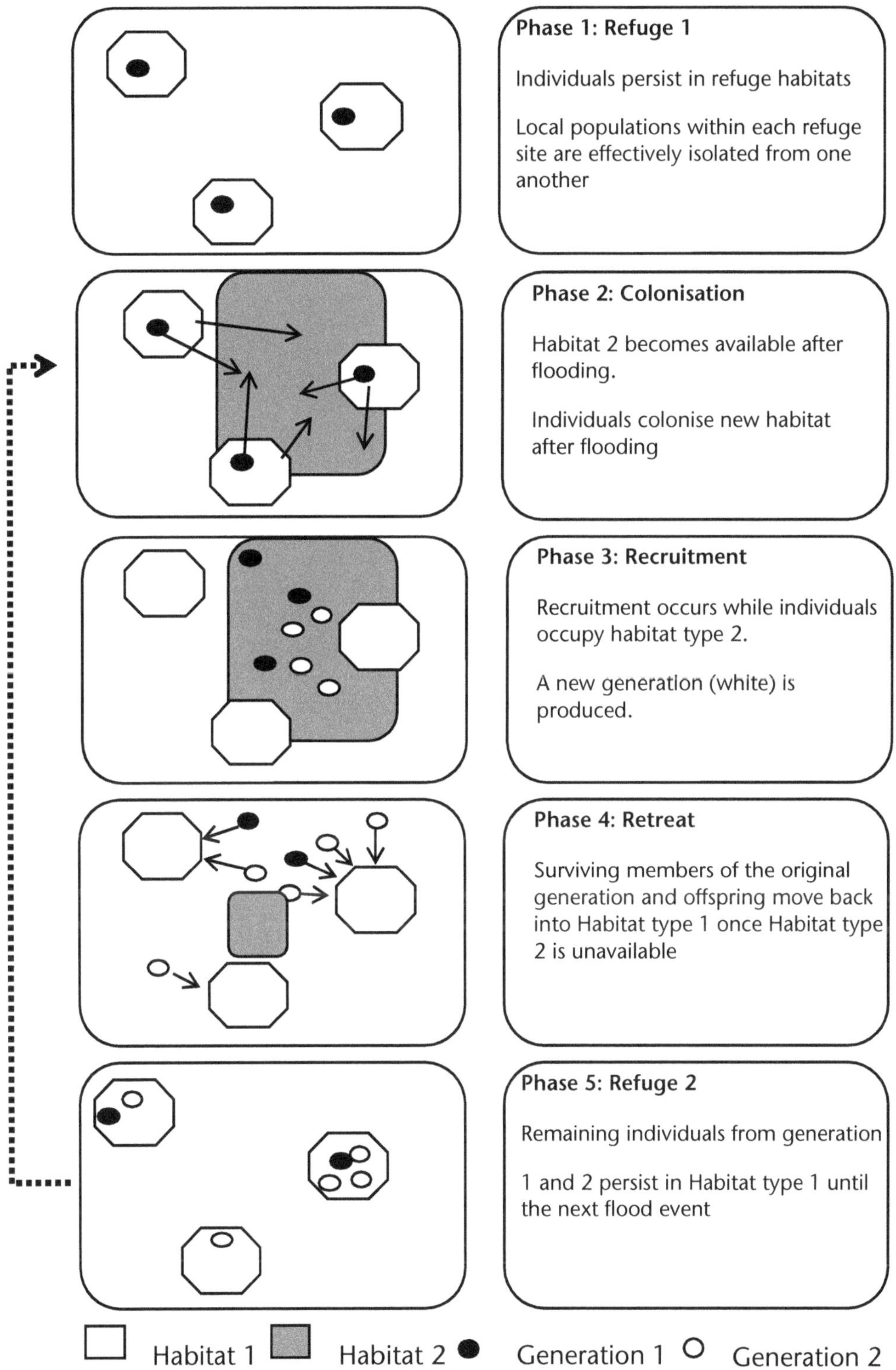

Figure 14.1: Simplified model of the response of non-burrowing wetland frog species to wetting and drying within a wetland mosaic.

Hydroperiod

Tadpole life-span (time from egg to complete metamorphosis) can range from as little as three weeks for burrowing species such as *L. rubella* to six months for some larger species (Anstis 2002). Hydroperiod therefore determines the length of time available for breeding activity and tadpole development, and hence which species are able to complete their development. Chronic reductions in wetland hydroperiod will exclude species with longer development times. The reverse is also true – increasing hydroperiod can lead to higher predator densities and can exclude species that are sensitive to predation (Adams 2000). Increasing water permanence can lead to reduced vegetation complexity (Casanova and Brock 2000; Warwick and Brock 2003), which may exacerbate predator effects by reducing the availability of cover. In the Murrumbidgee catchment, the conversion of seasonally flooded wetlands such as barren box swamp into permanent water storages coincided with the extinction of resident southern bell frog (*L. raniformis*) populations, probably because it led to increases in fish densities and reductions in vegetation cover (Wassens 2008).

Given the above factors, wetland hydrology can exert a strong influence on community composition and on waterbodies with differing hydroperiods typically supporting distinctly different frog communities. For example, a comparison of frog communities within minor stream systems in the Lachlan catchment showed that, while permanent and temporary waterbodies had similar numbers of species, the types of species present were significantly different. Of the species recorded, *L. latopalmata*, *L. rubella* and *Crinia parinsignifera* were more likely to occur in temporary waterbodies, while *L. peronii* favoured permanent sites (Wassens *et al.* 2007a).

Hydroperiod is important when managing environmental flooding and attempting to predict the impacts of flow regulation. Understanding the relationship between tadpole development times and hydroperiod can be useful when deciding how much water should be applied to a wetland and for what duration. Extreme care needs to be taken if setting lower limits for wetland hydroperiods, because the resulting juvenile frogs often need to stay close to the waterbody for some time to gain condition. For example, *L. raniformis* juveniles often stay close to their natal site for months after adults have abandoned the area (Wassens, unpubl. data). Similarly, tadpoles that accelerate their development as a result of decreasing water levels metamorphose at a smaller size, which can reduce their fitness as adults (Scott 1994).

Frequency of inundation

The frequency with which a wetland is inundated influences both the structure of frog communities and population persistence (Vignoli *et al.* 2007). There have been no specific studies on the impacts of altered flooding frequency on Australian frog species, however, the reduction in flooding frequency along the Murrumbidgee River since 1980 (Page *et al.* 2005) has coincided with local extinctions of *L. raniformis* and *L. caerulea*, both of which require regular flooding (Wassens 2008). The alteration of flooding frequency can lead to reduced recruitment opportunities for frogs or may force frogs into more permanent waterbodies which have higher predator densities, thus reducing recruitment outcomes (Vignoli *et al.* 2007). Direct mortality of resident frogs can occur where wetlands dry out completely for extended periods and no alternative waterbodies are available.

In principle, the maintenance of high-quality refuge habitats should mediate the impacts of reduced flooding frequency, although this has never been empirically tested. Maintaining high-quality refuge habitats might assist frogs in several ways – it can lower mortality by ensuring sufficient cover from predators, it can provide shelter sites and it can provide more constant food resources. High-quality refuge habitats may allow low levels of recruitment to occur during drought periods. For example, during the drought conditions experienced on the

River Murray in 2007 and 2008 small numbers of recently metamorphosed *L. peronii* and *Limnodynastes tasmaniensis* were observed around well-vegetated permanent waterbodies (Wassens, unpubl. data). These permanent habitats were not used for breeding when temporary waterbodies were available.

Timing of inundation

Many frogs have specific activity periods, and the timing of inundation can influence which species are able to respond. *C. signifera* and *C. parinsignera*, for example, can breed at just about any time of year including mid-winter. Both species respond well to winter flooding and are often the dominant frog species in wetlands during late autumn, winter and early spring. In contrast, *L. peronii* prefers higher temperatures. It is unlikely to respond during flooding in cooler months, but is often the dominant species in summer (Wassens, unpubl. data).

Vegetation

Both aquatic and terrestrial vegetation can influence frog community composition and response to flooding. Aquatic vegetation is used as an anchor point for egg masses (Anstis 2002), shelter from predators (Adams 1999) and a substrate for biofilms on which many tadpoles feed (Anstis 2002). Terrestrial and fringing vegetation is important for winter hibernation and shelter during dry conditions (Hamer *et al.* 2003). Some of the tree frogs, for example *L. peronii*, utilise hollows in standing timber for shelter and foraging, and may be absent from sites without trees. Loss of wetland vegetation is a common problem in areas affected by livestock grazing or hoofed feral animals. These factors can contribute to declines of frogs and poor breeding responses following flooding (Jansen and Robertson 2001; Jansen and Healey 2003).

Interactions with other species

Many frog species preferentially breed in waterbodies that temporarily or periodically dry out, because those wetlands have lower densities of fish than permanently inundated ones (Adams 2000). The presence of predatory fish can reduce frog recruitment via direct predation of eggs and tadpoles (Holomuzki 1995; Hamer *et al.* 2002; Teplitsky *et al.* 2003), refusal of females to deposit eggs (Holomuzki 1995) and changes in tadpole behaviour and fitness (Teplitsky *et al.* 2003). While many frog species use waterbodies that are almost exclusively fish-free, the species that utilise inland wetland systems are assumed to be relatively tolerant to the presence of fish because these larger wetland systems also support native fish populations. However, introduced fish species such as *Gambusia* sp. and European carp appear to have a far greater impact on tadpole behaviour and recruitment success than native fish species, potentially because they can occur at higher densities and are faster to colonise wetlands during flooding. European carp, which are common in many inland wetland systems, also reduce water quality and damage vegetation (Roberts *et al.* 1995; Pinto *et al.* 2005), which may affect tadpole survival.

Case study: flood management for species conservation

The southern bell frog *(L. raniformis)* is a vulnerable (EPBC Act 1999) wetland species, formerly common and abundant throughout the southern section of the Murray-Darling Basin. This species is now restricted to a few key floodplain habitats in the Lower Murrumbidgee and River Murray systems. Typical of non-burrowing frog species in the Murray-Darling Basin, *L. raniformis* generally utilises seasonally flooded wetlands as breeding sites and moves to permanent waterbodies when flooded habitats are not available (Wassens *et al.* 2008b). Consequently, the quality and proximity of permanent waterbodies can play a significant role in determining the occupancy patterns of *L. raniformis* and its response to wetland flooding.

The Lowbidgee region is an inland delta system fed by the Murrumbidgee River. This region contains many of NSW's most significant populations of *L. raniformis* (Wassens 2005). During surveys in 2001 and 2002 it was the dominant frog species in river red gum (*Eucalyptus camaldulensis*) wetlands and black box (*E. largiflorens*)/lignum (*Muelenbeckia florulenta*) wetlands (Wassens 2005). Since 2000, the volume of water entering the Lowbidgee floodplain has declined substantially. By 2004, *L. raniformis* was rare in most of its former habitats. The reductions in the volume of water entering the Lowbidgee floodplain has probably affected populations in two ways:

- direct loss of habitat, leading to reduced recruitment and increased mortality;
- loss of connectivity between wetlands, which occurs when the landscape becomes drier and the distances between waterbodies increase due to reductions in the area flooded.

Some former habitats have dried out completely, others are flooded less frequently and other sites have been reduced to small areas of permanent water which are maintained for stock and domestic purposes.

Emergency wetland watering 2007/08

In September 2007, lack of flooding and the drying of key wetlands formally occupied by *L. raniformis* raised the real concern that the species could be lost from the region unless there was intervention. Preliminary surveys of 21 sites during September 2007 failed to locate *L. raniformis* or wetlands in suitable condition to support a recruitment event (Wassens *et al.* 2008a). The NSW Minister for Climate Change, Environment and Water made available up to 10 GL of environmental water (initially 3 GL, with the additional water depending on surpluses and tributary inflows), which had been set aside in the water-sharing plan for the Murrumbidgee regulated river to water wetlands in the region (Wassens *et al.* 2008a). The aim of the environmental diversions was to inundate targeted wetlands within the Lowbidgee floodplain to create southern bell frog habitat.

Site selection

Dry conditions when wetlands were being selected for watering meant that *L. raniformis* individuals were not active and were therefore difficult to detect. Fortunately, comprehensive surveys of wetlands across the Lowbidgee floodplain had been conducted following the last large flood that occurred across the Lowbidgee floodplain, in 2000. Surveys had been conducted in October 2000 and January 2001, with follow-up surveys in January 2004 (Wassens 2005; fig. 14.2). The survey results were used as a guide to where *L. raniformis* had previously occurred and to define the key habitat characteristics of wetlands that supported *L. raniformis*. This information was used to generate a list of wetlands that were known to have supported *L. raniformis* populations in 2000 and 2004 or were likely to have supported *L. raniformis*, based on their habitat characteristics and proximity to known populations.

The challenge was to predict which wetlands were most likely to have maintained a viable population since 2004. Using the principles described above, we developed a decision-making framework to prioritise sites for watering (Box 14.1). As *L. raniformis* is known to utilise permanent waterbodies during dry periods, the proximity and quality of drought refuge habitats was a major consideration. Wetlands with no permanent water nearby (<2 km) were excluded from further consideration even if they were known to have supported *L. raniformis* in 2001. Wetland hydrology was considered both as a means of predicting whether *L. raniformis* was likely to have persisted since 2004 and as a means of determining the suitability of wetlands that had not been

Box 14.1: Decision-making framework for selecting sites for watering to maintain viable L. raniformis populations in the Lowbidgee region of NSW

1 The likelihood that *L. raniformis* previously occurred at the wetland (positive identification is weighted higher than anecdotal sightings)
   a Date *L. raniformis* last recorded
   b If not recorded at the site, distance to nearest known site
2 Drought refuge
   a Proximity of the site to drought refuge habitats (waterbodies that have held water consistently since late 2005)
   b Quality of drought refuge habitats if present
      - Aquatic and fringing vegetation cover
      - Predator density
   c Probability of individuals being able to access and persist in drought refuge habitats after the wetland dries
   d Presence of mortality sinks (e.g. regulators with an enclosed box base are known to trap frogs, snakes and turtles). *L. raniformis* is particularly vulnerable because individuals often move into the regulators because they hold water, but are unable to climb out when the water levels go down
3 Target wetlands (habitat suitability)
   a Quality and condition of target wetlands as breeding sites (estimate vegetation cover and complexity if possible)
   b Time since last flooded
   c Frequency of inundation
   d Estimated wetland hydroperiod
   e Wetland management (heavy grazing, cattle wallowing, water extraction, burning)
   f Landholder support (if applicable)
   g Representativeness of wetland plant communities (river red gum/black box–lignum)
4 Technical considerations
   a Estimate transfer losses
   b Volume of water required
   c Presence of water control structures (regulators, pipes, canals)
5 Landscape considerations: capability of the target wetland to act as a source population to allow for future recolonisation of other *L. raniformis* habitats in the region
   a Distance of wetland to other known populations
   b Position of wetland within the context of the known range of the species (edge, middle, centre of known range)
   c Dispersal pathways between wetlands

previously surveyed. Technical considerations, particularly transfer losses and the volume of water required to pool water in the wetland for extended periods, were also considered.

The framework attempted to assess the long-term viability of watering sites and their importance in maintaining the species across the whole floodplain. The initial allocation of just 3000 ML could save only two or three populations, thus it was important to ensure that those populations were well connected to former habitats so that recolonisation could occur during the next large flood.

Framework implementation

Technical considerations were used for the first cull, and any waterbody that could not be fully flooded with the available volume of water was excluded. As a result, river red gum and black box wetlands on the northern and southern sides of the Murrumbidgee River, which were more than 2 km from the Red Bank Weir or a suitable regulator structure, were excluded because they would have required more than the available 3 GL to fill. Likewise, waterbodies at the bottom of the Nimmie-Caria systems were not considered because of unacceptable transfer losses.

Of the remaining sites, three wetlands where *L. raniformis* had been recorded in 2001 were excluded because there were no longer any permanent waterbodies within 2 km of the target wetland, hence there was a high probability that *L. raniformis* was already extinct from those sites.

This left nine wetlands that fulfilled the selection criteria: they were known or were predicted to have supported *L. raniformis*, were near permanent water, could be filled with minimal transfer losses and would be connected to former *L. raniformis* habitats during a larger flood. As the volume of water was limited, these nine sites were ranked according to the selection criteria and the best six were selected for watering. Three were within black box/lignum wetlands in the Nimmie-Caira system (Avalon, Eulimbah and Warwaegae swamps) and three were within river red gum wetlands of Yanga National Park (Mercedes, Pococks and Two Bridges Swamp) (Table 14.1). We aimed to water clusters of wetlands in close proximity to each other and to water supply infrastructure. This ensured that there was some connectivity between watering sites in each system and minimised transfer losses.

Additional environmental water allocations were made available in early January and, after topping up the six initial watering sites, the remaining three wetlands were flooded. Paul Coates Swamp was in very close proximity to Mercedes Swamp and was known to have supported *L. raniformis* until 2004, but was prioritised below Mercedes Swamp because its large area made it difficult to flood to a sufficient depth with the volume of water initially available. Southern bell frogs were recorded at South Eulimbah Stock Dam, part of the larger Eulimbah Swamp complex, but the site was initially not considered because it already received stock and domestic water. It was topped up in March after southern bell frogs were found there, to ensure that resident individuals would survive through to 2008/09. The third site, Shaw's Swamp, was not predicted to support *L. raniformis* because it did not flood frequently enough. However, it filled via a leaking regulator and was later topped up to maintain the aquatic vegetation.

Outcomes

Initial response

L. raniformis responded to flooding at six of the nine sites watered. It was absent from Avalon Swamp where it had previously been recorded in 2004 and from Pocock's Swamp, which had not previously been surveyed. It was also absent from Shaw's Swamp in Yanga National Park, as predicted. Overall the abundance was very low; fewer than 20 individuals were recorded at each site during six days of searches. In comparison, 92 *L. raniformis* individuals were collected in a single night at Paul Coates Swamp in January 2001 (Wassens 2005).

As noted earlier, it may be possible to extend the length of time that populations persist between floods by providing persistent refuges of high quality and with low predator densities.

Table 14.1: General characteristics of wetlands targeted for environmental watering in 2007

Site	Description	Date *L. raniformis* last recorded	Time dry (months)	Distance to other known populations	Frequency of connection between target wetlands nearest population	Area flooded (ha)	*L. raniformis* present in 2008
Eulimbah Swamp	Large lignum wetland with deep canal on one side	Dec. 2007	8	<2 km Eulimbah Stock Dam	1–2 years via canal	<50 ha (2005 flooding extent 3000 ha)	Yes
South Eulimbah Stock Dam	S&D dam feed via lignum flood runner. Southern edge of Eulimbah Swamp	March 2008	0	<2 km Eulimbah	Yearly S &D via lignum flood runner	20	Yes
Warwaegae	Lignum/cumbungi permanent S&D dam and overflow into lignum	Never surveyed	5	6 km to Avalon	Yearly S &D via canal	6	Yes
Mercedes Swamp	Tall spike rush/river red gum wetland. Currently dry but flooded annually every year until 2006	Jan. 2001	6	< 1km Paul Coates	Yearly via weir (except 2006)	95	Yes
Two Bridges Swamp	Tall spike rush/river red gum on the flood runner to Piggery Swamp. Patches of damp soil remaining	Never surveyed	2	<2 km Mercedes	Yearly via weir (except 2006)	180	Yes
Paul Coates Swamp	Complex of open tall spike rush and forested short spike rush wetlands. Subject to forestry and some grazing	Jan. 2004	12	<1 km Mercedes	Yearly via weir (except 2006)	100	Yes
Avalon Swamp	Black box/lignum. Main section of waterbody permanent until 2006	Jan. 2004	8	25 km to Eulimbah	1–2 years via canal	34	No
Shaw's Swamp	Shallow forested, short spike rush wetland off Wagorah Creek in Yanga NP	Never surveyed	36	>5 km Two Bridges	>2 years?	100	No
Pocock's Swamp	Grassy/river red gum wetland connected to Mercedes Swamp during high flows	Never surveyed	36	<0.5 km Mercedes	>2 years?	65	No

S&D = sStock and domestic.

Poor-quality refuge habitats and chronic isolation may have contributed to the loss of the Avalon Swamp population. Avalon Swamp did contain a permanent waterbody but it was in poor condition with no aquatic or fringing vegetation. It was also later revealed to have dried out for a short period in 2006 and contained very high densities of common carp (J Spencer, DECCW, unpubl. data). The loss of the Avalon Swamp population highlights the sensitivity of *L. raniformis* to wetland drying and the importance of maintaining high-quality permanent waterbodies during droughts.

Recruitment outcomes

Calling activity commenced immediately after the waterbodies were flooded at all six sites where *L. raniformis* was present. Successful recruitment, evidenced by the presence of tadpoles and metamorphs (recently metamorphosed tadpoles), occurred at three sites – Mercedes Swamp, Two Bridges Swamp and South Eulimbah Stock Dam (Paul Coates Swamp was not resurveyed) (Fig. 14.3). Evidence of recruitment by *L. raniformis* was not observed at Eulimbah or Warwaegae swamps despite strong calling activity. Those sites also had comparatively low levels of recruitment by other frog species, indicating that they had lower suitability for frog breeding overall.

Differences in vegetation structure and high densities of juvenile carp, particularly at Warwaegae and Eulimbah swamp, are likely to have contributed to the poor breeding response at

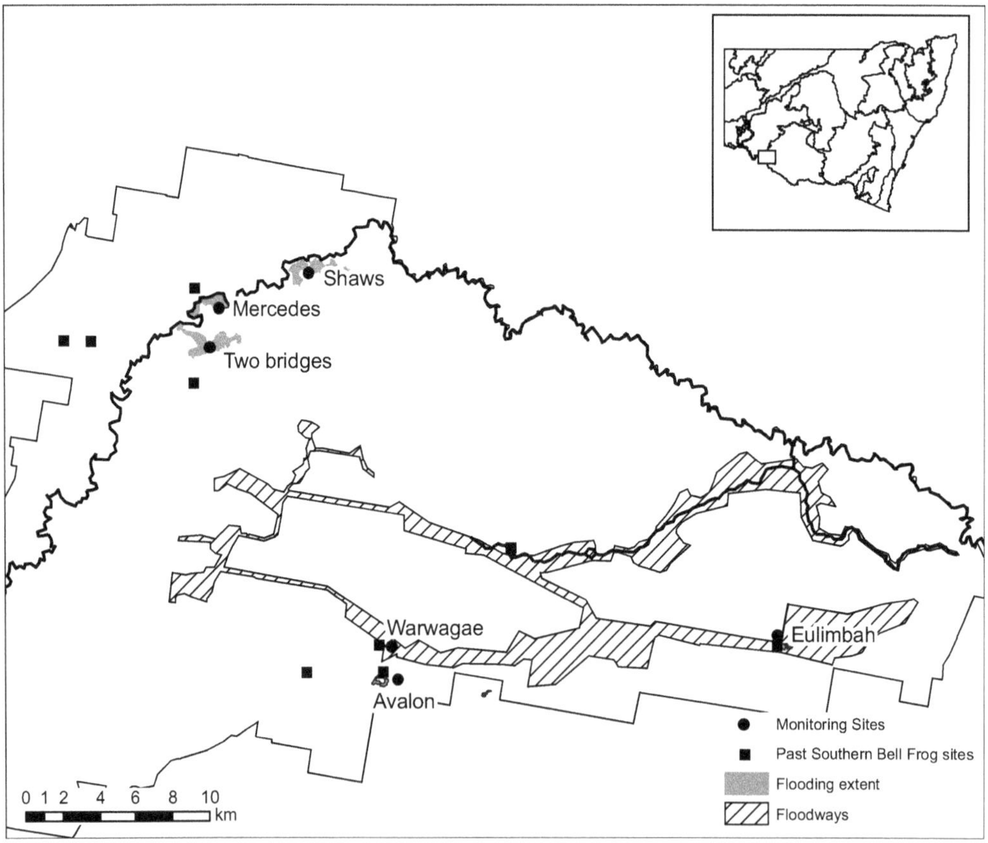

Figure 14.2: Location of wetland watering sites and previous records of *L. raniformis* within the Lowbidgee region of NSW. Flood extent and floodway layers supplied by State Water; extent of flooded wetlands by DECCW; previous records from Wassens (2005). From Wassens *et al.* (2008a).

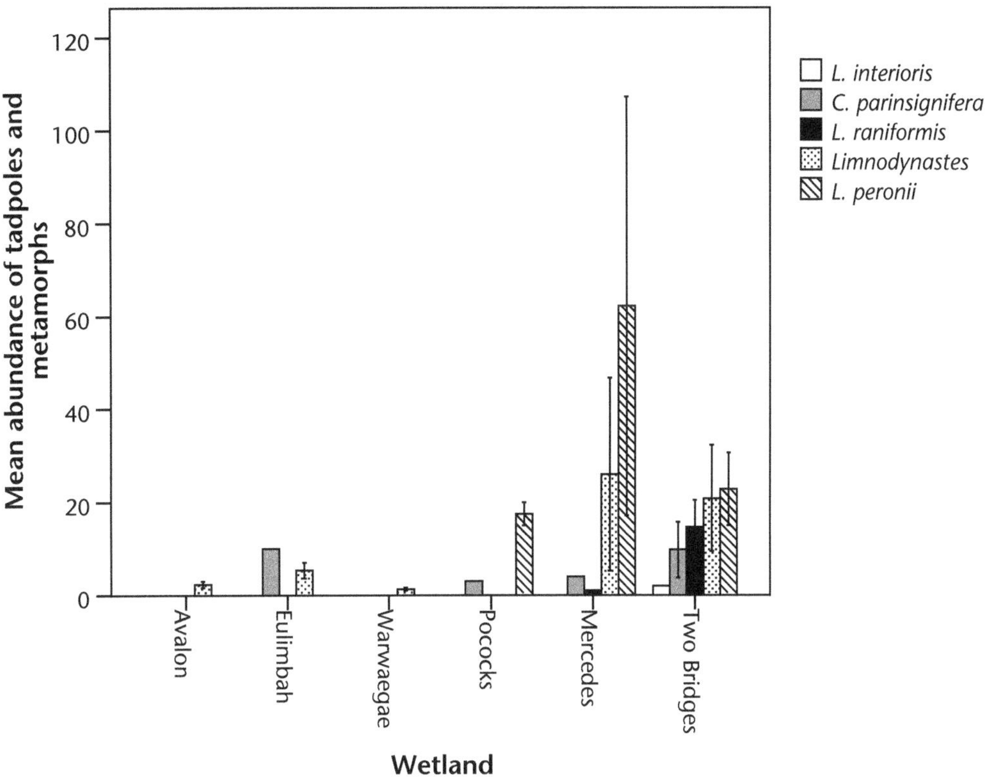

Figure 14.3: Mean (±SE) abundance of tadpoles and metamorphs at each wetland. *L. tasmaniensis* and *L. fletcheri* have been combined into a single group *Limnodynastes* because it is not possible to separate the species as tadpoles. Redbank Weir has been excluded because no tadpoles were collected. From Wassens *et al.* (2008a).

those wetlands. While all sites except South Eulimbah Stock Dam contained carp, their impact on *L. raniformis* recruitment appeared to be far more significant within the black box/lignum wetlands, probably due to differences in fish densities and aquatic vegetation communities. The structure of vegetation can reduce fish predation on tadpoles by supplying predator refuges (Lawler *et al.* 1999). It is likely that the dense emergent vegetation cover at Mercedes and Two Bridges swamps provided better cover and protection from European carp than the softer submerged and floating aquatic vegetation that dominated Eulimbah and Warwaegae swamps. The large number of metamorphs at South Eulimbah Stock Dam, which had similar habitat structure to Eulimbah and Warwaegae swamps but was free of carp, provides evidence for a potential link between carp and poor breeding outcomes.

Relationships with other frog species

Wetlands that supported successful recruitment of *L. raniformis* also contained the highest abundance and diversity of tadpoles and metamorphs of the other resident frog species: *Limnodynastes interioris, L. fletcheri, L. tasmaniensis, Litoria peronii* and *Crinia parinsignifera* (Fig. 14.2). This suggests that flooding to maintain *L. raniformis* will have positive benefits for other frog species. Subsequent fish surveys have shown Two Bridges Swamp, the site with the highest frog diversity and tadpole abundance, to have a high diversity of native fish (J Spencer, DECCW, unpubl. data).

Lessons learnt

Predicting outcomes to flooding

This case study represents the first time that flooding management had been used as an emergency measure to prevent the extinction of local populations of a threatened frog species. Success was influenced by our ability to predict which sites would still be occupied and to forecast recruitment outcomes. Overall, the decision-making framework was quite successful in predicting the persistence of *L. raniformis*. However, not all the selection criteria proved useful when making predictions. For example, it was difficult to describe aquatic vegetation communities at sites that had not been surveyed in 2001 and 2004. It was also difficult to predict wetland hydroperiod, as this largely depends on the volume of water entering the wetland. We relied heavily on the expert knowledge of landholders and State Water staff. Expert knowledge also played a key role in weighting the relative importance of the different criteria. For example, Shaw's Swamp conformed to the majority of selection criteria, except flooding frequency; expert knowledge was required to assess how important that factor was likely to be in relation to the others.

Frog populations are influenced by multiple factors at a range of spatial and temporal scales. The relative importance of the factors appeared to differ between wetland systems. This presented complications when applying the decision-making framework across different wetland types. The habitat requirements and ecology of *L. raniformis* in river red gum wetlands had been the focus of earlier research (Wassens 2005), which formed the basis for site selection strategy and made it easier to predict which wetlands would be occupied and would sustain recruitment. The response of populations to flooding in black box/lignum wetlands was harder to predict, particularly in terms of recruitment outcomes. The presence of carp appears to be far more important in black box/lignum wetlands than in river red gum wetlands. As a result, the relative importance of this factor (and possibly others) needs to be weighted differently in different types of wetlands.

Limited knowledge of past flooding frequencies presented a problem in some instances. For example, Pocock's Swamp was initially believed to have a similar flooding frequency to Mercedes Swamp (to which it is connected), but on further investigation was found to flood less frequently (every three to four years).

Monitoring outcomes

Surveys of calling activity are commonly used to assess frog populations and breeding response. This is quick and inexpensive, but calling activity frequently has little relationship with actual recruitment outcomes, as this case study highlights. The potential lag between the first failed recruitment event and the eventual disappearance of a population needs to be taken into account when planning monitoring. Focusing on outcomes such as recruitment, and measuring a wide range of environmental parameters which may indicate the causes of recruitment failures, are essential if intervention is to be employed.

Managing for stability in variable systems

Wetland frogs, *L. raniformis* in particular, typically exhibit boom and bust population dynamics. They are adapted to large-scale flooding which allows high levels of recruitment and high abundances during good periods; the high abundances help to sustain populations during periods of high mortality such as droughts. Population crashes and local extinctions are comparably common but are compensated by recolonisation and recovery (Alford and Richards 1999; Skelly *et al.* 1999; Richter *et al.* 2003). Our current flooding strategy is conservative and focuses on maintaining comparatively small areas of habitats in a stable state. Whether this type of conservative management can successfully maintain viable populations in the long term is open to question.

Conclusions

The response of frogs to flooding is rarely considered when planning environmental flow strategies. This may be due to the misconception that frogs are poorly understood and difficult to predict, or that their requirements are catered for when targeting other taxonomic groups. The continual decline of frog populations across the Murray-Darling Basin and the loss of species from wetland systems under active flooding management highlight the importance of considering frogs, along with other wetland taxa, when targeting environmental water.

Acknowledgments

This chapter was improved by the helpful comments provided by the two reviewers, David Hunter and Keith Ward. The following people assisted in the selection of sites for wetland watering and the development of the wetland watering strategy for the Lowbidgee floodplain case study (Wassens *et al.* 2008a): James Maguire, David Hunter, Mike Maher, Ross Hyne, Reinier Mann, Simone Stimpson, Paul McInnis and Narelle Jones (all from DECCW); Tom Davy from State Water; and Steve Blore, Michael Spinks and Frank Old from the Lowbidgee League. Sascha Healy and Omar Arnaiz assisted with field work for the case study.

References

Adams MJ (1999). Correlated factors in amphibian decline: exotic species and habitat change in western Washington. *Journal of Wildlife Management* **63**, 1162–1171.

Adams MJ (2000). Pond permanence and the effects of exotic vertebrates on anurans. *Ecological Applications* **10**, 559–568.

Alford RA and Richards SJ (1999). Global amphibian declines: a problem in applied ecology. *Annual Review of Ecology and Systematics* **30**, 133–165.

Anstis M (2002). *Tadpoles of South-eastern Australia: A Guide with Keys.* Reed New Holland, Sydney.

Babbitt KJ and Tanner GW (2000). Use of temporary wetlands by anurans in a hydrologically modified landscape. *Wetlands* **20**, 313–322.

Casanova MT and Brock MA (2000). How do depth, duration and frequency of flooding influence the establishment of wetland plant communities? *Plant Ecology* **147**, 237–250.

Hamer AJ, Lane SJ and Mahony MJ (2002). The role of introduced mosquitofish (*Gambusia holbrooki*) in excluding the native green and golden bell frog (*Litoria aurea*) from original habitats in south-eastern Australia. *Oecologia* **132**, 445–452.

Hamer AJ, Lane SJ and Mahony MJ (2003). Retreat site selection during winter in the green and golden bell frog, *Litoria aurea* Lesson. *Journal of Herpetology* **37**, 541–545.

Harrison S and Taylor AD (1997). Empirical evidence for metapopulation dynamics. In *Metapopulation Biology: Ecology, Genetics and Evolution.* (Eds I Hanski and ME Gilpin) pp. 27–39. Academic Press, San Diego.

Hazell D (2003). Frog ecology in modified Australian landscapes: a review. *Wildlife Research* **30**, 193–205.

Holomuzki JR (1995). Oviposition sites and fish-deterrent mechanisms of two stream anurans. *Copeia* **1995**, 607–613.

Jansen A and Healey M (2003). Frog communities and wetland condition: relationships with grazing by domestic livestock along an Australian floodplain river. *Biological Conservation* **109**, 207–219.

Jansen A and Robertson AI (2001). Relationships between livestock management and the ecological condition of riparian habitats along an Australian floodplain river. *Journal of Applied Ecology* **38**, 63–75.

Lawler SP, Dritz D, Strange T and Holyoak M (1999). Effects of introduced mosquitofish and bullfrogs on the threatened California red-legged frog. *Conservation Biology* **13**, 613–622.

Marsh DM and Trenham PC (2001). Metapopulation dynamics and amphibian conservation. *Conservation Biology* **15**, 40–49.

Marsh DM, Fegraus E and Harrison S (1999). Effects of breeding pond isolation on the spatial and temporal dynamics of pond use by the tungara frog, *Physalaemus pustulosus*. *Journal of Animal Ecology* **68**, 804–814.

Marsh DM, Rand AS and Ryan MJ (2000). Effects of inter-pond distance on the breeding ecology of tungara frogs. *Oecologia* **122**, 505–513.

Page K, Read A, Frazier P and Mount N (2005). The effect of altered flow regime on the frequency and duration of bankfull discharge: Murrumbidgee River, Australia. *River Research and Applications* **21**, 567–578.

Pinto L, Chandrasena N, Pera J, Hawkins P, Eccles D and Sim R (2005). Managing invasive carp (*Cyprinus carpio* L.) for habitat enhancement at Botany Wetlands, Australia. *Aquatic Conservation: Marine and Freshwater Ecosystems* **15**, 447–462. doi:10.1002/aqc.684|ISSN 1052-7613.

Richter SC, Young JE, Johnson GN and Seigel RA (2003). Stochastic variation in reproductive success of a rare frog, *Rana sevosa*: implications for conservation and for monitoring amphibian populations. *Biological Conservation* **111**, 171–177.

Roberts J, Chick A, Oswald L and Thompson P (1995). Effect of carp, *Cyprinus carpio* L, an exotic benthivorous fish, on aquatic plants and water quality in experimental ponds. *Marine and Freshwater Research* **46**, 1171–1180.

Schwarzkopf L and Alford RA (2002). Nomadic movements in tropical toads. *OIKOS* **96**, 492–506.

Scott DE (1994). The effect of larval density on adult demographic traits in *Ambystoma opacum*. *Ecology* **75**, 1383–1396.

Semlitsch RD (2000). Principles for management of aquatic-breeding amphibians. *Journal of Wildlife Management* **64**, 615–631.

Skelly DK, Werner EE and Cortwright SA (1999). Long-term distributional dynamics of a Michigan amphibian assemblage. *Ecology* **80**, 2326–2337.

Smith MA and Green DM (2005). Dispersal and the metapopulation paradigm in amphibian ecology and conservation: are all amphibian populations metapopulations? *Ecography* **28**, 110–128.

Snodgrass JW, Komoroski MJ, Bryan AL and Burger J (2000). Relationships among isolated wetland size, hydroperiod, and amphibian species richness: implications for wetland regulations. *Conservation Biology* **14**, 414–419.

Teplitsky C, Plenet S and Joly P (2003). Tadpoles' responses to risk of fish introduction. *Oecologia* **134**, 270–277.

Trenham PC, Koenig WD and Shaffer HB (2001). Spatially autocorrelated demography and interpond dispersal in the salamander *Ambystoma californiense*. *Ecology* **82**, 3519–3530.

Vignoli L, Bologna MA and Luiselli L (2007). Seasonal patterns of activity and community structure in an amphibian assemblage at a pond network with variable hydrology. *Acta Oecologica: International Journal of Ecology* **31**, 185–192.

Warwick NW and Brock MA (2003). Plant reproduction in temporary wetlands: the effects of seasonal timing, depth, and duration of flooding. *Aquatic Botany* **77**, 153–167.

Wassens S (2005). The use of space by the endangered southern bell frog (*Litoria raniformis*) in the semi-arid region of New South Wales, Australia. PhD thesis. Charles Sturt University, Wagga Wagga.

Wassens S (2008). Review of the past distribution and decline of the southern bell frog *Litoria raniformis* in New South Wales. *Australian Zoologist* **34**, 446–452.

Wassens S, Arnaiz OL and Watts RJ (2007a). Assessing the diversity, abundance and hydrological requirements of frog populations at 'Burrawang West' and 'Yarnel' Lagoons, two small wetlands on anabranch creeks of the mid-Lachlan River. Dept of Environment and Climate Change, Canberra.

Wassens S, Roshier DA, Watts RJ and Robertson AI (2007b). Spatial patterns of a southern bell frog (*Litoria raniformis*) population in an agricultural landscape. *Pacific Conservation Biology* **13**, 104–110.

Wassens S, Arnaiz OL, Healy S, Watts RJ and Maguire J (2008a). Hydrological and habitat requirements to maintain viable southern bell frog (*Litoria raniformis*) populations on the Lowbidgee floodplain: phase 1. Dept of Environment and Climate Change, Canberra.

Wassens S, Watts RJ, Jansen A and Roshier D (2008b). Movement patterns of southern bell frogs (*Litoria raniformis*) in response to flooding. *Wildlife Research* **35**, 50–58.

Withers PC and Thompson GG (2000). Cocoon formation and metabolic depression by the aestivating hylid frogs *Cyclorana australis* and *Cyclorana cultripes* (Amphibia: Hylidae). *Journal of the Royal Society of Western Australia* **83**, 39–40.

Chapter 15

Linking river red gum condition to hydrological change at Yanga National Park

Li Wen, Neil Saintilan, Kerrylee Rogers and Joanne Ling

Introduction

Yanga National Park is located on the Lowbidgee floodplain of the Murrumbidgee catchment (Fig. 15.1/colour plate, p. 314). It is classified as warm, persistently dry grassland under the modified Köpen classification system (Stern *et al.* 2000). It has a semi-arid climate with low annual rainfall and hot summers. Mean summer maximum temperatures range between 30.9°C and 33.0°C, and mean winter minimum varies from 3.5°C to 6.8°C. Multi-year mean annual rainfall is 320 mm with considerable interannual variation (BOM 2008).

Yanga National Park is renowned for its large stands of river red gum (*Eucalyptus camaldulensis*) forests and woodlands that are critical fish and waterbird habitats and refuge for biodiversity in arid and semi-arid Australia (Maher 1990; Kingsford and Thomas 2001; Watts *et al.* 2003; Gilligan 2005). In the natural state, many of the river red gum forests in the region are characterised by variable and unpredictable patterns of high and low flows and water levels.

Over the past 50–100 years, the wetlands on the Lowbidgee floodplain have changed significantly from their natural state (Kingsford and Thomas 2001). The ecosystem is experiencing degradation due to in-stream regulation structures (e.g. dams and weirs), flood prevention structures (e.g. levees), deterioration of water quality, clearing of riparian vegetation and grazing, and the presence of exotic species (Hillman *et al.* 2000). Of these factors, altered flow regime is believed to be an important contributor (Kingsford and Thomas 2001; Wen *et al.* 2009). Major restoration effort is therefore directed to environmental water allocation aiming to re-establish the ecological integrity of the river-floodplain ecosystem.

Condition of river red gum forest and flooding

River red gum (*E. camaldulensis*) is a widely distributed, iconic eucalypt that often dominates watercourses and their floodplains throughout semi-arid inland Australia. With its striking height (up to 45 m; Brooker *et al.* 2002), river red gum forest forms a landscape distinct from its surrounding dryland shrublands and woodlands. The biomass of river red gum forests can match those of forests in high-rainfall regions (Bren 1990). River red gum forests also play an important role in nutrient cycling between floodplains and rivers (Baldwin 1999) and provide habitat for plants and animals (MacNally *et al.* 2001).

In the semi-arid Murray-Darling Basin, river red gum depends on overbank flows for recruitment and survival, and established river red gum forests also rely on periodic floods to maintain a healthy state (Bren and Gibbs 1986; Bacon *et al.* 1993; Cunningham *et al.* 2007; VEAC 2008). The unregulated flooding regime in this part of the Murry-Darling Basin consisted of peak flows in late winter and spring with low flows in summer and autumn (Dalton 1990; Page *et al.* 2005; Wen 2009). However, regulation of river through dams, weirs, levees and diversion has drastically altered the flow regime and led to increased river flow capacity, decreased total annual flow, reduced extent and depth of winter flooding and reduced frequency of flooding (Stefano 2001; Maheshwari *et al.* 1995) for lowland floodplains. These changes have produced major deterioration in much of the riparian forest, including reduced tree growth rate, accelerated mortality and minimal regeneration (Bacon *et al.* 1993). For example, in a survey of river red gum along the River Murray floodplains below Euston, the Murray-Darling Basin Commission (2003) reported that up to 80% of trees were stressed.

Quantitative exploration of the relationships between flooding duration and frequency and river red gum conditions is impeded by the paucity of long-term observations of river red gum condition and overbank flow pattern, and by the strongly non-linear and higher-order interactions between cause and response variables (Tockner and Stanford 2002). These cause commonly used modelling methods to fail to find meaningful ecological patterns (De'ath and Fabricius 2000).

Multivariate time series model and hydrological alteration in Yanga

Climate and hydrological data

Annual rainfall and annual maximum summer temperature from 1963 to 2005 at Balranald were acquired online from the Bureau of Meteorology website (BOM 2008). Daily river discharge data from 1936 to 2007 were obtain from the NSW Department of Water and Energy hydrological network gauges No. 4100040 (Murrumbidgee River downstream of Maude) and No. 4100041 (Murrumbidgee River downstream of Redbank) (Fig. 15.1, p. 314). The discharge records were obtained online from Department of Water and Energy (DWE Waterinfo 2008). The Redbank gauge record, however, has a long-term gap from March 1993 to late August 1996. To fill the data gap, a time series regression model was fitted using upstream gauge (at Maude) daily discharge data and its lags and climate data (see below). The record at Maude has only short-term (days) gaps that were filled by either simple regression or auto-regression and moving average (ARMA) models.

Short-term hydrological gaps at Maude

The method used to fill the gaps in hydrological records at Maude depended on their position in the hydrograph. When the gaps were located on the rising or falling limb of the hydrograph, a simple linear regression was used:

$$Y = a \times D + b \qquad (1)$$

where Y is river discharge, D is the number of days away from previous flow peak or valley, and a and b are constants.

When the gaps are located near the flow peak or valley, an ARMA model which included an auto-regressive term, a moving-average term and an error term was used:

$$X_t = \varepsilon_t + \sum_{i=1}^{p} \phi_i X_{t-i} + \sum_{i=1}^{q} \theta_i \varepsilon_{t-1} \tag{2}$$

where $\phi_1, \phi_2 \ldots \phi_p$ are auto-regressive parameters, $\theta_1, \theta_2 \ldots \theta_q$ are moving average parameters and ε_1 is the model error.

The completed time series of daily river discharge downstream of Maude was then used as an input variable to model the river flow downstream of Redbank. The results and accuracy of the models are not reported in this chapter due to the number of missing values.

Long-term hydrological gap at Redbank

The general expression of the time series regression model takes the form of:

$$\left(1 - \sum_{i=1}^{n} \alpha_i B^i\right) Y_t = C + \sum_{j=1}^{m} \left(\sum_{k=0}^{l} \beta_{jk} B^k X_{jt} \right) + N_t \tag{3}$$

where B^i *and* B^k is backshift operator for responsible and explanatory variables (such as B^i for Y means $Y_t = Y_{t-1}$, α and β are coefficients, C is constant, X_{jt} is the value of observation for explanatory variable X_j at time t, Y_t is the value of observation for dependent variable Y at time t, and N_t is the error term of the model:

$$\phi(B) N_t = \theta(B) \varepsilon_t \tag{4}$$

where $\phi(B)$ and $\theta(B)$ are polynomals in B and ε_t is white noise, a normal distributed time series with mean zero and constant variance.

The model was fitted using 40 years of daily river discharge from 1936 to1975, and tested using 10 years (1976–85) of the hydrological record. The verified model was used to forecast the daily river discharge from 1986 to 2007 (22 years), including the 13 years with missing data (1993–2006).

The model fitted is presented below:

$$Ln(D_r + 1) = 1.245 + \sum_{i=0}^{9} \alpha_i \times Ln(D_{mi} + 1) + \sum_{i=1}^{2} \beta_i \times Rain_i \tag{5}$$

where $\alpha_1, \alpha_2 \ldots a_9 = 0.066, 0.161, 0.236, 0.163, 0.083, 0.040, 0.018, 0.029, 0.016, 0.011$; $\beta_1, \beta_2 = 0.006, 0.007$; D_r is river discharge at Redbank (ML/day), D_{mi} is river discharge at Maude (ML/day) and $Rain_i$ is rainfall (mm/day).

With an adjusted $R^2 = 0.99$, the model is considered very accurate. Furthermore, as the slopes in the trend lines are close to 1 (1.0013 and 0.9868 for model building data set and model testing data set, respectively; Fig. 15.2), no evidence of over- or underestimation was indicated.

Hydrological alterations

Hydrological alteration before and after 1970 was investigated using HIA V_7, software developed by the Nature Conservancy (Nature Conservancy 2007). The year 1970 was chosen as a dividing point because the reduction in river flow became significant after the late 1960s at

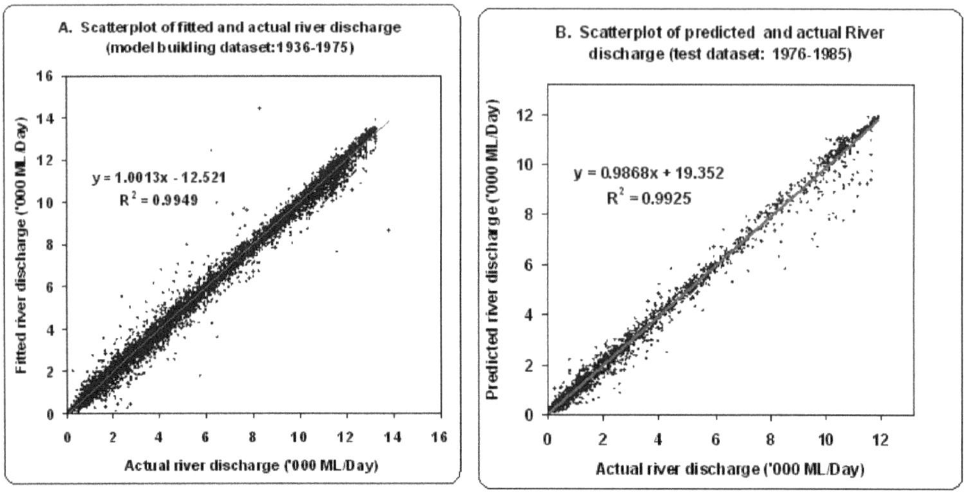

Figure 15.2: Graphs show the accuracy of time series regressive model. (a) Model fitting. (b) Model testing. As the slopes of both trend lines are close to 1, there is no evidence to suggest over- or underestimation. The R^2 is also very close to 1 for both model fitting and model testing, suggesting the model captures almost all variance in river discharge.

Hay (Wen 2009) and because 1970 separates the two major floods (1956 and 1974) during the period, to reduce bias.

Overbank flow events were defined as daily river discharge greater than the bankfull discharge of 9500 ML/day downstream of Redbank, a figure estimated on the base of field observation (Watts *et al.* 2003).

There have been dramatic changes in hydrology downstream of Redbank Weir after 1970 (Table 15.1, Fig 15.3). The annual median flow decreased more than 60% from 2727.0 ML/day to 1048.5 ML/day (Table 15.1), and the reduction was evident for every month since 1970 (Fig. 15.3). The biggest reduction (up to 90%) was in November, when the natural spring peak flow occurs. The altered and reduced peak occurred in September, when the demand for irrigation is at its lowest level.

Both the 30-day maximum and minimum flows decreased between periods. The duration and frequency of extreme low flows increased 33.3% and 300%, respectively. One direct impact of the decreased flow is the reduction of floodplain inundation, which depends on the overbank flows. The frequency of bankfull river discharge declined 53.0% from 18.5%

Table 15.1: Changes in selected hydrological indicators in Murrumbidgee downstream of Redbank Weir

Indicator[a]	1936–70	1971–2007	Change (%)
Annual median flow (ML/day)	2727.0	1048.5	-61.6
30-day maximum flow (ML/day)	10368.0	7430.4	-28.3
30-day minimum flow (ML/day)	691.2	259.2	-62.5
Extremely low flow[b] duration (days/year)	7.5	10.0	33.3
Extremely low flow frequency	1.0	4.0	300.0
Change rate (cubic metre/sec/day)	1.6	0.6	-62.5
Frequency of bankfull flows (%)	18.5	8.7	-53.0

a: Median values except for overbank flow frequency.

b: Flows fall below 10% during the corresponding period.

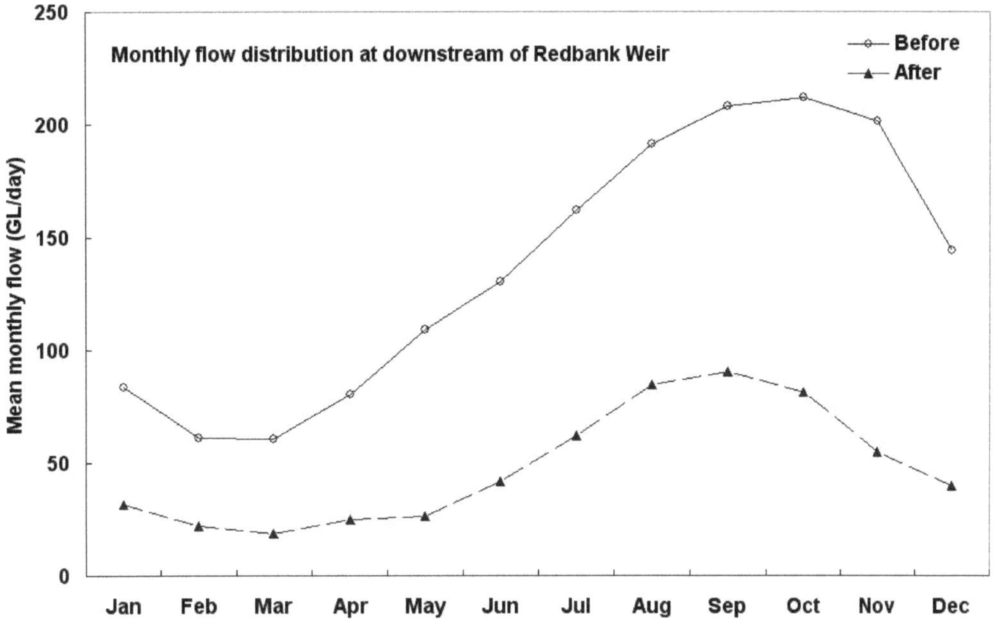

Figure 15.3: Monthly flow distribution at downstream Redbank weir before (solid line with circles) and after (dashed line with triangles) 1970.

(approximately one in five years) to 8.7% (less than one in 10 years) (Table 15.1), which may be the single most important factor in the continuous deterioration of river red gum condition in Yanga National Park.

Measuring the condition of river red gum forest

Historical vegetation mapping

Historical extent and condition of vegetation communities in Yanga National Park was mapped by McCosker (2008) for the NSW Department of Environment, Climate Change and Water using aerial photographs taken in 1965, 1973, 1997 and 2005. River red gum data for this study were extracted from these maps. Detailed methods including ortho-rectification of aerial photos, classification and identification of vegetation communities and ground-truthing can be found in McCosker (2008). Table 15.2 presents the four river red gum communities identified in Yanga. Crown condition classification is given in Table 15.3.

Six levels of crown condition were assigned to each of the defined polygons (Table 15.3). The crown condition rating scheme was derived from the method described by Walker and Hopkins (1984) for determining projective foliage cover, a measure of crown density or the amount of light the plant is able to intercept. Projective foliage cover was assessed from aerial photography using stereoscope. Trees with dark dense crowns have a high projective foliage cover and are considered as being in good condition. Trees with light coloured transparent crowns have low projective foliage cover and are considered as being stressed, or in poor condition.

River red gum extent and condition trends

The total area of river red gum in Yanga remained constant at 21 000 ha over the sampling period (Fig. 15.4). For each community type, the riparian river red gum (type A) remained relatively stable at about 1000 ha. There was noticeable increase in type C community

Table 15.2: River red gum (*E. camaldulensis*) communities identified in Yanga National Park

Community type	Description	Characteristic groundcover	Typical environment
A	River red gum riparian open forest	*Juncus* spp. *Phragmites australis, Eleocharis acuta*	Distributed along the river channel and main waterways
B	River red gum woodland with sedge groundcover	*E. acuta, Paspalidium jubiflorum, Wahlenbergia fluminalis, Senecio quadridentatus, Carex tereticaulis*	Surrounding lakes and swamps and the lower slopes along the river and channels
C	River red gum woodland with grassy/chenopod shrub groundcover	*Sclerolaena muricata, Salsola kali, E. sphacelata*	Occurs on higher ground on the floodplain
D	River red gum forest with shrubby understorey	*Muehlenbeckia florulenta, Acacia stenophylla, Chenopodium nitrariaceum*	Drainage depressions and flood-outs of major watercourses on the floodplains

Modified from McCosker (2008).

(chenopod understorey) and a corresponding decrease in type D river red gum (non-chenopod shrub understorey). A similar incursion of chenopod shrub community due to the lack of inundation was also observed in the Macquarie Mashes (Ch. 9, this volume) and the Chowilla floodplain (Ch. 21, this volume).

The crown conditions declined for all community types over the sampling period (Fig. 15.5). It is noticeable that type A (riparian river red gum open forest) and type B (river red gum woodland with sedge groundcover) had better condition than type C (river red gum with grassy/chenopod shrub groundcover) and type D (river red gum with shrub understorey). The condition of type A was steadier than the others. The condition of type B and D communities deteriorated most during the sampling period.

Relationships between crown condition and flooding

Classification and regression trees

A classification and regression tree (CART) is a relatively new analytic procedure for predicting the values of a continuous or categorical response variable from continuous or categorical predictors. In classification trees the dependent variable is categorical, whereas in regression

Table 15.3: Vegetation crown condition ratings

Class	PFC	Condition interpretation
1	>75%	Excellent condition, dense crowns
2	50–75%	Good condition, healthy open crowns
3	25–50%	Moderate condition, sparse foliage, dieback evident
4	5–25%	Poor condition, dieback evident in most trees
5	1–5%	Very poor condition, many dead trees
6	<1%	Most trees either dead or almost totally defoliated

PFC = projective foliage cover.
From Walker and Hopkins (1984) cited by McCosker (2008)

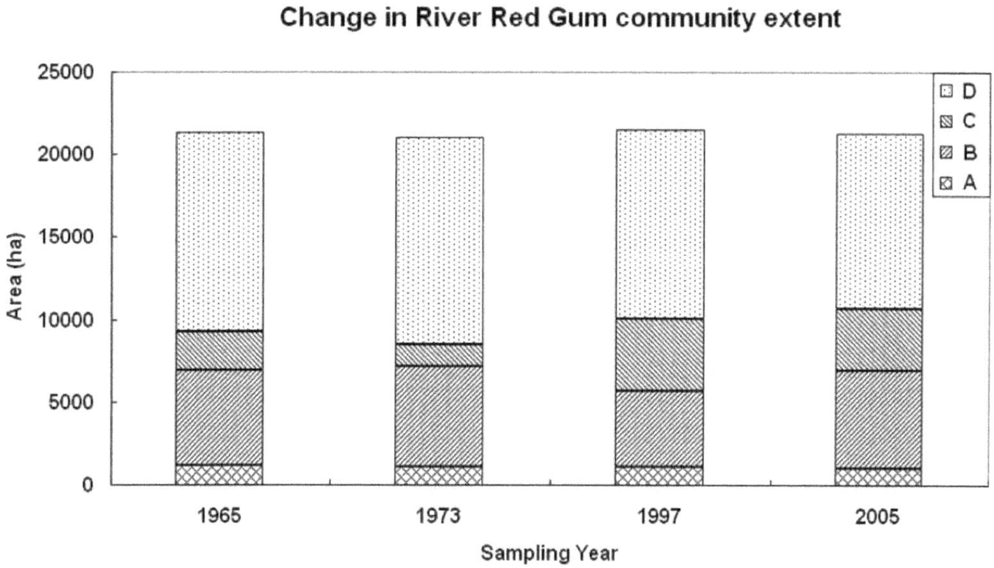

Figure 15.4: Trends of river red gum (*E. camaldulensis*) extent in Yanga. The total area of river red gum remained at 21 000 ha over 40 years of sampling period. The extent of type C (river red gum with grassy/chenopod shrub groundcover increased at the cost of type B (river red gum with sedge groundcover) and D (river red gum with shrub understorey).

trees the dependent variable is quantitative. Regression trees parallel regression/ANOVA (analysis of variance) modelling, while classification trees parallel discriminant analysis. CART has a number of advantages over classic regression methods, that are ideal for analysing and modelling complex ecological data. These features include (Hansen *et al.* 1996; De'ath and Fabricius 2000):

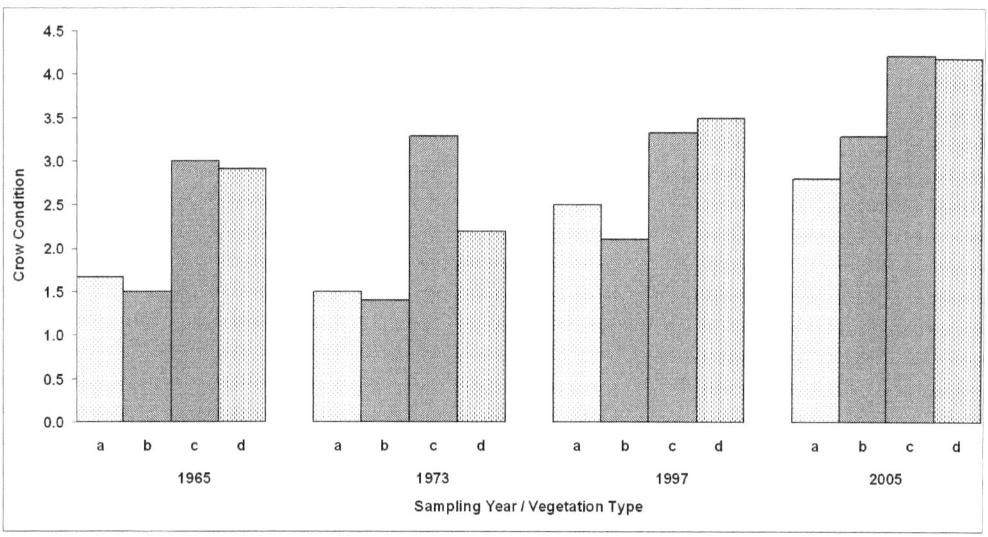

Figure 15.5: River red gum crown condition against sampling year for each community type shows the general declining trends (the higher the rating, the poorer the condition) except in 1973.

- trees are non-parametric and therefore do not require normality assumptions of the data;
- trees can handle data of different types – continuous, categorical, ordinal and binary. Transformations of the data are therefore not required;
- trees can be useful for detecting important variables and interactions and identifying outliers;
- trees cope with missing data by identifying surrogate splits in the modelling process. Surrogate splits are splits highly associated with the primary split. They can be used in situations when data cannot be collected using the primary split variable;
- construction of CART is straightforward and robust. CART is a relatively automatic 'machine learning' method – compared to other multivariate modelling methods, relatively little input is required;
- trees provide simple and meaningful interpretations to non-statisticians.

The classic computational algorithms for classification and regression trees were popularised by Breiman *et al.* (1984; see also Ripley 1996; Hastie *et al.* 2001; Therneau and Atkinson 2008). The CART algorithm involves a recursive optimisation through binary splitting of the data based on the Gini (classification) or impurity (regression) criteria. At each step, the program will find a logical split condition to assign observations to the two child nodes. For continuous predictors, these logical conditions are usually of the type *If x > Value then Node ID=k*. For categorical predictors, the logical split conditions are usually of the type *If x=Category I then Node ID=k*.

The *rpart* package developed by Therneau and Atkinson (2008), which closely follows the algorithm of Breiman *et al.* (1984), was used to perform the CART modelling in R environment (R development core team 2008). The data set is described in Table 15.4. The responsible variable is river red gum crown condition derived from the aerial photography. Five groups (a total of 17) of explanatory variables are included in the modelling process:

1 Veg is the river red gum community type identified from the aerial photography;
2 $Flood_i$ is overbank flow history defined as the accumulated bankfull flow days for one to six years prior to sampling, i= 1, 2, ... 6;

Table 15.4: Variables included in CART modelling

Variable	Type	Range	Description
CC	Numeric	1–6[a]	Response variable. River red gum crown condition, n = 128
Veg	Categorical	A–D[b]	Community type n = 128
Floodi	Numeric	0–122	Accumulated bankfull flow days at ith year prior to sampling. (i = 1, 2, ... 6), e.g. for 1997 data, Flood6 is the total flooding days in the period 1991 to 1996
Fi	Logical	Yes, No	Occurrence at ith year prior to sampling. (i = 1, 2, ... 6)
Tempi	Numeric	27.5–30.0	Maximum summer temperature at ith year prior to sampling, oC. (i = 1, 2)
Raini	Numeric	201.5–517.0	Annual rainfall at ith year prior to sampling, mm/year. (i = 1, 2)

a: See Table 15.2
b: See Table 15.3.
Modified from Wen *et al.* (2009).

3 F_i is flood occurrence for the ith year prior to sampling, i =1, 2, ... 6;
4 $Rain_i$ is the annual rainfall at the ith year prior to sampling, i = 1, 2;
5 $Temp_i$ is the maximum summer temperature at the ith year prior to sampling, i =1, 2.

Interpreting CART

The classification tree constructed by processing the entire data set is presented in Figure 15.6a. The tree has a root node, non-terminal and terminal nodes. Root and each non-terminal node have a question that asks whether a particular attribute (variable) satisfies a given specification generated by the algorithm. The terminal nodes are the assigned classes. For tree A (Fig. 15.6a), the root node divided the data set into two groups based on community type. The first group comprises type A and B communities; types C and D form the other group. The tree indicates that types A and B are in better condition than types C and D, which is consistent with aerial photograph assessment (Fig. 15.5). The groups have distinct flooding history and hence different flooding requirements.

Although climate parameters were included in the analysis none emerged as important, indicating they are not useful predictors of condition. This may suggest that local rainfall is not adequate to sustain river red gum community health, and that floodwater from the Murrumbidgee is vital for the survival of river red gum.

Tree A (Fig. 15.6a) shows the need to construct an individual tree for each river red gum type.

For riparian river red gum forests (type A), which are located along the Murrumbidgee River channel and main watercourses, the modelling process did not produce a classification tree. It gave a root node with a median value of 2, indicating that the selected explanatory variables were not important predictors for the observed river red gum crown conditions. As this community is close to the river channel and hence has access to a permanent water resource, overbank flows may play only a minor role in maintaining its condition.

For river red gum forest with sedge groundcover (type B), occurring at the edge of lakes and swamps and flood runners within the floodplain, 64 days of overbank flow in the previous five years is critical (Fig. 15.6b). For river red gum with grassy/chenopod shrubby groundcover, which is located at higher elevations of floodplain, 194 days of overbank flow in the previous five years is required to maintain a crown condition of class 3 (Fig. 15.6c). If there is less than 64 days of overbank flow accumulated in the previous five years, the community would be in poor to very poor condition. For river red gum communities located at drainage depressions further away from the river channel (type D, river red gum with shrubby understorey), more frequent overbank flow of 59 days in the previous three years is required to maintain a fair crown condition (class 3) (Fig. 15.6d).

The modelling results of river red gum flooding requirements are at the lower end of field observations. For example, in a survey of the major river red gum forest ecosystems in Victoria, VEAC (2008) reported that the forest requires flooding every two to five years. Roberts and Marston (2000) observed that the healthy stands of river red gum in Barmah State Forest, Victoria, had historical flood durations of one to seven months. In this study, the minimum flooding requirement would be an underestimation as the irrigation history was not included in the model and floodwater from Maude overbank flows was also excluded. During major flooding events, floodwater from Maude (Fig. 15.1) can reach Yanga National Park (DWR 2001). Nevertheless, the volume of irrigation water was relatively small and had regular flow paths, therefore the effects were likely to be limited and locally restricted. In addition, flooding at Maude was less frequent than at Redbank. When flooding did occur at Maude, there were overbank flows at Redbank in most cases (from 1936, there are only 18 days when the discharge at Maude reached full bank but Redbank did not). Therefore, it is reasonable to assume that using flooding at Redbank alone is adequate for the purposes of this study.

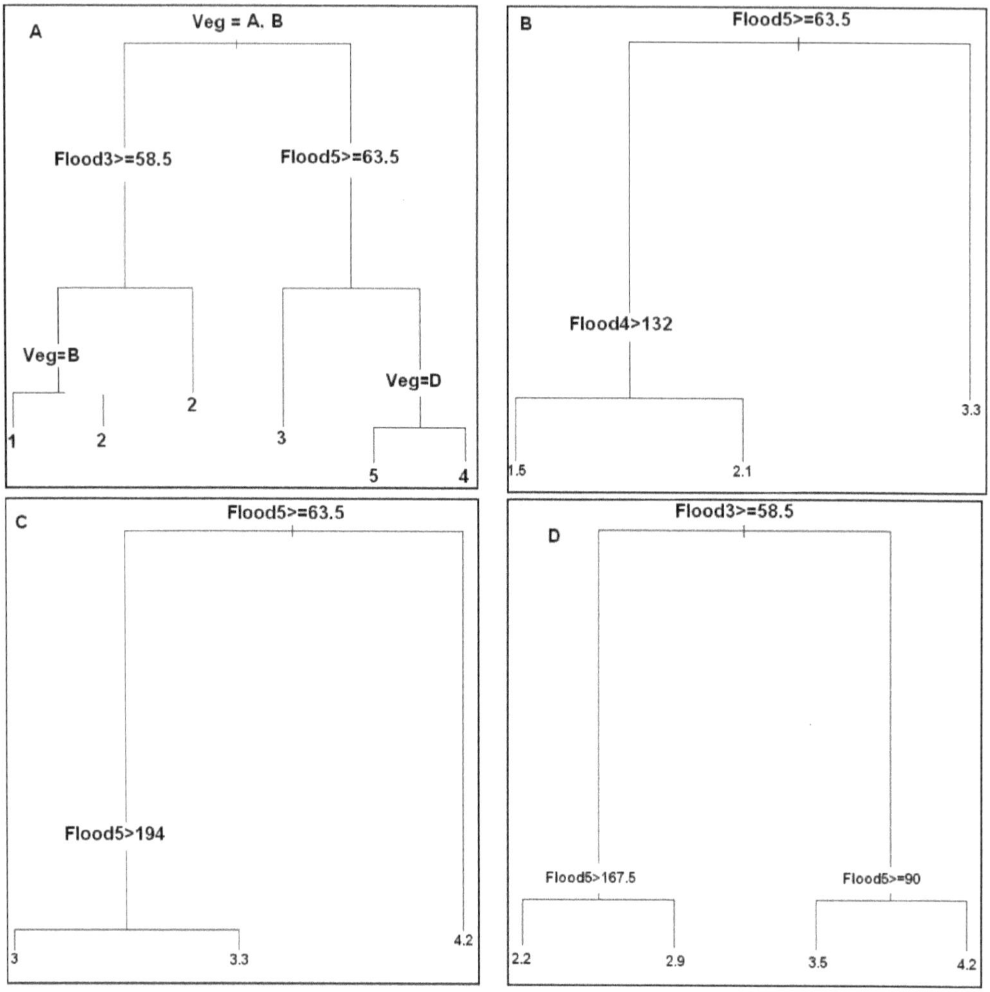

Figure 15.6: Classification and regression trees obtained for (a) the entire data set including all four river red gum communities, (b) river red gum with sedge groundcover, (c) river red gum with grassy groundcover and (d) river red gum with shrubby understorey. The length of each branch indicates the importance of the splitting. The river red gum condition is rated at six levels, where one is excellent and six is poorest. Three is the threshold which divides the community into health and stress states.

Conclusions

The hydrological investigation showed dramatic reductions in river flows after 1970 in the lower reaches of the Murrumbidgee River. The reductions are clear for high, medium and low flows. A direct result of the reductions is that the overbank flows into Yanga National Park have decreased in frequency by more than 50% since 1970, which is proven to be the single most important factor contributing to the poor river red gum condition (Wen *et al.* 2009).

The study also showed that river red gum communities in Yanga National Park have a distinct flooding history and thus require different flooding regimes to maintain and/or restore their condition.

1 The condition of riparian river red gum is not related to overbank flows, and remained in a healthy state over the sampling period.

2 River red gum with sedge groundcover requires about two months of overbank flow in five years to stay in good condition.
3 River red gum forest with grass/chenopod shrub groundcover and river red gum forest with shrubby understorey require much longer overbank flow (more than six months in five years) to be maintained in good condition.

Because the future water availability for Yanga National Park is likely to be limited (CSIRO 2008), it is implausible to provide environmental water for all river red gum communities. The restoration priorities should be directed to the river red gum with sedge groundcover as this community type is generally in better condition than other community types given the same frequency and duration of overbank flows.

In summary, our results demonstrated that CART analysis is a simple yet powerful technique to determine links between river red gum condition and hydrological change with significant potential for river and environment flow management; river red gum community in Yanga National Park should be managed as hydrological units given the limitation of environmental water allocations.

Acknowledgments

The study was performed under the framework of the NSW Rivers Environmental Restoration Program for ecological investigation on Lowbidgee floodplain. The authors acknowledge support from the Commonwealth Water for the Future program and the NSW Government.

References

Bacon PE, Stone C, Binns D, Leslie D and Edwards J and D (1993). Relationships between water availability and *Eucalyptus camaldulensis* growth in a riparian forest. *Journal of Hydrology* **150(2–4),** 541–561.

Baldwin DS (1999). Dissolved organic matter and phosphorus leached from fresh and 'terrestrially' aged river red gum leaves: implications for assessing river–floodplain interactions. *Freshwater Biology* **41**, 675–685.

BOM (Bureau of Meteorology) (2008). http://www.bom.gov.au/climate/averages/tables/cw_049002.shtml. Accessed December 2008.

Bren L (1990). Red gum forests. In *The Murray*. (Eds N Mackay and D Eastburn) pp. 230–242. Murray-Darling Basin Commission, Canberra.

Bren LJ and Gibbs N (1986). Relationships between flood frequency, vegetation and topography in an *E. camaldulensis* forest. *Australian Forest Research* **16**, 357–370.

Breiman L, Friedman J, Olshen R and Stone C (1984). *Classification and Regression Trees*. Wadsworth, Belmont.

Brooker M, Connors J, Slee A and Suffy S (2002). EUCLID: *Eucalypts of Southern Australia*. CSIRO Publishing, Melbourne.

CSIRO (2008). Water availability in the Murrumbidgee. Report to the Australian Government from the CSIRO Murray-Darling Basin Sustainable Yields Project. CSIRO, Canberra. Available at www.clw.csiro.au/publications/waterforahealthycountry/mdbsy/pdf/Murrumbidgee-Report.pdf. Accessed December 2008.

Cunningham S, Nally R, White M, Read J, Baker P, Thomson J and Griffioen P (2007). Mapping the current condition of *E. camaldulensis* (*Eucalyptus camaldulensis* Dehnh.) stands along the Victorian Murray River floodplain. Report to the North Victoria Catchment Management Authorities and the Victorian Dept of Sustainability and Environment.

Dalton K (1990). *Managing our River Red Gums.* Soil Conservation Service of New South Wales, Sydney.

De'ath G and Fabricius K (2000). Classification and regression trees: a powerful yet simple technique for ecological data analysis. *Ecology* **81(11),** 3178–3192.

DWE Waterinfo (2008). http://www.waterinfo.nsw.gov.au/river-provisional-sites.shtml. Accessed December 2008.

DWR (2001). *Lowbidgee Management Plan Stage 2: Land and Water Management 1992–1997.* Compiled for the Lowbidgee Stage 2 Steering Committee, NSW Dept of Water Resources, Sydney.

Gilligan D (2005). Fish communities of the Murrumbidgee catchment: status and trends. Final Report Series No. 75. NSW Dept of Primary Industries – Fisheries. Available at http://www.dpi.nsw.gov.au/research/areas/systems-research/aquatic-ecosystems/outputs/2005/545. Accessed December 2008.

Hansen M, Dubayah R and Defries R (1996). Classification trees: an alternative to traditional land cover classifiers. *International Journal of Remote Sensing* **17(5),** 1075–1081.

Hastie T, Tibshirani R and Friedman JH (2001). *The Elements of Statistical Learning: Data Mining, Inference and Prediction.* Springer Science and Business Media, New York.

Hillman TJ, Koehn JD, Mitchell D, Thompson D, Sobels JD and Woodside D (2000). The Murrumbidgee: assessing the health of a 'working river'. Report to Irrigated Agribusiness Taskforce and NSW Dept of Land and Water Conservation.

Kingsford RT and Thomas RF (2001). *Changing Water Regimes and Waterbird Habitats on the Lower Murrumbidgee River in Arid Australia.* NSW National Parks and Wildlife Service, Sydney.

Kingsford RT and Thomas RF (2004). Destruction of wetlands and waterbird populations by dams and irrigation on the Murrumbidgee River in arid Australia. *Environmental Management* **34(3),** 383–396.

MacNally R, Parkinson A, Horrocks G, Conole L and Tzaros C (2001). Relationships between terrestrial vertebrate diversity, abundance and availability of coarse woody debris on south-eastern Australian floodplains. *Biological Conservation* **99**, 191–205.

Maher P (1990). Bird survey of the Lachlan/Murrumbidgee confluence wetlands. NSW National Parks and Wildlife Service, Sydney.

Maheshwari BL, Walker KF and McMahon TA (1995). Effects of river regulation on the flow regime of the River Murray, Australia. *Regulated Rivers: Research and Management* **10**, 15–38.

McCosker R (2008). Yanga vegetation mapping: historical community extent and condition. Technical report to NSW Dept of Environment and Climate Change.

MDBC (2003). Preliminary investigations into observed *E. camaldulensis* decline along the River Murray below Euston. Technical Report 03/03. Murray-Darling Basin Commission. Available at http://publication.mdbc.gov.au/product_info.php?products_id=151. Accessed December 2008.

Nature Conservancy (2007). Indicators of hydrologic alteration Version 7 user's manual. Available at http://www.nature.org/initiatives/freshwater/files/ihav7.pdf. Accessed December 2008.

Overton I and Doody T (2010). Ecosystem response modelling in the Chowilla floodplain, Lindsay and Wallpolla Islands icon site. In *Ecosystem Response Modelling in the Murray-Darling Basin.* (Eds N Saintilan and I Overton) pp. 357–372. CSIRO Publishing, Melbourne.

Page K, Read A, Frazier P and Mount N (2005). The effect of alternated flow regime on the frequency and duration of bankfull discharge: Murrumbidgee River, Australia. *River Research and Application* **21**, 1–12.

R Development Core Team (2008). http://www.r-project.org/. Accessed December 2008.

Ripley BD (1996). *Pattern Recognition and Neural Networks.* Cambridge University Press, Cambridge.

Roberts J and Marston F (2000). *Water Regime of Wetland and Floodplain Plants in the Murray-Darling Basin.* CSIRO Land and Water, Canberra.

Stefano J (2001). River red gum (*Eucalyptus camaldulensis*): a review of ecosystem processes, seedling regeneration and silvicultural practice. *Australian Forestry* **65(1),** 14–22.

Stern H, de Hoedt G and Ernst J (2000). Objective classification of Australian climates. *Australian Meteorological Magazine* **49**, 87–96.

Therneau T and Atkinson E (1997). An introduction to recursive partitioning using the RPART routines. Technical report. Mayo Foundation, Rochester. Available at http://cran.r-project.org/web/packages/rpart/rpart.pdf. Accessed December 2008.

Therneau TM and Atkinson B (2008). The rpart package. Available at http://cran.r-project.org/web/packages/rpart/rpart.pdf. Accessed December 2008.

Tockner K and Stanford J (2002). Riverine flood plains: present state and future trends. *Environmental Conservation* **29**, 308–330.

Thomas R, Bowen SM, Simpson S, Cox SJ, Sims NC, Hunter SJ and Lu Y (2010). Inundation response of vegetation communities of the Macquarie Marshes in semi-arid Australia. In *Ecosystem Response Modelling in the Murray-Darling Basin.* (Eds N Saintilan and I Overton) pp. 137–150. CSIRO Publishing, Melbourne.

VEAC (2008). River red gum forests investigation. Final report. Victorian Environmental Assessment Council, Melbourne. Available at http://www.veac.vic.gov.au/riverredgumfinal/VEAC_RRGF_final_report-all.pdf. Accessed December 2008.

Walker J and Hopkins MS (1984). Vegetation. In *Australian Soil and Land Survey: Field Handbook.* (Eds RC McDonald, RF Isbell, JG Speight, J Walker and MS Hopkins) pp. 46–67. Inkata Press, Sydney.

Watts R, Read A, Page K, Crook D, Frazier P, Hardwick L, Jansen A, Lowe B, Lugg A, Roshier D and Ryder D (2003). Ecological assessment of flow scenarios in the Murrumbidgee River (Zone J) for the Living Murray Initiative. Final report to the Murray-Darling Basin Commission for the Living Murray Initiative. Available at http://www.mdbc.gov.au/subs/information/mfat/downloads/reg_zonej.pdf. Accessed December 2008.

Wen L (2009). Reconstruction of natural flow in a regulated system, the Murrumbidgee River, Australia, using time series analysis. *Journal of Hydrology* **364(3–4),** 216–226.

Wen L, Ling J, Saintilan N and Rogers K (2009). Investigation of the hydrological requirements of the river red gum (*Eucalyptus camadulensis*) forest using classification and regression tree (CART) modelling. *Ecohydrology* **2**, 143–155.

Chapter 16

Ecosystem response modelling in the River Murray

Ian Overton, Dave Penton and Tanya Doody

Introduction

The River Murray provides water to Australia's largest irrigation region and water supply to the city of Adelaide and other towns (Fig. 16.1). The river supports internationally recognised wetlands and a floodplain area in excess of 600 000 ha. Since the early 1920s, the River Murray has been regulated by weirs and reservoirs to reduce the impact of large floods on town infrastructure and to provide a regular water supply. Over the last century, this regulation, in conjunction with overextraction and climate change, has caused environmental degradation to the River Murray. Concern about the river's continuing deterioration in health has brought attention to the river's management. In particular, scientists and managers are proposing new environmental flow strategies for releasing and managing flows to provide environmental benefits to the floodplain, wetlands and in-stream water quality. This chapter describes efforts to model the ecosystem response to proposed and existing environmental flow strategies.

Regulation, overextraction and climate change have caused environmental degradation to the River Murray over the last century and the rate of loss of ecological function is increasing. The River Murray is a highly regulated river. It has 14 weirs and locks and several storage facilities in the stretch from Hume Dam to the mouth. Consumptive users extract approximately two-thirds of natural flows (CSIRO 2008). These extractions have severely degraded the environment and in the current drought the Murray Mouth is closed, causing a potential environmental disaster in the Coorong.

Naturally saline groundwater has risen close to the floodplain surface in many areas as a result of locks, broad-scale clearing of native forests and irrigation water applied to surrounding lands. Rising groundwater has led to large-scale floodplain salinisation and further decline in ecological condition. Flooding has reduced, leading to the River Murray floodplain suffering severe decline in health of its dominant eucalyptus tree species, river red gum (*Eucalyptus camaldulensis*) and black box (*E. largiflorens*). A survey of red gum and black box health below Euston in 2003 found 70% of trees were poor or dead (MDBC 2003).

The Murray-Darling Basin Commission has established six icon sites along the River Murray: Barmah-Millewa Forest, Gunbower-Koondrook-Perricoota, Hattah Lakes, Chowilla Floodplain and Lindsay Wallpolla Islands, Lower Lakes, Coorong and Murray Mouth and the river channel. The condition of these icon sites was reported in 2008 (MDBC 2008). In general, across the system, severe drought conditions continue to affect waterbird communities and limit the availability of other wetland, floodplain and riverine habitats. Most floodplain or shallow icon sites were dry or almost dry and supported few waterbirds (MDBC

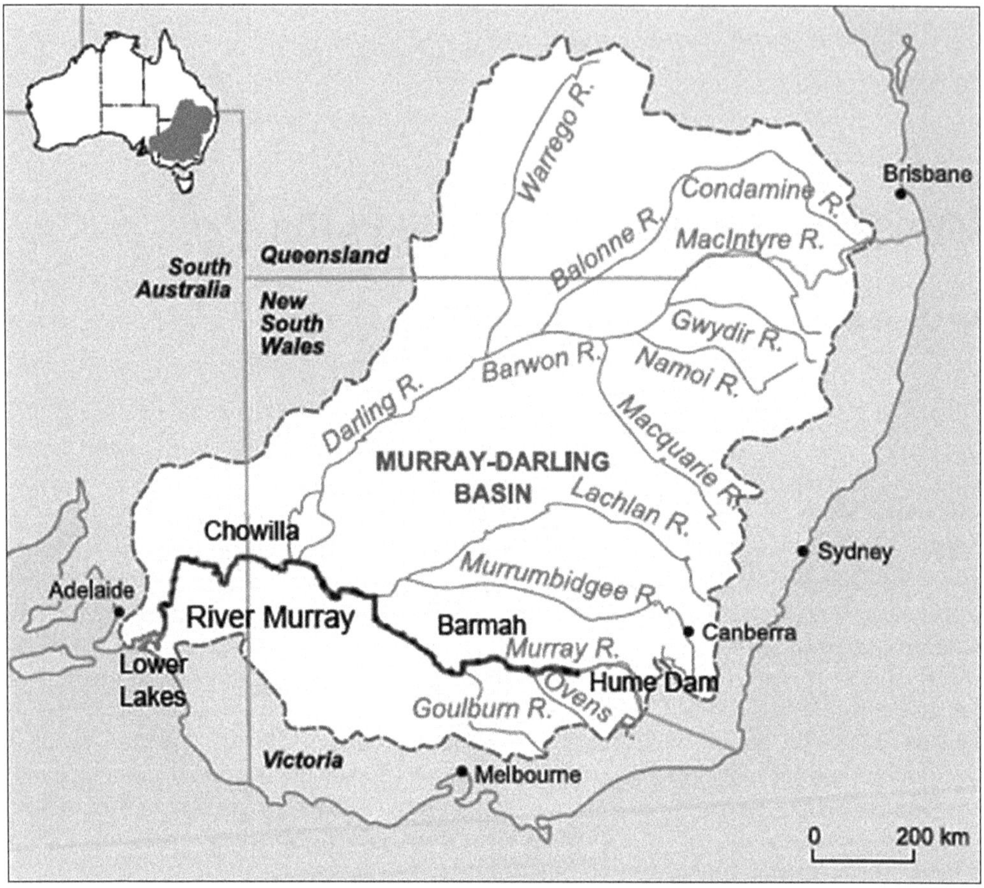

Figure 16.1: The Murray-Darling Basin within Australia, showing the River Murray. Chowilla (Zone 16) and Barmah (Zone 3) are shown in the lower and upper parts of the River Murray.

2008). The dominant tree species on the floodplain, river red gum and black box, continued to decline across all icon sites, with understorey vegetation suffering a loss of biodiversity (MDBC 2008).

Climate change is reducing the rainfall in south-east Australia (Chew *et al.* 2002). This reduced rainfall, combined with elevated groundwater levels from natural upstream drainage, has led to greater health decline in the Lower River Murray. The shallow saline groundwater, coupled with reduced flooding frequency, has led to soil salinisation which further reduces water availability to vegetation (Overton *et al.* 2006a). The upper River Murray regions have traditionally been considered safe from vegetation decline, however, Brett Lane and Associates (2005) found that 25% of red gums and 14% of the grey box maintained less than 50% of their original canopies in Gunbower Forest. Other ecosystem communities are also suffering from increased summer and permanent inundation. In contrast, some wetlands are suffering from reduced flooding connectivity.

Many studies have examined the longitudinal (upstream and downstream) dimension of the river, including the ecological impacts of river regulation on native flora and fauna and the physical changes occurring in the littoral zone. It has been established that long-term modification of flow rates, changes to the frequency of flooding events and alteration to flow timing has caused degradation beyond the littoral zone into both the lateral (floodplain extent) and vertical (soil profile) dimensions. Such human-induced changes contribute to land degradation throughout the region and increased accession of saline groundwater to the river (Jolly 1996). The semi-arid nature of the River Murray floodplain exacerbates these human-induced ecological imbalances. In semi-arid regions, floods are more extreme and less predictable and flow timing is more variable compared to other riverine environments, which may intensify the salinity problems (Walker and Thoms 1993).

Consequently, regulation of the River Murray has severely affected the balance of physical and ecological processes which maintain this unique riverine environment. Managing the river system to balance social, economic and environmental requirements, when resources are scarce, requires quantitative analytical tools (Young *et al.* 2000). However, no quantitative tool exists for measuring the environmental benefits of flood inundation or costs of flow reductions or balancing these with social or economic factors. Primarily, managers need a detailed flood extent prediction tool and an accurate vegetation map. With these, managers could determine the extent, depth and frequency of vegetation inundation from given flow management options and infer environmental impacts.

A floodplain inundation model has been developed for the River Murray region of the Murray-Darling Basin, that predicts the extent and depth of a range of flow regimes (Overton 2005, Penton and Overton 2007). This model was used to assess the water volumes required to sustain river red gum communities along the River Murray floodplain in Victoria (Overton and Doody 2007). The results indicated that 74% of the floodplain area was at high risk, with flooding frequency levels below those that would likely sustain healthy tree populations. Volumes of annual environmental flow requirements were calculated based on health indices for flooding frequency.

Previous ecosystem response models

Relationships between vegetation condition and flooding in the Barmah-Millewa Forest icon site are described in Chapter 20 (this volume). The original study (Bren and Gibbs 1986) linked red gum health and distribution to flood extents, and was the first spatial analysis of ecohydrology on the River Murray.

In 2002, the Murray-Darling Basin Ministerial Council called for social, economic and ecological assessments of the costs and benefits of returning additional water to the River Murray as environmental flows. The Ministerial Council chose three flow volumes – 350, 750 and 1500 GL per year – as reference points for assessment. The Murray Flows Assessment Tool (MFAT) (Young *et al.* 2003) was developed to assist in this ecological assessment.

The tool was used to identify the impacts of proposed environmental water allocations on the River Murray icon sites. This approach allowed incorporation of a range of ecosystem components where knowledge was available and provided the ability to assess the likely condition of those components for a given flow regime.

MFAT contains five ecological models including floodplain vegetation, wetland vegetation, waterbirds, native fish and the risk of algal blooms, to assess habitat condition based on modelled daily river flows. The models are driven by ecological information from selected localities (where an assessment is made) along the river. Results for localities can then be grouped to provide habitat condition scores for river zones and the whole river system.

MFAT scores the potential condition of habitat for floodplain and wetland vegetation, waterbirds and native fish for any given flow pattern and provides a score of the risk of blue-green algal blooms. Scores range from 0 (unsuitable) to 1 (ideal) habitat conditions. Preference curves summarise the ecological information that underpins the assessment models (Fig. 16.2),

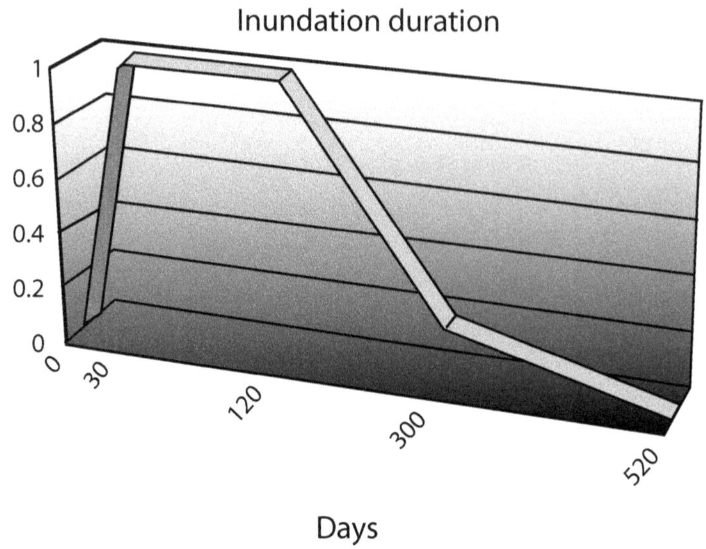

Figure 16.2: Empirical model of river red gum response to flooding, relating health to inundation duration (Young *et al.* 2003).

describing the preferred habitat condition (e.g. flood timing) for selected species. For example, the preference curve in Figure 16.2 describes how many days adult vegetation of a particular species prefers to be inundated for, in order to maintain its condition (1 = most preferred, 0 = least preferred). Preference curves are developed for adults as well as recruitment and can be combined to give an overall habitat condition score.

The MFAT is a decision support tool and not a true ecosystem model of the interaction between river ecosystem components, flow regimes and other drivers (Davies and Acreman 2003). Estimates of ecological response during historical periods should not be interpreted as ecological outcomes but as highly qualified indicators of relative potential, based on flow-related habitat only (Davies and Acreman 2003).

MFAT provides a framework for combining basic information on river flows, interactions with adjacent floodplains, billabongs and wetlands, and ecological responses based on habitat suitability. However, the level of complexity of MFAT is low. Development of better floodplain modelling tools, using GIS, digital elevation and hydraulic models etc., is fundamental for understanding the ecosystem response to flow regime (Davies and Acreman 2003).

In large lowland floodplain rivers, like the River Murray in Australia, detailed understanding of channel flows and their connection with floodplain inundation is necessary when describing the ecological outcomes that might result from different environmental water allocations (Oliver *et al.* 2007). This requires detailed floodplain elevation modelling, flood extent, vegetation mapping and condition assessment over the river floodplain.

Floodplain inundation model

Historic records of flow on the River Murray are limited to particular gauging stations that were in operation over different periods. It was necessary to use modelled flow data to analyse changes in river flows due to river regulation in each defined section of the River Murray over long periods. The Murray-Darling Basin Commission's MSM-BigMod river flow model provided 116 years of flow data from 1891 to 2007 for both predicted flows under 'current' conditions and under 'natural' conditions (i.e. if the current infrastructure were not in place)

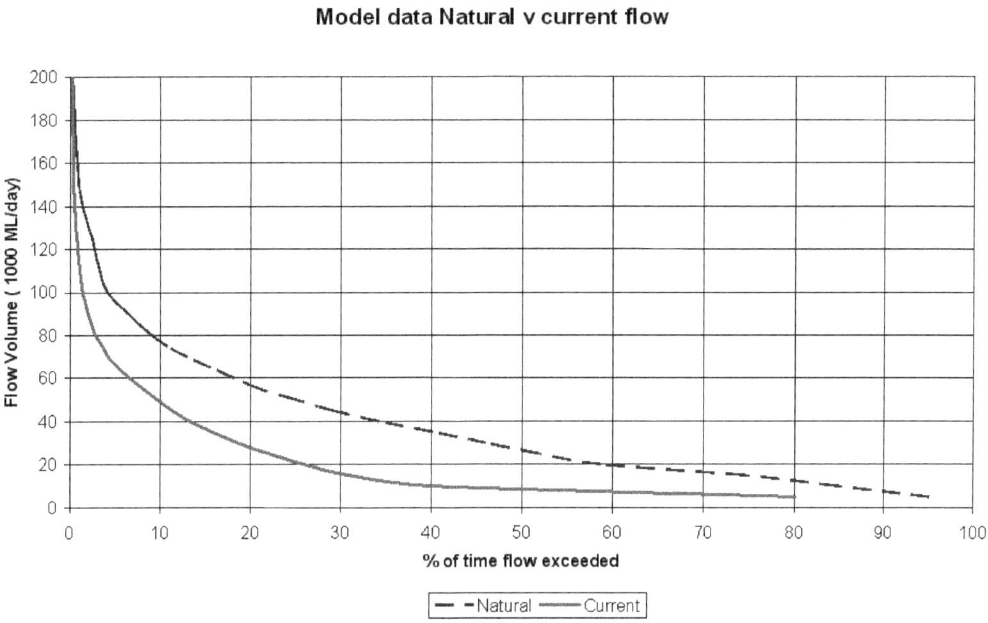

Figure 16.3: Flow exceedence curves for the River Murray for 'natural' (pre-regulation) and 'current' (modelled actual flows) conditions.

(Close 1986). Figure 16.3 (colour plate, p. xx) shows the change in the percentage of time exceeded for flows in the River Murray for 'natural' and 'current' conditions.

To predict potential changes to the floodplain, two scenarios were considered. First, predicted flows for the next 50 years were modelled by repeating the last 50 years of flow from 1957 to 2007. Second, future flows were modelled by considering potential climate change, using a prediction of a 33% reduction in river flows estimated by Jones *et al.* (2002). Current modelling of climate and river flows within the Murray-Darling Sustainable Yield (CSIRO 2007) project suggests this value is likely to be conservative. Future flows were modelled by gradually increasing a flow reduction until the maximum of 33% reduction in 2057 was achieved.

The River Murray floodplains are complex environments where small elevation differences across the floodplain create a range of different flooding regime habitats and dictate the need for precise and detailed hydraulic models. Hydraulic models of this detail require either a high-resolution digital elevation model or ground-surveying methods, which would be prohibitively expensive for the River Murray. High-resolution digital elevation models are available for the River Murray in the form of airborne laser scanning or LiDAR (Light detecting and ranging and laser altimetry) for the upstream section and digital photogrammetry (from high-resolution photography) for the downstream section. Modelling the movement of water across large floodplains is a complex task and complete data for hydraulic model development and calibration is often limited.

A satellite image-derived flood extent model, the River Murray Floodplain Inundation Model (RiM-FIM) (Overton *et al.* 1999, 2006a, 2006b, 2009; Overton 2005), provides a predictive capability for flood extents, depths, connectivity and volumes. RiM-FIM was created as a research and decision support tool for environmental flow management in the River Murray. It encompasses the River Murray floodplain from Hume Dam at Albury (Victoria) to Lake Alexandrina at Wellington (South Australia).

The model was developed using a Geographical Information System (GIS), remote sensing, LiDAR elevation data and hydrological modelling. Floodplain inundation extents were detected from satellite imagery for a range of flows and interpolated to model flood growth patterns. RiM-FIM predicts flooding extent on the River Murray floodplain from a range of river flows and weir levels and is useful for predicting the extent of inundation on the River Murray floodplain, including the flow regimes of wetlands and floodplain vegetation. It allows spatial and quantitative analysis of the flood extents to be used as an input into the management decision process.

Other studies on modelling river flow and inundation have been undertaken for wetlands on the Darling River in NSW (Shaikh *et al.* 2001), floodplains on Roanoke River in North Carolina (Townsend and Walsh 1998) and Murrumbidgee River (Frazier *et al.* 2003).

River Murray floodplain zones

The extent of the River Murray floodplain is defined by the 1956 flood, the largest in the 20th century. The Murray-Darling Basin Commission defined the 1956 flood boundary from historic records. Gaps in the boundary occurred where ground data on the extent of the flood were not available. The original 1956 flood boundary was edited by overlaying the boundary onto 1996 aerial photography and identifying the new floodplain boundary, by the extent of the visible floodplain edge indicated by vegetation on aerial photography.

River Murray flows are dependent on releases from Hume Dam, as well as flows into the river from major tributaries and outflows from irrigation and water storage extractions. Flow in each reach is dependent on management of locks and weirs, meaning a single flow input at the top of the system cannot be used to predict area of inundation for the whole system. To account for this variation, the River Murray and its floodplain were divided into 22 zones from Hume Dam in Victoria to Wellington in South Australia (Fig. 16.4/colour plate, p. 315).

Flow travel time for each zone was calculated from Commission data (Crabb 1997). The data present an approximation of the time a flow wave (a change in flow rate) takes to get from one zone to the next. Wave speed decreases at high (overbank) flows as water spreads out onto the floodplain, due to increased resistance to flow and storage effects (Jacobs 1990).

Floodplain inundation mapping

The development of RiM-FIM required analysis of flood extent, linking extents to LiDAR DEM heights, interpolation between flood events and linking extents to river heights to predict impacts of weir manipulation. The steps involved in developing this model shown in the flow diagram in Figure 16.5.

Remote sensing is particularly useful for monitoring flood extents because it provides basic data more cheaply and efficiently than ground-based methods (Whitehouse 1989). Studies to determine the aerial extent of flooding have commonly involved optical satellite image analysis (Usachev 1985; Walker *et al.* 1986; Townsend and Walsh 1998, Shaikh *et al.* 2001; Sheng *et al.* 2001; Frazier *et al.* 2003), radar remote sensing (Townsend and Walsh 1998) or an integration of remote sensing and GIS (Brivio *et al.* 2002). These methods have proved to be very useful and economical for large area flood analysis. More detailed studies have used digital elevation models to create a floodplain surface that can be inundated at certain river heights (Townsend and Walsh 1998). Elevation methods are particularly useful for predictive studies of changing flow paths through the floodplain by manipulating flow barriers. However, surface modelling may not give the best representation of floodplain inundation, as there are numerous impediments and small channels across a predominantly flat floodplain. Modelling of floodplain inundation from surface elevation also requires detailed information on stage heights, backwater curves, flow impedances and roughness coefficients – simple height levels are insufficient in this dynamic environment.

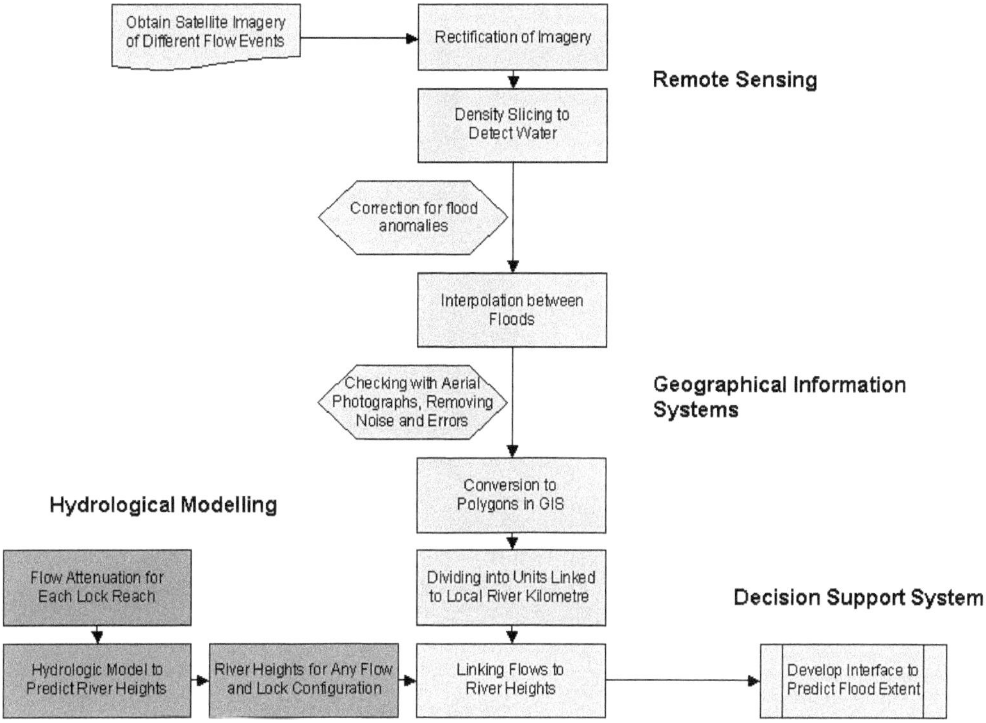

Figure 16.5: The process of building the floodplain inundation model for South Australia (Overton *et al.* 2006b).

Landsat TM and ETM+ imagery was used to provide good coverage and resolution of the study area with image resolution of 30 × 30 m. Landsat TM Band 5 (mid-infrared) was chosen as the band to use for a density slicing process, which was used to determine the presence of surface water and conducted in a GIS (Fig. 16.6). Water is detected by isolating the pixels that have a very low reflectance value, indicating that the light was absorbed and not reflected, for this band.

Seventy-eight satellite images were used, based on the criteria that:

- they were located on a rising limb of the hydrograph;
- there must not have been a greater sized flood within the last six months (this would leave standing bodies of water);
- they must not have contained any upper- or lower-level cloud cover that shadowed any portion of the River Murray floodplain.

The satellite images were registered to map grid co-ordinates so that flood masks (flood extents) would adjoin and be superimposed upon each other within the GIS. Registration was performed using map co-ordinates, digital data and image-to-image rectification

Integrating flood events

The images were classified as inundated or dry. The inundated area was classified according to the flow through the upstream gauging station. The collection of inundated areas were combined such that the greatest observed flow was assigned to each cell. For example, if the

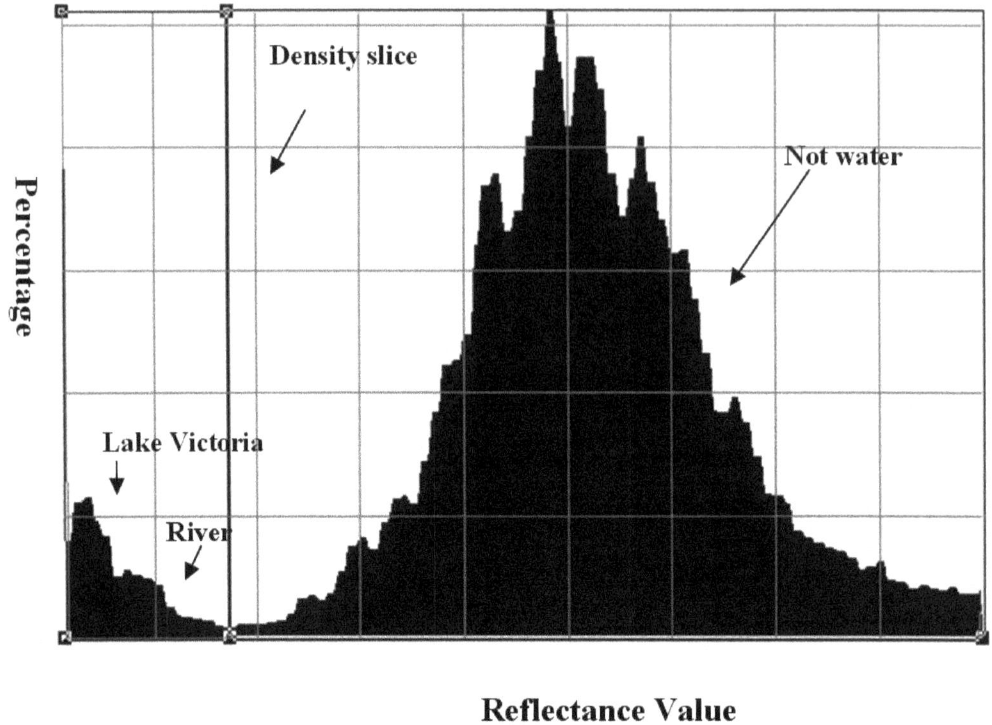

Figure 16.6: Histogram of a 102 GL/day flow with the central line indicating the point at which the reflectance value is split to indicate water or no-water. Two low reflectance value peaks represent Lake Victoria (with deep water) and the River Murray (with shallow turbid water) (Overton *et al.* 2006b).

lower flow scene showed the surface to be wet at its maximum extent then the lower value was used; otherwise, the current value was used. The floodplain inundation images were used to determine the exact extent of flooding that would occur at a givenWhile satellite images could provide a range of flood events (flow intervals), it was necessary to consider flows between those events to provide a more continuous predictive model. Interpolation between the discrete flow intervals using a digital elevation model produced finer intervals of flood extent. This flood growth map was linked to a hydrological model of the River Murray to determine the commence-to-flood river flow levels (discharge) and river heights. The RiM-FIM model provided a series of GIS layers showing areas inundated at different flow. Further, the high-resolution digital elevation model was used to produce improved inundation maps and estimates of depths.

Integrating LiDAR model with flood elevation

The technique for producing inundation maps involves calculating spot heights for inundations observed in satellite imagery, predicting the water height across the entire satellite image and generating inundation maps for observed and unobserved inundations. The result is a high-resolution set of images that describe the water height for different flow scenarios.

The initial objective of integrating a digital elevation model with flood masks was to calculate the depth of water in flooded areas. Depth is not directly observable from such a model or from a flood mask. Digital elevation models of the River Murray describe the elevation of the landscape relative to the Australian height datum (AHD), an estimate of height above sea level. Flood masks indicate whether a particular location is inundated. Depth can be calculated from these data sets given two assumptions. The first is that at the edge of an inundated area the depth of water will be zero and the water will be the same height as the landscape. The second assumption is that in a local region water maintains approximately the same height, which allows the water height at a point in the landscape to be predicted. The depth of an inundated location is calculated as the water height less the landscape height.

Since the water height points cannot be determined for every part of every inundation observed in a flood mask, a water height layer was generated using an interpolation technique.

An interpolated surface was fitted to the water heights using a kriging technique with linear trend removal. This surface needed to match the decline in heights along the river and out into the floodplain. Once the interpolated surface had been fitted to the observed water heights, the flood height and a revised inundation map were produced. Figure 16.7a shows a 23 000 ML/day flow and Figure 16.7b shows a 30 000 ML/day flow.

Producing a continuous commence-to-fill surface for comparison to hydrographs required the prediction of unseen flows. The depth of the observed floods is equal to the water height less the landscape height (Fig. 16.7c), with the edge of flooding defined as the point where water height equals the landscape height. New depths were created by linearly interpolating between observed water heights. For example, consider a point in the landscape (Fig. 16.8/colour plate, p. 316) where the water height of the 23 000 ML/day flow was 93 m AHD and the water height of the 30 000 ML/day flow was 94.4 m AHD. The water height of a 27 000 ML/day flow for that location was predicted as 93.8 m AHD. The surfaces were generated at 1000 ML/day intervals from the smallest regular flow to the highest recorded flow. A flood mask defines the outer boundary of the River Murray floodplain derived from aerial photography. The predicted height surface for a 27 000 ML/day flow is shown in Figure 16.7e.

The commence-to-fill at each point was calculated as the greatest predicted flow to inundate each point. Hydrographs from gauging stations along the River Murray show the frequency at which each flow has been observed. These frequencies were mapped directly onto the commence-to-fill layer to produce an inundation frequency layer for the river. A set of colour aerial photographs of a 70 000 ML/day flood were used to validate the growth behaviour of floods.

Hydrological modelling for South Australia

Hydrological flow models were integrated with RiM-FIM to enable the output data from the hydrological models to feed directly into RiM-FIM. The GIS model queries look-up tables for the required hydrological parameters, which are then linked to the spatial layers in the GIS.

This type of loose coupling allows existing hydrological models to be run without reprogramming within the GIS (Karami and Houston 1996; Sui and Maggio 1999) and proved to be an effective and economical approach, as the hydrological parameters are fixed values.

River Murray flow model

The channel flow component of the floodplain inundation model in South Australia was developed from:

- river height measurements at certain distances from the mouth of the river for certain flows;
- river flow measurements at gauging stations.

There were insufficient measuring points along the river to establish a relationship between river height and flow. To offset this, a modelled series of backwater curves was produced, consisting of a series of water surface elevation curves between locks for a steady uniform flow. A

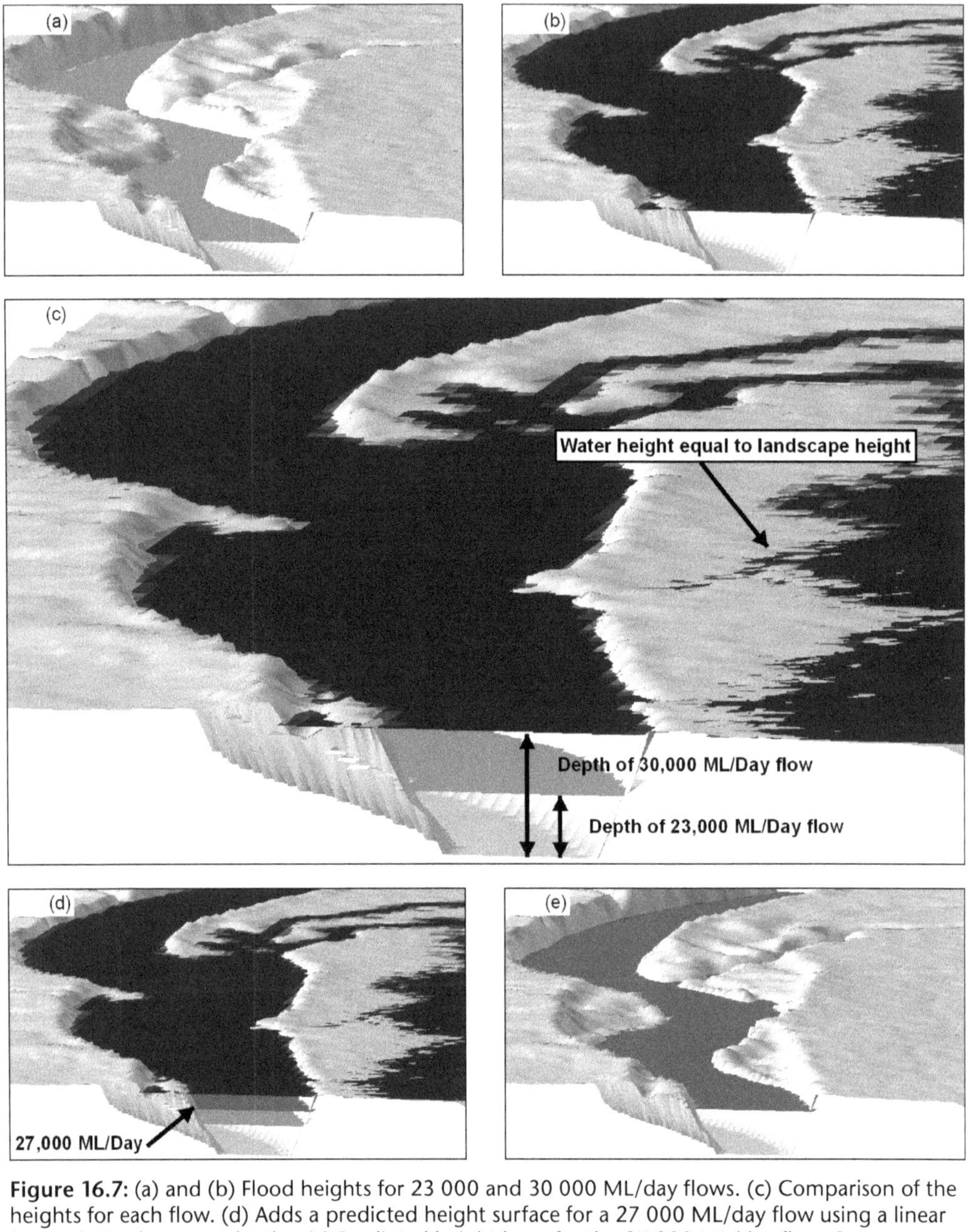

Figure 16.7: (a) and (b) Flood heights for 23 000 and 30 000 ML/day flows. (c) Comparison of the heights for each flow. (d) Adds a predicted height surface for a 27 000 ML/day flow using a linear interpolation between depths. (e) Predicted height layer for the 27 000 ML/day flow. Source: Overton *et al.* (2009).

relationship was established between flow rate and river height, where flow rate equates to a river height at weir pool level.

The locking of the river has resulted in water level being held at higher levels than under unregulated (natural) conditions. The use of backwater curves describes the behaviour of locked rivers compared to a pre-locking scenario where flow is downhill and unrestrained. River heights near the mouth of the river are much lower than those near the South Australian and Victorian border due to downfall flow of the river and its attenuation. Once flow levels have reached 60 000 ML/day, the river is above the maximum height of the locks and so backwater curves become straight.

Floodplain inundation response units

As river heights can be manipulated by the six locks, predominantly independent of each other, the floodplain needed to be divided into regions that could be related to local conditions in the river. The floodplain was segmented into a number of floodplain inundation units, regions on the floodplain that flood in response to a particular point along the river referred to as the trigger point. These floodplain inundation units were identified by considering the potential flood behaviour and discrete regions were generated by river morphology.

As a height difference of several metres can occur between the upper and lower pool levels of a weir, behaviour of the flow path around the weir is complex. A weir manipulation trial at Lock 5 demonstrated that the area around the lock responded to changes in the river height below the lock. There is very little difference between the river height within 2 km intervals in the modelled backwater curves, which implies there is probably an error margin of approximately two triggers up or downstream in the assignment of a unit to a trigger. Each polygon in the vector data set for a particular reach now contained attributes for reach, floodplain inundation response unit, flow and trigger.

Once the data had been converted to polygons in the GIS, it was clear that each floodplain unit contained small areas of floodplain that flood at the same particular flow. These areas were termed 'Wet at Same Height' (WaSH) areas, as they will inundate at the same river height for a given trigger point due to their flat elevation. The WaSH areas are likely to have the same or similar vegetation, as the distribution of floodplain vegetation communities is strongly related to flooding frequency.

Hydrological inputs to the model

Different flow rates at the NSW border produce certain river elevations along the river. It is predominantly the local river height that determines whether the river will break the banks and flood an adjacent area. In a hot dry summer period, a flow rate at the border will be further reduced downstream than it is in wetter months, as water is evaporated and extracted from the river and less is replaced by rainfall. Curves were generated which show the attenuation of flow at each weir for a given month.

The river hydraulic model was used to generate backwater curves for a range of weir manipulations and flow values. For a nominated water level and discharge at the lock, the backwater model defines the upstream water levels to the next lock. In this way, discharge relationships can be derived between the water levels at the downstream and upstream ends of a river segment. The precision of the model is 5 cm in the river height predictions.

GIS development

A raster grid of cell values that represent the commence-to-flow was based on the flows on the day of the satellite images. The raster data (pixel-based data) were converted to vector data (area- or shape-based data) to reduce the data redundancy and allow for easy retrieval, updating and generalisation of graphics and attributes that are especially important if equations are to be used to generate models. The polygons in the RiM-FIM are areas of the floodplain that commence to flood at the same river height or river flow (WaSH areas).

Decision support system for South Australia

The height of the river at trigger points are affected by river regulation, satellite-derived inundation and the interpolation process. Originally, flow values derived from the satellite images were regressed against the range of river heights for each trigger point. This produced a spread of data points across the range of actual flow values. However, because area inundated is extremely sensitive to slight alterations in river height, it was decided that a relationship between river height and area of floodplain inundation should be used in each floodplain inundation response unit. The model is limited to the spatial resolution at which it was acquired, in this instance 30 m. If a change in the height of the river results in less than a 30 m change at the flood boundary, the area of inundation will not increase.

A user interface was designed with input fields to prompt the user for flow values at the border. The result is a GIS map of the floodplain showing the areas flooded by that flow regime. Figure 16.8 (colour plate, p. 316) shows the output of the GIS floodplain inundation model after user input.

The initial stages of the project provided flood units and river heights at particular triggers. The hydrological model parameters made it possible to simulate a flow event from the border under different weir configurations for certain months. The GIS provided the map of the area inundated. The decision support system contained layers such as riparian vegetation, transport infrastructure, water infrastructure and major wetlands.

The inputs to the decision support system include the user-defined flow at the border, weir configurations of all six weirs, and month of year. The outputs of the model are river heights for each trigger kilometre and area of inundation.

Discussion on flood mapping

Prior to the RiM-FIM becoming available, this type of analysis was not possible. The commence-to-flood flow for the whole floodplain allows analysis of flood extent and flood return periods given defined flow hydrographs. We can thus examine current and previous flood regimes on the floodplain and predict flood regimes from future flow scenarios such as environmental flow strategies and climate change.

Figure 16.9 shows the results of integrating a high-resolution digital elevation model into the RiM-FIM. The resolution is significantly improved from a contour-based interpolation, and localised errors in the image classification are removed. The result is a high-resolution set of images that describe the water height for different flow scenarios. The surfaces provide river managers, conservationists, scientists and others with a predictive model of floodplain inundation for the River Murray floodplain.

Applications of the floodplain inundation model for ecosystem response modelling

Changes in flooding frequency

A flood index has been developed annually for the whole floodplain which scores the number of 'natural return periods' that part of the floodplain or wetland is experiencing. An area not flooded for six years, that once flooded every three years, has a flood index of 2. The effect of a flood is to reduce the flood index by 1 and not to reset the flood index to zero directly after a flood.

Red gums require flooding approximately once every 18 months. They will show signs of stress when the interval between floods is more than five years (approximately three times the natural return period). Evidence is seen on the Chowilla floodplain, where creeks and wetland areas not flooded for five years began to decline in 2003. Black box naturally flooded once

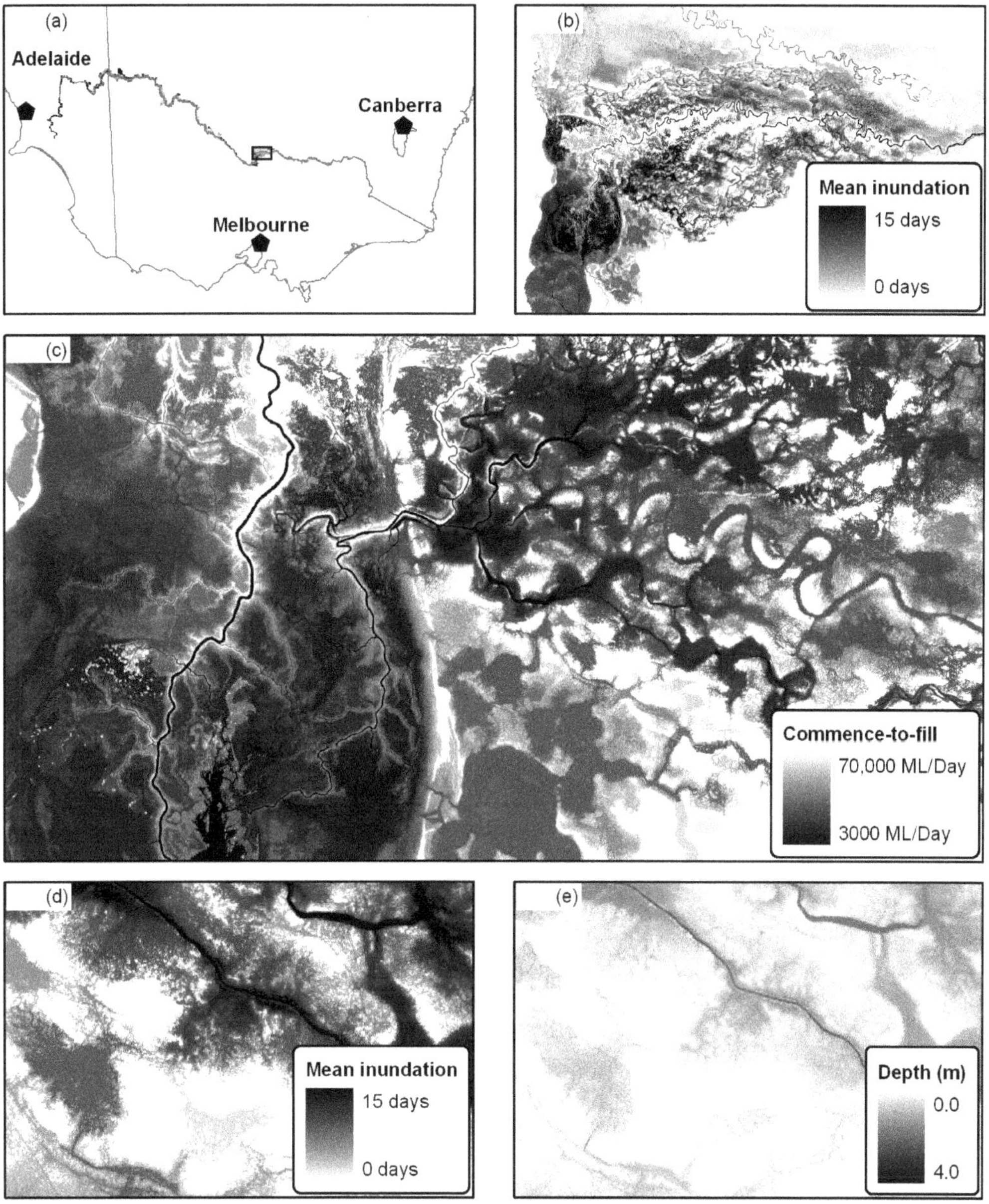

Figure 16.9: (a) Barmah-Millewah state forest on the River Murray. (b) Mean days connected in August for the region. (c) Commence-to-fill using the digital elevation model-based interpolation. (d) Mean days connected in August for a smaller region. (e) Depth of a 113 000 ML/day flood for that region. Source: Overton *et al.* (2009).

every four years, but under current conditions the recurrence of such floods is once every 12 years, therefore these trees are showing signs of severe stress.

A comparison of the vegetation health on the Chowilla floodplain with the derived flood index found a good correlation between healthy vegetation occurring in areas of a flood index of 3 or less (Overton and Doody 2007). Healthy vegetation was also found to occur within 50 m of

Table 16.1: Area of floodplain in good health (flood index £3) under different flow regimes

Area (ha)	1977	2007	2057 flow repeated	2057 climate change
Zone 3	78 500 ha (100%)	38 300 ha (49%)	36 600 ha (47%)	32 600 ha (42%)
Zone 16	31 200 ha (90%)	9500 ha (28%)	9500 ha (28%)	9360 ha (27%)
Whole river	598 000 ha (92%)	260 000 ha (40%)	247 000 ha (38%)	221 000 ha (34%)

Overton and Doody (2007).

permanent water bodies (Overton and Doody 2007). The flood index is therefore linked to vegetation decline (high risk) when the index is 3 or greater (three times the natural return period since last flood). This flood index of 3 roughly equates to a flow of approximately 1/3 'natural'.

The RiM-FIM was used to derive flooding frequency across the River Murray floodplain for the last 100 years. A flood index was derived for two periods, 1977 and 2007. The flood index for 1977 was derived to predict the general health of the floodplain in a period of less water extraction and wetter climate, but recent enough to include the major regulation infrastructure. A major flood in 1956 meant that a period of time sufficiently far away to not be influenced by that one-in-100 year event was required. The flows in the period leading up to 1977 can be considered a base case for good floodplain health.

The results show that the flood index values below 3 (low risk) were present over most of the floodplain in 1977, suggesting that the volume of flow in the river during the previous 20 years was sufficient to maintain vegetation health (Figs 16.5–16.9). This is supported by historical observations of healthy floodplain vegetation in that period. Further evidence for using 1977 flood volumes as a potential for analysis of base case was reported by Telfor and Overton (1999). Mapping floodplain health on the River Murray floodplain from 1945 to 1998 using field assessment and historic aerial photography, they found the floodplain was in relatively good condition in 1972.

Results for the flood index for 2007 indicated up to 60% of the floodplain is at high risk from reduced flooding frequency. These areas are 49% good for the Barmah area and 28% good for the Chowilla area (Table 16.1).

Effects of climate change and increased development within the Murray-Darling Basin, including plantations, farm dams and groundwater extraction, were also modelled to determine the impact on the floodplain ecosystem for a 50-year prediction, with repeated historic flow and a predicted climate change reduction in flows. The results show that up to 38% of the floodplain will remain in good health under a repeated flow and down to 34% in good health under a climate change scenario.

The flood index changes with different flows as the flooding history and river management changes. The flood index versus river flow for Zone 3 (Barmah) is shown in Figure 16.10 and Zone 16 (Chowilla) in Figure 16.11. Most of the flow range was within a flood index of 3 in 1977 for both areas. Future predictions with climate change give the highest flood indexes, with some flow bands having a flood index of 28. The flood index is generally higher for Zone 16 than it is for Zone 3 due to the pressure of water extractions as the river flows downstream.

Figures 16.12 and 16.13 illustrate the results for area of floodplain versus flood index for the two zones. They indicate a movement of floodplain area to higher flood indexes over time.

Figure 16.14 (colour plate, p. 317) highlights the area of Zone 3 (Barmah) in 1977 and 2007, with areas of flood index 3 or less in yellow and areas over 3 in red. These show that substantial

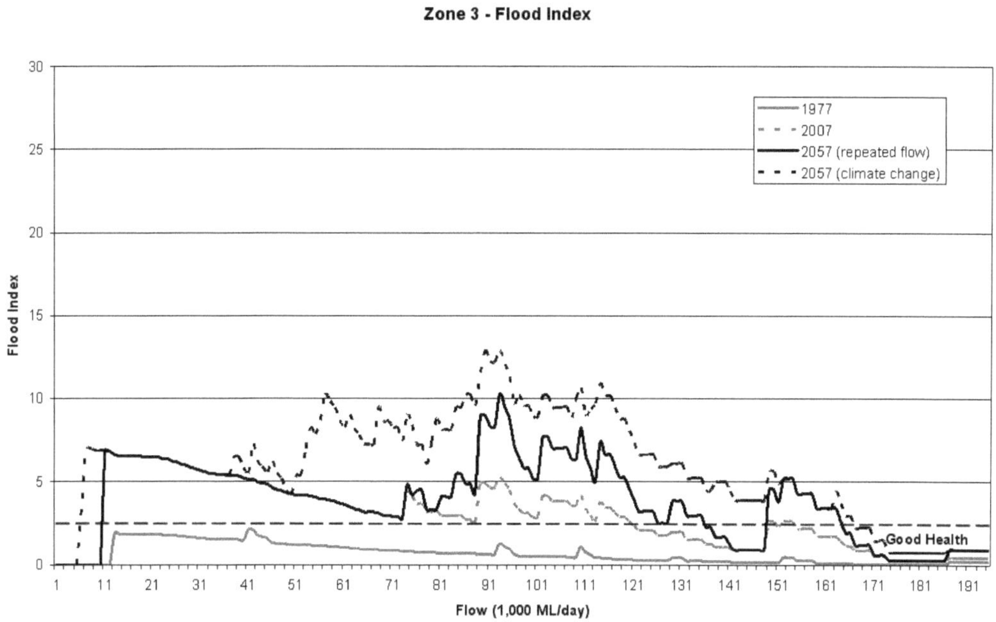

Figure 16.10: Flood index for different river flows for Zone 3 at Barmah in the Upper River Murray (Overton and Doody 2007).

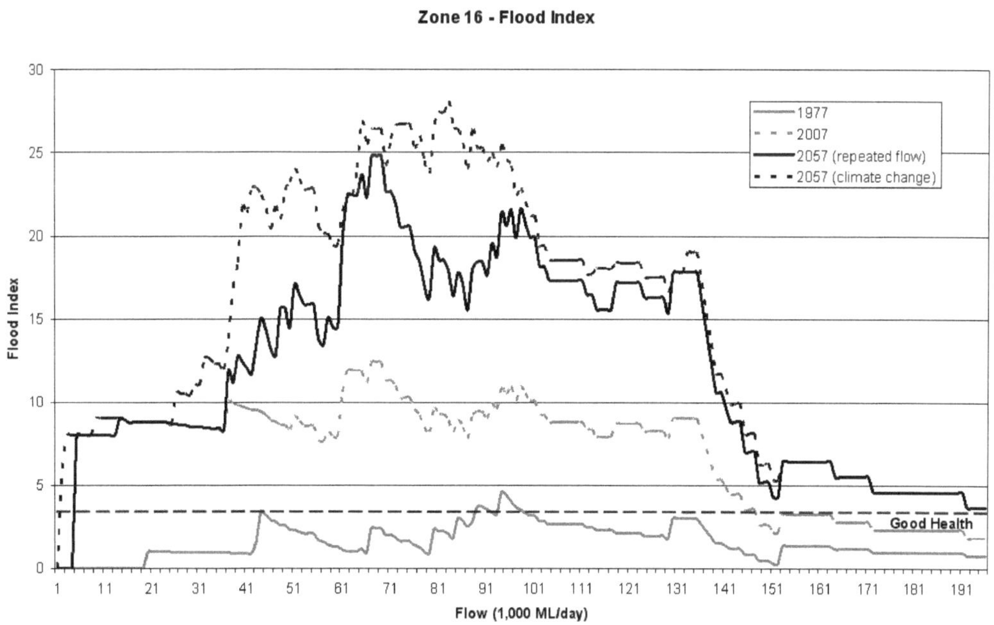

Figure 16.11: Flood index for different river flows for Zone 16 at Chowilla in the Lower River Murray (Overton and Doody 2007).

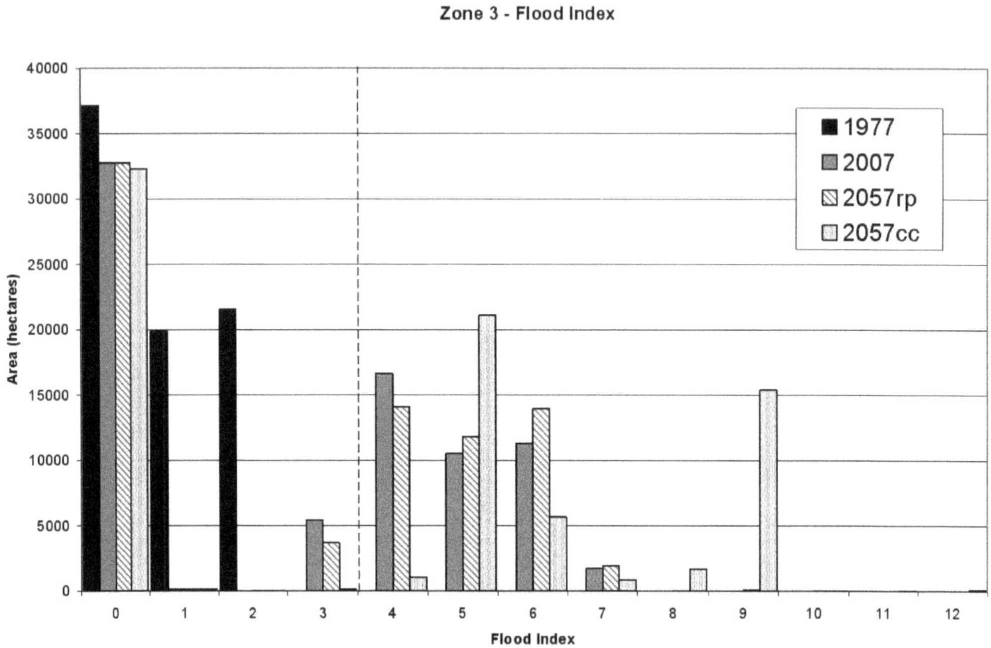

Figure 16.12: Flood index by area of floodplain for Zone 3 at Barmah in the Upper River Murray (Overton and Doody 2007).

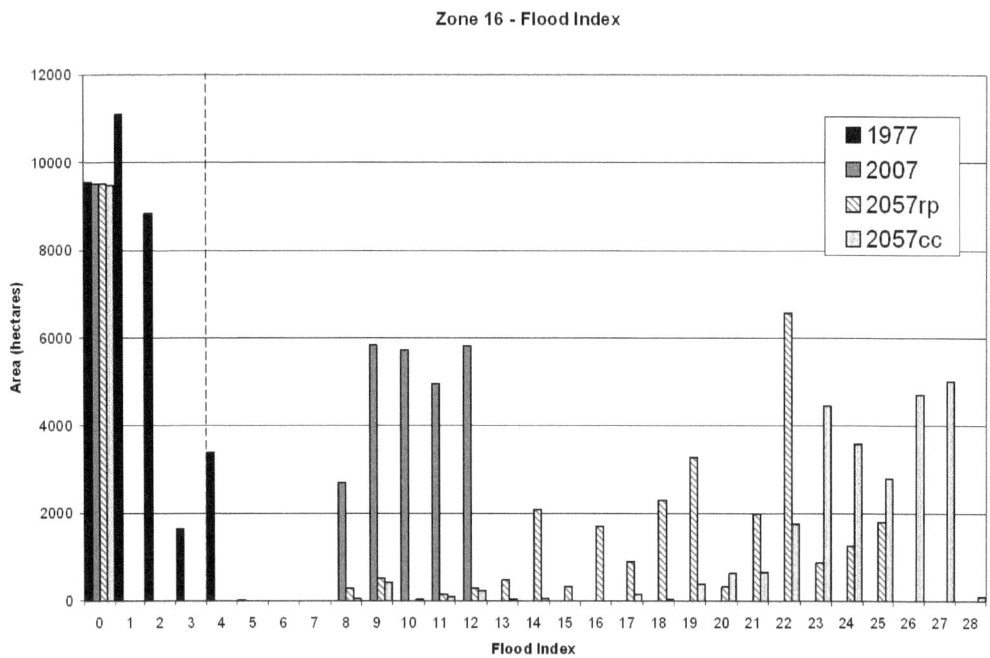

Figure 16.13: Flood index by area of floodplain for Zone 16 at Chowilla in the Lower River Murray (Overton and Doody 2007).

Table 16.2: Flow requirements for different vegetation types based on flooding frequency of base case (1970s) conditions

Vegetation class	Average flow (ML/day)	Days in 10 yrs (1977)	Volume required (GL/yr)	Volume last 10 years (GL/yr)
Floodplain woodland	70 000	108	756	70
Floodplain forest	34 000	386	1313	160
Woodland	53 500	189	1011	86
Woodland/lignum	127 250	8	102	0
Riverine forest	38 000	316	1201	141

Overton and Doody (2007).

areas of floodplain are now at risk as a result of a lack of flooding due to river regulation and the current drought period.

By considering that generally healthy floodplain existed in 1977, it is possible to calculate the flow volumes that would be required to maintain vegetation health. Vegetation classes for each zone of the River Murray were summarised by area and flow band. For each zone, the number of days of inundation for each flow band was calculated for a 10-year period before 1977 and 2007. The number of days multiplied by the average flow rate for each vegetation type provided an estimate of the volume of water required (GL/yr) over that 10-year period to maintain vegetation health (the same as for 1977). The same calculation was made using the period of reduced flow rates from 1997–2007, to estimate the volume of water that was received in that 10 years. The required number of days for floodplain woodland vegetation (black box) is given as 220 for Zone 16, which is approximately 2.5 three-month floods in 10 years, or a flood one in four years. For red gum areas (riverine forest) the number of days is 335, which equates to four three-month floods in 10 years, or one in 2.5 years. Both these flood returns are consistent with flooding regimes proposed for the Chowilla floodplain in the asset environmental water management plan for the icon site, determined using a complex vegetation health process model for water availability (Overton *et al.* 2006a).

Table 16.2 indicates the different volumes required across the vegetation types compared to those received in 2007. Water of this volume needs to be supplied in different ways to achieve the desired outcome in all the vegetation types. Some will need long small-volume floods, while others will need shorter but larger-volume floods.

Changes in wetland connectivity and other measures

The RiM-FIM provides a research tool for assessing the changing extent of flood habitats within the River Murray corridor. This volume has provided an analysis of wetland connectivity and macrophyte diversity (Ch. 18, this volume). A similar analysis could be done using the RiM-FIM for all River Murray wetlands. The RiM-FIM also provides quantification of changes to floodplain areas and has been used by Stokes *et al.* (this volume) to assess changes to invasive species occurrence related to flow regimes.

Framework for ecosystem modelling

The RiM-FIM provides a spatial framework for coupling ecosystem responses with hydrological drivers and other factors, such as soil type and land use, using the cell-based structure for modelling (such as the WINDS modelling described in Ch. 21, this volume). Within this spatial framework it is possible to analyse vegetation diversity, species composition and health with changing flow regimes. Cells within the model can be isolated to identify histories of flooding at individual sites such as wetlands or key floodplains.

Conclusions

A floodplain inundation model has been derived and linked to historic and potential future river flow records. The results indicate a decline in the flooding frequency of most of the floodplain and wetlands. A flood index was derived that related the current flooding frequency to the 'natural' frequency to predict floodplain health. Evidence showed a declining floodplain health with a flood index over 3 (three times the natural return interval). This index is a highly simplistic approach but is useful to give an overall trend in floodplain health.

Results have shown that over 50% of the floodplain is at risk of decline from lack of flooding. This area increases under the future scenarios to 60–70%, which includes areas of permanent inundation and areas of elevated floodplain which survive on rainfall. The area of floodplain that is inundated less than once every 13 years (the average return period for most of the floodplain habitat) is drastically shrinking.

Volumes of flow to maintain floodplain health can be derived for different flow bands by analysing a period when the health of the floodplain was maintained. This period was found to be prior to 1977.

The derived volumes appear to be realistic and compare well with a more detailed process modelling exercise for the Chowilla floodplain, which determined the same flooding return periods for the same vegetation types (Overton *et al.* 2006a).

The research indicates human activity and water resource overuse has had a significant impact on the River Murray floodplain over the last 100 years, as highlighted by the declining vegetation health. Future water management plans need to provide for allocation of environmental water to improve the floodplain ecosystem health, balancing the allocations with water requirements for irrigators and other water users along the river.

The change in flooding frequency has been particularly felt in the Lower River Murray, where saline groundwater is close to the surface, leading to soil salinisation. RiM-FIM has been used to help model the groundwater–surface water interactions affecting vegetation health (Doody and Overton 2008).

Previous exercises in putting a broad flow target for environmental sustainability (Jones *et al.* 2002) had a target of 1/2 to 2/3 for moderate to high probability of having a healthy working river. This study has suggested that 1/2 to 2/3 provides a flood index of 1.5 to 2, which has been shown to represent a broad healthy condition of the floodplain in 1977. When the index gets to 3 the degradation of the floodplain will be clear. The flow volume of the period prior to 1977 averaged between 1/3 and 2/3 of the natural flow regime.

The River Murray is Australia's largest river and is an essential economic and environmental resource. Human intervention to manage the River Murray over the past 100 years has involved regulating the river and providing water storages for regular water supply to irrigators and domestic water users. Due to the impact of river regulation, resource overallocation and reduced rainfall, the River Murray floodplain is suffering severe decline of its dominant eucalypt tree species. Flooding regimes along the Murray are only a fraction of what they were before regulation, and many floodplain ecosystems are suffering from the decline in water availability. The changed flooding regimes on the River Murray floodplain from Hume Dam to the Lower Lakes have been assessed using the GIS-based RiM-FIM. A flood index has been developed annually for the whole floodplain, which scores the number of natural return periods that part of the floodplain or wetland is experiencing. Natural return periods are based on recorded river flows prior to regulation and extraction 100 years ago. The flood index is linked to vegetation decline (high risk) when the index is 3 or greater (three times the natural return period since last flood). The results show that up to 74% of the floodplain is at high risk from reduced flooding frequency. Flood index values below 3 (low risk) were present over most of the floodplain in 1977, suggesting that the volume of flow in the river during the previous 20 years was sufficient to maintain vegetation health. This is supported by historical observations of healthy floodplain vegetation in that period. Effects of climate change and increased development within the Murray-Darling Basin, including plantations, farm dams and groundwater extraction, were also modelled to determine the impact on the floodplain ecosystem for a 50-year prediction, with repeated historic flow and a predicted climate change reduction in flows. The research indicates human activity and water resource overuse has had a significant impact on the River Murray floodplain over the last 100 years.

References

Bren LJ and Gibbs N (1986). Relationships between flood frequency, vegetation and topography in a river red gum forest. *Australian Forest Research* **16,** 357–370.

Brett Lane and Associates (2005). Gunbower Forest flood enhancement monitoring program implementation: sentinel wetland and understorey surveys. North Central Catchment Management Authority, Canberra.

Brivio PA, Colombo R, Maggi M and Tomasoni R (2002). Integration of remote sensing data and GIS for accurate mapping of flooded areas. *International Journal of Remote Sensing* **23(3),** 429–441.

Chiew FHS and McMahon TA (2002). Modelling the impacts of climate change on Australian streamflow. *Hydrological Processes* **16(6),** 1235–1245.

Close AF (1986). Computer modelling of the River Murray. In *Hydrology and Water Resource Symposium*. November 1986. Griffith University, Brisbane.

Crabb P (1997). Murray-Darling Basin Resources. Murray-Darling Basin Commission, Canberra.

CSIRO (2007). Overview of project methods. Report to the Australian Government from the CSIRO Murray-Darling Basin Sustainable Yields Project. Available at http://www.csiro.au/partnerships/MDBSY.html.

CSIRO (2008). Water availability in the Murray-Darling Basin. Report to the Australian Government. CSIRO, Canberra. Available at http://www.csiro.au/files/files/po0n.pdf.

Davies P and Acreman M (2003). Review of the interim report of the SRP, the Murray Flow Assessment Tool and the SRP-REG scientific process for the Murray-Darling Basin Commission. Report prepared for the Murray-Darling Basin Commission, Canberra.

Doody TM and Overton IC (2008). Riparian vegetation changes from hydrological alteration on the River Murray, Australia: modelling the surface water–groundwater dependent ecosystem. In *Proceedings of the HydroChange'08 Conference*. Kyoto, Japan.

Frazier PS, Page KJ, Louis J, Briggs S and Robertson A (2003). Relating wetland inundation to river flow using Landsat TM data. *International Journal of Remote Sensing* **24(19),** 3755–3770.

Jacobs T (1990). River regulation. In *The Murray*. (Eds N Mackay and D Eastburn). Murray-Darling Basin Commission, Canberra.

Jolly ID (1996). The effects of river management on the hydrology and hydroecology of arid and semi-arid floodplains. In *Floodplain Processes*. (Eds MG Anderson, DE Walling and PD Bates). Wiley, New York.

Jones R, Whetton P, Walsh K and Page C (2002). Future impacts of climate variability, climate change and land use change on water resources in the Murray-Darling Basin. CSIRO, Canberra.

Karami HA and Houston BH (1996). Evaluating strategies for integrating environmental models with GIS: current trends and future needs. *Computers, Environment and Urban Systems* **20(6),** 413–425.

Murray-Darling Basin Commission (2003). Preliminary investigation into observed river red gum decline along the River Murray below Euston. Technical report 03/03. Murray-Darling Basin Commission, Canberra.

Murray-Darling Basin Commission (2008). The Living Murray: Icon Site Condition Report, October 2008. Murray-Darling Basin Commission, Canberra.

Oliver R, Bickford S, Byrne G, Gallant JC, Matveev V, Overton IC, Penton D, Ryan S and Watson G (2007). *A Biophysical Framework for Assessing Environmental Water Allocation and Delivery in the Murray River. Proceedings of the International Conference on Environmental Flows*, Brisbane, September 2007.

Overton IC (2005). Modelling floodplain inundation on a regulated river: integrating GIS, remote sensing and hydrological models. *River Research and Applications* **21(9),** 991–1001.

Overton IC and Doody TM (2007). Flooding frequency and vegetation health relationships for environmental flows in the River Murray in Victoria. Report to the Victorian Environmental Assessment Council. CSIRO, Adelaide.

Overton I, Newman B, Erdmann B, Sykora N and Slegers S (1999). Modelling floodplain inundation under natural and regulated flows in the Lower River Murray. In *Proceedings of the 2nd Australian Stream Management Conference: The Challenge of Rehabilitating Australia's Streams.* Adelaide. (Eds I Rutherfurd and R Bartley). CRC for Catchment Hydrology, Melbourne.

Overton IC, Jolly ID, Slavich PG, Lewis MM and Walker GR (2006a). Modelling vegetation health from the interaction of saline groundwater and flooding on the Chowilla floodplain, South Australia. *Australian Journal of Botany* **54(2),** 207–220.

Overton IC, McEwan K, Gabrovsek C and Sherrah J (2006b). The River Murray floodplain inundation model (RiM-FIM). Technical report. Water for a Healthy Country. CSIRO, Adelaide.

Overton IC, Siggins A, Gallant J, Penton DJ and Byrne G (2009). Flood modelling and vegetation mapping in large river systems. In *Laser Scanning for the Environmental Sciences.* (Eds G Heritage and A Large) ch. 2. Wiley Publishing, Oxford.

Penton DJ and Overton IC (2007). Spatial modelling of floodplain inundation combining satellite imagery and elevation models. In *MODSIM 2007: International Congress on Modelling and Simulation.* December 2007. (Eds L Oxley and D Kulasiri) pp. 1464–1470. Modelling and Simulation Society of Australia and New Zealand.

Shaikh M, Green D and Cross H (2001). A remote sensing approach to determine the environmental flows for wetlands of the Lower Darling River, New South Wales, Australia. *International Journal of Remote Sensing* **22(9)**, 1737–1751.

Sheng Y, Gong P and Xiao Q (2001). Quantitative dynamic flood monitoring with NOAA AVHRR. *International Journal of Remote Sensing*, **22(9),** 1709–1724.

Sui DZ and Maggio RC (1999). Integrating GIS with hydrological modelling: practices, problems and prospects. *Computers, Environment and Urban Systems* **23**, 33–51.

Telfor A and Overton IC (1999). Assessment of the impact of the Bookpurnong/Lock 4 Irrigation District on floodplain health and implications for future options. Report prepared for the Loxton to Bookpurnong Local Action Planning Committee. PPK/Mapping and Beyond, Adelaide.

Townsend PA and Walsh SJ (1998). Modelling floodplain inundation using an integrated GIS with radar and optical remote sensing. *Geomorphology* **21**, 295–312.

Usachev VF (1985). Evaluation of flood plain inundations by remote sensing methods. In *Hydrological Applications of Remote Sensing and Remote Data Transmission.* (Ed. BE Goodison) pp. 235–243. Publication No. 145, International Association Hydrological Science, Washington DC.

Walker KF and Thoms MC (1993). Environmental effects of flow regulation on the Lower River Murray, Australia. *Regulated Rivers: Research and Management* **8,** 103–119.

Walker RD, Aubrey MC and Fraser SJ (1986). *Remote Sensing in Australia.* Canbec Press, Melbourne.

Ward K and Colloff M (2010). Ecosystem response modelling for the Barmah-Millewa Forest: the interface between science and management. In *Ecosystem Response Modelling in the Murray-Darling Basin.* (Eds N Saintilan and I Overton) pp. 345–356. CSIRO Publishing, Melbourne.

Young WJ, Lam DC, Ressel V and Wong IW (2000). Development of an environmental flows decision support system. *Environmental Modelling and Software* **15,** 257–265.

Young WJ, Scott AC, Cuddy SM and Rennie BA (2003). Murray Flow Assessment Tool: a technical description. CSIRO Client Report for the CRC for Freshwater Ecology, October 2003.

Chapter 17

Flow and metabolic activity in the channel of the River Murray

Rod Oliver and Zygmunt Lorenz

Introduction

The energy supplies for riverine food webs come from organic materials that are created in the channel through plant production or transported into the river from the catchment (Vannote *et al.* 1980; Thorp and Delong 1994). In general, the quantity of organic material determines the overall biomass of secondary producers that can be supported. The organic materials are delivered in many different forms: as intact organisms, as non-living particulate organic matter (detritus) and as dissolved organic matter. Each category contains an array of organic compounds of different composition, energy content and nutrient status. The mix of organic materials, and their utilisation, plays a major part in determining the relative roles of herbivores, carnivores, detritivores and decomposers in transferring organic material and nutrients through the food web (Moore *et al.* 2004; Dodds and Cole 2007). As a result, community composition is influenced by the types of organic material available, and in turn influences the pathways for the transformation and transferral of energy and nutrients (Moore *et al.* 2004). Since the supply of organic material shapes food webs by influencing ecosystem structure and function, it also affects their resilience and resistance to disturbance.

In the River Murray, iconic species such as fish and waterbirds have been used to indicate ecosystem condition and to assess the effectiveness of natural resource management strategies (MDBC 2005). Sustaining populations of these higher organisms often underpins management goals and achieving this depends on supplying their requirements for habitat and food resources. Current research efforts are largely focused on relating characteristics of habitat structure to population dynamics. Much less effort is directed at identifying the influence of changing food supplies on ecosystem structure, or understanding the links between habitat and food resources.

In lowland floodplain rivers like the Murray, it is expected that a significant proportion of river channel food resources will be supplied by organic materials transported from the floodplain (Vannote *et al.* 1980; Oliver and Merrick 2006). However, like many rivers that are regulated to supply irrigation, industrial and urban users, flooding in the River Murray has been significantly reduced and the channel increasingly isolated from the floodplain (Maheshwari *et al.* 1995). Despite this disconnection being widely recognised, its effects on supplies of organic material to the river, and on primary production and respiratory activity within the channel, have only recently been investigated in any detail (Oliver and Merrick 2006; Gawne *et al.* 2007). Indeed, for rivers in the Murray-Darling Basin and in fact across Australia, little is known of the importance of different sources of organic material to river channel energy supplies, or their influence on riverine food web structures. In particular, there is scant information on the effects of hydrological fluctuations on supplies of organic materials, or on the food web connections critical in transferring energy to iconic groups such as fish and birds (Bunn *et al.* 2003; Vink *et al.* 2005). Consequently, the sources and pathways of energy transfer that are required to meet current management targets for biota are largely unknown. This makes it difficult to assess the need for environmental flows to deliver materials to support these management objectives, or to predict the impacts of climate change on energy supplies and to plan for transition.

River metabolism

Photosynthesis and respiration are the metabolic processes responsible for the formation and breakdown of organic material. The balance between these two processes within the river channel provides insight to the fluxes of energy through the food web (Odum 1956). Gross primary production (*GPP*) is the total carbon fixed into organic material by photosynthesis; in the river channel contributions come from phytoplankton, periphyton and submerged macrophytes. Community respiration (*CR*) is the loss of carbon due to the respiratory metabolism of organic materials to provide cellular energy and nutrients. All living organisms within the river channel contribute to this process. Subtracting *CR* from *GPP* gives the net ecosystem production (*NEP*), which describes the balance between the production and breakdown of organic material (Odum 1956).

In single organisms, the interpretation of metabolic activity is relatively straightforward. In plants, net production (*NP*) is generally the difference between photosynthesis and respiration. In secondary producers, positive and negative net production is determined by the balance between the intake of organic material as a food source and the amount respired due to cellular metabolism. Net production indicates whether the organic content has increased, in which case *NP* is positive, or decreased, in which case *NP* is negative. Positive *NP* indicates growth. Negative *NP* indicates a drawdown of cellular carbon reserves and cannot be sustained indefinitely, leading to death or the formation of resting stages.

The analysis of whole system metabolism is more complex to interpret because of the many interactions involved. Complications include variations in the rates of primary production and respiration over space and time, and the range of different sources of organic materials available to drive respiration. For example, organic material can be formed by primary producers at a sampling site or it can be imported from upstream, from the catchment or from the atmosphere. As a result, heterotrophic respiratory activity at a site can exceed *GPP* if the quantity of metabolisable organic carbon is subsidised by imported material. In this case a negative *NEP* does not necessarily indicate cellular energy limitation. Conversely, organic materials produced by primary production in one part of the river can be transported by flow and respired in a different part, or can be exported totally from the river system to coastal waters. In these cases, the export of primary production prior to it being broken down reduces the respiration rate at the site of production, resulting in an apparently larger *NEP* at that site.

To assess the balance of metabolic activities along a river system, measurements need to be made at spatial scales that describe the influence of import, export and transport of organic materials, and at temporal scales that capture the changing balance in *NEP* in response to factors such as daily and seasonal weather conditions. The studies reported here form part of a series of investigations aimed at describing metabolism within the 1400 km section of the River Murray from Hume Dam to Lindsay-Walpolla (Fig. 17.1). Although the river is regulated,

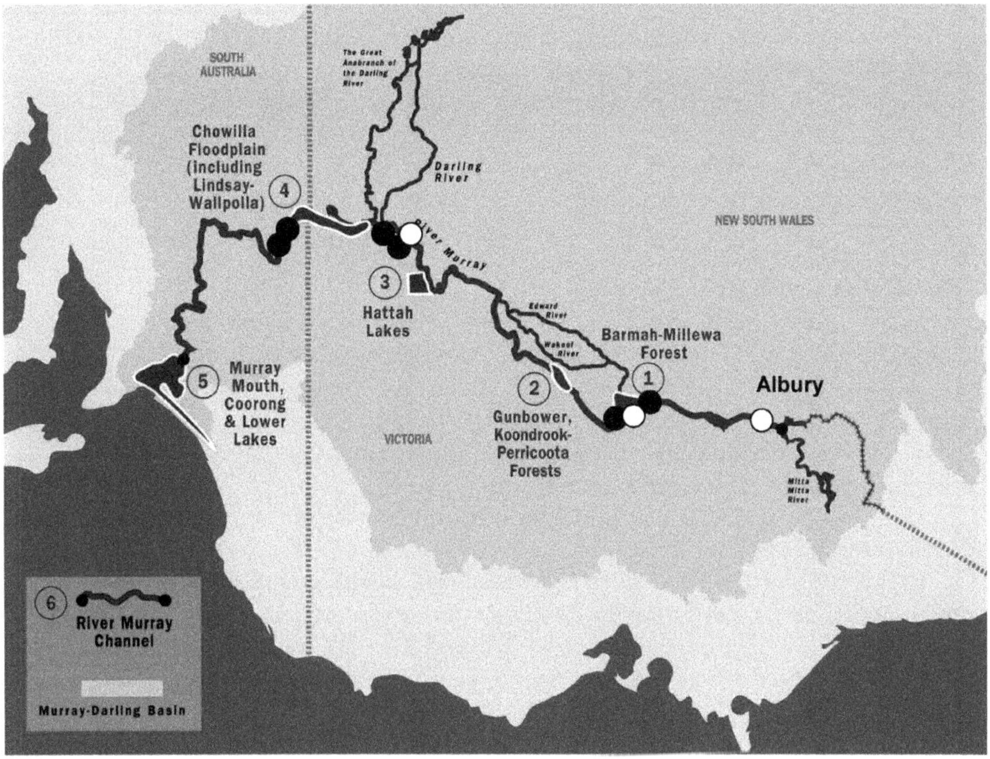

Figure 17.1: Sampling sites on the Murray River for the first (○) and second (●) sampling campaigns. Map modified from Murray-Darling Basin Commission webiste: Ministerial Council communique 2003 showing icon sites.

long flowing sections remain between the controlling weirs. A major objective of this study was to assess the contribution of internal and external sources of organic material to river metabolism and to estimate the supply of organic carbon moving through the food web to support higher trophic levels (Oliver and Merrick 2006).

Sampling sites and data integration

As only a certain number of site visits was possible, there was an unavoidable trade-off between spatial and temporal coverage of measurements. The number of sampling sites that could be investigated along the river during any single field trip needed to be balanced against the frequency of trips. Sampling sites were selected to provide information on longitudinal differences in river metabolism and to cover a range of flow velocities that might reflect different channel habitats. Two sampling campaigns were required: the first was in 1999–2000 (Oliver and Merrick 2006) and the second in 2006–07. The first campaign used three sampling sites in flowing river reaches spread along the length of the system (Fig. 17.1), sampled quarterly or monthly. The second campaign consisted of six river reaches forming contrasting pairs, three nominated as flowing reaches and overlapping the sites from the first campaign, and three sites associated with in-stream weirs that extended the coverage into low-flow environments as well as further downstream. In this campaign, sampling was focused only on the growing season.

A potential problem with this disjointed approach was the inherent difficulty of combining field measurements that had been made across large spatial scales and over different seasons, and in this case including a seven-year gap between data sets. The most suitable way of handling the dilemma was to link biological responses to environmental variates, so that modelling could be used to extend interpretations of measurements over intervening river reaches and across disjunct time periods. Although difficult, it is important to establish these quantitative linkages as they help to identify potential tools and options for river management. In this study, parallel measurements were made of environmental variates relevant to metabolic processes, particularly underwater irradiance, water temperature and river flow characteristics.

Measuring river metabolism

A widely used technique for measuring river metabolism is the analysis of diel time series of dissolved oxygen concentrations (Odum 1956; Young and Huryn 1996; Oliver and Merrick 2006). Oxygen is generated during photosynthesis and consumed during respiration, so oxygen concentration time series measured during daylight provide an estimate of *NEP* because photosynthesis and respiration are both occurring. Overnight measurements are expected to show the decline in oxygen concentration due to respiration and so provide an estimate of *CR*. The *GPP* is calculated from these two measurements by correcting the daytime *NEP* measurement for the estimated reduction due to *CR*.

The River Murray channel in general contains few submerged macrophytes, so measurements were focused on estimating the total channel metabolism and the contributions from planktonic and benthic micro-organisms (Oliver and Merrick 2006). Contributions can also be made by organisms attached to in-stream structures such as large woody debris, but this was expected to be small in the River Murray as historically it was desnagged for navigation and contains relatively little woody material (Treadwell *et al.* 1997; Gawne *et al.* 2007).

Changes in the dissolved oxygen concentration of the open water were measured using oxygen sensors fitted to recording sondes. These were deployed in the water column at three locations along a 5 km river reach for 36–48 hours. Planktonic metabolism was measured by enclosing water samples in clear, tubular, perspex incubation chambers that extended over the depth of the illuminated surface layer (the euphotic zone). Periphytic production in the illuminated littoral zone was measured using cylindrical benthic chambers fitted with clear perspex lids. These were pushed into the bottom sediments to enclose an area of 700 cm^2 overlain with a 15 cm height of water. Both the planktonic and benthic chambers were fitted with recording oxygen probes and mixed using submersible recirculating pumps. Diel chamber incubations were run in parallel with the open water measurements. In the absence of other contributors, the difference between total system and planktonic rates should be similar to the direct estimates of benthic metabolism. This provided a means of checking for unexpected contributions.

Metabolic rates in closed chambers were calculated directly from concentration changes in the oxygen time series. In contrast, estimating metabolic rates from open water measurements is less straightforward because oxygen concentration changes depend not only on the rates of photosynthesis and respiration, but on gas exchange at the air–water interface. If respiration reduces the oxygen concentration in the water below that balancing the atmospheric concentration (the saturation oxygen concentration), oxygen diffuses into the water column. Conversely, if photosynthesis increases oxygen concentrations above the saturation concentration then oxygen diffuses out of the water column. The open water oxygen measurements can also be influenced by groundwater exchange if volumes are large and dissolved oxygen concentrations differ significantly between the river and the groundwater. There was no evidence of substantial groundwater effects at the sampling sites used in this study.

Oxygen data analyses

Changes in dissolved oxygen concentrations over time in the open water (dO/dt) were considered due to the combined influence of photosynthesis, respiration and oxygen exchange at the air–water interface (Equation 1) (Young and Huryn 1996; Oliver and Merrick 2006):

$$dO/dt = GPP(E_t) + kD + CR \tag{1}$$

$GPP(E_t)$ describes the dependence of integral gross photosynthetic production (*GPP*) on light intensity, E_t being the incident, photosynthetically active radiation (*PAR*) at time t (Kosinski 1984; Young and Huryn 1996). Light measurements were obtained from a recording light sensor deployed on location.

Atmospheric gas exchange was estimated as the product of a re-aeration coefficient k (time^{-1}) and the oxygen deficit D. The deficit is the difference between the calculated saturation oxygen concentration and the measured oxygen concentration (Odum 1956; McCutchan *et al.* 1998). Theoretical saturated oxygen concentrations were calculated from water temperatures measured at five-minute intervals using a formula from the International Oceanographic Tables (1973), but without salinity correction.

A three-dimensional curve fitting routine (Sigma Plot) was applied to Equation 1 using measured rates of oxygen concentration change (dO/dt), irradiance (E_t) and deficit (D) to estimate *CR*, k and *GPP* (Oliver and Merrick 2006).

Annual system metabolism

The regular measurements made at three sites during the first campaign were integrated over the annual cycle by assuming smooth transitions from one measurement to the next. The river was found to be moderately productive across sites with *GPP* ranging between 775 and 1126 $gO_2m^{-2}yr^{-1}$ (Fig. 17.2) and community respiration between 872 and 1284 $gO_2m^{-2}yr^{-1}$ (Oliver and Merrick 2006). As annual rates of *GPP* were of similar magnitude to annual rates of *CR*, *NP* was not significantly different from zero (Fig. 17.2). This was interpreted as meaning that, along the length of the river, external supplies of organic material were not playing an

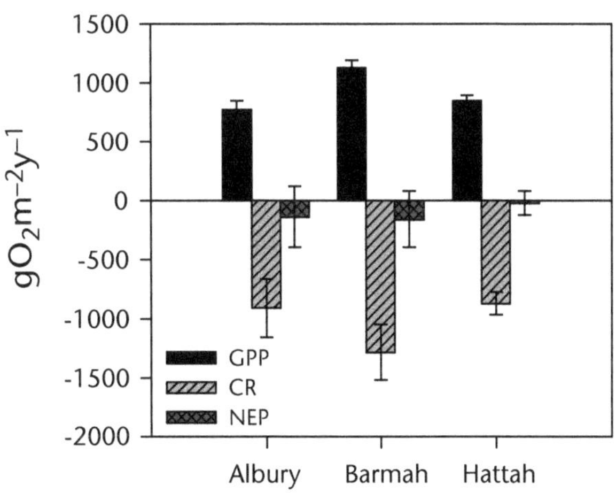

Figure 17.2: Annual metabolism at three sites along the Murray River, estimated from the first sampling campaign.

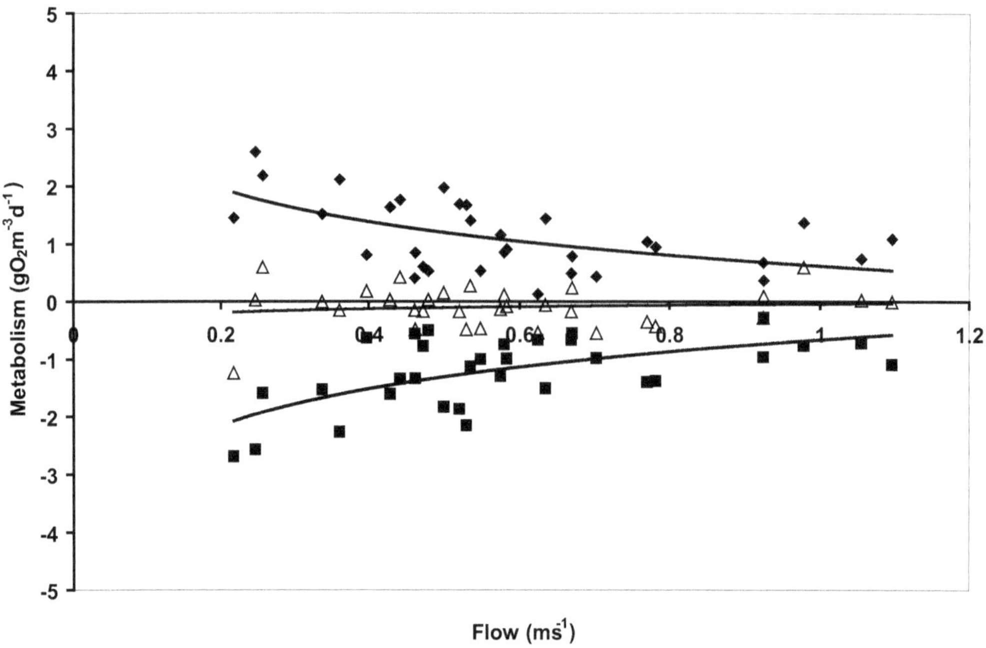

Figure 17.3: Volumetric community *GPP* (♦), *CR* (■) and *NEP* (△) displayed on a flow velocity axis for sites with velocities >0.2 m s^{-1}. Trend lines are included to highlight the response pattern.

important role in metabolism, otherwise respiration rates would be expected to exceed primary production (Oliver and Merrick 2006). Two approaches were used to further assess these findings and to extend the analysis across all sites and sampling times.

Metabolism at river sites

In the first approach, metabolic rates at river sites were plotted on a water velocity axis (Fig. 17.3). River sites were defined as those having velocities greater than 0.2 m s^{-1}, as this velocity marks a change in river conditions (discussed below). Note that the trend lines in Figure 17.3 are indicative only and show the average change in the rates of metabolism. Although *GPP* and *CR* increased as water velocities declined, *NEP* was zero when averaged across all sites and sampling times, although positive and negative values were observed on specific sampling occasions (Oliver and Merrick 2006). This supports the suggestion that external sources of organic material were playing only a small part in river metabolism. The findings imply that the riverine food webs are supported by organic material generated within the channel, and that this production must be fully utilised for *NEP* to be zero.

Partitioning metabolism at river sites

To further investigate these interactions, metabolic activity was partitioned between compartments (water column and sediments) and organisms (primary producers and heterotrophs) using results from incubation chambers and numerical analyses. This was done to identify the major sites of metabolic activity and to relate metabolic activities to appropriate environmental parameters. For example, the light conditions experienced by micro-algae mixing vertically through the water column are different from those experienced by micro-algae on the bottom sediments.

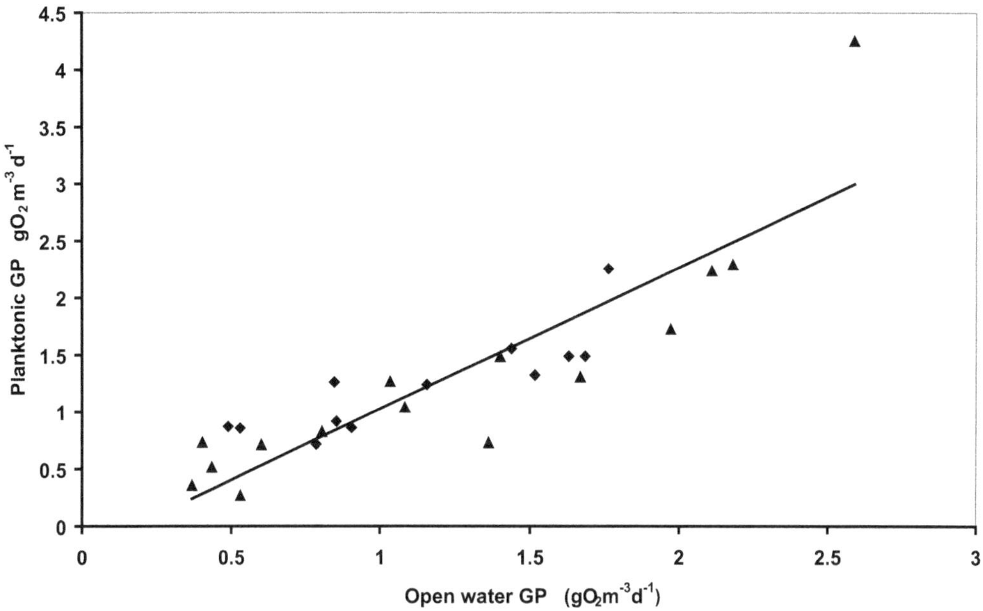

Figure 17.4: Comparison of planktonic *GPP* and total system *GPP* for all river sites. Sampling set 1 = ▲, sampling set 2 = ♦.

Partitioning photosynthesis

A comparison between the total system *GPP* and planktonic *GPP* at river sites showed that these measurements were equivalent, indicating that most of the photosynthesis was carried out by phytoplankton (Fig. 17.4). Linear regression indicated a ratio not significantly different from 1.0 and an intercept not significantly different from zero ($y = 1.14x - 0.07$, $r^2 = 0.76$, $Pslope < 0.0001$, $Seslope = 0.13$, $Pconst = 0.68$). This result was consistent with the findings from benthic chambers, which showed that benthic micro-algae contributed only 5–6% of *GPP* at the river sites (Oliver and Merrick 2006). The findings confirmed that macrophytes and other attached organisms were not making a significant contribution at these sites. It was concluded that, as phytoplankton photosynthesis accounted for most of the primary production and that production matched the total system respiration to give zero *NP*, phytoplankton were likely to be the major source of organic material supporting the food web.

Light and photosynthesis

Light availability is a major environmental driver of phytoplankton photosynthesis. Its influence was investigated by comparing phytoplankton biomass-specific rates of gross primary production (GPP_B) with the mean irradiance of the water column (Oliver and Merrick 2006). To estimate GPP_B, all planktonic photosynthesis was attributed to phytoplankton and standardised to the planktonic chlorophyll-*a* concentration. This was considered reasonable, as there are unlikely to be other contributors to planktonic photosynthesis. A significant linear relationship was found using data from all river sites for both sampling periods (Fig. 17.5), with regression yielding the relationship $y = 0.34x + 4.4$ ($r^2 = 0.81$, slope $P < 0.0001$, intercept $P = 0.56$). Considering that these are field measurements, this regression indicates a surprisingly consistent relationship between the mean irradiance and GPP_B within the River Murray. There is little indication from this relationship that nutrients are influencing rates of photosynthesis.

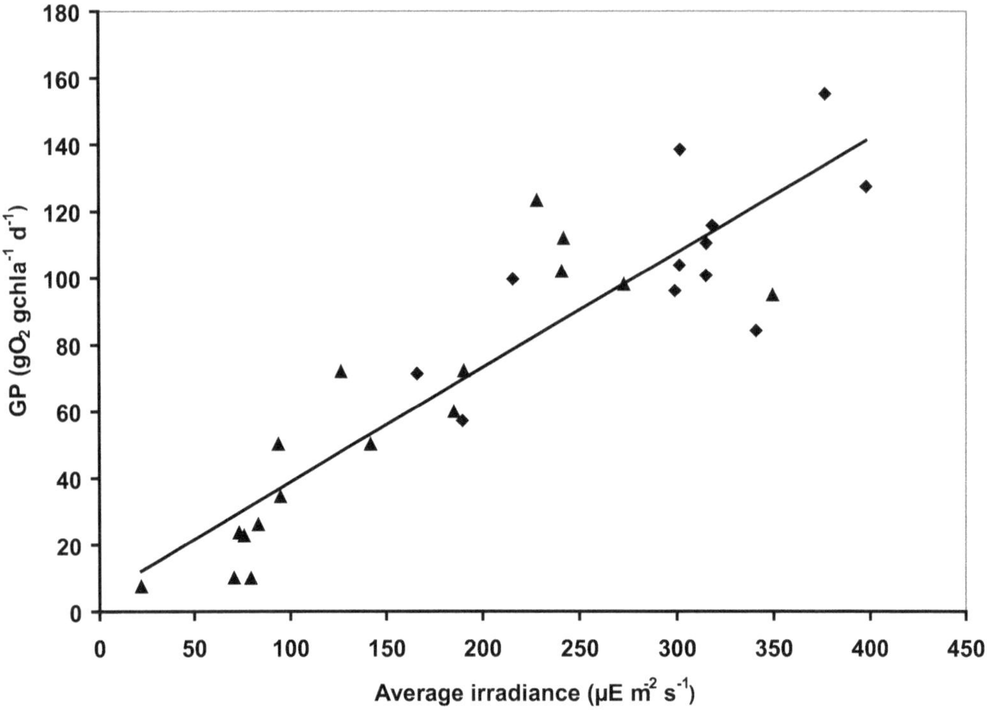

Figure 17.5: Chlorophyll-specific planktonic *GPP* for all river sites (sampling set 1 = ▲, sampling set 2 = ♦) as a function of the mean irradiance encountered in the mixed water column.

Partitioning respiration

A comparison of total system *CR* and planktonic *CR* indicated that plankton accounted for most of the respiration during the first sampling campaign. Two-factor ANOVA without replication (n = 14, $P = 0.03$) indicated a significant difference, with planktonic respiration on average contributing 88% of the total. Plankton contributed significantly less during the second campaign, averaging about 78% of the total respiration (two-factor ANOVA without replication, n = 12, $P < 0.008$). As the difference between the total channel metabolism and planktonic metabolism is attributed to benthic activity, this suggests an enhanced benthic respiration during the second period of sampling. Calculations suggested that the increased benthic respiration was due to increased sedimentation losses of phytoplankton because of decreases in river depth as a result of drought. This is consistent with the *GPP* measurements that indicate phytoplankton are the major source of organic material driving respiration.

Phytoplankton share of planktonic respiration

Unlike planktonic photosynthesis that can be unequivocally attributed to phytoplankton, planktonic respiration can result from the activity of a range of organisms including phytoplankton, zooplankton, bacteria and fungi. Regression analysis was used to try to partition the planktonic respiration, and identify the contribution of phytoplankton from an estimate of the potential respiratory carbon pool associated with the phytoplankton.

In general, phytoplankton respire recently fixed photosynthate, but if this is insufficient then cellular reserves are utilised. In the first instance, specific cellular carbon stores are respired; as these are drawn down then structural elements are sacrificed to sustain cellular function (Beardall and Raven 1990). The potential respiratory carbon pool was calculated on a daily basis as the sum of *GPP* and the phytoplankton biomass, both expressed in units of carbon (gC). Measurements of *GPP* estimate the amount of recently fixed photosynthate, while the phytoplankton biomass provides an estimate of structural and long-term storage components that are utilised in metabolism following extended low-light periods (Beardall and Raven 1990; Oliver and Merrick 2006). Although the calculated total stock of phytoplankton respiratory carbon is likely to overestimate the cellular respiratory carbon pool, it does help account for the buffering of respiration by cellular reserves.

The total daily stock of phytoplankton respiratory substrate was significantly correlated with planktonic respiration for all river sites over all sampling times (Fig. 17.6), accounting for 87% of the variation and giving an intercept not significantly different from zero ($y = -1.12x - 0.1$, $r^2 = 0.87$, $Pslope < 0.001$, $Pconst = 0.18$) (Oliver and Merrick 2006). This suggests that phytoplankton account for the planktonic respiration at the river sites. It could be argued that the numbers of heterotrophs, such as bacteria and fungi, also correlate linearly with the phytoplankton respiratory substrate and contribute substantially to the respiration, but this seems unlikely. Samples were taken during periods when phytoplankton populations were increasing in number and when they were decreasing in number. During periods of phytoplankton decline, when the number of moribund phytoplankton cells increases relative to viable cells, heterotrophic respiration might be expected to increase relative to that of the intact phytoplankton. This would cause significant deviations from the observed regression, but as this was not observed most of the respiration is attributed to viable phytoplankton. Of course there will be some contribution from planktonic heterotrophs, but it is considered to be small. These findings suggest that the respiratory breakdown of phytoplankton cells due to heterotrophic consumption is occurring largely in the sediments.

Modelling phytoplankton NP at river sites

In the preceding analyses, planktonic *GPP* and *CR* were attributed to phytoplankton metabolism. Consequently, the derived relationships between *GPP* and the mean irradiance of the water column (Fig. 17.5), and *CR* and the respiratory substrate (Fig. 17.6), can be used to describe characteristics of phytoplankton metabolism, including estimating phytoplankton *NEP* separately from the total system *NEP* (Oliver and Merrick 2006). In the calculations, units were standardised to carbon by assuming photosynthetic and respiratory quotients are both equal to 1, and that the phytoplankton carbon to chlorophyll ratio is 20 (Oliver and Merrick 2006).

The linear relationship describing the reliance of biomass-specific *GPP* on mean irradiance ($\overline{E}$) can be expressed as:

$$GPP^b = a\overline{E} \qquad (2)$$

where a is the regression coefficient and GPP^b is in gC gC^{-1} d^{-1}. If b is the phytoplankton concentration in gC m^{-3}, then the rate of *GPP* in gC m^{-3} d^{-1} is:

$$GPP = ba\overline{E} \qquad (3)$$

The reliance of planktonic community respiration (*CR*) on the phytoplankton respiratory carbon pool can be expressed as:

$$CR = -c(b + GPP) \qquad (4)$$

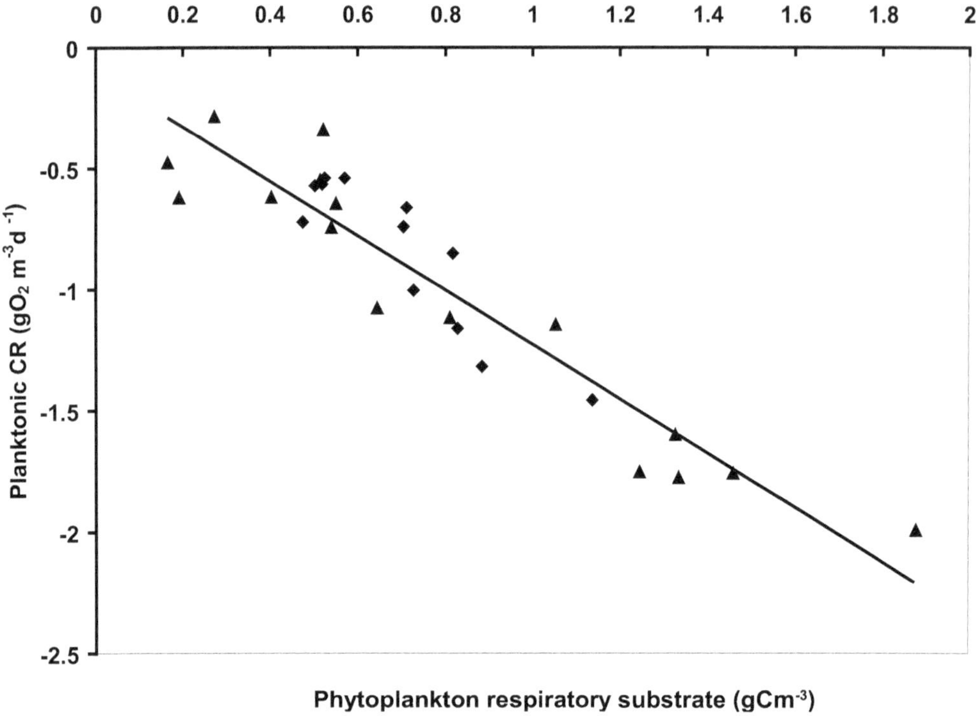

Figure 17.6: Planktonic community respiration versus phytoplankton respiratory substrate (calculated as the sum of *GPP* and the chlorophyll-based biomass) for all river sites. Sampling set 1 = ▲, sampling set 2 = ♦.

where c is the regression coefficient and CR is in gC m^{-3} d^{-1}. By substituting GPP from Equation 3:

$$CR = -c(b + ba\overline{E}) \tag{5}$$

The NP of phytoplankton can be estimated by substituting Equation 3 and Equation 5 into the expression $NP = GPP + CR$ and simplifying to give:

$$NP = b(a\overline{E}(1 - c) - c) \tag{6}$$

Annual rates of phytoplankton *NEP* were estimated from application of this form of expression, but as derived in detail from the linear regression equations and ranged from 25 to 41 gC m^{-2} yr^{-1} for the different sampling sites and periods (Oliver and Merrick 2006). At the river sites, this phytoplankton *NEP* appears to be the annual supply of organic carbon available to support the food web.

Metabolism at weir pool sites

Sampling sites within weir pools experienced reduced water velocities. As velocities declined below 0.2 m s^{-1} increased fluctuations were observed in total system metabolism, especially in *CR* and *NEP*, with *NEP* more frequently and increasingly negative (Fig. 17.7). Note that the trend lines in Figure 17.7 are indicative only of average changes in the rates of metabolism.

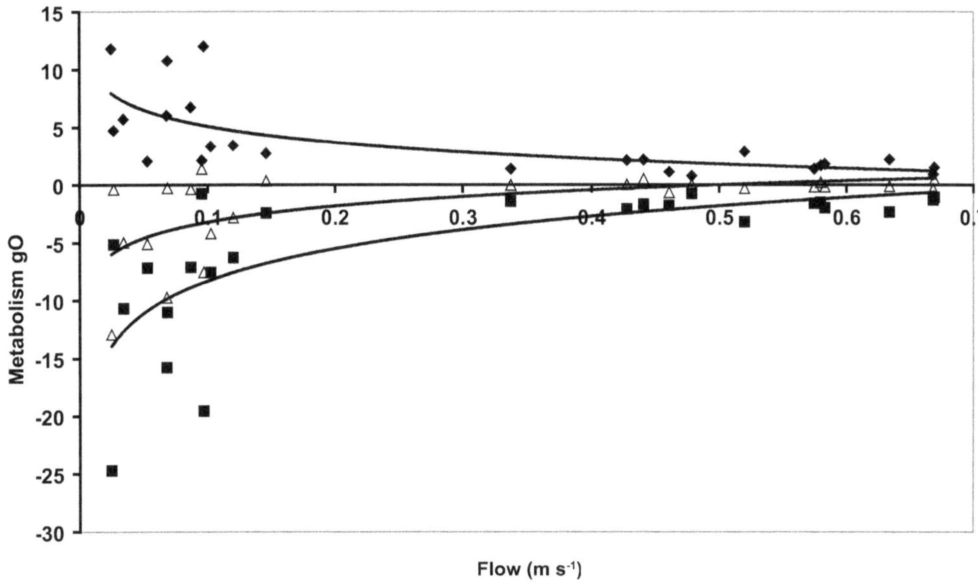

Figure 17.7: Areal community *GPP* (♦), *CR* (■) and *NEP* (△) displayed on a flow velocity axis for all sites from the second sampling period.

The occurrence of negative *NEP* values in the weir pools suggests that they receive external supplies of organic material. This organic material could have been derived from sources external to the river as a result of weir pools having more substantial floodplain connections or larger amounts of fringing emergent vegetation, or it could come from sedimentation of phytoplankton or detritus delivered continuously from upstream. An alternative explanation is that, because of the limited number of measurements in the weir pools, the cycles of production and respiration have not been properly captured. Unfortunately, the weir pool data are too sparse to draw conclusions on the metabolic balance and likely sources of organic material.

Partitioning metabolism in weir pools

Planktonic metabolism in the weir pools was often substantially less than the total metabolism, suggesting a major contribution from benthic sources. Rates of benthic metabolism were estimated as the difference between total and planktonic rates for all sites from the second sampling set (Fig. 17.8). In general, benthic metabolism accounted for most of the system metabolism in the weir pools (compare Figs 17.7 and 17.8). Comparison with river sites indicated a very large increase in benthic activity as flow velocities declined below 0.2 m s^{-1} (Fig. 17.8), with benthic *GPP* making substantial contributions to the total weir pool production.

Despite the increased metabolic activity observed in the weir pools, flowing river sections upstream and downstream of these sites did not show signs of enrichment and metabolic rates matched the patterns described generally for river sites. This suggests that the weir pools have only local effects, with little influence on metabolism over most of the river length. However, because of their increased metabolic activity they might be expected to play a significant role in supporting populations of organisms associated with the pools or perhaps reliant on them during critical periods of their life-cycles.

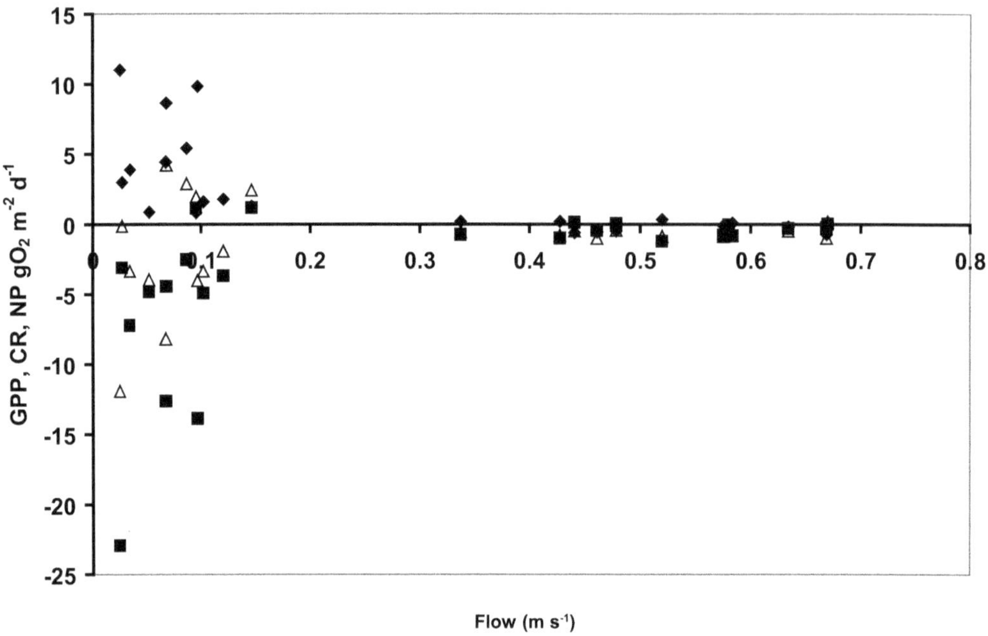

Figure 17.8: Areal benthic *GPP* (♦), *CR* (■) and *NEP* (△) displayed on a flow velocity axis for all sites from the second sampling period.

Conclusions

Measurements made seven years apart at sites spread over 1400 km of river channel gave a consistent picture of metabolism in flowing sections of the River Murray. Phytoplankton photosynthesis accounted for most of the riverine primary production and supplied net production of 25–41 gC m^{-2} yr^{-1} to the food web. Annual net production for the system was close to zero, indicating that channel primary production was fully utilised and that there was little supply of usable, external organic material to the channel. Although this might be considered an unsuitable result for a floodplain river, it is not unexpected.

The River Murray has been increasingly isolated from its floodplain as a result of massive increases in water diversion for commercial uses (MDBMC 1995; Maheshwari *et al.* 1995). In addition, the floodplain has been substantially cleared of native vegetation and replaced with a range of land uses, mainly agriculture. Riparian vegetation that once fringed the river channel has been removed over large areas. As a consequence, the hydrological connections important for carrying organic material between the floodplain and river channel have been broken and the sources of terrestrial organic material have been substantially altered and reduced. These are the likely causes for the small supply of external organic carbon that is contributing to metabolism in the River Murray.

This situation is not unique. Jones *et al.* (2003) analysed dissolved inorganic carbon concentrations in rivers across the US for a 22-year period from 1973 to 1994. They found a decline that they attributed to large-scale reductions in terrestrial organic material production and import into aquatic ecosystems. It is likely that this type of system degradation is widespread, but the extremely small external supply of organic carbon to the River Murray seems extraordinary. It is not possible to estimate what part this plays in the demise of populations of higher organisms such as fish and waterbirds that depend on the aquatic system for food resources, but it is likely to be substantial.

To address these problems, management strategies must reinstitute at least some of the connectivity supplying food resources to the river. Determining how much area to flood, and what forms of terrestrial habitat it should contain, will require measurements on rivers less affected than the Murray so that the influence of more contiguous supplies of organic material can be assessed. It appears that movement of water across the floodplain will be a critical component of any effort to rehabilitate the supply of organic materials. The importance of floodplain passage has been demonstrated in many rivers (Junk *et al.* 1989; Hein *et al.* 1999). In a number of these studies the organic material returning from the floodplain was found to contain a large algal component contributing to the organic carbon supply. During an experimental manipulation of the Yolo Bypass floodplain of the Sacramento River (Lehman *et al.* 2008), water returning from the floodplain had a high concentration of chlorophyll-*a* made up of diatoms and green algae. This enabled the floodplain to export 14–37% of the downstream combined floodplain plus river load of algal biomass even though it received only 3% of the river stream flow. The significance of such organic carbon subsidies to the River Murray are unknown, but are suspected to be important. Algal material exported from the floodplain could provide a readily available food supply to channel food webs whereas returning terrestrial organic material might be more refractory, providing a longer-term supply of material requiring extended processing.

To identify the most appropriate forms of organic material to be returned to the river, and the most suitable way of doing this in order to enhance food webs, will require a far better understanding of trophic structure and function than we currently have. In particular, improved understanding of the links between microbial metabolism and higher trophic levels, and their responses to the supply of organic materials, will be critical to developing the predictive capacity necessary to underpin management strategies.

References

Beardall J and Raven JA (1990). Pathways and mechanisms of respiration in microalgae. *Marine Microbial Food Webs* **4**, 7–30.

Bunn SE, Davies PM and Winning M (2003). Sources of organic carbon supporting the food web of an arid zone floodplain river. *Freshwater Biology* **48**, 619–635.

Dodds WK and Cole JJ (2007). Expanding the concept of trophic state in aquatic ecosystems: it's not just the autotrophs. *Aquatic Sciences* **69**, 427–439.

Gawne B, Merrick C, Williams DG, Rees G, Oliver R, Bowen PM, Treadwell S, Beattie G, Ellis I, Frankenberg J and Lorenz Z (2007). Patterns of primary and heterotrophic productivity in an arid lowland river. *River Research and Applications* **23**, 1070–1087.

Hein T, Heiler G, Pennetzdorfer D, Riedler P, Schagerl M and Schiemer F (1999). The Danube Restoration Project: functional aspects and planktonic productivity in the floodplain system. *Regulated Rivers: Research and Management* **15**, 259–270.

International Oceanographic Tables 1973. Vol. 2. Prepared under the supervision of the Joint Panel on Oceanographic Tables and Standards. National Institute of Oceanography of Great Britain, Wormley, Surrey/Unesco, Paris.

Jones JB, Stanley EH and Mulholland PJ (2003). Long-term decline in carbon dioxide supersaturation in rivers across the contiguous United States. *Geophysical Research Letters* **30**, 1–4.

Junk WJ, Bayley PB and Sparks RE (1989). The flood-pulse concept in river-floodplain systems. In *Proceedings of the International Large River Symposium*. (Ed. DP Dodge). *Canadian Special Publication Fisheries and Aquatic Science* **106**, 110–127.

Kosinski RJ (1984). A comparison of the accuracy and precision of several open-water oxygen productivity techniques. *Hydrobiologia* **119**, 139–148.

Lehman PW, Sommer T and Rivard L (2008). The influence of floodplain habitat on the quantity and quality of riverine phytoplankton carbon produced during the flood season in San Francisco Estuary. *Aquatic Ecology* **42,** 363–378.

Maheshwari BL, Walker KF and McMahon TA (1995). Effects of regulation on the flow regime of the river Murray, Australia. *Regulated Rivers: Research and Management* **10**, 15–38.

McCutchan JH, Lewis WM and Saunders JF (1998). Uncertainty in the estimation of stream metabolism from open-channel oxygen concentrations. *Journal of the North American Benthological Society* **17**, 155–164.

MDBC (2005). The Living Murray Foundation report on the significant ecological assets targeted in the First Step Decision. Murray-Darling Basin Commission, Canberra.

MDBMC (1995). An audit of water use in the Murray-Darling Basin. Murray-Darling Basin Ministerial Council, Canberra.

Moore JC, Berlow EL, Coleman DC, de Ruiter PC, Dong Q, Hastings A, Johnson NC, McCann KS, Melville K, Morin PJ, Nadelhoffer K, Rosemond AD, Post DM, Sabo JL, Scow KM, Vanni MJ and Wall DH (2004). Detritus, trophic dynamics and biodiversity. *Ecology Letters* **7,** 584–600.

Odum HT (1956). Primary production in flowing waters. *Limnology and Oceanography* **1,** 102–117.

Oliver RL and Merrick C (2006). Partitioning of river metabolism identifies phytoplankton as a major contributor in the regulated Murray River (Australia). *Freshwater Biology* **51,** 1131–1148.

Thorp JH and Delong MD (1994). The riverine productivity model: an heuristic view of carbon sources and organic processing in large river ecosystems. *Oikos* **70,** 305–308.

Treadwell SA, Campbell IC and Edwards RT (1997). Organic matter dynamics in Keppel Creek, southeastern Australia. *Journal of the North American Benthological Society* **16,** 58–61.

Vannote RL, Minshall GW, Cummins KW, Sedell JR and Cushing CE (1980). The river continuum concept. *Canadian Journal of Fisheries and Aquatic Sciences* **37,** 130–137.

Vink S, Bormans M, Ford PW and Grigg NJ (2005). Quantifying ecosystem metabolism in the middle reaches of Murrumbidgee River during irrigation flow releases. *Marine and Freshwater Research* **56,** 227–241.

Young RG and Huryn AD (1996). Interannual variation in discharge controls ecosystem metabolism along a grassland river continuum. *Canadian Journal of Fisheries and Aquatic Sciences* **53,** 2199–2211.

Chapter 18

Allocating water to the wetlands of the Murray Valley to maximise aquatic plant species diversity

George Ganf, Sean White and Rod Oliver

Introduction

Between 1922 and 1937 a series of weirs and locks were built along the length of the River Murray. The purpose was to provide for navigation and create a constant pool level to facilitate pumping for irrigation and water supply (http://www.mdbc.gov.au/rmw/river_murray_system). Between 1935 and 1940 a series of barrages was constructed across the southern end of the Lower Lakes to impound freshwater, improve domestic and irrigation supplies and reduce salinity levels in the Lower Murray.

In the early 1990s there was increasing concern about the health of the river and its associated wetlands as more and more water was removed for consumptive purposes. In 1996 the Murray-Darling Basin Ministerial Council introduced a permanent cap on diversions. The Council agreed that the cap be defined as 'the volume of water that would have been diverted under the 1993/94 levels of development'. The average long-term monthly inflow to the Murray Basin peaks in September at 1600 GL, but since 2002/03 the peak has just exceeded 400 GL per month and 2006/07 had the lowest flows on record due to the extended drought.

In 2001 the Murray-Darling Basin Ministerial Council called for social, economic and ecological assessments of the costs and benefits of returning additional water to the River Murray as environmental flows. The Council recommended that three flow volumes – 350, 750 and 1500 GL per year – be considered. This recommendation was examined using the Murray Flow Assessment Tool (MFAT, http://www.mdbc.gov.au/subs/information/mfat/index.htm), a decision support system that relates river flow to potential habitat condition. The results of the floodplain and wetland vegetation studies suggested that across the River Murray system 1500 GL per year would provide a significant improvement to the habitat and 750 GL would provide some improvement. The study did not investigate volumes of >1500 GL per year.

Modelling water allocation to wetlands

It is against this background that this study was undertaken. The methodology and results are reported in this chapter.

The primary aim was to relate the water regime preferences of wetland plants to discharge or river flow throughout the regulated section of the River Murray and to predict the volume of additional water required to maximise aquatic plant species diversity at a regional scale. The

Table 18.1: Studied area and number of wetlands from the 16 reaches of the Murray River from Albury to Renmark, including the Edward-Wakool anabranch system

Reach	Gauging station ID	Number of wetlands	Area (ha)
Edwards River at Deniliquin	409003	186	8554.8
Wakool River at Stoney Crossing	409013	122	638.5
Edwards River at Moulamein	409014	91	1179.2
Wakool River DS off-take regulator	409019	32	489
Yallakool Creek DS off-take regulator	409020	50	307.2
Edwards River DS Steven's Weir	409023	145	1544.6
Colligen Creek DS off-take regulator	409024	23	506.1
Murray River DS Yarrawonga Weir	409025	137	643.9
Edwards River at Liewah	409035	3	5.3
Murray River at Tocumwal	409202	485	8521.1
Murray River at Swan Hill	409204	575	13 321.1
Murray River at Torrumbarry	409207	65	2518
Murray River below Wakool Junction	414200	183	7249.8
Murray River at Euston	414203	631	7520.4
Rufus River at Lake Victoria outlet	426502	1	5.1
Murray River US Lock 6	426510	17	298.9
Total		2746	53 303

DS = downstream, US = upstream.
Data from Green and Alexander (2006).

project deals with 2746 wetlands with a total area of 53 303 ha, spanning 16 hydrologically distinct reaches (Table 18.1) within the regulated main stem of the River Murray from Albury to Renmark including the Edward-Wakool anabranch. Geographically, these wetlands are restricted to NSW and Victoria because no commence-to-fill (CTF) data were available for wetlands in South Australia. This subset is approximately one-third of the wetlands spread through the NSW–Victorian portion of the river. A statistical modelling approach was used to analyse the relationship between discharge, CTF, wetland connectivity and probability of occurrence of wetland plant species. A key assumption was that the amount of water required for wetlands is that which maximises plant species diversity at the regional spatial scale (the 16 defined reaches in Table 18.1). This approach was adopted in preference to analysing individual wetlands because detailed data on individual wetlands were scarce. Furthermore, a common characteristic among all wetlands was needed to be able to make predictions across over 2700 individual wetlands. Connectivity was chosen as the common characteristic because it relates wetlands to the hydrograph for individual river reaches. It is recognised that although the distribution and occurrence of wetland plant species is strongly influenced by water regime it is not the only factor that influences the probability of occurrence. Salinity, soil type, nutrient availability, turbidity, flow and grazers all modify the influence exerted by water regime.

Connectivity

The probability of occurrence of a wetland plant was modelled as a function of water regime. This was estimated as the wetland–river connectivity, which is the number of days per year a wetland receives water from the river expressed as a proportion. The connectivity of two wetlands for 1992–96 at the Swan Hill Reach (#409204) with CTF of 10 000–20 000 ML d^{-1} are

Table 18.2: Number and area of wetlands and their CTF for 575 wetlands in the Swan Hill Reach

CTF (ML d–1)	Number of wetlands	Cumulative number of wetlands	Area of wetlands (ha)	Cumulative area of wetlands
<6000	8	8	444	444
6000	11	19	6983.6	7427.6
6800	1	20	0.6	7428.2
7590	2	22	25.7	7453.9
11 000	1	23	1	7454.9
12 500	1	24	24.3	7479.2
13 700	6	30	47	7526.2
16 000	31	61	290.8	7817
17 000	1	62	3	7820
17 280	10	72	1422.1	9242.1
19 000	3	75	16.6	9258.7
22 100	1	76	9.6	9268.3
49 100	360	436	3353.4	12 621.7
58 700	138	574	698.9	13 320.6
68 300	1	575	0.5	13 321.1

Data from Green and Alexander (2006).

shown in Figure 18.1a alongside their average connectivities. The 575 wetlands in the Swan Hill Reach (Table 18.2) have CTF between <6000 and 68 300 ML d^{-1}. The hydrograph (Fig. 18.1b) shows that only wetlands with a CTF <35 000 ML d^{-1} received water from the main channel in a 10-year period. For those wetlands, connectivity was influenced by the duration and magnitude of the discharges (Fig. 18.1c).

Connectivities were allocated to 10 categories and the wetlands within each range summed for each reach, and across all reaches to estimate the valley-wide wetland connectivities. The exact relationship between discharge and connectivity differs among river reaches, reflecting the differences in their annual hydrographs, but the relationship can be described by a logarithmic decrease in connectivity with increasing commence to fill (Fig. 18.2).

Analysis of pre- and post-regulation wetland connectivity

Three 20-year data periods (1900–20, 1957–77 and 1979–99) were used to estimate connectivity under natural and current conditions using modelled hydrographs from BigMod (MDBC). Under natural conditions, data are modelled without water extraction or regulatory structures impeding flow through the river; under current conditions, modelled data include the effects of water extraction and regulatory structures. A total of 2641 wetlands were analysed along the main channel of the River Murray (Table 18.3). Comparisons of natural and current within a period demonstrate the effect of water extractions. Comparisons between periods illustrate the effect of water extractions and fluctuations in climate and flow patterns. The most significant change caused by river regulation is that the number of wetlands that never or only rarely receive water from the main channel increased on average for these periods from 1320 to 1985. The number of wetlands falling into the categories 0.1 to 0.4 fell from 978 to 376 and those in categories 0.4 to 1 fell from 343 to 279. This last decrease is of particular concern because it is these wetlands that harbour the greatest species and functional group diversity (Figs 18.3 and 18.4).

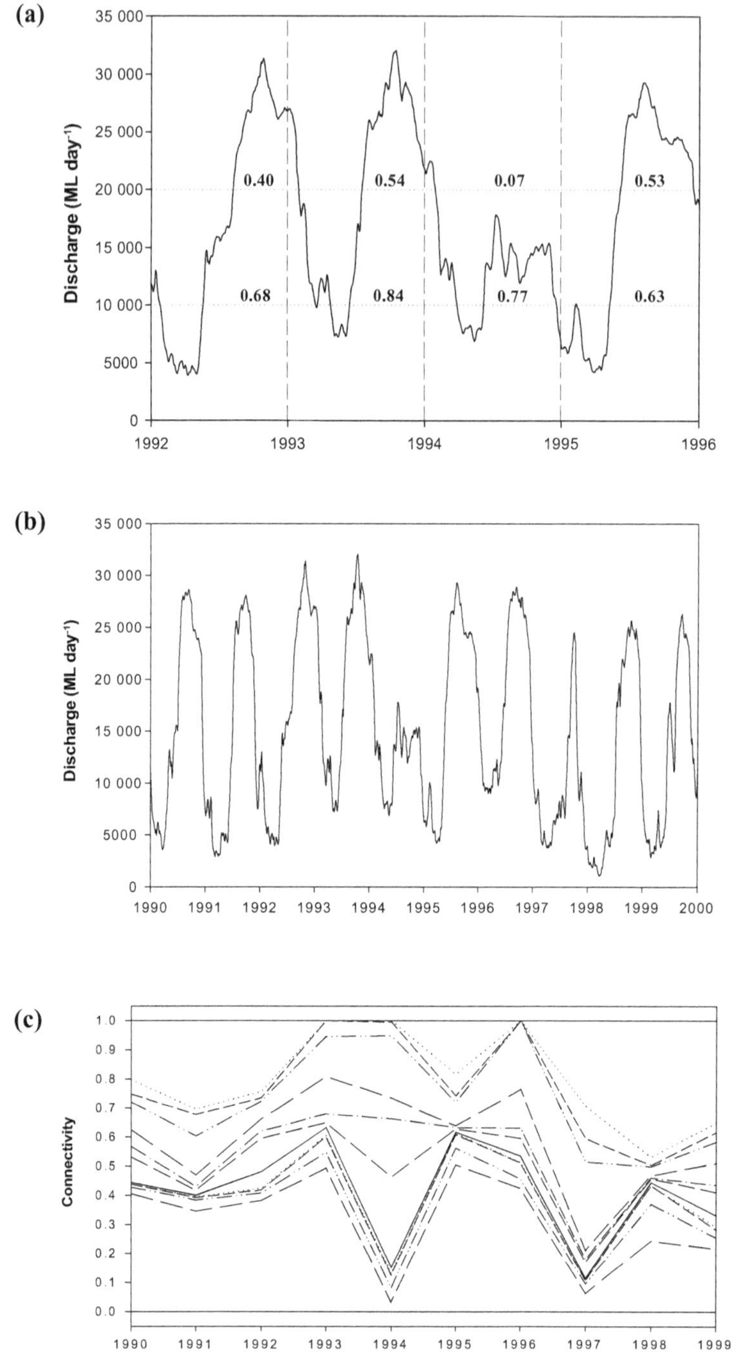

Figure 18.1: (a) Annual hydrograph and attendant measurement of wetland–river connectivity for two wetlands in the Swan Hill Reach with CTF values of 10 000 and 20 000 ML d^{-1}. Connectivity is presented above each dotted line. The averages are 0.31 and 0.58. (b) The annual hydrograph for the Swan Hill Reach 1990–2000. (c) Annual variation in the connectivity of the wetlands in the Swan Hill Reach that received water from the main channel 1990–2000. The upper graph represents wetlands with CTF of 6000 ML d^{-1} and the lower one wetlands that have CTF of 22 100 ML d^{-1}.

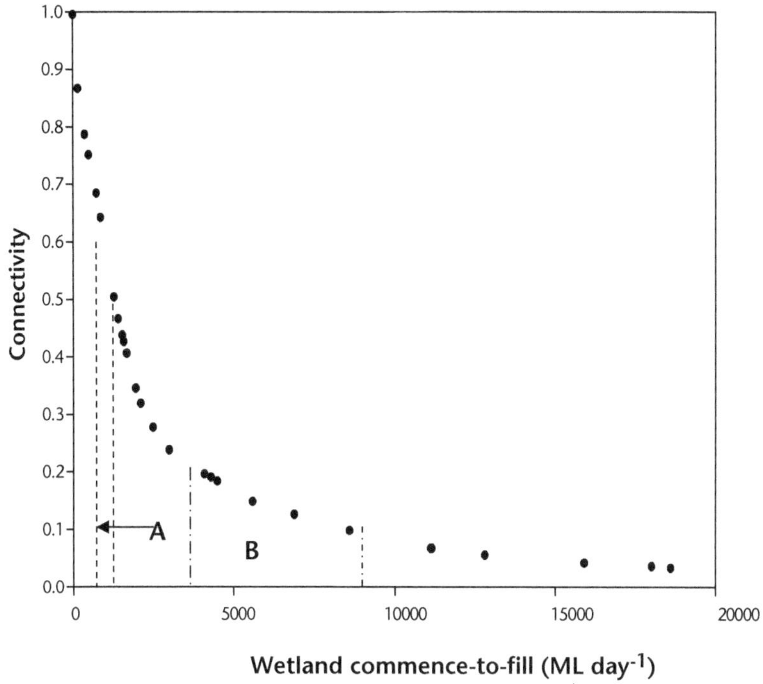

Figure 18.2: Relationship between wetland CTF and the wetland–river connectivity (dots represent individual wetlands with similar CTFs). The dotted vertical lines represent two 0.1 connectivity bands. (a) A wetland with a 0.5–0.6 connectivity has a very narrow CTF range of c. 278 ML d^{-1}. (b) For a wetland to fall within the 0.1–0.2 connectivity category it must fall within a wider CTF range of c. 4271 ML d^{-1}. There is a high degree of redundancy of wetlands in low-connectivity categories but low redundancy in high-connectivity categories.

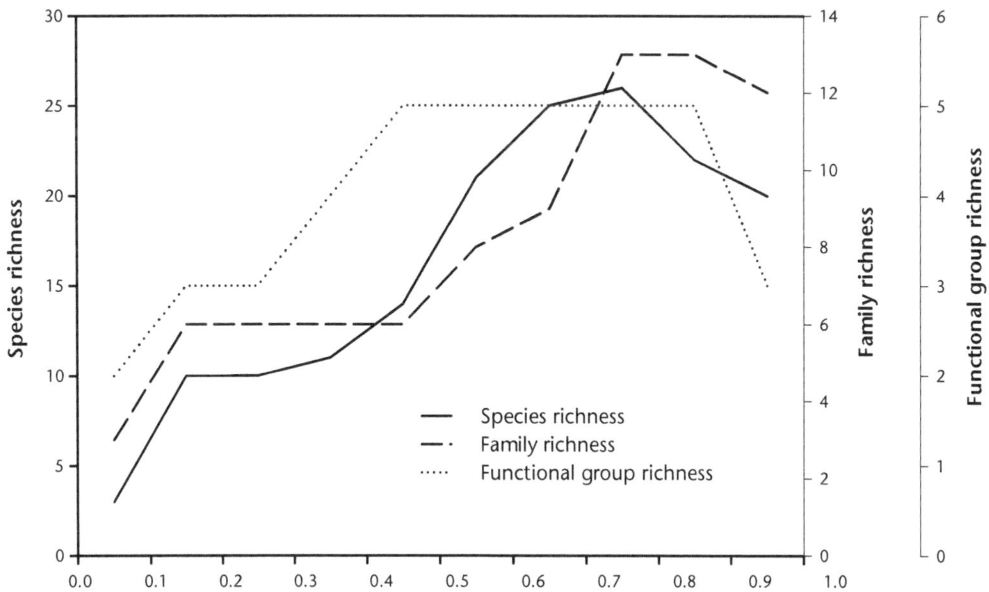

Figure 18.3: Effect of wetland–river connectivity on species richness, family richness and functional group richness.

Table 18.3: Number of wetlands falling into 10 connectivity categories (0.1 intervals) pre-regulation (natural) 1900–20 and post-regulation (current), 1957–77 and 1979–99

		0–0.1	**>0.1–0.2**	**>0.–0.3**	**>0.–0.4**	**>0.–0.5**	**>0.–0.6**	**>0.–0.7**	**>0.7-0.8**	**>0.–0.9**	**>0.9**
1900–20	Natural	1339	677	226	79	112	27	41	51	47	42
	Current	2052	240	60	26	32	53	28	58	30	62
1957–77	Natural	1311	693	104	167	120	50	25	30	51	90
	Current	1967	286	47	52	57	33	44	67	23	65
1979–99	Natural	1311	619	182	187	120	38	21	67	8	88
	Current	1937	312	52	54	54	34	49	68	22	59

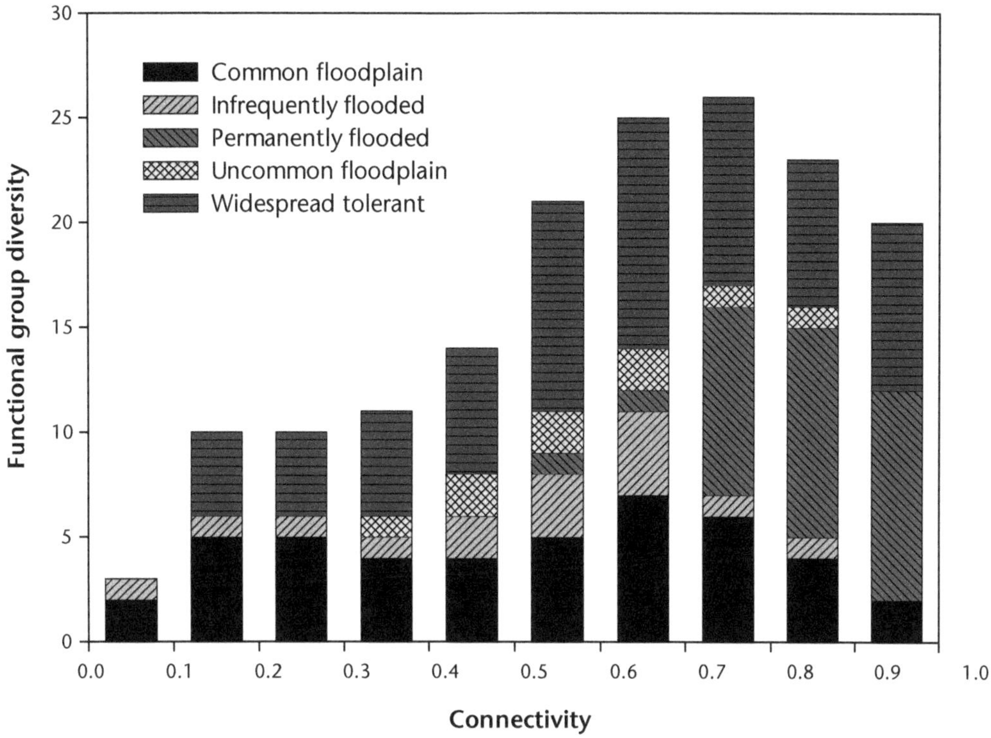

Figure 18.4: Effect of wetland–river connectivity on functional group diversity.

Wetland plants and connectivity

Species models were constructed from field survey data drawn from four sources (Ward 1994; Blanch *et al.* 1999; Siebentritt 2003; South Australian Baseline Wetland Survey). These data reflect the probability of occurrence of each species in the field across a connectivity gradient and are dependent upon knowledge of the relationship between the species and the connectivity over the previous two years. Observations of this type are very scarce, which restricted the total number of species to 36. These data were used to construct Gaussian (bell-shaped) curves describing the probability of presence of 36 species of wetland plants (Table 18.4) across a gradient of wetland–river connectivity (see Fig. 18.5 for an example of the distribution curves). Although implicit in the species distribution models, the effects of stress (e.g. salinity), resource availability (e.g. nutrients) and biotic interactions (e.g. competition) are not explicitly delineated in the species models.

The 36 species included submerged macrophytes, emergent macrophytes and floodplain species from 17 families. Most of the species have a wide distribution through Australia (Table 18.4). Each of the species can be allocated to a functional group which describes the species' tolerance to aspects of water regime – flooding, exposure and so on. Blanch *et al.* (1999, 2000) identified five functional groups in mature wetland plants (although more exist for germinants): widespread tolerant, common floodplain, permanently flooded, uncommon floodplain and infrequently flooded.

Community richness (species, family and functional group) based on these 36 model species varies across the connectivity gradient (Fig. 18.4). The highest richness occurs at connectivities of around 0.7. Because of the exclusive distribution of some species, no single connectivity value

Table 18.4: Model species

Species	Family	Functional group	Distribution
Azolla filiculoides Lam.	Azollaceae	Permanently flooded	NSW, SA, Tas., Qld, Vic.
Bolboschoenus caldwellii (V.J.Cook) Soják	Cyperaceae	Infrequent flooded	Qld, NSW, Vic.,Tas., WA, SA, NT
Bolboschoenus medianus (V.J.Cook) Soják	Cyperaceae	Widespread tolerant	NSW, Vic., Tas., SA
Brachyscome basaltica F.Muell.	Asteraceae	Infrequent flooded	NSW, Qld, Vic., SA, Tas.
Carex tereticaulis F.Muell.	Cyperaceae	Common floodplain	NSW, Vic., Tas., WA, SA
Centipeda cunninghamii (DC.) A. Braun & Asch.	Asteraceae	Widespread tolerant	Qld, NSW, Tas., WA, SA, NT
Cotula coronopifolia L.	Asteraceae	Common floodplain	Qld,^ Vic.,^ NSW,^ Tas.,^ WA,^ SA^
Cynodon dactylon (L.) Pers.	Poaceae	Widespread tolerant	NSW, Vic., Tas., WA, SA, NT
Cyperus bulbosus Vahl	Cyperaceae	Infrequent flooded	Qld, WA, SA, NT, NSW
Cyperus exaltatus Retz.	Cyperaceae	Widespread tolerant	Qld, Vic., WA, SA, NT
Cyperus gymnocaulos Steudel	Cyperaceae	Widespread tolerant	Qld, Vic., WA, SA, NT
Eleocharis acuta R.Br.	Cyperaceae	Uncommon floodplain	Qld, Vic., WA, SA, NT
Glycyrrhiza acanthocarpa (Lindley) J. Black	Fabaceae-Faboideae	Widespread tolerant	Qld, Vic., NSW, WA, SA
Juncus aridicola L. Johnson	Juncaceae	Permanently flooded	Qld, Vic., WA, SA, NT, NSW
Juncus usitatus L. Johnson	Juncaceae	Common floodplain	Qld, Vic., WA,* SA*
Lemna minor L.	Lemnaceae	Permanently flooded	
Ludwigia peploides subsp. *Montevidensis* (Sprengel) Raven	Onagraceae	Permanently flooded	NSW, Qld, Vic., SA
Muehlenbeckia florulenta Meissner	Polygonaceae	Common floodplain	Qld, Vic., NSW, SA, NT
Myriophyllum salsugineum Orch.	Haloragaceae	Permanently flooded	Vic., Tas., WA, SA, NSW
Paspalidium jubiflorum (Trin.) Hughes	Poaceae	Common floodplain	
Paspalum distichum L.	Poaceae	Widespread tolerant	Qld, Vic., Tas., WA, SA, NT
Persicaria prostratum R.Br.	Polygonaceae	Widespread tolerant	Qld, Vic., Tas., WA, SA, NT, NSW
Phragmites australis (Cav.) Trin. ex Steudel	Poaceae	Widespread tolerant	Qld, Vic., Tas., SA, WA, NT, NSW

Species	Family	Functional group	Distribution
Phyla canescens (L.) Greene	Verbenaceae	Infrequent flooded	Qld,* Vic.,* WA,* SA,* NSW
Potamogeton crispus L.	Potamogetonaceae	Permanently flooded	Qld, Vic., WA, SA, NT, NSW
Pratia concolor (R.Br.) Druce	Lobeliaceae	Common floodplain	Qld, Vic., WA, SA, NT, NSW
Pseudoraphis spinescens (R.Br.) Vick.	Poaceae	Uncommon floodplain	Qld, Vic., NSW, WA, SA, NT
Rorippa nasturtium-aquaticum (L.) Hayek	Brassicaceae	Common floodplain	Qld,* Vic.,* Tas.,* WA,* SA,* NT,* NSW*
Rumex bidens R.Br.	Polygonaceae	Widespread tolerant	Vic., Tas., SA
Schoenoplectus validus (Vahl) Á. Löve & D. Löve	Cyperaceae	Permanently flooded	Qld, Vic., Tas., WA, SA, NSW
Senecio cunninghamii DC.	Asteraceae	Widespread tolerant	Qld, Vic., NSW, SA
Sporobolus mitchellii (Trin.) C.E. Hubb ex. S.T. Blake	Poaceae	Common floodplain	Qld, Vic., WA, SA, NT, NSW
Stemodia florulenta W.R. Barker	Scrophulariaceae	Common floodplain	Qld, Vic., WA, SA, NT, NSW
Triglochin procerum R.Br.	Juncaginaceae	Permanently flooded	NSW, Qld, Vic., SA
Typha domingensis (Pers.) Steudel	Typhaceae	Permanently flooded	Qld, Vic., Tas., WA, SA, NT, NSW
Vallisneria americana Graebner	Hydrocharitaceae	Permanently flooded	Qld, Vic., Tas., WA, SA, NT, NSW

Distribution: * = naturalised, ^ = introduced.
Authority, life history and distribution from Black (1986). Functional group based on Blanch *et al.* (1999).

can support all 36 modelled species. For example, a wetland with a connectivity of 0.5–0.7 is unlikely to support both *Phyla canescens* and *Potamogeton crispus* (Fig. 18.5). Maximum diversity is therefore achieved within a reach only by having wetlands with a range of complementary connectivity values.

Across the gradient in connectivity there is a transition from a permanently flooded system to an ephemeral system, which corresponds to a transition in the habitat specialisation and life-cycle of the flora. The permanently flooded community is dominated by perennial species such as *Vallisneria americana*, *Azolla filiculoides* and *Typha domingensis.* These species commonly rely heavily on vegetative reproduction as the primary means of spread and temporal persistence. The intermediate wetlands have a strong seasonal change in water level. They contain annual species which can complete their life-cycle during drawdown (e.g. *Cynodon dactylon*), terrestrial invaders (e.g. *Rumex bidens*) and floodplain species which require occasional wetting (e.g. *Muehlenbeckia florulenta*). Species from intermediate wetland include a number of annual species which rely on seedbanks as a means of year-to-year persistence. Seed longevity is often short (less than five years), in the expectation that flooding will occur regularly. The ephemeral wetland community contains species which have adaptations to periods (years) of dry conditions such as *Carex tereticaulis*, *Pratia concolor*, *Paspalidum*

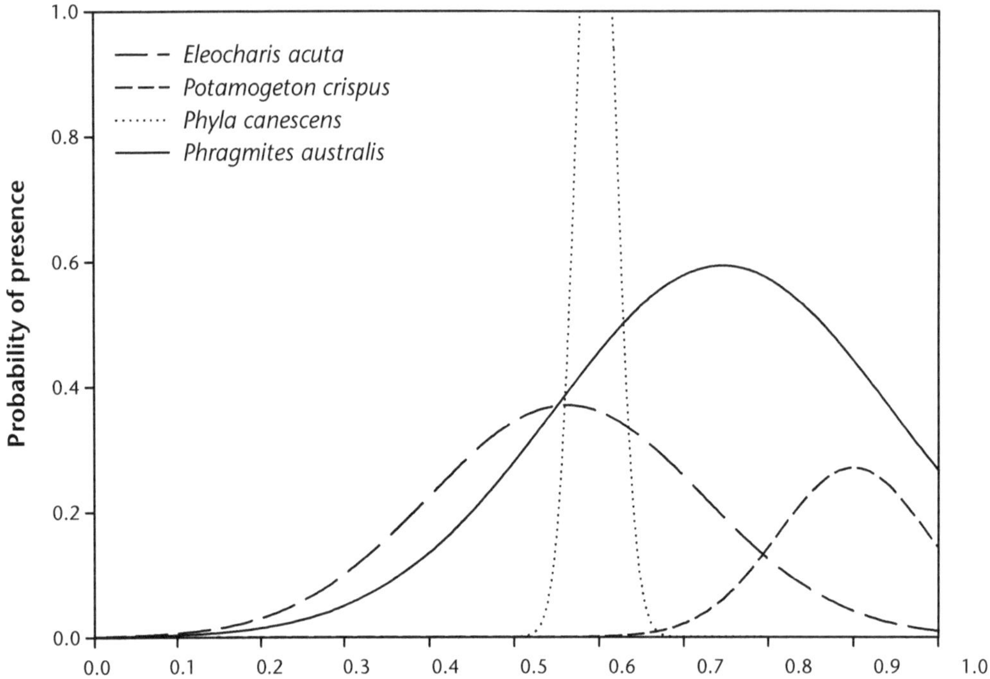

Figure 18.5: Gaussian distribution curves describing the probability of presence for four species across a gradient in wetland–river connectivity.

jubiflorum and *Brachyscome basaltica*. These species often have long-lived seeds which can persist in the sediment for years until conditions are conducive to germination. This transition is reflected in a change in the prevalence of the functional groups (Fig. 18.4). The groups 'widespread tolerant' (e.g. *Phragmites australis*) and 'common floodplain' (e.g. *Juncus usitatus*) are present across the whole gradient in wetland–river connectivity. Species from the 'permanently flooded' (e.g. *Potamogeton crispus*) functional group are most prevalent at connectivities >0.7 and are not found at connectivities below 0.5. The functional groups 'uncommon floodplain' (e.g. *Eleocharis acuta*) and 'infrequently flooded' (e.g. *Brachyscome basaltica*) are most prevalent at connectivities of 0.4 to 0.6. All functional groups are supported by connectivity values between 0.5 and 0.9.

Effect of changing discharge on community composition through the River Murray basin

Because each river reach has a unique hydrograph the following procedure was adopted.

- An annual cycle of mean daily flows was calculated from BigMod modelled hydrographs for 1989–2000 for each river reach under natural and current conditions (Table 18.1). This period covered both wet and drier years.
- The nominal daily hydrograph for each reach was used to calculate the connectivity for each wetland from its CTF value. This gave an estimate of the average number of days per year a wetland received water from the main channel over the period.

The presence of each of the 36 species was predicted in each wetland CTF for each reach. To examine how changes in discharge might affect this, the daily nominal hydrograph for each reach (averaged over the 10-year period 1990–2000) was multiplied by factors ranging from 0.5

Table 18.5: Average daily discharges for various reaches with minimum and maximum discharge values used in modelling

	Average discharge (ML d^{-1})		
Reach	**Lower (×0.5)**	**Nominal**	**Upper (×1.5)**
409003	2769	5538	8307
409013	9825	19 652	29 477
409014	1345	2690	4035
409019	122	245	367
409020	206	412	618
409023	1667	3334	5002
409024	628	1256	1883
409025	7637	15 275	22 912
409035	1248	2497	3745
409202	7610	15 221	22 831
409204	4755	9510	14 264
409207	5886	11 772	17 658
414200	8107	16 215	24 322
414203	9826	19 652	29 477
426502	2024	4048	6071
426510	9978	19 957	29 935

(simulating drought) to 1.5 (simulating flood) at intervals of 0.1. This approach maintained the annual pattern in discharge but allowed a summary change to be made in the annual discharge specific to each reach (Table 18.5). The connectivity was estimated for each hydrograph and the presence of each of the 36 model species predicted within each wetland CTF value for each reach under natural and current conditions.

Three indicators were chosen to describe the community, based on the prediction of the presence of the 36 model species. These were species richness, family richness and functional group richness (Begon *et al.* 1990).

Community indicators were estimated for each discharge interval from 0.5 to 1.5 to investigate how changes in discharge would affect the wetland plant community. The minimum discharge required to support maximum richness was collated for each reach and summed across the Murray-Darling Basin under natural and current conditions. The cumulative number or area of wetlands in which the indicator was maximised were ranked by discharge and plotted. The water return from flooded wetlands was arbitrarily set at 100%. To provide a context for the results, the discharge required to restore natural wetland–river connectivity was estimated (Fig. 18.6). These data suggest up to 4000 GL yr^{-1} above the 10-year mean would be needed to bring all wetlands in the study area to natural levels of wetland–river connectivity.

Under average nominal discharge for current conditions, species richness was maximised in only 700 of the 2746 wetlands studied, a small proportion suggesting suboptimal conditions and reduced species richness in the remaining 2046 wetlands (Fig. 18.7a). Incremental increases in discharge through the system would lead to proportional increases in the area and number of wetlands in which species richness was maximised. An additional 2100 GL yr^{-1} would maximise species richness in the majority of wetlands. Further increases in the discharge would have little benefit.

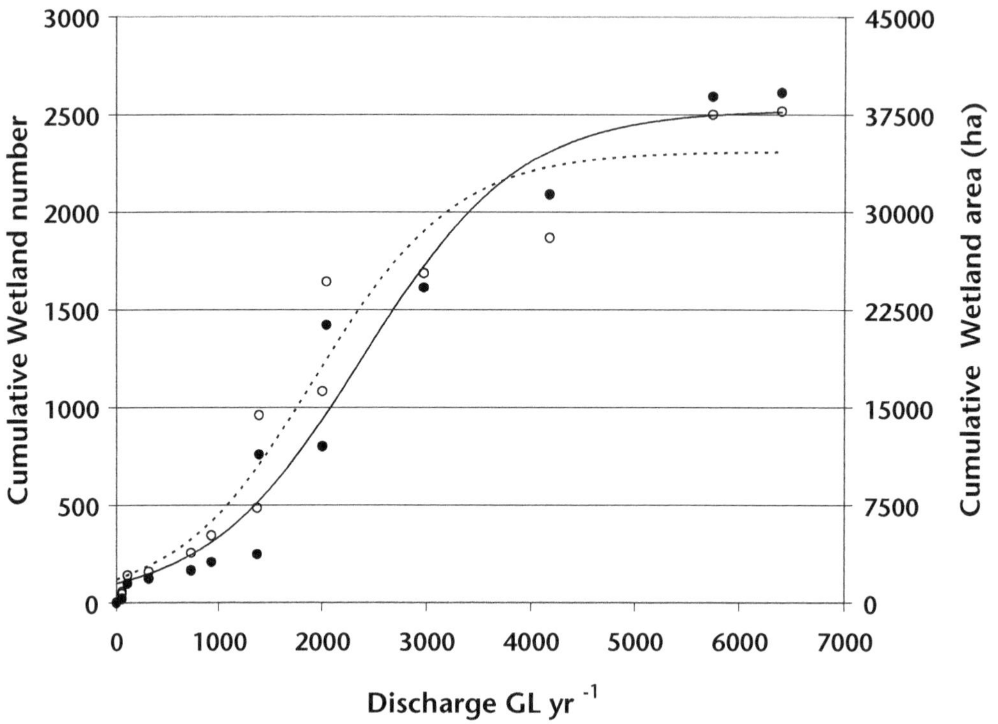

Figure 18.6: Additional discharge (GL yr^{-1}) above nominal average required to restore pre-regulation wetland–river connectivity through reaches of the Murray River.

Under average nominal discharge for natural conditions, species richness was maximised in almost 2300 wetlands (Fig. 18.7a). The only benefit to increasing discharge was in a single reach, which increased the number of wetlands by approximately 200. Provided that peak flows were retained, the annual discharge could be reduced by up to 3000 GL yr^{-1} before any drop in the area or number of wetlands was detected. Moreover, the slope of the drop in wetland area or number with incremental decreases in discharge was lower than for the current system. Overall, this suggests that natural flows created a system more resilient to change in discharge. The reason for the difference between the current and natural systems is unclear and warrants further investigation, but is probably related to the wider range of flood peaks under natural conditions.

Similar results were obtained for family richness (Fig. 18.7b) and functional group richness (Fig. 18.7c). Under current conditions with nominal average discharge, family richness was maximised in c. 1100 wetlands and functional group richness was maximised in all 2746 wetlands. The difference in the three richness indicators (species, family and functional group) is likely due to the different number of members each indicator included. Species richness is measured out of 36, family richness out of 17 and functional group richness out of five. Species richness will therefore be the most sensitive indicator and the most difficult to maximise, and functional group the least sensitive.

At nominal average discharge, the number of wetlands in which a given indicator was maximised was consistently smaller under current conditions than natural conditions (with the exception of functional group richness). The difference in terms of species richness, for example, was c. 1400 wetlands (Fig. 18.7a). To close this gap, the current system would need an additional 1200 GL yr^{-1}. To close the gap for the other indicators would require the addition of between 800 GL yr^{-1} for sum areal cover and 1800 GL yr^{-1} for family richness. To maximise

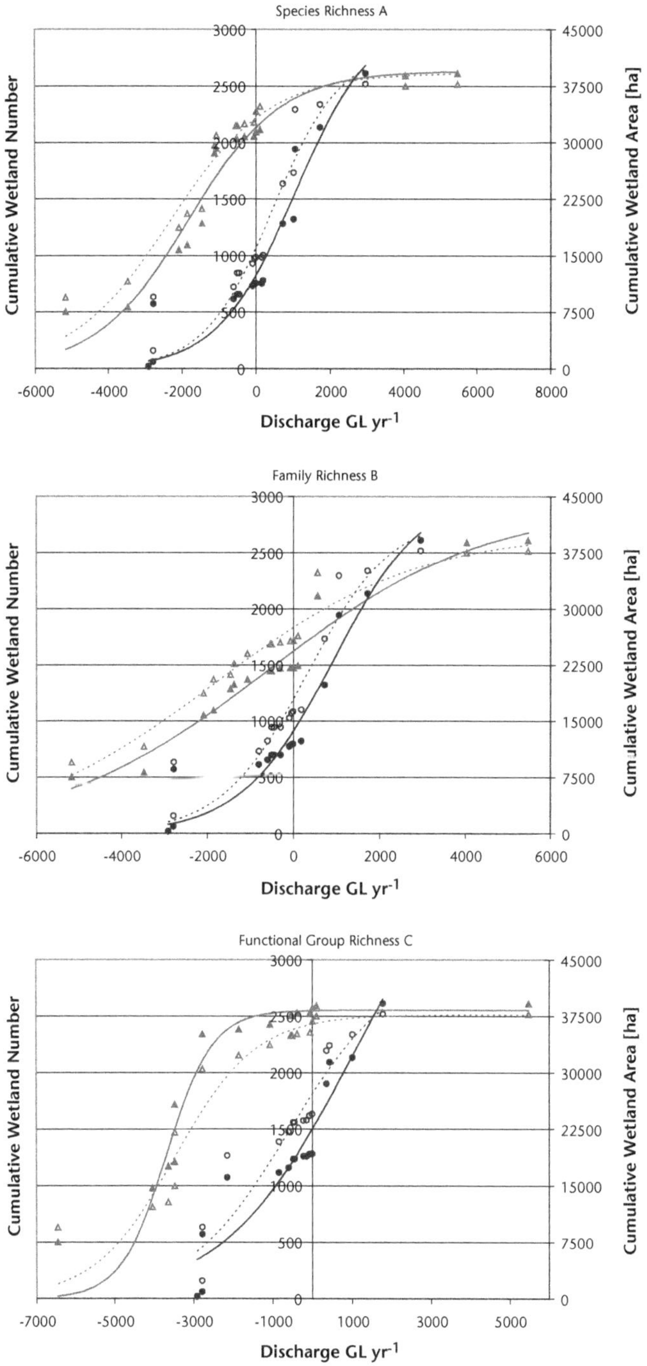

Figure 18.7: Effect of changing discharge (GL yr^{-1}; zero is nominal average discharge) on the cumulative area (ha) and number of wetlands in which the following indicator is maximised: (a) species richness, (b) family richness, (c) functional group richness. Natural conditions = triangle and dashed regression line. Current conditions = circle and solid regression line. Wetland area = closed symbols. Wetland number = open symbols. Sigmoidal regressions describe the relationship between discharge and wetland area. Calculations of discharge assume the water return from wetlands equals 100%. Note the different *x*-axis scales – the dotted vertical line in each figure marks nominal discharge.

the indicators across the whole basin would require the addition of 1800–3600 GL yr^{-1} (Fig. 18.7). The number of wetlands in which indicators were maximised would reach zero across the system with losses of discharge from the current nominal average of between 500 GL yr^{-1} for sum areal cover and 3800 GL yr^{-1} for functional group richness.

Concordance among indicators and reaches

Data were analysed to test for concordance or correlation among reaches and indicators. This analysis examines the similarity of rankings of the community indicators. For example, we can consider the relationship between discharge and either the area or number of wetlands in which species richness is maximised (Fig. 18.7a). For this indicator, there is a particular ranking of reaches from those that respond (have the indicator maximised) at low discharges to those that respond at high discharges. This ranking is the same as the order in which reaches are plotted from left to right along the *x*-axis. The analysis asks how similar that ranking is to the ranking of another community indicator. The data can be weighted to test more specifically for concordance among the uppermost rankings (top-down concordance) or the bottom-most rankings (bottom-up concordance). This tests the concordance at low levels of additional discharge (or negative levels) or high levels of additional discharge, respectively.

A significant concordance was found among indicators in the natural system ($W = 0.32$; Table 18.6), suggesting that a response in one indicator was likely to be associated with a response in the other indicators. The overall concordance was mainly the result of a high correlation ($C_B = 0.38$) at low (or negative) levels of additional discharge. The response of reaches was also significantly correlated ($W = 0.21$) in the natural system. Concordance was significant at both ends of the discharge spectrum but a higher correlation occurred at low levels of additional discharge ($C_B = 0.58$).

No significant concordance among reaches or indicators was found under current system ($P > 0.05$). To illustrate this, consider the ranking of reaches that respond to changing discharge in terms of species richness (Fig. 18.7a). Under natural conditions there is significant concordance among the reaches, meaning that the ranking or order in which the different reaches respond to changing discharge is similar whether the response is measured by species richness or by some other indicator. In the current system, the order or rank in which the different reaches respond to changing discharge varies among the indicators. Hence, under current conditions a reach might require a small amount of additional discharge to maximise species richness but a large amount to maximise species equitability, whereas under natural conditions the reach would require roughly the same amount of additional discharge to maximise both indicators. The lack of concordance among reaches suggests that regulation and water extraction have fragmented the plant communities along the river, creating a number of discrete units that respond discordantly (in different indicators) to changes in discharge.

The reciprocal, which assesses concordance among indicators, can be similarly interpreted. Where there is concordance among the indicators, as there is in the natural system, it suggests the ranking in which the various indicators are maximised with changing discharge is similar among the different reaches. That is, with incremental increases in discharge a given indicator (e.g. species richness) is most often maximised first and another indicator (e.g. species equitability) is maximised last. This allows prediction of a hierarchy of community indicators, from the indicator which requires the smallest amount of additional water to the indicator that requires the most additional water (Table 18.7). While there is a predictable order in the ranking of indicators in the natural system, there is no similar concordance for the current system, suggesting the ranking of indicators is reach-specific.

Table 18.6: Concordance among indicators and reaches (based on Kendall's χ^2 test) under current and natural conditions.

	Concordance		
	Overall	**Top-down**	**Bottom-up**
Among indicators			
natural	$W = 0.3248^a$	$C_T = 0.2688$	$C_B = 0.3780^b$
current	$W = 0.2721$	$C_T = 0.2599$	$C_B = 0.2544$
Among reaches			
natural	$W = 0.2064^a$	$C_T = 0.1763^b$	$C_B = 0.5813^c$
current	$W = 0.0923$	$C_T = 0.1537$	$C_B = 0.0056$

Top-down and bottom-up concordance were tested on data after substitution of rank for Savage score.
Significance: (a) $P \leq 0.05$. (b) $P \leq 0.01$. (c) $P \leq 0.001$.

In summary, under current conditions, wetlands of the River Murray system are characterised by hydrological conditions suggestive of low species richness, low species evenness (measured as either equitability or species diversity index) and low areal cover but high functional group richness. Wetland community richness shows comparatively low resilience to changing discharge. Consequently, the system would benefit from any additional disbursement of water. The various indicators suggest the River Murray would benefit from, at most, an additional 1800–3600 GL yr^{-1}.

Regulation has more strongly affected richness indicators (species, family and functional group) than evenness (equitability, Simpson's diversity index) or area indicators (sum areal cover; Table 18.7). Compensating for the effect of extraction and regulation on community indicators would require the addition of 800–1800 GL yr [1] (Table 18.10).

Overall concordance was found to occur only in the natural system, suggesting that water extraction and, more especially, regulation, has fragmented the river system into a number of discrete systems (reaches) which respond discordantly to changes in discharge (Table 18.6).

Wetland elimination study

A study was undertaken into the importance of individual wetlands in supporting reach-level community richness. Richness indicators were estimated for the whole reach with sets of wetlands in each CTF removed in turn (Table 18.8). Any drop in reach-level richness indicated a wetland CTF which supports a suite of species (or families or functional groups) unique to the reach, and as such deserves priority preservation.

Wetlands making a unique contribution to reach-level diversity

Wetlands which support a unique flora had CTF values of 8000 ML d^{-1} (2920 GL yr^{-1}) or less (Table 18.9). Under current conditions, 15 wetland CTF values, spread among 11 reaches, make a unique contribution to the reach-level community. In contrast, only nine CTF wetlands were found to be unique under natural conditions, suggesting a greater degree of overlap or redundancy of wetland type in the natural condition than in the current condition. Seven wetland CTF values were common to both natural and current conditions.

There was general agreement among the richness indicators: species, family and functional group. Species richness quantifies the presence of 36 species and as such is the most sensitive to changes in the reach. Next is the family richness, which includes 17 groups. Functional group was the least sensitive indicator, with only five groups. Thus, the loss of a single functional group was likely to be accompanied by a loss of three or four families and six or seven species.

Table 18.7: Ordinal ranking, weighted ranking and expected (ideal) ranking of reaches and indicators in the natural system in response to changing discharge

	Reach			Community indicator		
Rank	**Ordinal ranking**	**Weighted ranking**	**Expected ranking**	**Ordinal ranking**	**Weighted ranking**	**Expected ranking**
1	409013	17	6	Functional group richness	20	16
2	409207	34	12	Family richness	22	32
3	409202	35	18	Species richness	30	48
4	414203	38	24	Species equitability	45	64
5	409025	40	30	Diversity index	46	80
6	409204	40	36	Sum areal cover	52	96
7	414200	46	42			
8	409035	50	48			
9	409023	53	54			
10	409024	59	60			
11	426502	61	66			
12	409020	63	72			
13	409019	65	78			
14	426510	66	84			
15	409003	74	90			
16	409014	75	96			

Expected weighting assumes perfect correlation and is calculated as the rank multiplied by six (the number of indicators) or 16 (the number of reaches). The ranking of reaches and indicators in the current system were discordant and so are not listed.

Correlation coefficients among the three indicators were greater than 0.9, suggesting the three indicators were responding in a similar fashion (auto-correlation).

Hydrological explanation

The hydrological basis for high redundancy of wetland type under natural conditions can be explained on the basis of the relationship between CTF and wetland connectivity. Commonly,

Table 18.8: Example of the procedure for the elimination study

		Area of distribution (ha)			
CTF ($ML\ d^{-1}$)	**Wetland-river connectivity**	***Azolla filiculoides***	***Bolboschoenus caldwellii***	***Bolboschoenus medianus***	***Brachyscome basaltica***
<6000	1.00	7.1	0	7.1	0
6000	0.84	0	5.3	5.3	0
10 400	0.51	0	58.3	58.3	0
15 770	0.28	0	4.2	4.2	4.2
23 000	0.13	0	129.9	0	129.9

The total number of species found within this hypothetical reach is four (*Azolla filiculoides*, *Bolboschoenus caldwellii*, *Bolboschoenus medianus* and *Brachyscome basaltica*). When wetlands within the first CTF value are excluded (0 $ML\ d^{-1}$), richness drops to three (*Azolla* is missing). Excluding any other single CTF does not affect the reach-level richness.

Table 18.9: Details of wetlands which contribute a unique suite of species to reach-level richness (species, family and functional group)

Reach	Number of wetlands	CTF (ML d^{-1})	Species richness		Family richness		Functional group richness	
			Current	Natural	Current	Natural	Current	Natural
409003	64	<6000	15	15	8	9	2	2
409003	4	6000		2				
409013	1	8000	4	8	1	1		1
409014	1	1100	8					
409019	20	<6000	12		8		1	
409019	1	50	11		2			
409020	6	<6000		8		1		1
409020	1	152	4		1			
409024	1	275	1					
409204	8	<6000	14	1	8	1	1	
409204	11	6000	1					
409207	20	<6000	2	11	7			1
409207	45	6000	8	14	1		1	1
414200	1	<6000	15	1	8	1	2	
414203	2	5000	12		7		1	
414203	3	8000	2		1			
426501	1	6000	12	3	6	3	2	2

Table 18.10: Additional discharge (GL yr^{-1}) required to restore wetland community indicators to pre-regulation levels and maximise community indicators across the whole basin, under current conditions

	Discharge (GL yr^{-1})	
Community indicator	**Pre-regulation**	**Maximise across basin**
Species richness	1200	2100
Family richness	1800	1800
Functional group richness	0	0
Species equitability	1200	3000
Species diversity index	800	3500
Sum areal cover	1600	3600

the relationship between CTF and connectivity is non-linear but can be approximately described by a transformed regression (x-axis log.-transformed) (Fig. 18.2). For demonstration purposes, consider two connectivity categories: 0.1–0.2 and 0.6–0.7. Hypothetically, each of these categories supports one or more species not found at any other connectivity. To maximise reach-level species diversity it is therefore desirable to have a wetland within each of these two connectivity categories. In the low-connectivity category (0.1–0.2), a wetland must have a CTF value of 3937–8208 ML d^{-1}, which is a wide range. On the other hand, for a wetland to have a connectivity of 0.6–0.7 it must have a CTF value of 657–935 ML d^{-1}. This is a much narrower range, and therefore the probability of a wetland occurring in this range is less likely.

The spread of CTF values is shown in Figure 18.2 where, despite the even spacing of wetlands across the CTF gradient (shown by the points), obvious redundancy occurs at low-connectivity categories while a number of higher categories are missed. Within the category of 0.5–0.6 there is one wetland group, but there are six wetland groups with connectivity of 0.1–0.2. Within that first reach, the wetland would be of particular importance in supporting a unique suite of species and loss of the wetland will have a significant impact on the reach-level species richness.

Summary

Only a small number of wetlands (190 out of 2750) make a unique contribution to reach-level richness, but these wetlands contribute as many as 15 unique species to reach-level richness. The wetlands are spread among different reaches and have CTF values of no more than 8000 ML d^{-1}. Despite their low CTF values, the connectivity of these wetlands ranges from 0.37 to 1.0, indicating that some management of the wetlands may be possible (necessary) to preserve reach-level community richness.

Data and model evaluation

To consider the Murray Valley as an entity we took a broad-brush approach. This necessitated the use of data that were not specifically collected for this study; indeed, none of the data used were primary information.

Roberts and Marston (2000) reviewed the sparse and scattered information on the response of Murray-Darling Basin wetland and floodplain plants to water regime. Although information for some species permitted their inclusion in the suite of 36 species there was insufficient information on other key species such as *Bolboshchoenus fluviatilis*, *Eleocharis sphacelata*, *Juncus ingens* and *Potamogeton ochreatus*. Furthermore, the data represent a fraction of the species present in the Murray Valley and do not represent the distribution of the species across multifarious gradients of water regime, nutrients, salinity and soil type etc. This is a fertile area for future research integrating the two disciplines of hydrology and aquatic plant ecology that, as Roberts and Marston (2000) state, will serve persons and organisations involved in ecological aspects of river flow and wetland management.

The CTF data collected by Green and Alexander (2006) used a range of techniques ranging from satellite imagery to discussion with local landholders. The data are limited in some areas and only recently have data from South Australia become available. Direct verification of wetland sill heights is needed, as these may change as river banks slump and levees are constructed. The model is incapable of predicting the depth and duration of the wetting cycle – it reveals only the frequency of the inundation – but with appropriate data these water regime characteristics could be included.

The hydrological data span the period from 1900 to present. We used an average hydrograph for the period 1989–2000 to predict the influence of additional flow on aquatic plant diversity. Whether it is best to average flow over this 10-year period and use the mean hydrograph to drive probability responses, or whether it is best to accumulate responses with time, is problematic. For example, a wetland with a constant connectivity of 0.6 is assumed to behave in a similar manner to a wetland with five years at 0.3 and five years at 0.9.

Whether it is possible to deliver environmental water to individual reaches can only be resolved by the application of the two computer-based models: the Monthly Simulation Model (MSM) which models the river from Dartmouth to the South Australian border, and feeds into BigMod (a daily simulation model) that models the river from Dartmouth to the Lower Lakes. These models rely on different state river flow models. Uncertainties in the application of these models to the South Australian section of the river are problematic (Fernando *et al.* 2007) and the complexity of the models makes scenario testing difficult.

Despite these shortcomings, it is encouraging that the predicted additional flows required to maximise aquatic plant species diversity are similar to those suggested by others.

Conclusions

Regulation and water extraction have had a significant and diverse impact on the wetlands of the River Murray. First, community indicators showed a negative effect by regulation and/or water extraction at nominal discharge, affecting the community in c. 400–1750 of the 2746 wetlands studied (Fig. 18.7). Second, the way in which the indicators respond to changes in water allocation has been altered as a result of regulation, the indicators showing decreased resilience to dropping discharge. Third, regulation has segmented the river into a number of discrete community units (reaches) which have discordant community responses to incremental changes in water allocation (Table 18.6), departing from the predictable response hierarchy (Table 18.7) that might have otherwise guided restoration and management.

To be effective, restoration or management of the River Murray system must occur at the reach scale. This conclusion is based on three observations. First, due to the exclusive distribution of some species along the wetland–river connectivity gradient, no single wetland (or CTF group) can support all the modelled species. Second, remarkably few individual wetlands (190 of 2746) support a unique suite of species (Table 18.9). It is in the range of wetlands, spread through the landscape, that maximum community richness is attained. Third, flow regulation has affected the manner in which the wetland community responds to incremental increases in discharge. In the unregulated river system, there was a predictably ordered response by the indicators to increasing discharge within a reach: functional group richness was maximised first, then family richness and species richness (Table 18.9). Regulation has abolished this concordance (Table 18.6) such that different indicators are ultimately more important in different reaches. Management must therefore be reach-specific.

A system-wide restoration of community indicators to pre-regulation levels would require an additional 800–1800 GL yr^{-1} (2192–4932 ML d^{-1}; Table 18.10). The addition of any smaller discharge would have a proportional but reach-level benefit to the system that could be measured in all community indicators except functional group richness. A more ambitious goal would be to maximise community indicators across the river valley – a goal that would require the addition of 1800–3600 GL yr^{-1}. These data can be placed in the context of the need for an additional 4000 GL yr^{-1} to restore pre-regulation levels of wetland–river connectivity (Fig. 18.7a).

Further losses in community indicators are possible. For indicators to reach zero in *all* reaches would only require a drop in discharge of 500–3800 GL yr^{-1} (1370–10 411 ML d^{-1}; Fig. 18.7) below current average nominal flows. At that point it is likely that groups, starting with species, then families and finally functional groups, will be permanently lost from the River Murray system.

Various experts advised the Murray-Darling Basin Ministerial Council that additional discharges of 350–1500 GL per year were needed restore habitat condition in the Murray Valley. The estimates reported here to maximise wetland species diversity at a regional scale are remarkably close to the upper figure.

Acknowledgments

The authors wish to thank the following people and organisations who, by contributing data, made this project possible: Murray Wetlands Working Group (wetland CTF data), Keith Ward (field survey data) and the Murray-Darling Basin Commission (model hydrological data). The Murray-Darling Basin Commission websites are an invaluable source of information and their use and data gleaned is gratefully acknowledged.

References

Begon M, Harper JL and Townsend CR (1990). *Ecology: Individuals, Populations and Communities.* 2nd edn. Blackwell Scientific, London.

Black JM (1986). *Flora of South Australia.* 4th edn. South Australian Government Printing Division, Adelaide.

Blanch SJ, Walker KF and Ganf GG (1999). Tolerance of riverine plants to flooding and exposure indicated by water regime. *Regulated Rivers: Research and Management* **15**, 43–62.

Blanch SJ, Walker KF and Ganf GG (2000). Water regimes and littoral plants in four weir pools of the River Murray, Australia. *Regulated Rivers: Research and Management* **16**, 445–456.

Fernando TMKG, Maier HR, Dandy GC and Croke BFW (2007) Assessing prediction uncertainty in the BIGMOD model: a shuffled complex evolution metropolis algorithm approach. In *MODSIM 2007: International Congress on Modelling and Simulation.* December 2007. (Eds L Oxley and D Kulasiri) pp. 1499–1505. Modelling and Simulation Society of Australia and New Zealand.

Green D and Alexander P (2006). River Murray Wetland database: NSW, Victoria wetland commence-to-flow levels June 2006. Funded and published by NSW Murray Wetlands Working Group, Murray-Darling Basin Commission, NSW Dept of Natural Resources and Natural Heritage Trust.

Roberts J and Marston F (2000) Water regime of wetlands and floodplain plants in the Murray-Darling Basin: a source book of ecological knowledge. Technical report 30-00. October 2000. CSIRO Land and Water, Canberra.

Siebentritt MA (2003). The influence of water regime on the floristic composition of Lower River Murray wetlands. PhD thesis. Adelaide University, Australia.
SKM (2004). River Murray Wetlands baseline survey – 2004. SA Murray-Darling Basin Natural Resources Management Board, Berri.
SKM (2006). River Murray Wetlands baseline survey – 2005. SA Murray-Darling Basin Natural Resources Management Board, Berri.
Ward KA (1994). Flood requirements of the wetland flora in Barmah Forest, Victoria. Draft report. Floodplain Ecology Group, Dept of Conservation and Climate Change.

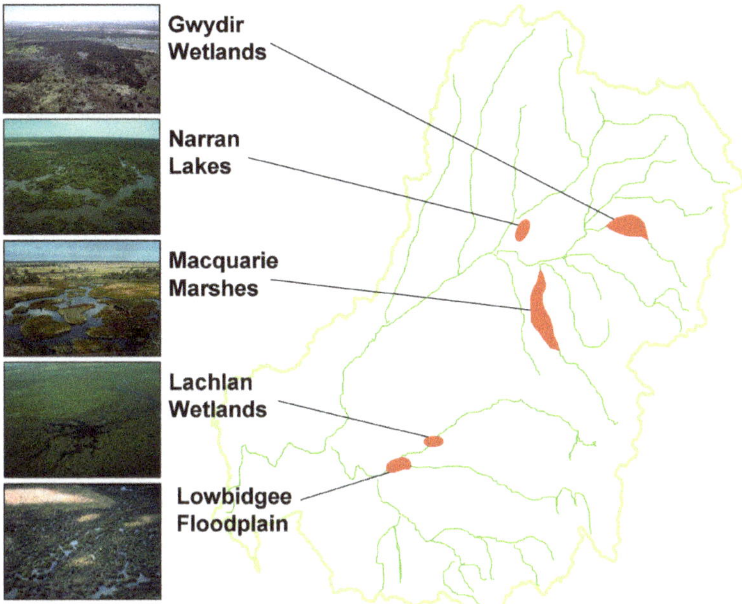

Figure 4.1: Location of wetlands targeted by the Rivers Environmental Restoration Program.

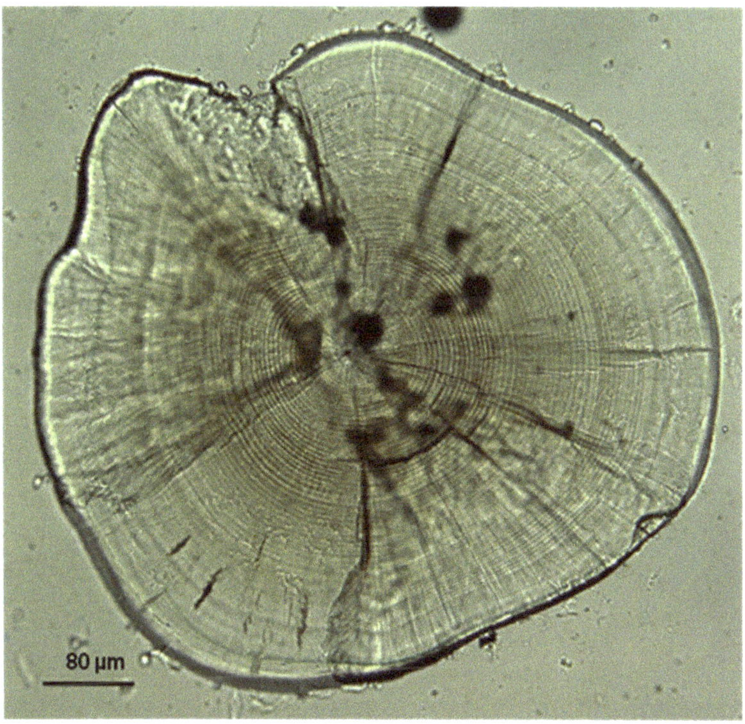

Figure 7.3: Sagittal plane section of the sagitta from a 25.5 mm standard length Australian smelt (*Retropinna semoni*) (200x magnification) showing readable microstructure from the core to the edge of the otolith. Reproduced from Fowler (2008).

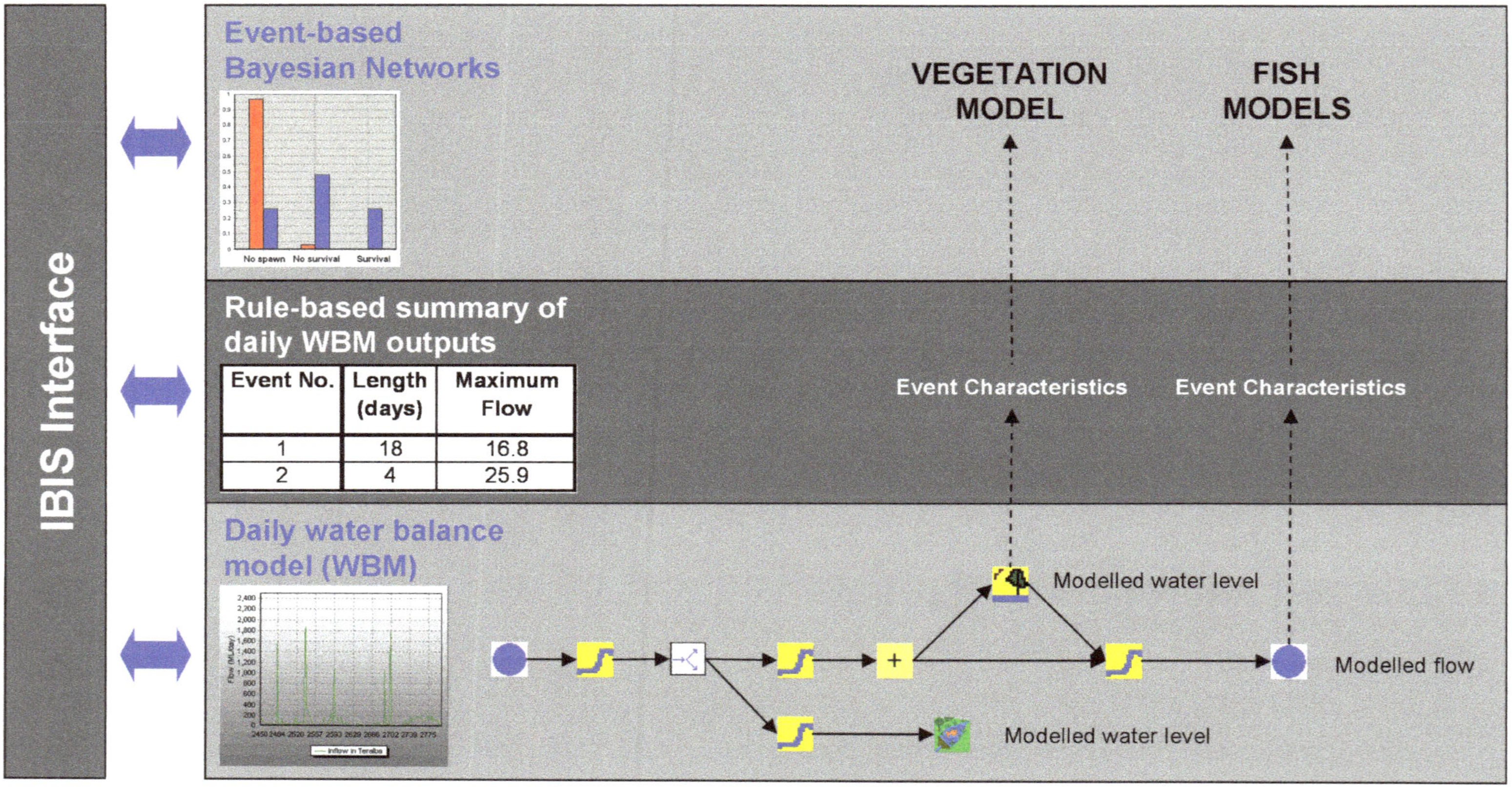

Figure 8.2: Structure of the Gwydir application of the IBIS DSS. The integrated model couples a daily water balance model to event-based Bayesian networks representing aspects of the ecology. Users interact with the models through the IBIS interface.

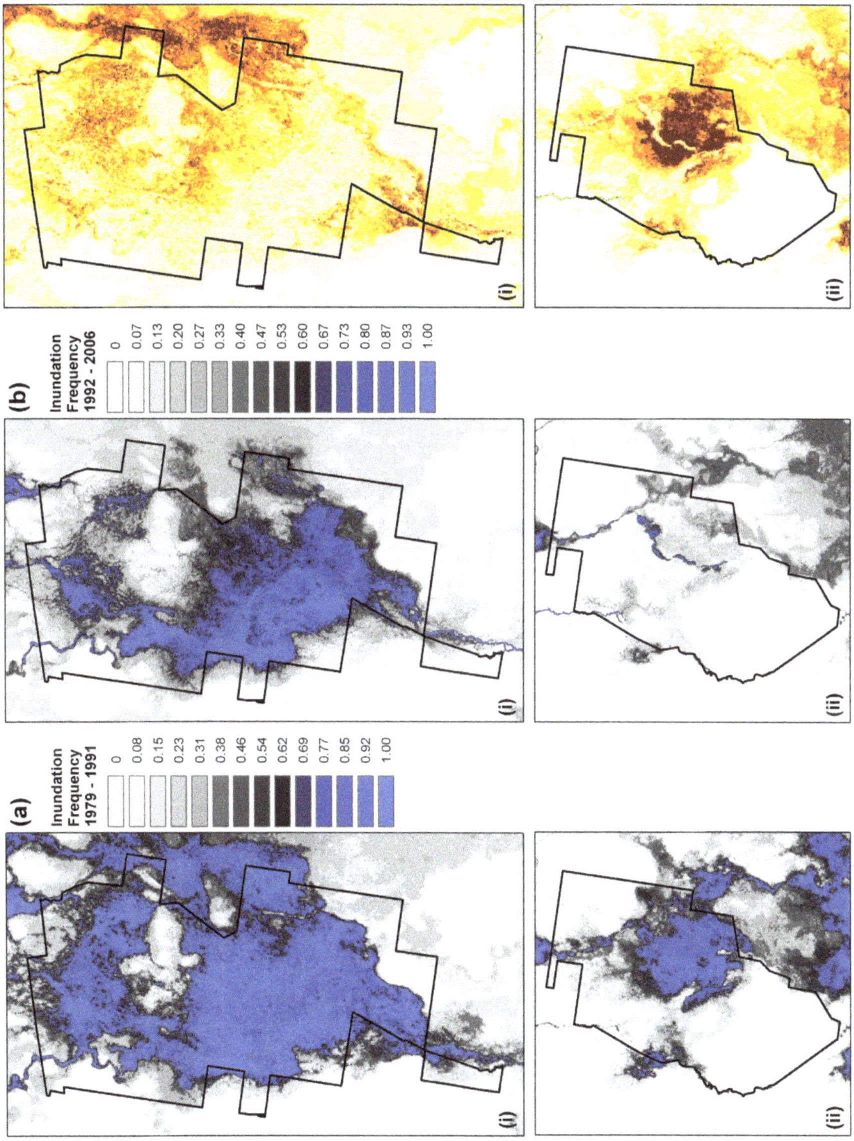

Figure 9.2: Inundation frequency index derived from the count of spring flooding as a proportion of the number of years in each period: (a) 1979–1991 (13 years) and (b) 1992–2006 (15 years). (c) Inundation frequency change zones at 0.07 intervals indicative of the proportional decrease (negative) or increase (positive) of frequency from the first period, for the (i) northern and (ii) southern Macquarie Marshes nature reserve sections.

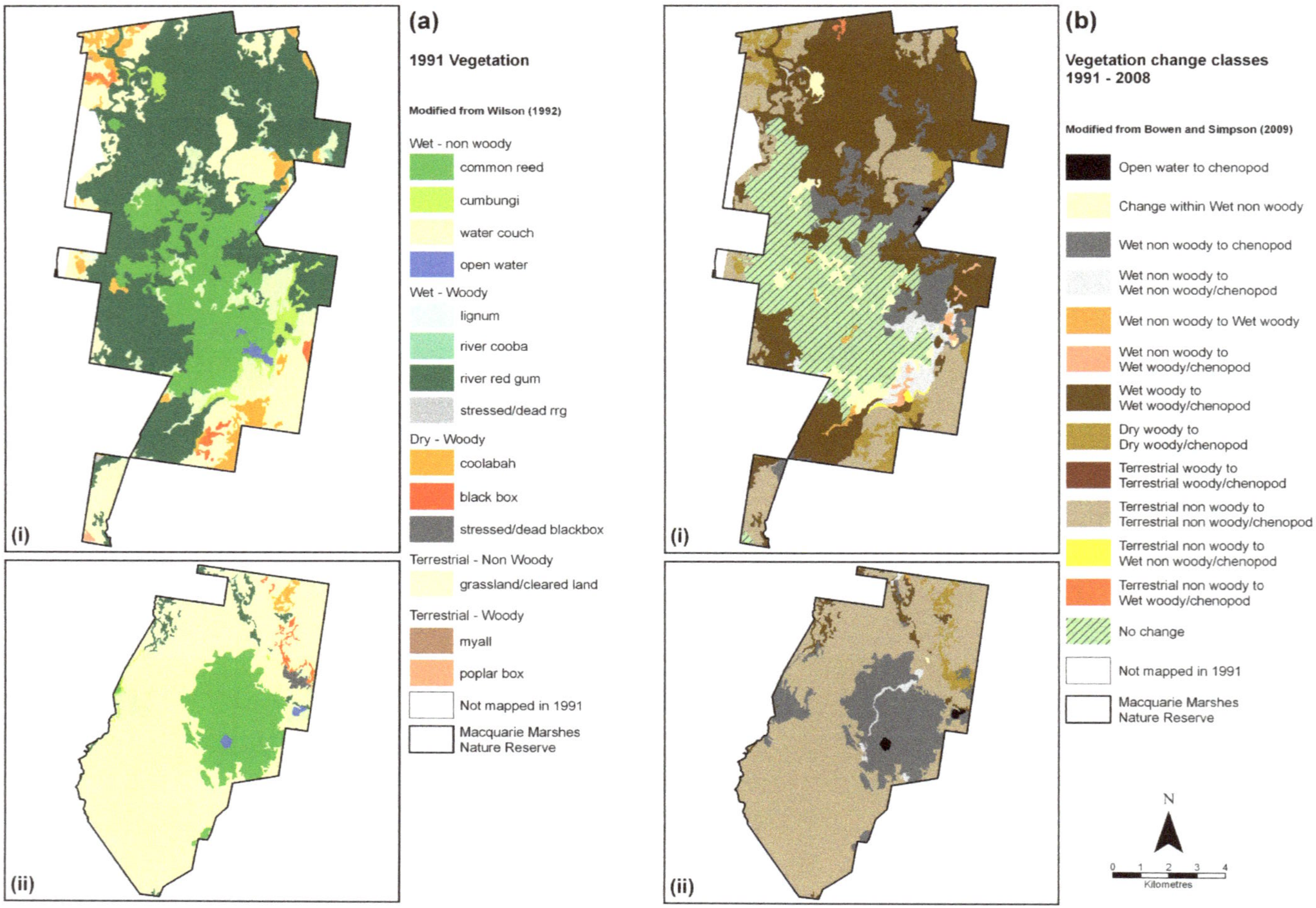

Figure 9.3: (a) Vegetation community groups and dominant canopy species mapped in 1991. (b) Vegetation change classes derived from a comparison of the vegetation communities mapped in 1991 and 2008, for the (i) northern and (ii) southern Macquarie Marshes nature reserve sections.

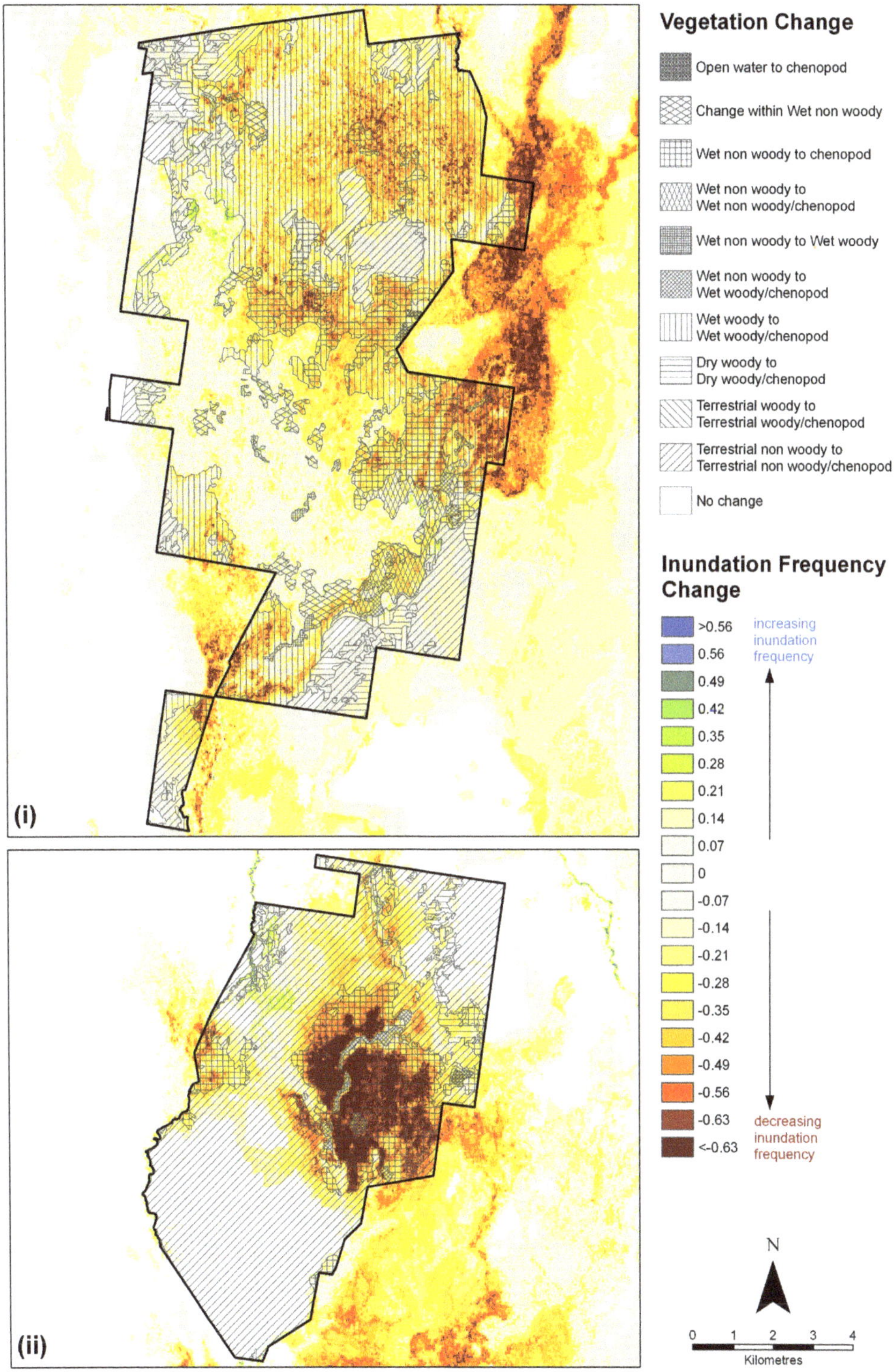

Figure 9.4: Overlay of the vegetation change classes (see Table 9.1) and inundation frequency change zones between the two periods 1979–91 and 1992–2006, for the (i) northern and (ii) southern Macquarie Marshes nature reserve sections.

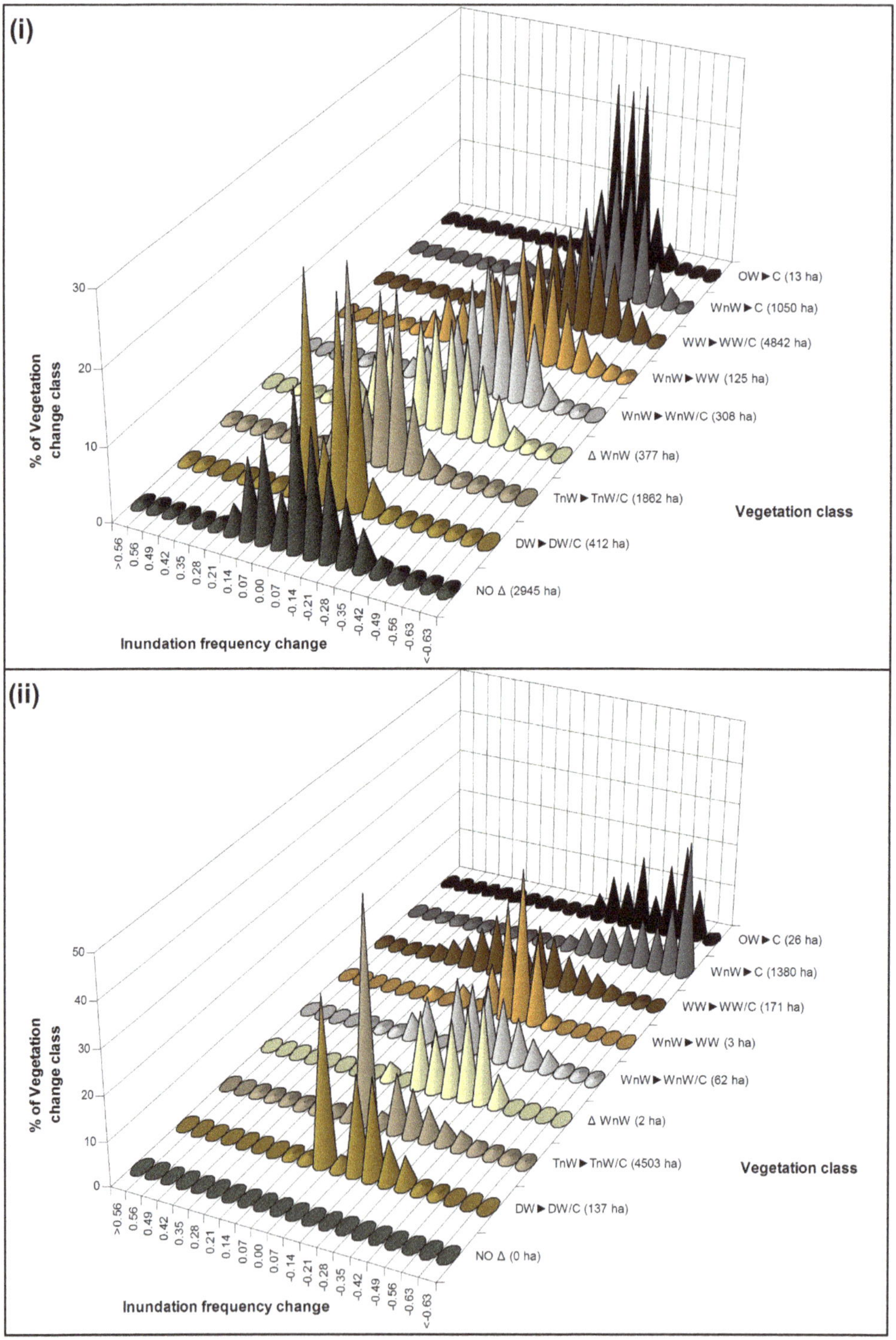

Figure 9.5: Percentage area (y axis) of each vegetation change class (z axis) (see Table 9.1 for codes) within each inundation frequency change zone (x axis) at 0.07 intervals indicative of the proportional loss (negative) or gain (positive) of floods from the first period for the (i) northern and (ii) southern Macquarie Marshes nature reserve sections.

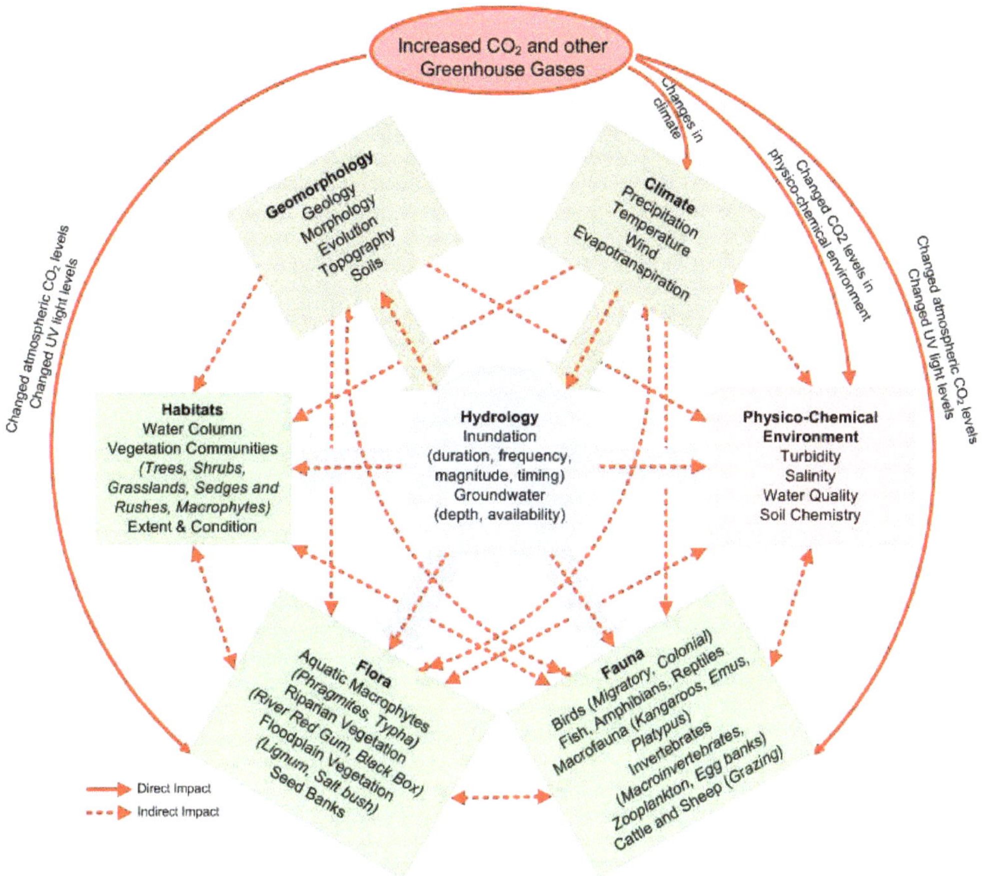

Figure 10.1: Conceptual model of the potential direct and indirect impacts of climate change on components of the Macquarie Marshes.

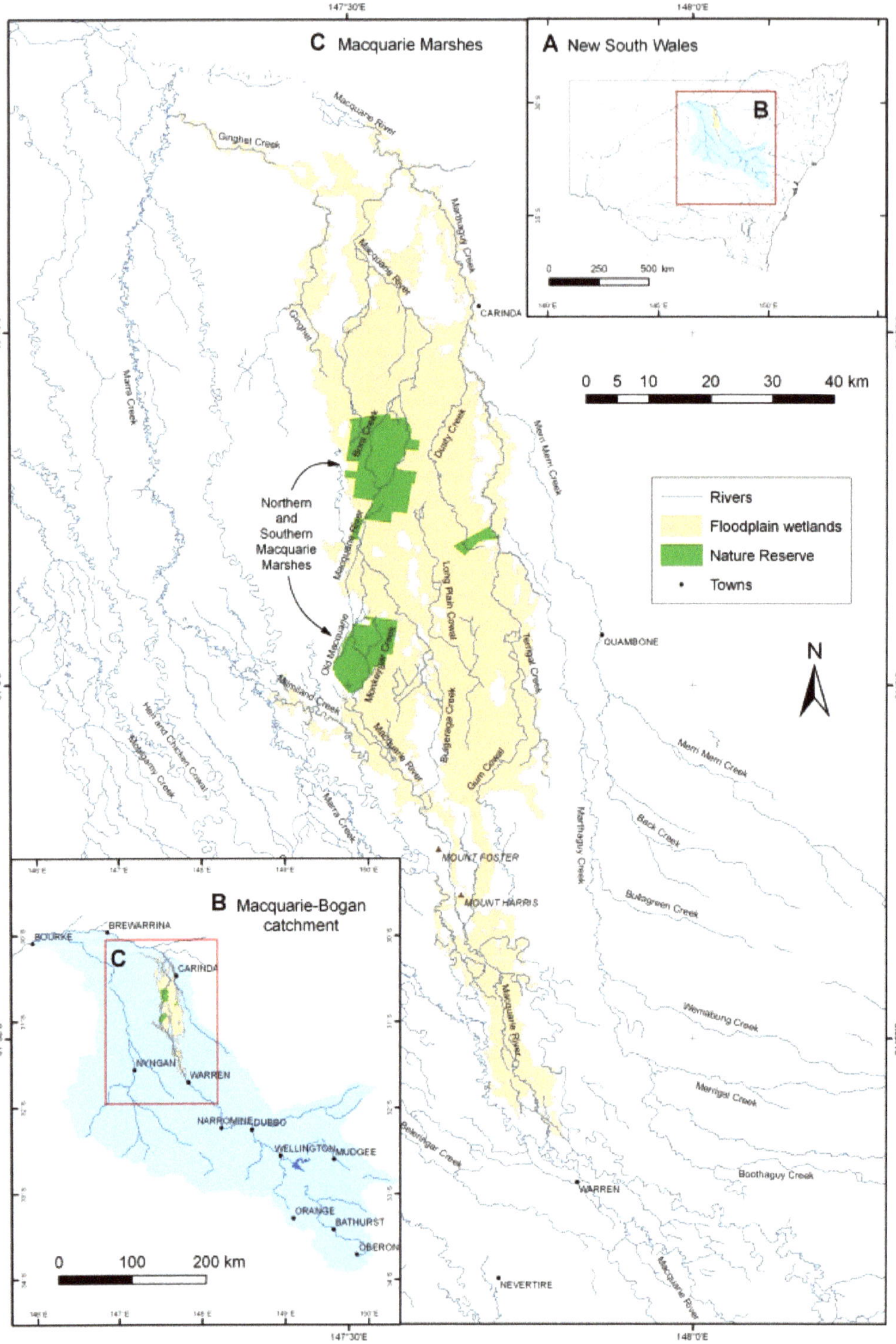

Figure 10.2: Location of the Macquarie Marshes.

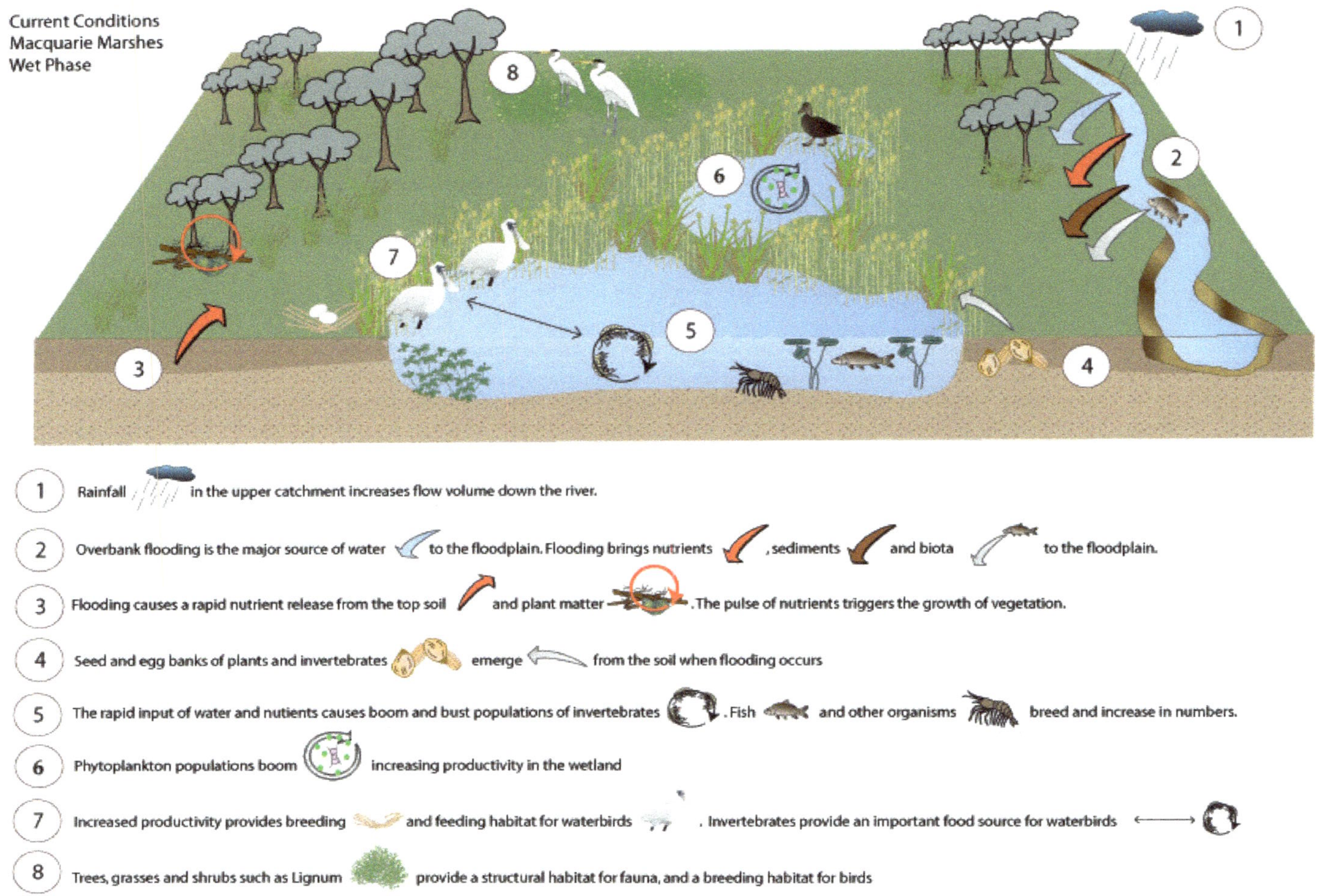

Figure 10.4: Conceptual model of a floodplain, such as the Macquarie Marshes under current climatic conditions during the wet phase.

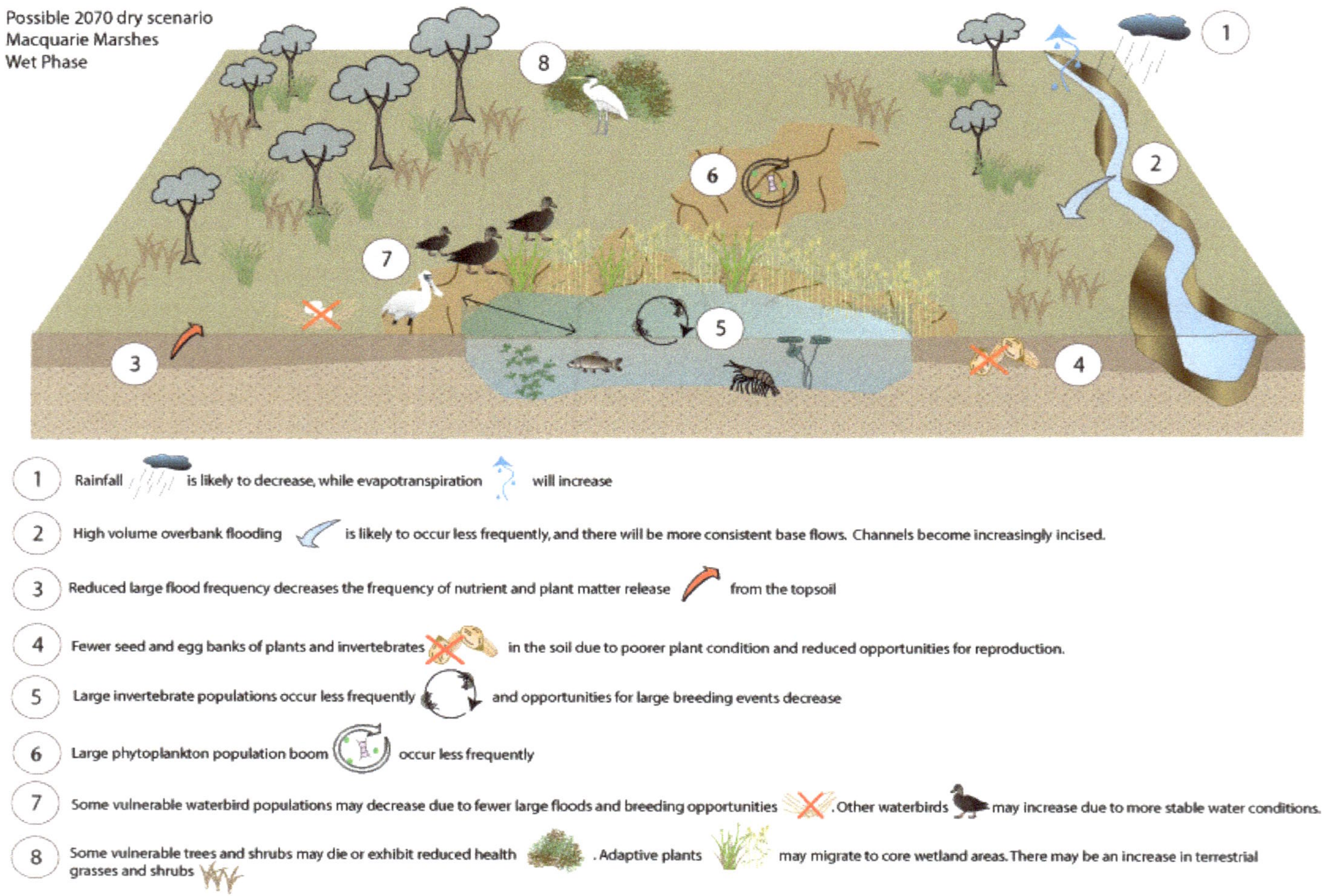

Figure 10.5: Conceptual model of a floodplain, such as the Macquarie Marshes under possible 2070 dry scenario conditions during the wet phase.

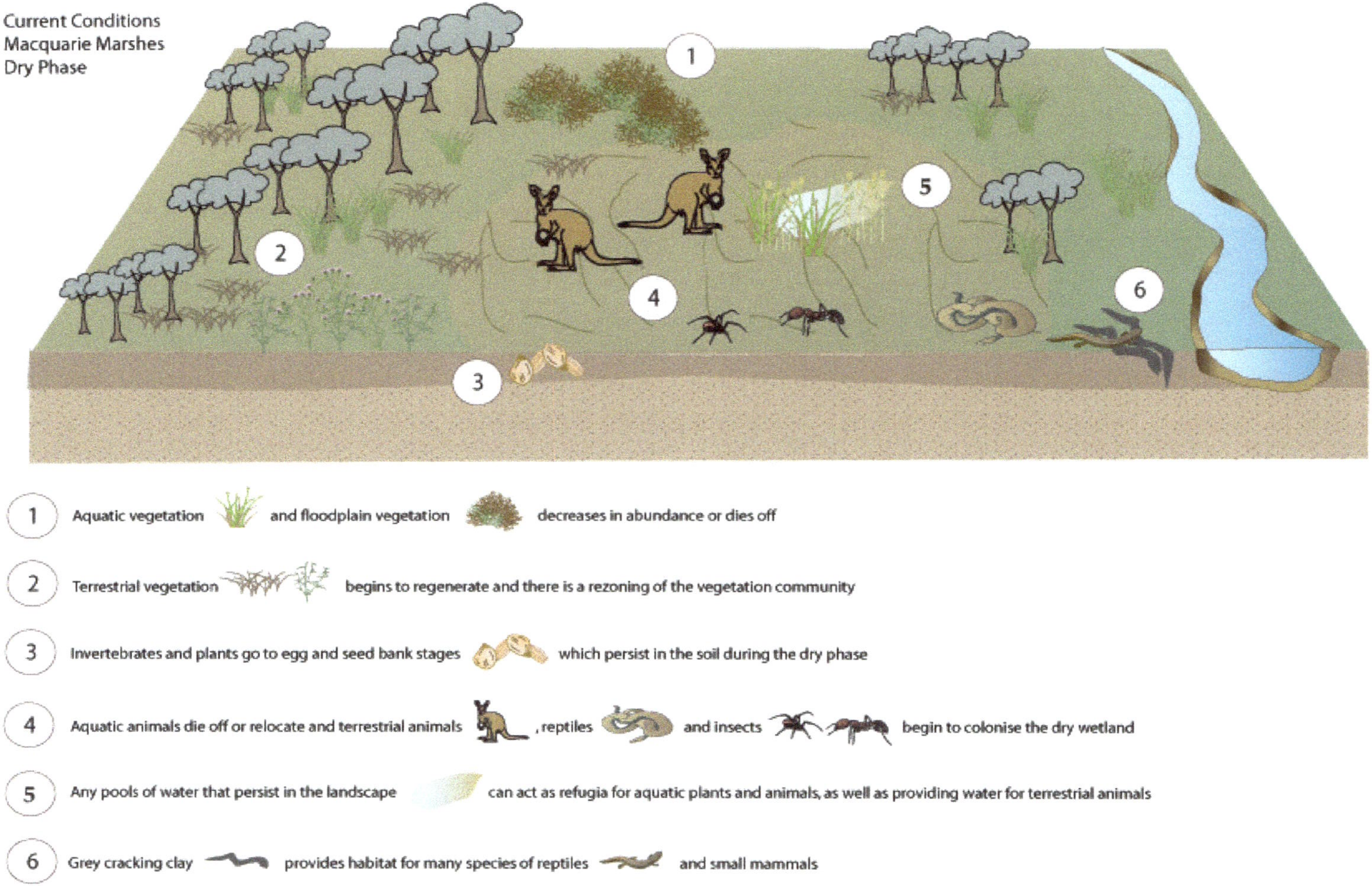

Figure 10.6: Conceptual model of a floodplain, such as the Macquarie Marshes under current climatic conditions during the dry phase.

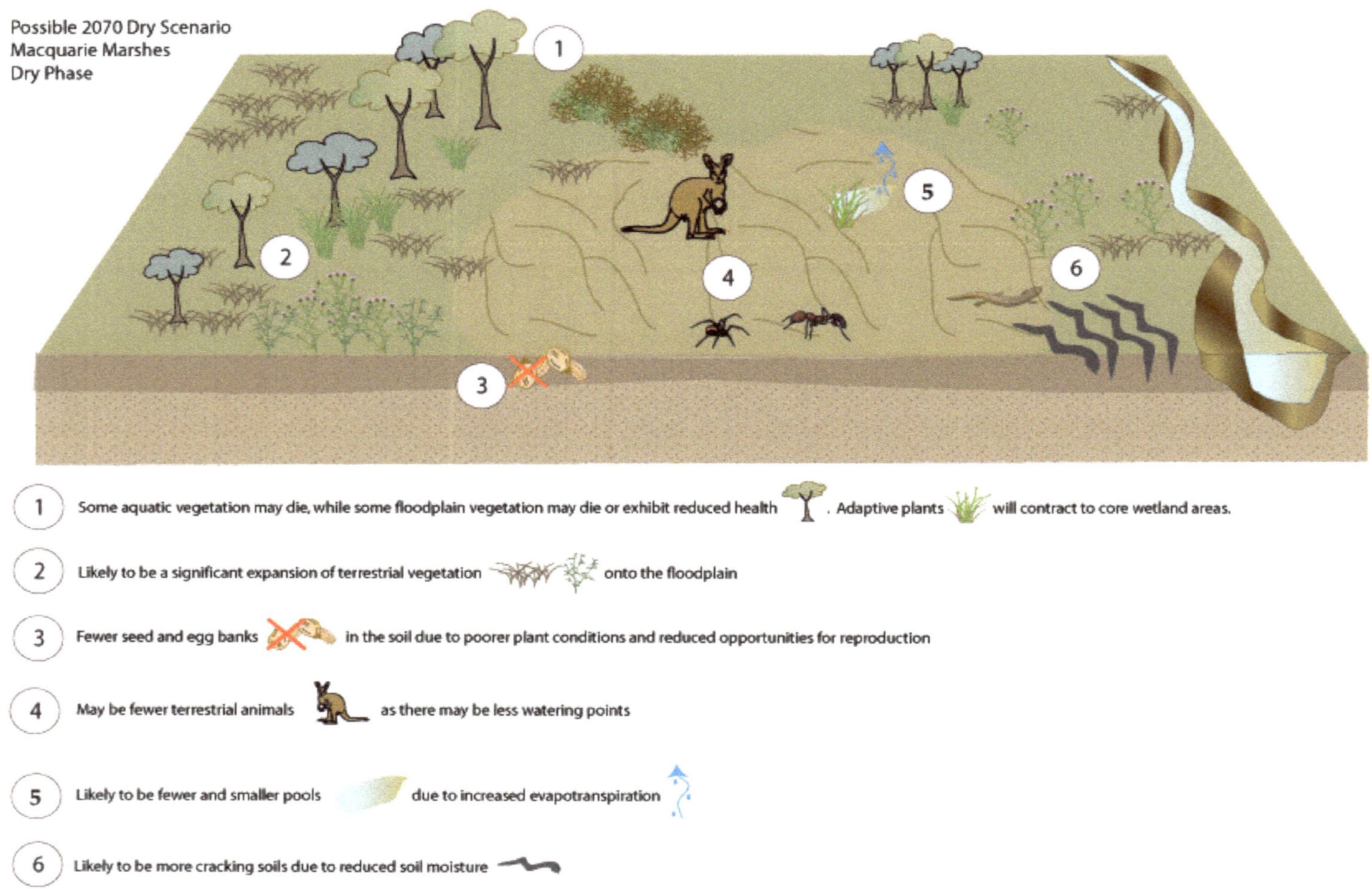

Figure 10.7: Conceptual model of a floodplain, such as the Macquarie Marshes under possible 2070 dry scenario conditions during the dry phase.

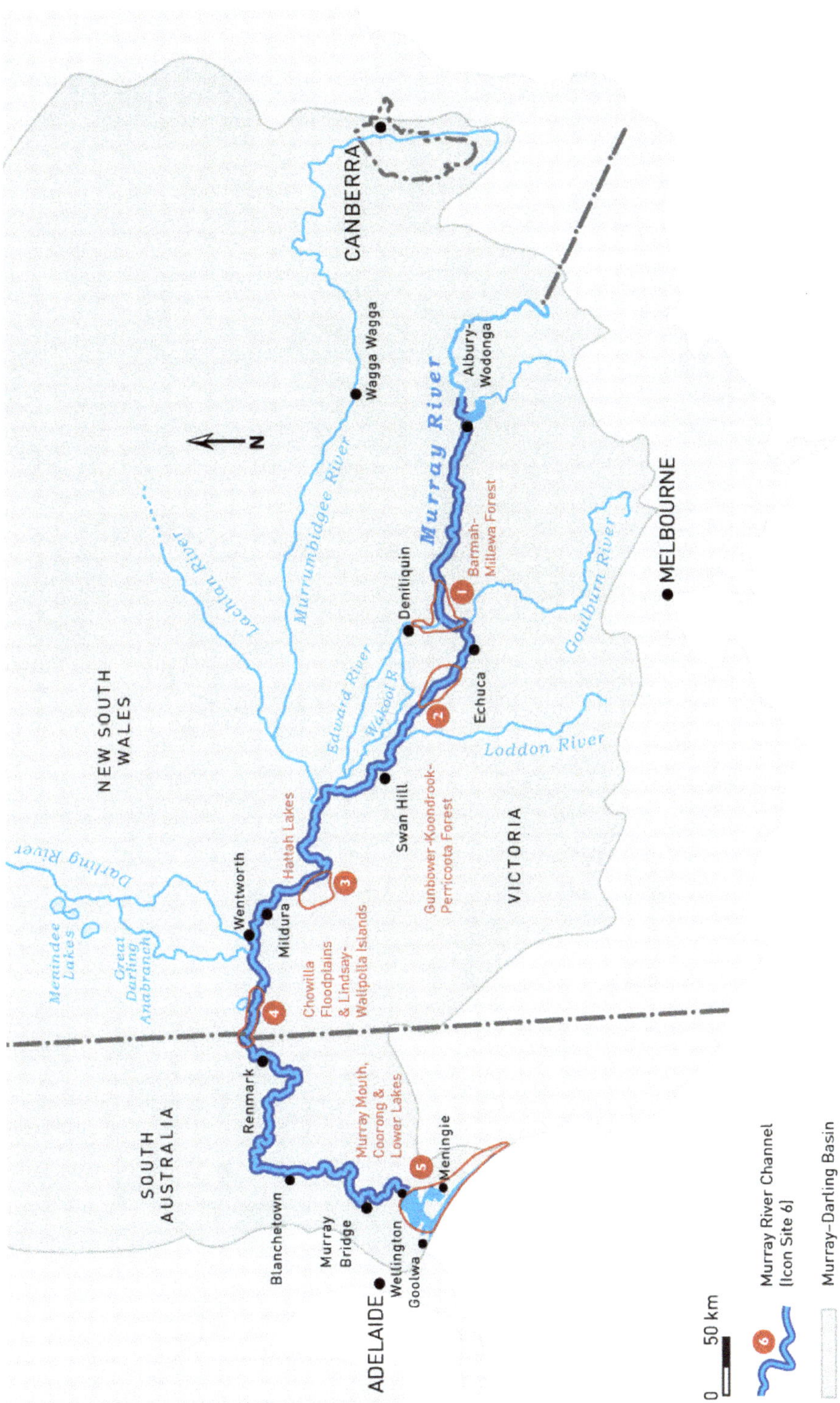

Figure 11.1: Icon sites in The Living Murray.

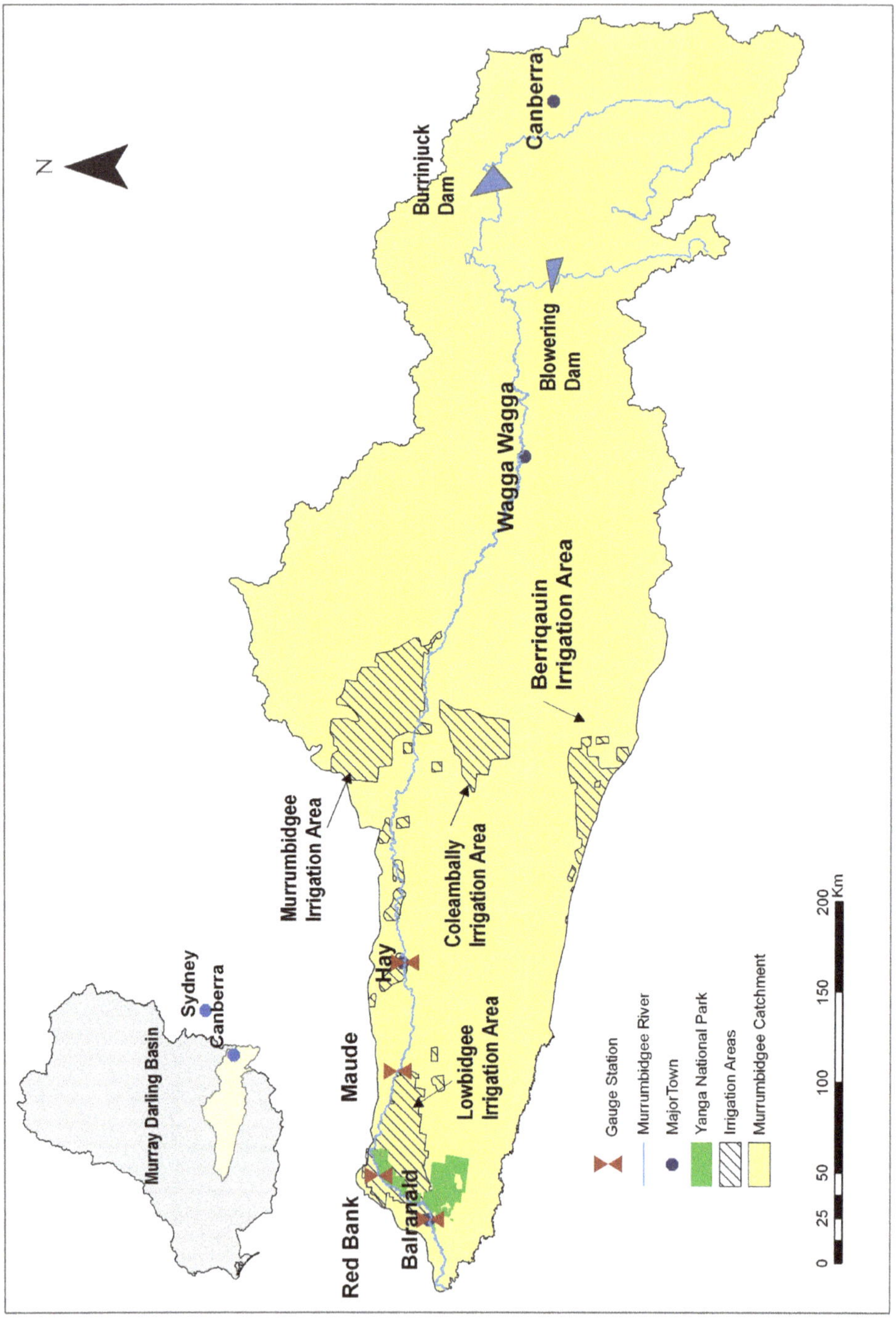

Figure 15.1: Location of Yanga National Park within the Murrumbidgee catchment. Major river developments (weirs and irrigation areas) are shown. Redbank and Maude gauges were used to investigate the hydrology of the park.

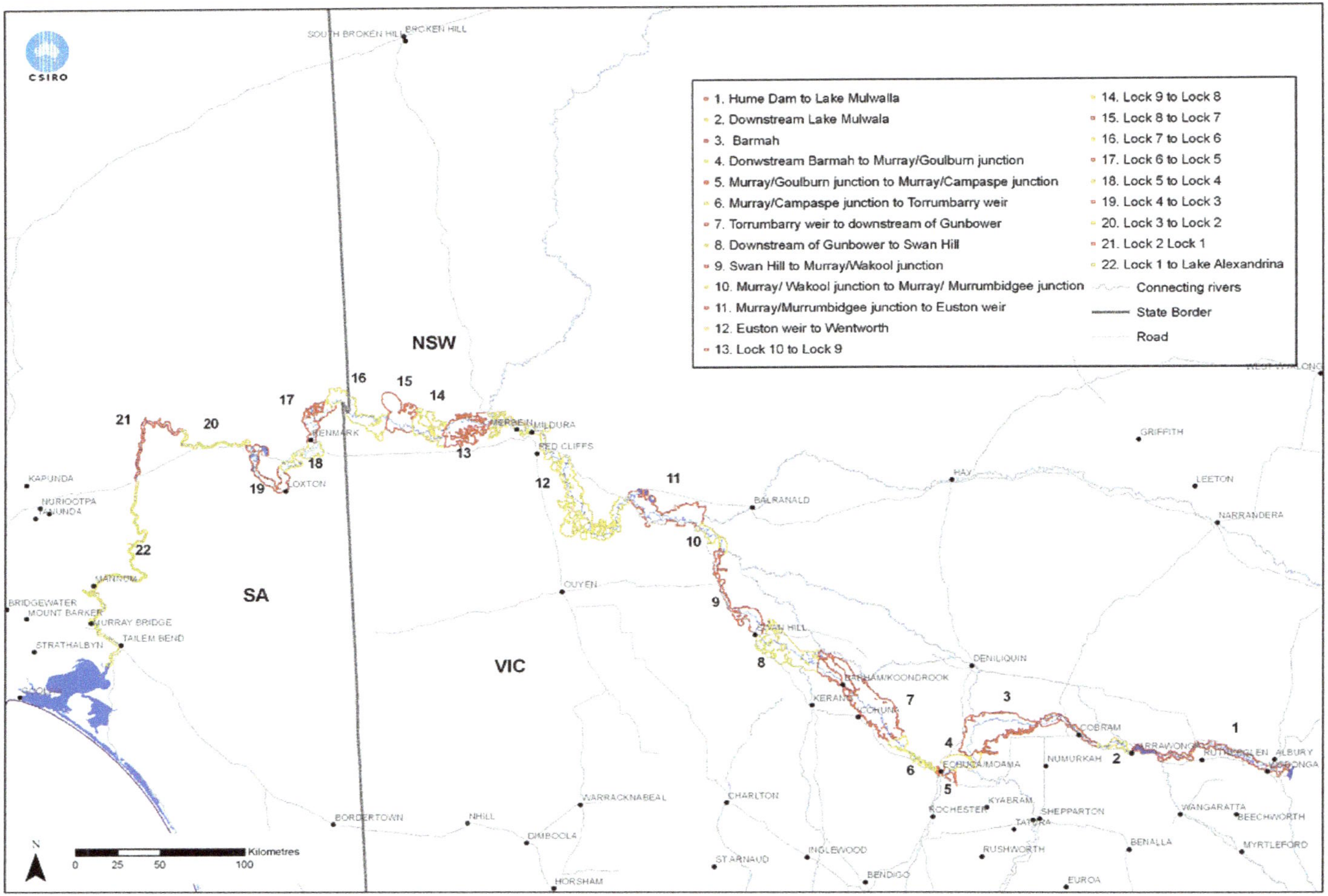

Figure 16.4: The 22 zones of the Murray River floodplain, from Hume Dam in Victoria to Wellington in South Australia (Overton *et al.* 2006b).

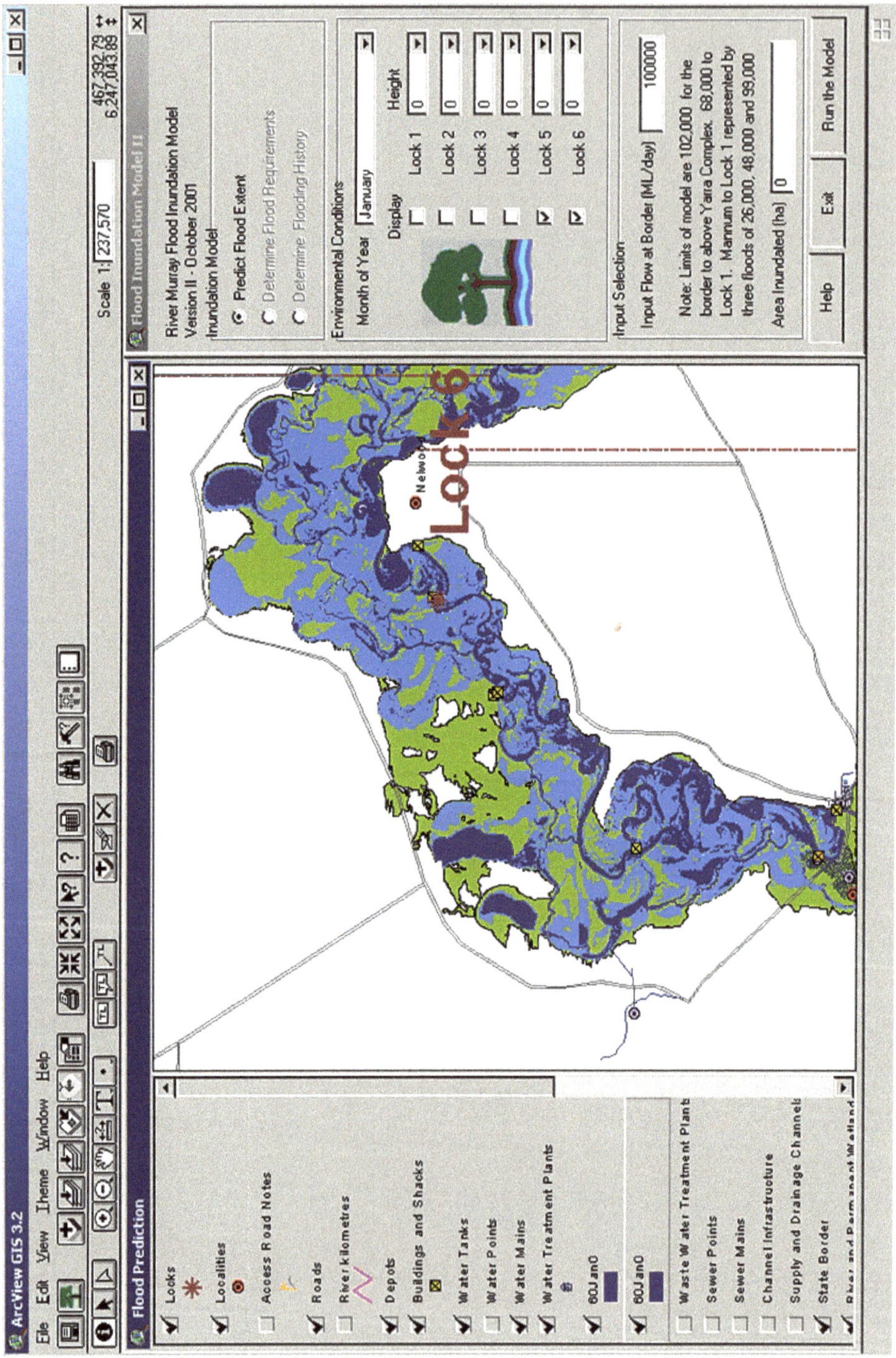

Figure 16.8: Example output of the GIS floodplain inundation model showing an area around Lock 6 at Chowilla with two flood predictions of 60 000 ML/day (dark) and 100 000 ML/day (light) (Overton *et al.* 2006b).

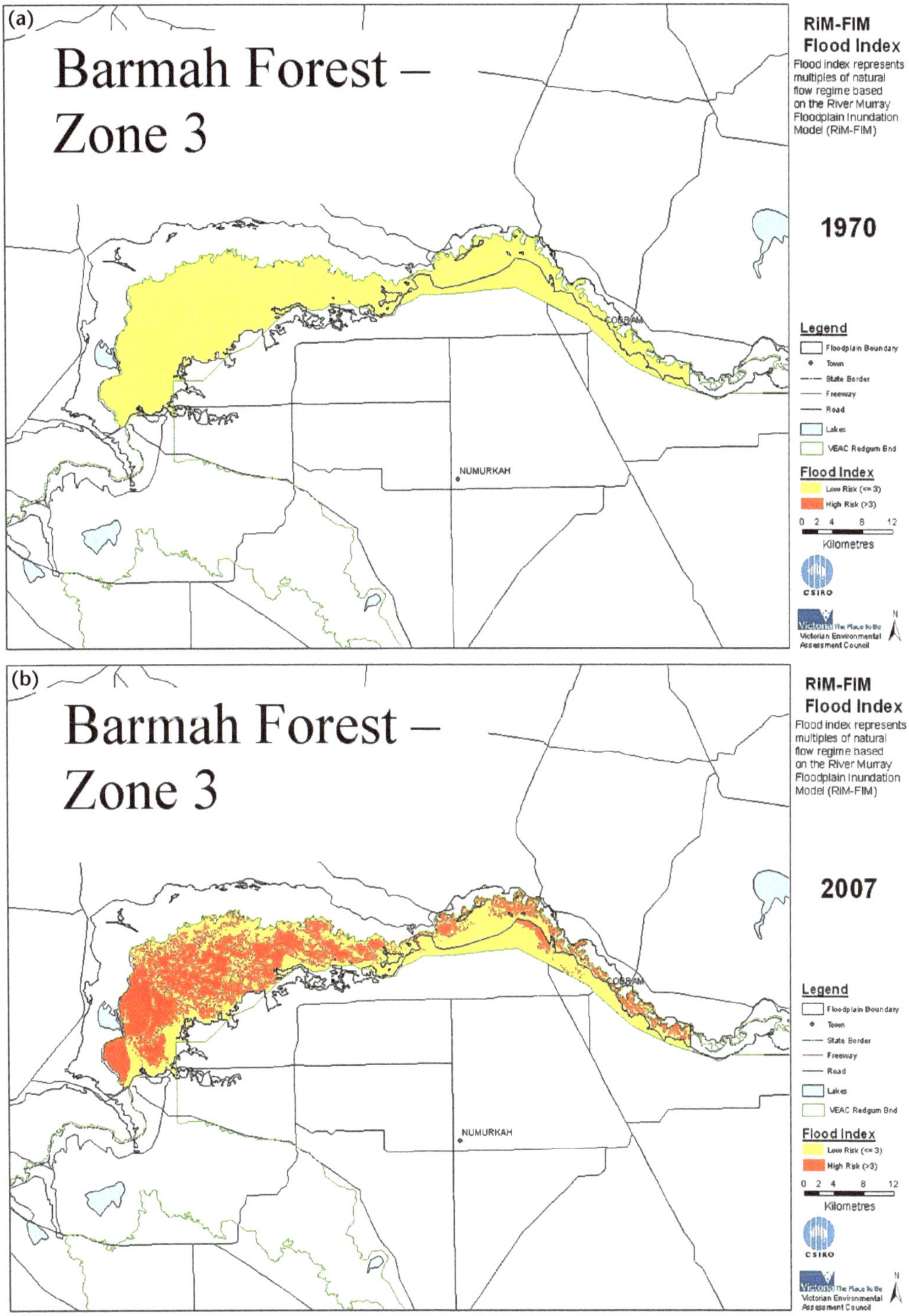

Figure 16.14: (a) Zone 3 at Barmah showing areas of good health (yellow) and poor health (red) for 1970. (b) Zone 3 at Barmah showing areas of good health (yellow) and poor health (red) for 2007 (Overton and Doody 2007).

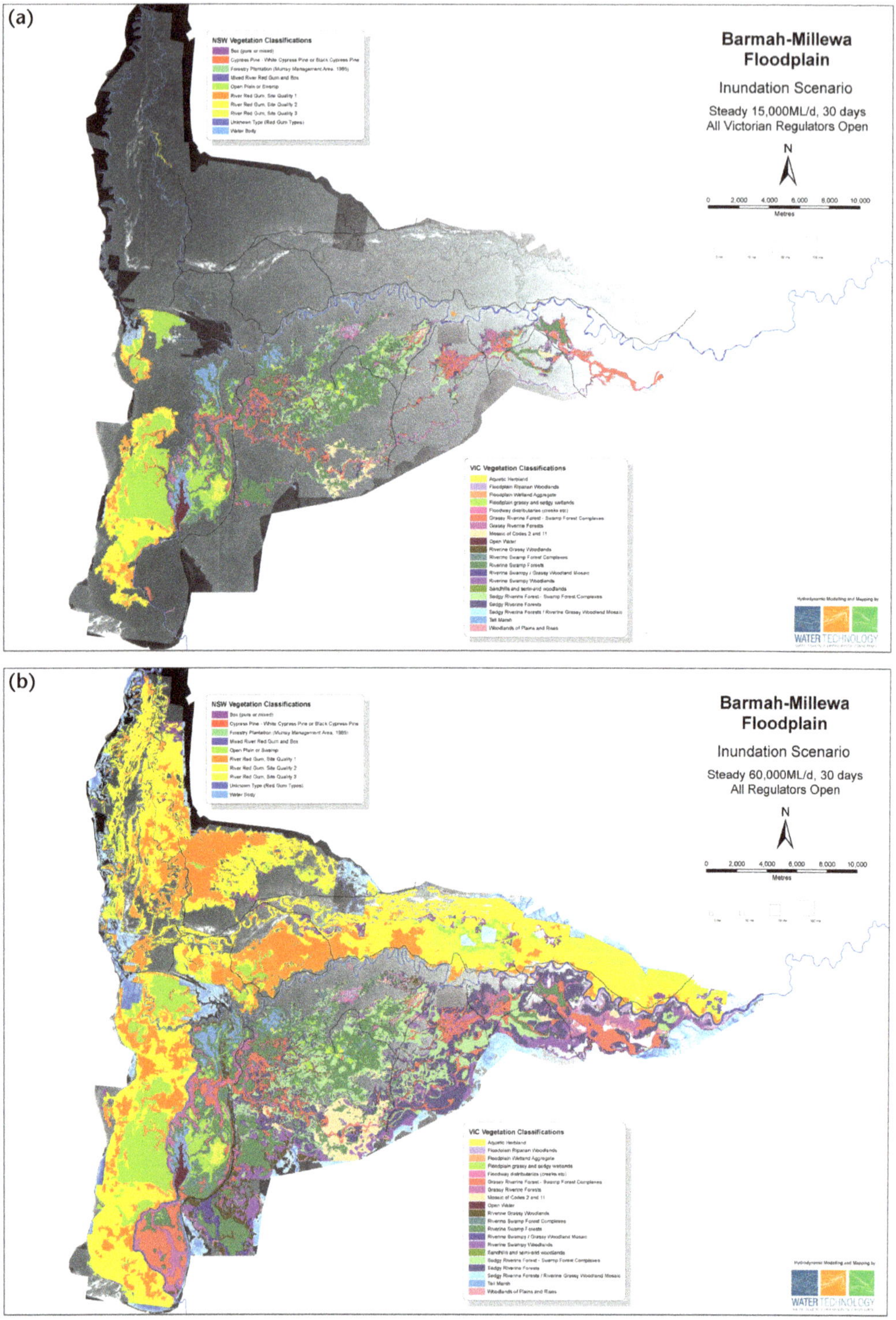

Figure 20.2: Areas of different vegetation classes inundated in relation to two contrasting flow scenarios, (a) 15 000 ML d^{-1} and (b) 60 000 ML d^{-1}, derived from the Water Technology MIKE FLOOD model (with permission of Water Technology).

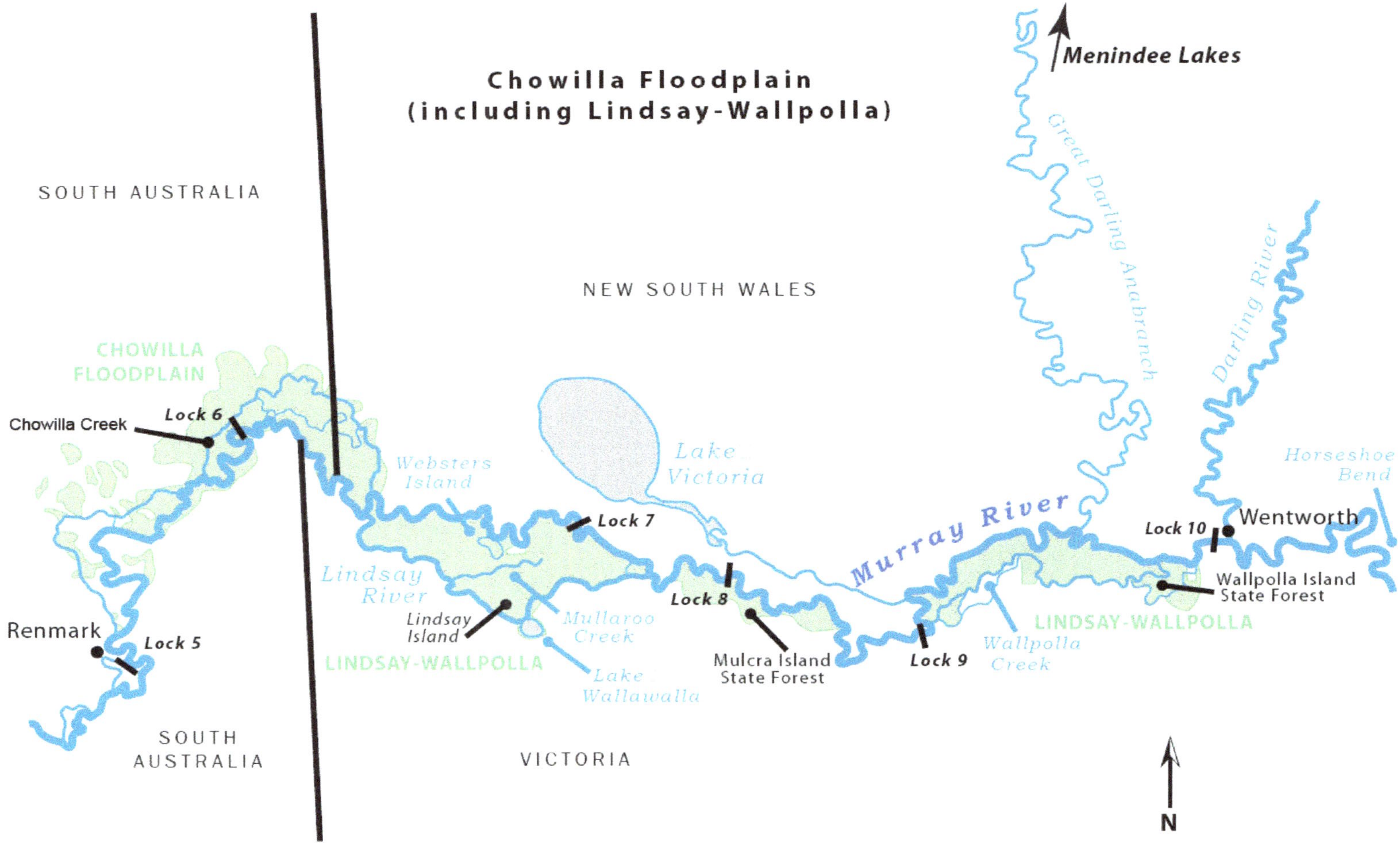

Figure 21.2: Chowilla floodplain, Lindsay and Wallpolla Islands icon site (MDBC 2006).

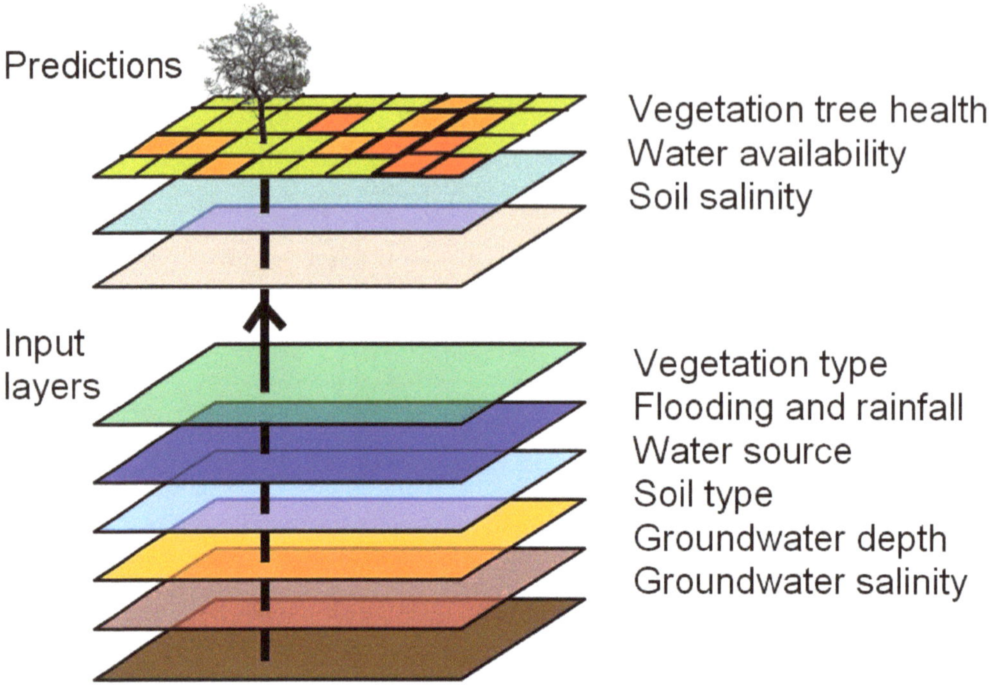

Figure 21.3: WINDS model input layers and layer predictions.

Figure 21.4: (a) Health decline of black box high on the floodplain due to soil salinisation. (b) Health decline of red gum along ephemeral creeks due mainly to a lack of flooding.

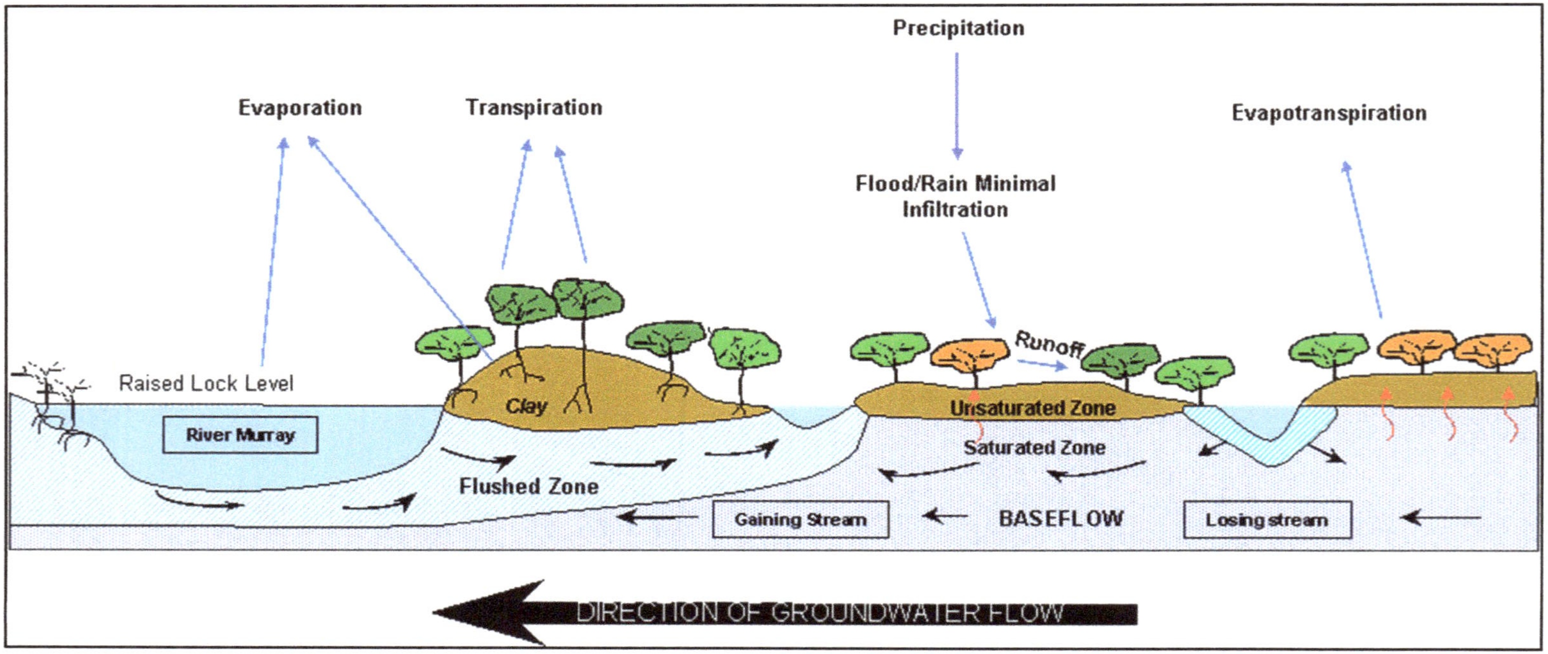

Figure 21.5: Conceptual diagram of floodplain processes.

Figure 21.6: (a) Losing creek with good fringing vegetation, lower part of Chowilla Creek. (b) Gaining creek with poor salinised edges, upper part of Salt Creek.

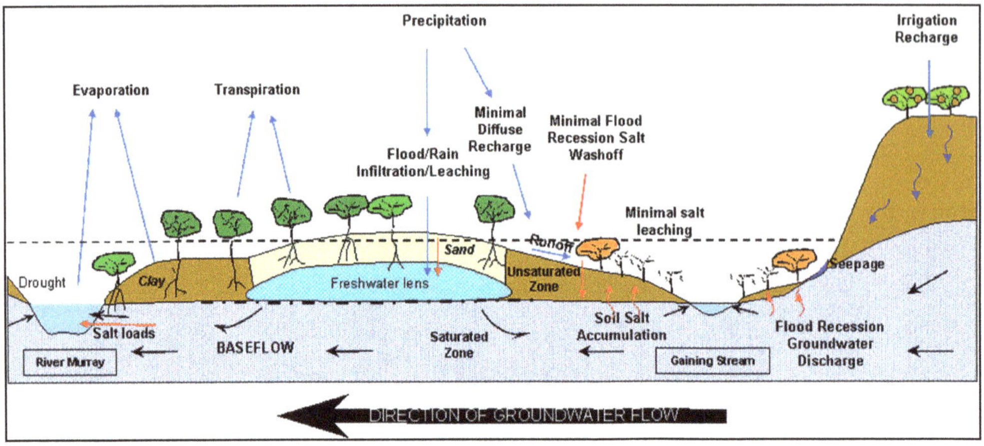

Figure 21.7: Conceptual diagram of floodplain processes.

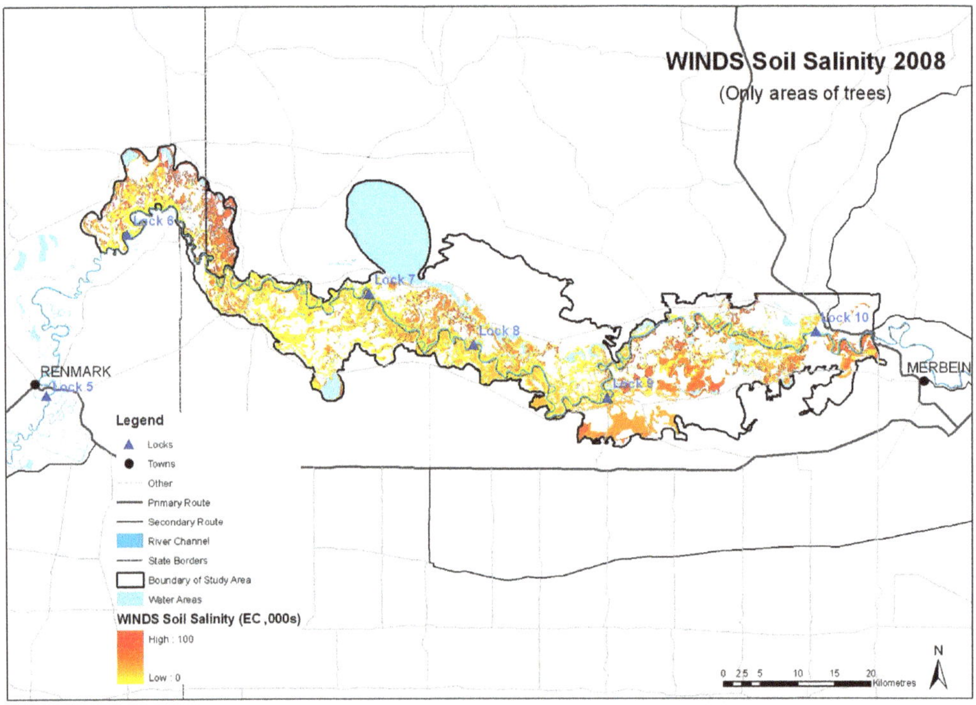

Figure 21.8: WINDS model output showing predicted soil salinity for the Chowilla floodplain in 2008 for the area of the floodplain with tree vegetation. Dark colours represent high salinities of up 100 dS m^{-1}.

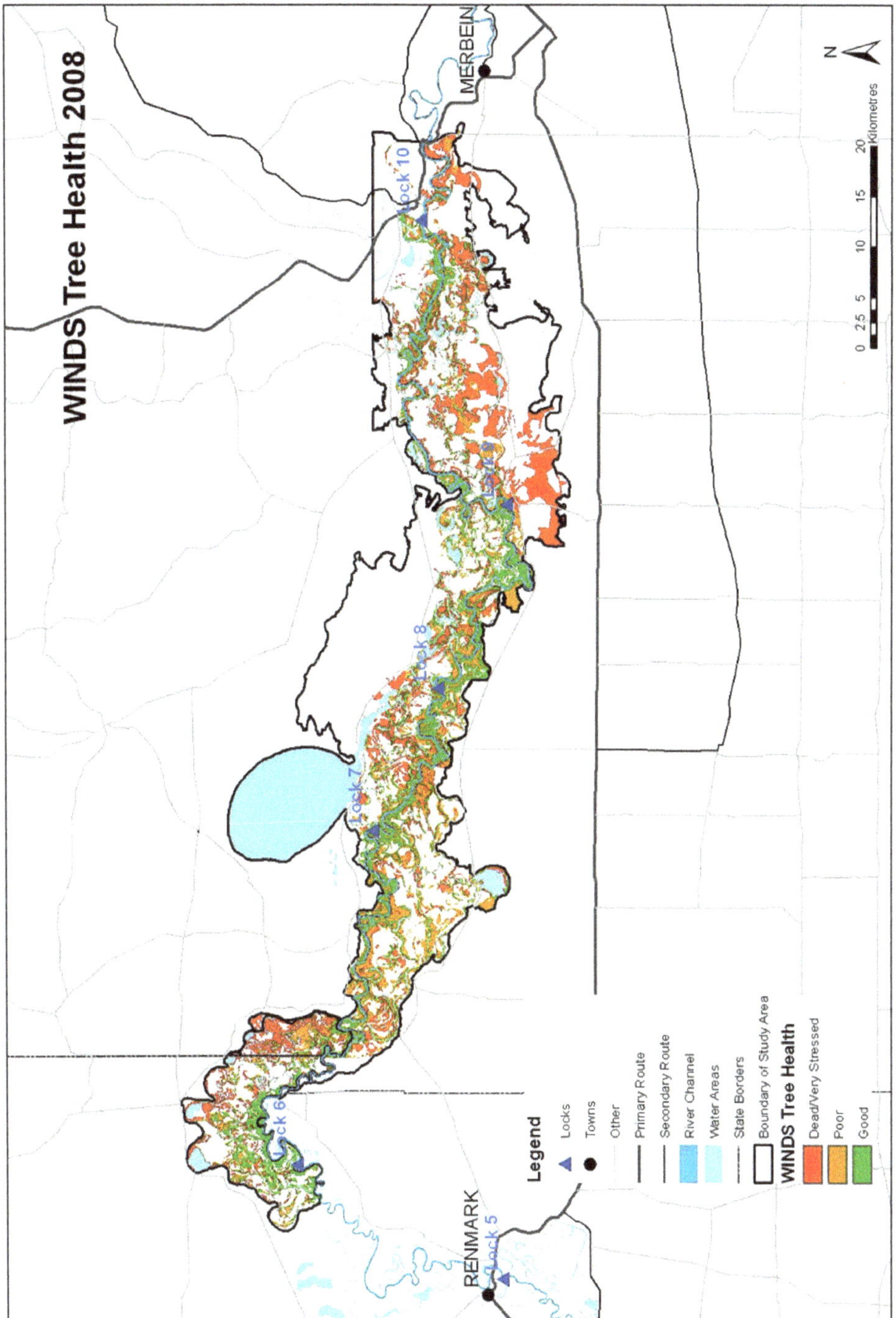

Figure 21.9: WINDS model output showing predicted tree vegetation health and understorey condition for 2008, using a three-class classification.

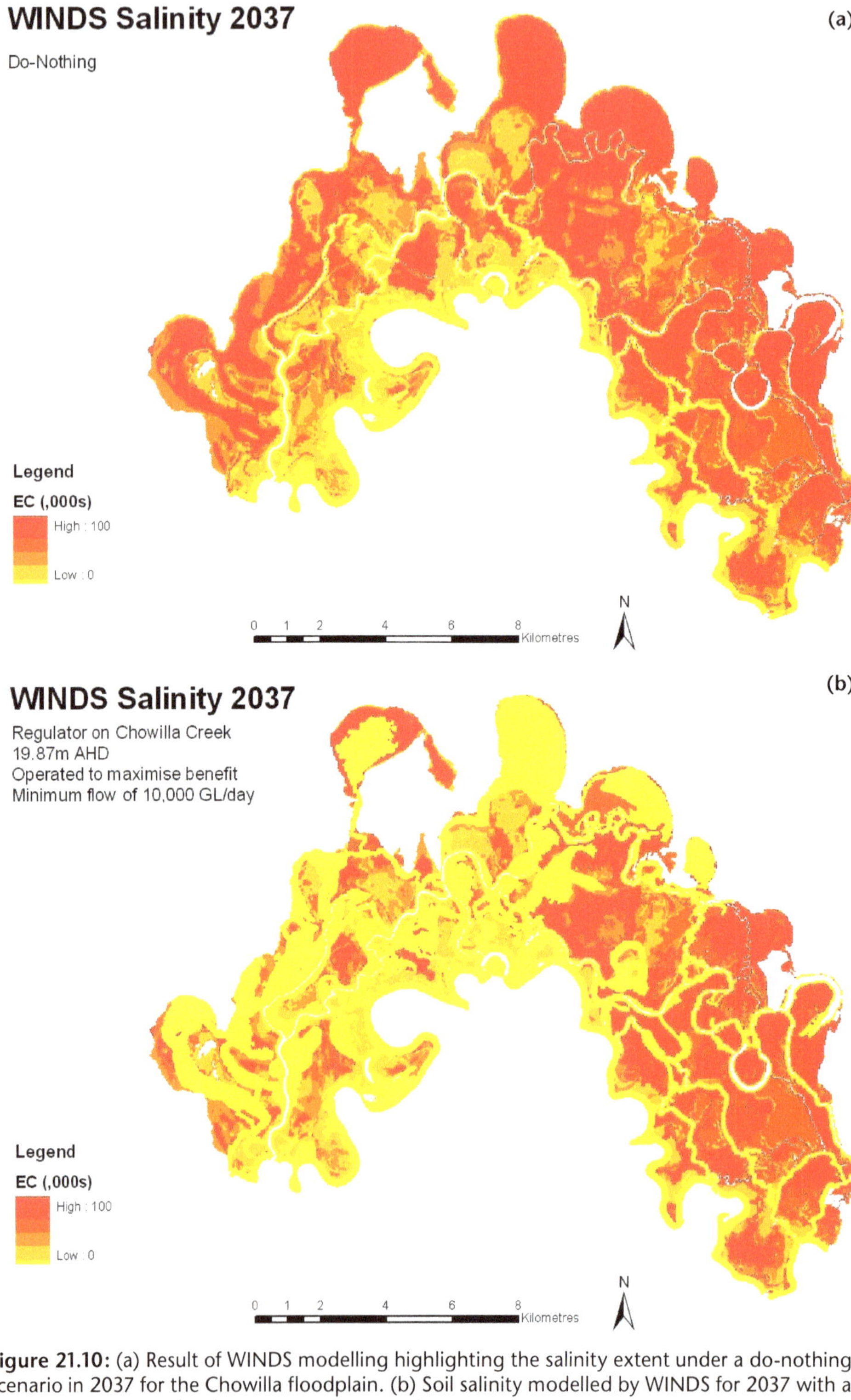

Figure 21.10: (a) Result of WINDS modelling highlighting the salinity extent under a do-nothing scenario in 2037 for the Chowilla floodplain. (b) Soil salinity modelled by WINDS for 2037 with an operational scenario for the environmental regulator.

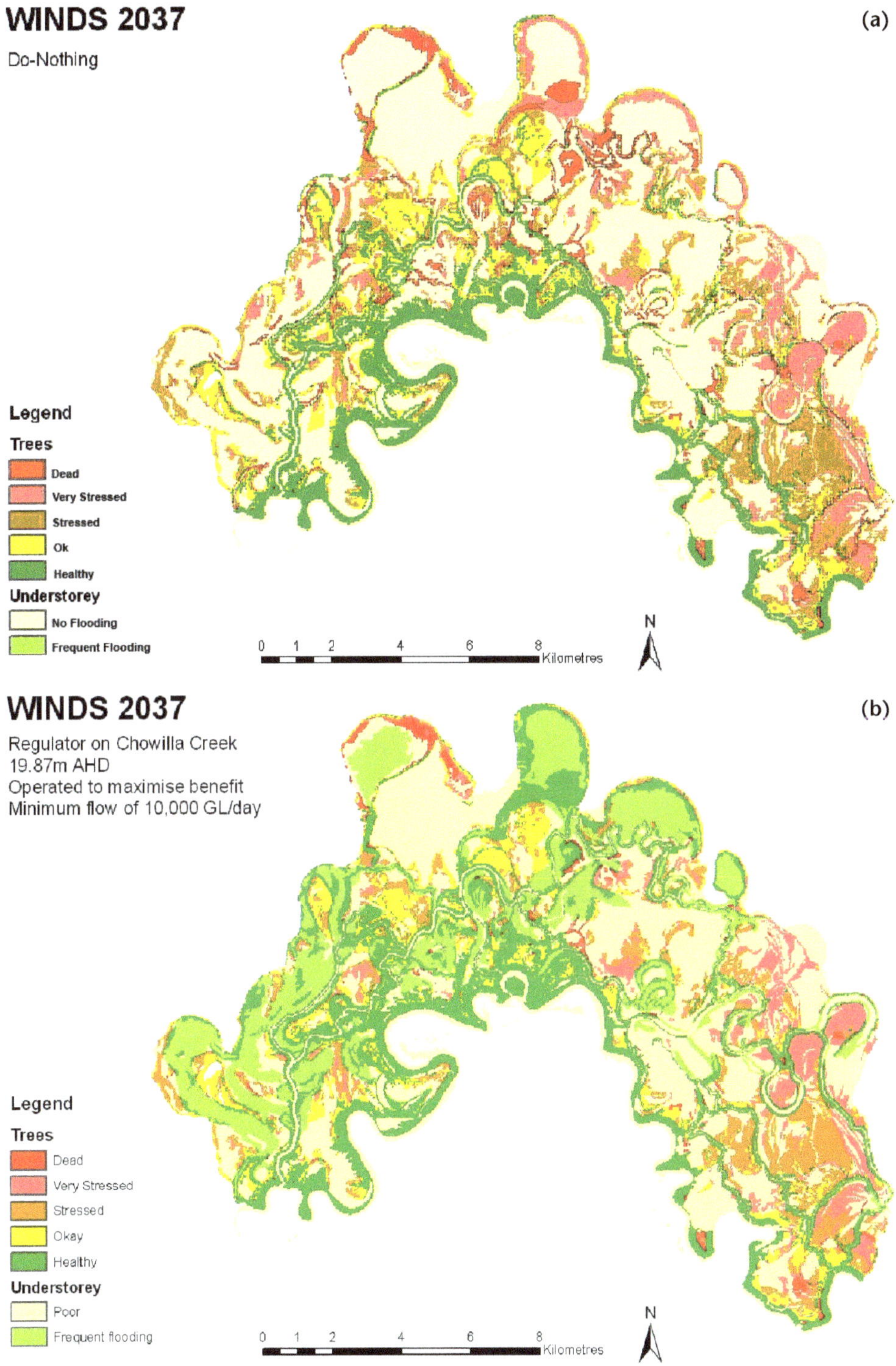

Figure 21.11: (a) WINDS prediction of Chowilla floodplain tree health in 2037 under a do-nothing scenario. (b) WINDS prediction of Chowilla floodplain tree health in 2037 with regulator use.

Figure 21.12: (a) Red gum regeneration in areas previously occupied by reed beds within a wetland. (b) Red gum surviving in lowlying areas of an ephemeral creek, with old red gums on the banks in decline.

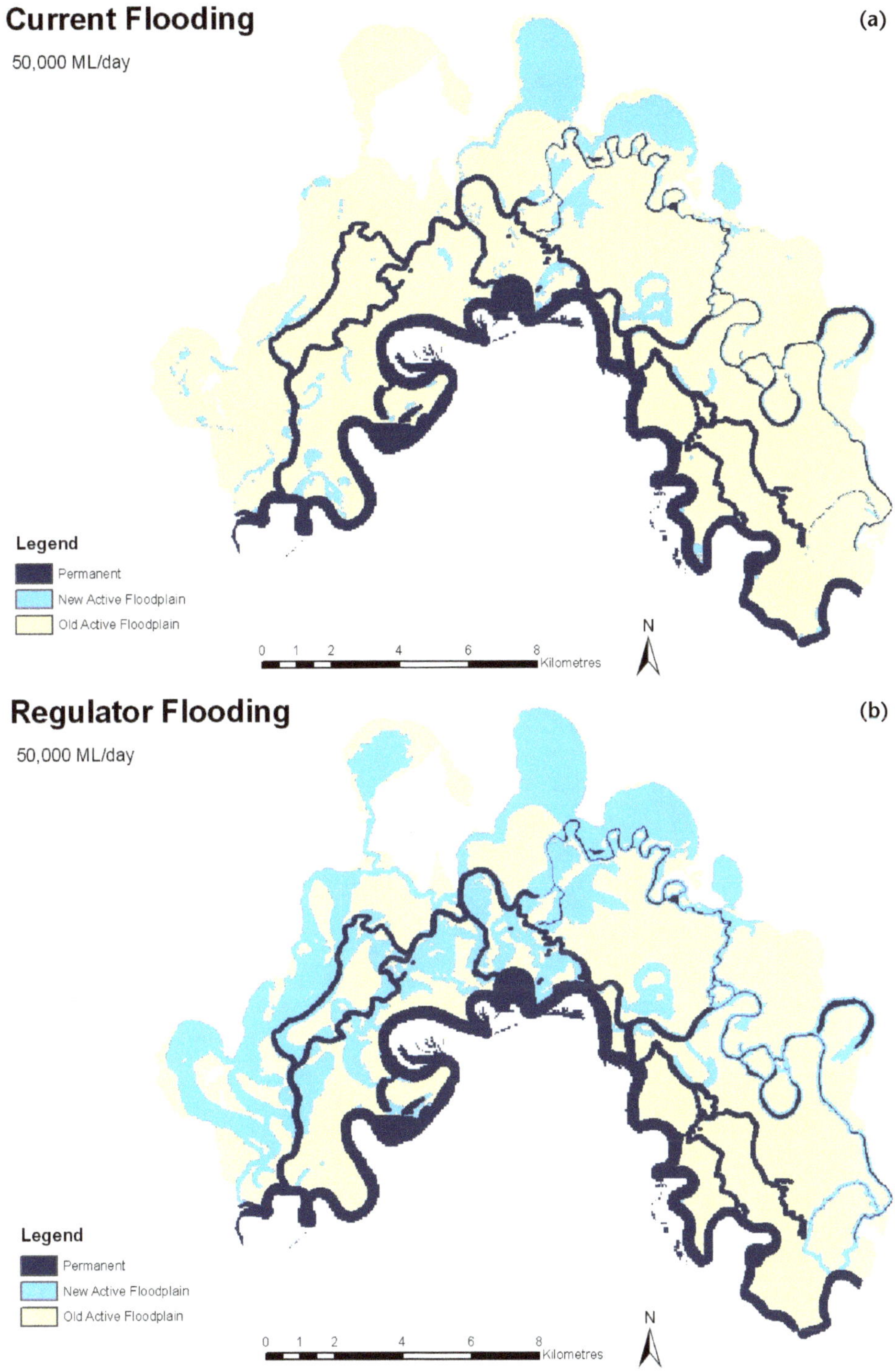

Figure 21.14: (a) Extent of the current 50 000 ML d^{-1} flood event. (b) Extent of a 50 000 ML d^{-1} flood event with the proposed regulator in operation.

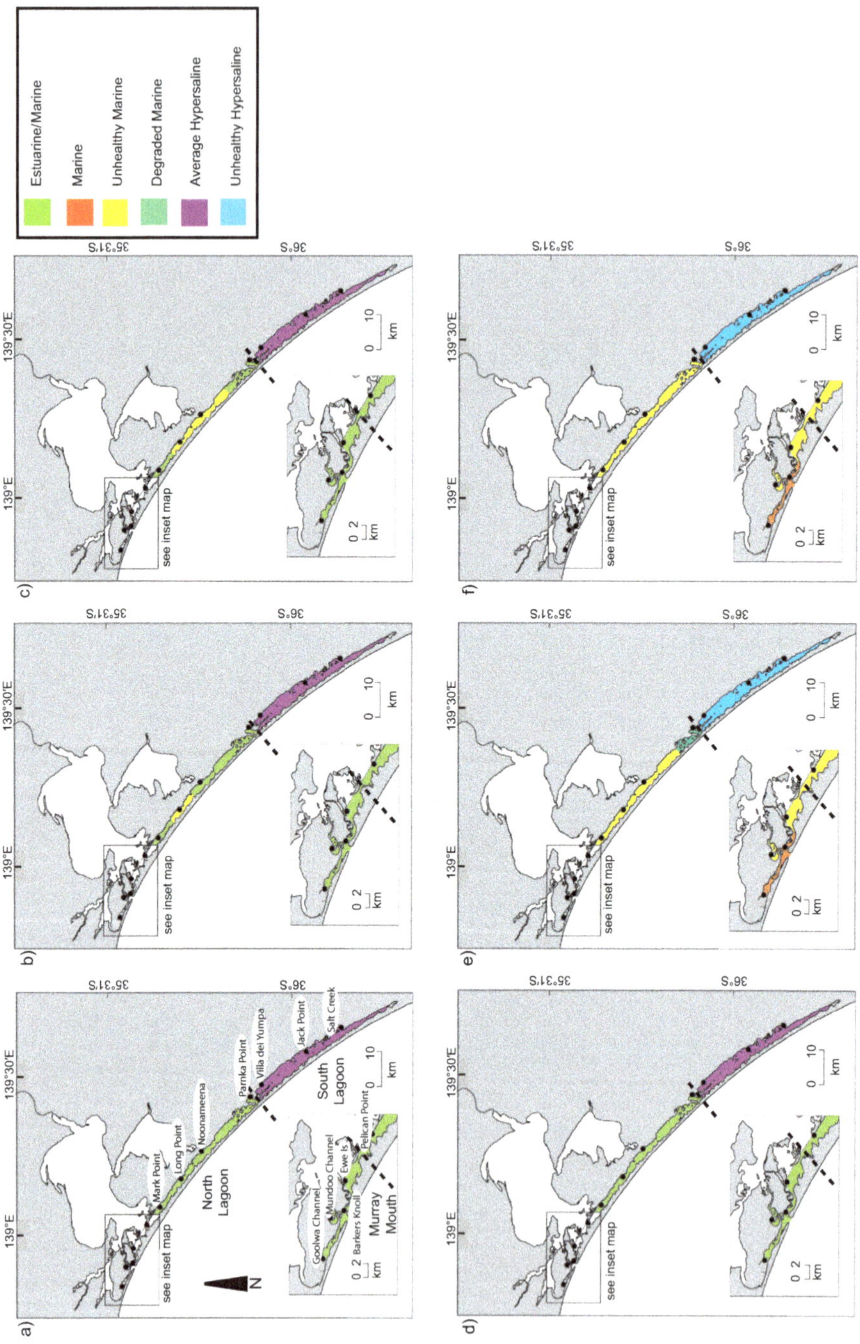

Figure 22.1: Distribution of ecosystem states along the Coorong (a) in 1999 to 2001, (b) in 2002, (c) in 2003, (d) in 2004–05, (e) in 2006 and (f) in 2007. Dotted lines indicate boundaries between the three regions. Dots indicate the locations of the 12 focal sites used in CLLAMMecology. Names of regions and sites are only listed in panel (a), but apply to all other panels as well. Source of map: Craig Noell, SARDI Aquatic Sciences, South Australia.

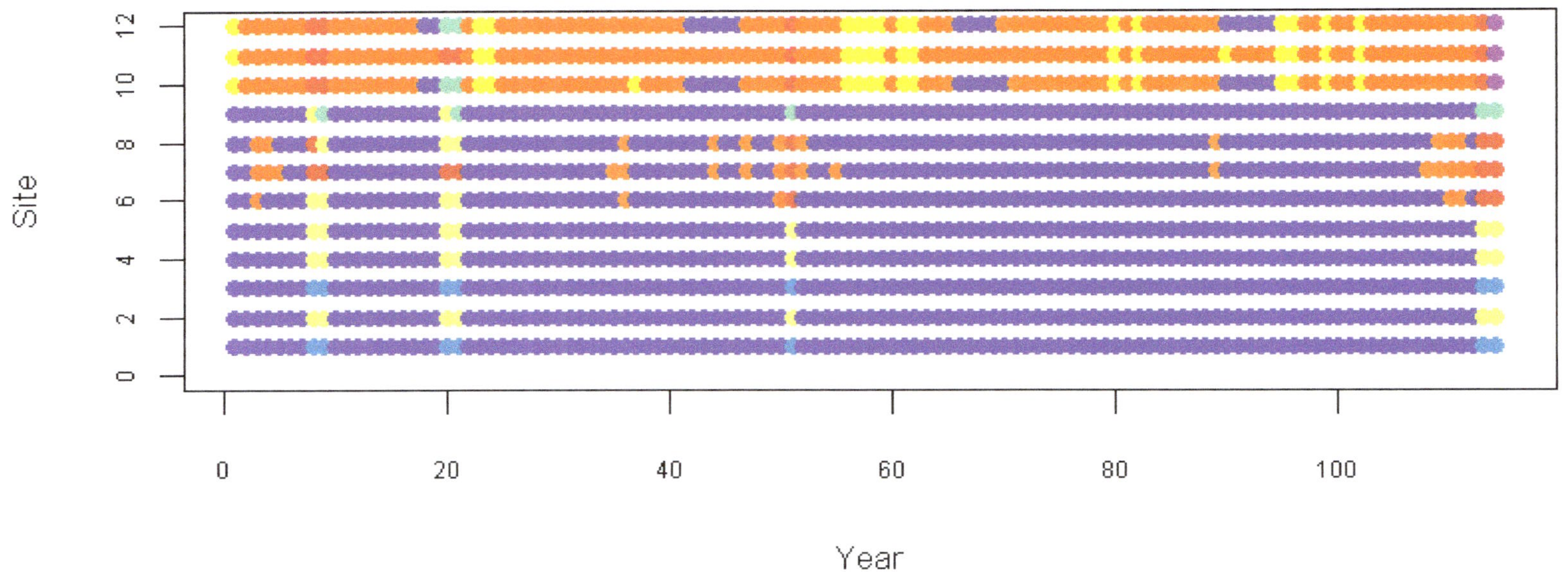

Figure 22.3: Distribution of ecosystem states in the Coorong under the baseline scenario. Each bar shows the distribution of the states within each site across the 114-year model run. Sites are numbered from north to south (e.g. Goolwa Channel = site 1 and Salt Creek = site 12). The changes in the bar colours represent the transitions between states. For each bar, colours represent the following states: dark blue = Estuarine/Marine, light blue = Marine, light green = Unhealthy Marine, dark green = Degraded Marine, yellow = Healthy Hypersaline, orange =Average Hypersaline, red = Unhealthy Hypersaline and purple = Degraded Hypersaline.

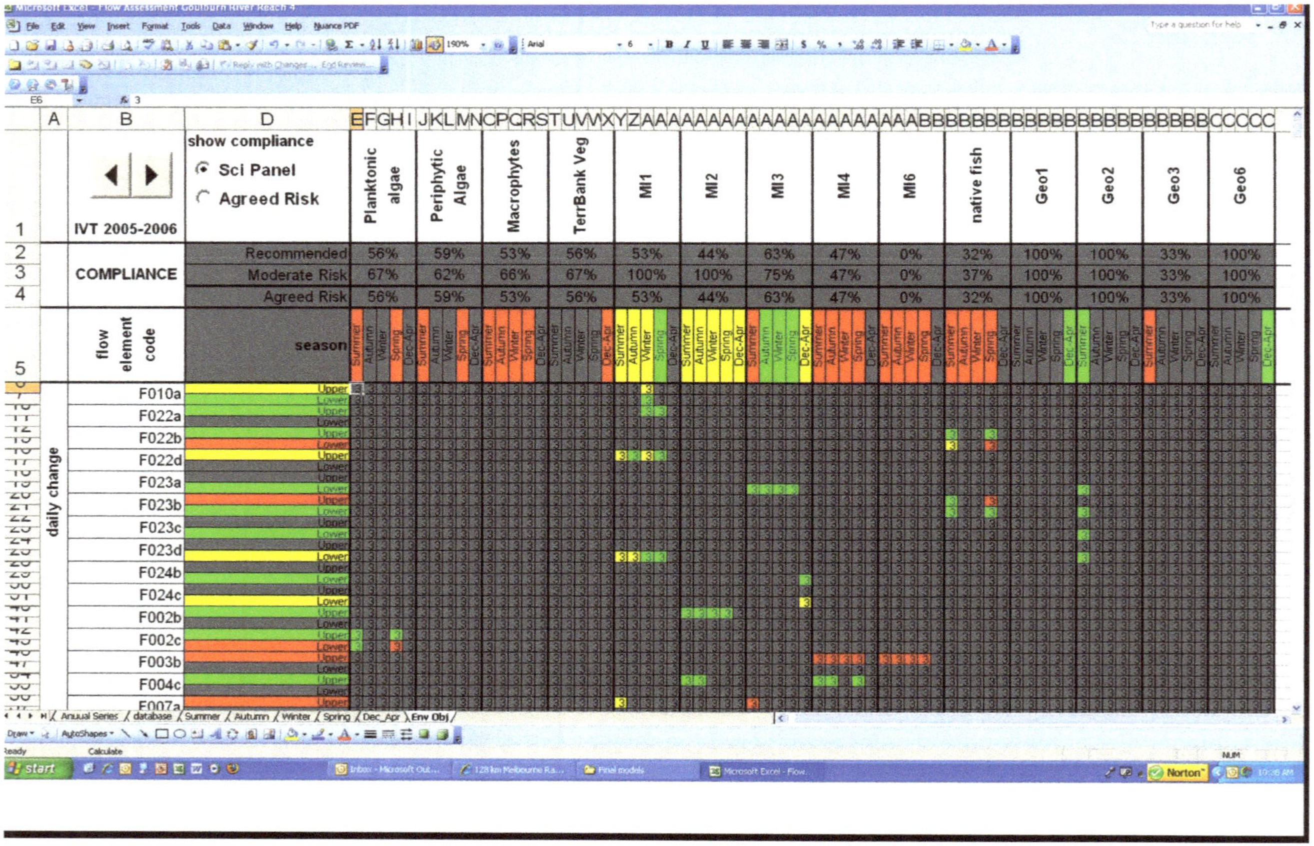

Figure 23.2: Decision support tool output for the Murchison reach. Red lights in columns or rows indicate that a flow stressor (or one of its elements) has exceeded a threshold representing a high risk to stated ecological objectives in a particular season. Yellow lights indicate a moderate risk to ecological objectives and green lights represent low risk. Grey shading indicates where the stressor is not relevant. The number or nature of red or green lights acceptable for a given scenario can be discussed between stakeholders. Important questions are whether different weightings should be given to different types of stressors, or whether different weightings should be given to individual objectives.

Chapter 19

Modelling invasive plants in relation to flooding and drying: implications for ecosystem functions

Kate Stokes, Keith Ward, Paula Ward and Matthew Colloff

Introduction

Floodplains are known to support assemblages of plant species that are adapted to disturbance (Lytle and Poff 2004), with vegetation zones determined by the tolerance of species to flooding (Keddy 1984; Silvertown *et al.* 1999; van Eck *et al.* 2004). Anthropogenic alteration of floodplains can dramatically alter the frequency and magnitude of flood events, and is considered a key driver of shifts in composition from native to introduced plants (Poff *et al.* 1997; Tickner *et al.* 2001; Nilsson and Svedmark 2002; Richardson *et al.* 2007), potentially resulting in negative alterations in the functional responses of communities (e.g. services in pollination, biological control or biomass production). Increased awareness of negative ecosystem consequences has triggered attempts to restore flood regimes via environmental water allocations. However, significant uncertainty remains regarding the nature and extent of ecological responses following provision of environmental water, including potential increases in the abundance and diversity of invasive weeds. We use the River Murray (Fig. 19.1) of south-east Australia as a case study, and compare historic vegetation surveys with present community composition to evaluate whether specific hydrological variables can predict exotic weed distributions. We aim to assess whether increases have occurred in the abundance and diversity of exotic species, whether alterations in exotic species abundance are associated with alterations in flood patterns at specific sites and whether exotic species possess similar or different life-history attributes from those characterising the native community. Finally, the potential threat of weed distribution resulting from flow restoration for biodiversity benefits is discussed.

What makes a site vulnerable to invasion?

A combination of observational, experimental and theoretical studies reviewed by Fridley *et al.* (2007) examined linkages between species richness of native plants and community resilience to invasion by exotics. The review indicated that positive associations between various attributes of native biodiversity and exotic species tend to predominate at larger spatial scales (≥1 km^2) and negative associations at finer scales (≤10 m^2). At large spatial scales, dispersal processes and abiotic stresses are believed to drive diversity gradients across different habitats, whereas at small scales biotic resistance mechanisms (e.g. competition and recruitment limitation) may control the extent of invasion. Therefore, risk indicators for weed invasion will be

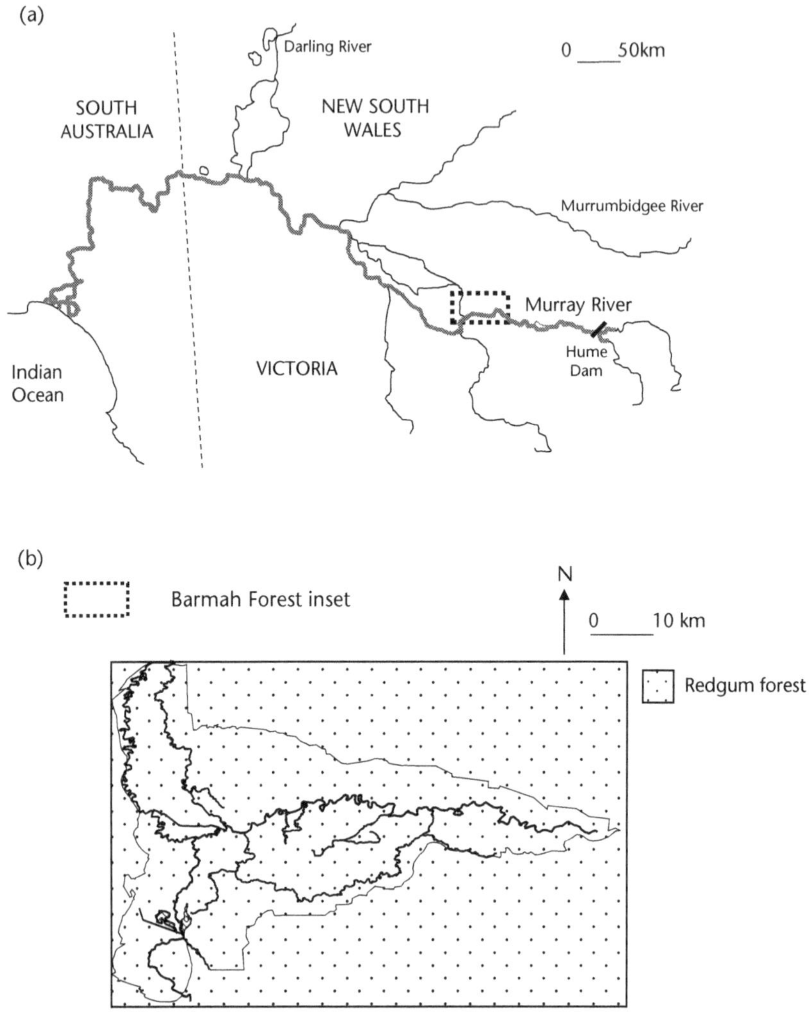

Figure 19.1: Location map showing (a) the Murray River corridor and (b) the course of the Murray River within Barmah Forest.

scale-dependent and result from interrelated factors including the reproductive and physiological characteristics of the invaders, the characteristics of the original plant community, the availability of propagules and the nature and frequency of any disturbance. This study adopts a multi-scale approach, focusing on floodplains of the River Murray, an area of approximately 600 000 ha at the regional scale. At the smaller site scale we chose the Barmah-Millewa forest, the largest river red gum forest in Australia and an internationally listed Ramsar wetland covering approximately 66 000 ha of floodplain.

Plant functional traits

Plant functional traits relate to universal plant functions of growth (e.g. light and nutrient acquisition, water use efficiency) and persistence (e.g. recruitment, dispersal, defence against herbivores and other disturbances). They are important ecological tools because they can be compared over large geographic regions and temporal scales, as well as enabling comparisons between separate data sets and studies (Westoby 1998; Weiher *et al.* 1999). Plant functional traits are not direct measures of ecosystem function, but have been used to infer underlying ecological processes and predict the response of ecosystems to fire (Keith *et al.* 2007), grazing (McIntyre *et al.* 1995), inundation (Boutin and Keddy 1993) and the invasibility of habitats (Pysek and Richardson 2007). Additionally, in some disturbed ecosystems feedback mechanisms operate so that plant traits themselves strongly influence the probability or severity of disturbances. For example, the introduction of invasive grasses to abandoned farmland in Spain resulted in increased fuel load and fire frequency, leading to powerful feed-forward processes (Grigulis *et al.* 2005).

Classification of species into functional groups based on hydrological preferences

Data on the germination, growth and reproduction of plants in wetlands of NSW was used by Brock and Casanova (1997) to classify species into functional groups based on plant trait responses to variable water regimes. Five groups were identified:

1 terrestrial dry (Tdr);
2 terrestrial damp (Tda);
3 amphibious fluctuation tolerators (Ate);
4 amphibious fluctuation responders (morphologically plastic) (Arp);
5 amphibious fluctuation responders (floating leaves) (Arf).

Each contained a mixture of annual and perennial species. The amphibious groups are the most versatile and can tolerate or respond to a range of wetting and drying patterns. Morphological fluctuation responders include species such as *Myriophyllum variifolium*, which produces different leaf types for underwater and above-water photosynthesis (Brock and Casanova 1997). Another mechanism of altered morphology is to grow laterally, putting down roots at nodes when stranded above the flood line and extending vertically suspended nodes and leaves when flooded, such as the heterophyllous *Limnosella australis*.

Changes in flood patterns on the River Murray

With completion of the Hume (1936) and Dartmouth dams (1979) the Murray has become a highly regulated river, driven by the demand for irrigation water. In the Upper Murray flood patterns have changed from semi-annual winter/spring floods of four to five months duration, supplied by snowmelt and inflows from the upper catchment, to shorter overbank floods during the summer months when irrigation water swells the river close to channel capacity. Reductions in flood frequency, duration, depth and extent have been apparent over the last 20 years, coinciding with lower rainfall and inflows. Floods now occur too late in the year for native plant communities that are adapted to post-flood growth and reproduction during late spring and early summer (Bren 1992; Kingsford 2000). At Barmah Forest, natural flood events occurred annually from 1990–94 and in 1996 and 1997 (the last major natural flood). Subsequent floods of shorter duration in 1998, 2001 and 2005 were floods supplemented by environmental water allocations.

Flow patterns in the Lower Murray have altered since the construction of 10 low-level weirs in 1922–35. Flows remain variable, but by the early 1990s low flows (100–300 GL per month) had decreased five-fold and moderate flows (500–1500 GL per month) had increased two-fold since weir construction (Walker and Thoms 1993). For the Lower Lakes, Coorong and Murray Mouth water resource development has increased the average period between floods required to flush the river mouth and help sustain the lake and estuarine ecosystems from 1.2 years to 2.2 years (CSIRO 2008). Flood volumes have also been greatly reduced – the average annual flood volume is only a fifth of the volume under without-development conditions (CSIRO 2008).

River Murray floodplain inundation model

The River Murray Floodplain Inundation Model (RiM-FIM) (Overton *et al.* 2006) was created as a research and decision support tool for environmental flow management, developed using a Geographical Information System (GIS), remote sensing and hydrological modelling. RiM-FIM predicts the extent of flooding on the River Murray floodplain, from a range of river flows and weir levels in South Australia, enabling spatial and quantitative analysis of flood extents to contribute to management decision processes. The model contains 22 different zones demarcated by regulation structures. RiM-FIM is a steady state model that predicts the extent of flooding from a given flow on the first day of the flood; it does not currently consider antecedent conditions or the effect of flood duration (Overton *et al.* 2006).

Regional-scale vegetation data sets

The regional-scale data set consists of point distribution records obtained from herbaria and floristic surveys for species recorded to occur within the River Murray floodplain corridor, as defined by the 1956 flood boundary (the maximum flood recorded in recent history). The data set contains 144 170 records from state-based surveys and 149 929 records from herbarium collections (Bickford *et al.* 2006). Earliest collection records were obtained in 1836 and the most recent in 2005. Herbarium records constitute nearly all pre-1980 collections, whereas the majority of survey records in the data set (95%) were collected post-1990, providing a record of both past and present species distributions (Table 19.1).

Site-scale vegetation data sets

At smaller spatial scales, floristic surveys were conducted at Barmah Forest to improve understanding of the flood requirements and flood response of flora. Previous surveys were conducted in 1990 and 1991–94 (Ward 2007) at five wetland locations throughout the forest. These sites were relocated and understorey vegetation resurveyed seasonally during 2006/07 and 2007/08, providing a comprehensive multi-site data set where the maximum distance between site locations is approximately 10 000 m. At each of five sites, two 150 m transects

Table 19.1: Vegetative composition of the Murray Floodplain point record data set

Category	No. of species recorded from herbaria	No. of species recorded in floristic surveys	Total no. of species/ category	Total species/ category (%)
Native vascular plants	1459	1201	1462	62
Exotic plants	499	445	503	21
Bryophytes/ algae	163	50	164	7
Naturalised plants	231	188	233	10

Total = total number of species recorded between the two data accumulation methods.

were re-established within discrete vegetation zones ascending from a low to a relatively higher elevation. Three 20 × 20 m quadrats were located along each transect, marking the end points and mid position. Topography of each quadrat was surveyed to assist in estimating individual quadrat flood regimes. Cover abundance of all species present within ten 1 m^2 random sample units confined within each 20 × 20 m quadrat was recorded seasonally. Additional parameters measured include the proportion of bare ground, leaf litter, open water and water depth. However, there was little rainfall and no flooding during 2006/07 and most study sites remained dry during this monitoring period (Ward 2007).

Predicting the spatial location of exotic plants using hydrological parameters

Initially, point location data for all exotic species recorded within the spatial boundaries of the River Murray floodplain since 1950 were mapped in GIS and overlain against a hydrological parameter produced by the RiM-FIM model. The commence-to-fill value (CTF) was used, which indicates the river flow (ML d^{-1}) required to inundate a specific spatial location on the floodplain. No spatial associations between exotic species and particular CTF values were observed. Therefore all species (natives and exotics) were segregated into functional groups to determine whether specific functional groups are associated with spatial locations that flood at a particular river flow. At Barmah Forest, species classified as amphibious fluctuation responders were clearly associated with floodplain locations inundated at low river flow values, but in general spatial segregation is weak and distributional overlap is observed between different functional groups (Fig. 19.2). Therefore data were segregated to examine the distribution of individual species in relation to CTF values. The hydrological niches of 21 common exotic species and 21 common native species were mapped by matching each spatial point occurrence for a particular species with its corresponding range of CTF values within each of Zones 1 to 9. An example is given for Zone 9 (Fig. 19.3). Representation of exotic and native species between different functional groups was intended to be equal but unfortunately that aim was constrained by the frequency of individual species within the large-scale data set. Only species occurring at high frequency (>50 records) were analysed and there were no exotic species recorded at that level of abundance for the amphibious fluctuation responder groups. The resulting minimum, maximum and mean CTF values show extremely wide variation for almost all species in all zones (e.g. Zone 9, Fig. 19.3).

Common species tend to be generalists but it is these species that are likely to prove the most severe threat to floodplain systems, precisely because they are able to thrive in a variety of hydrological conditions. Any marginal differences between CTF values of native and exotic species could indicate a threshold value that managers could exploit to create unsuitable hydrological conditions for exotics. However, there was no clear indication that the exotic species group responded negatively to particular river flows, and exotic and native species occupied zones experiencing similar inundation conditions. This analysis is based on presence-only data and does not provide information on the health and vigour of plants, or their relative abundance. A complete investigation of life-history patterns is required to fully understand the mechanisms structuring plant zonation patterns along flooding gradients. Furthermore, lower and upper elevation boundaries may be determined by different mechanisms. For example, hydrological preferences may dictate the lower elevation boundaries of a species, whereas competition may play a role in structuring upper boundaries. In general, it is difficult to identify these different drivers using the large-scale data set because the data suffers from several biases, including identification problems in the field, accessibility of sampling sites,

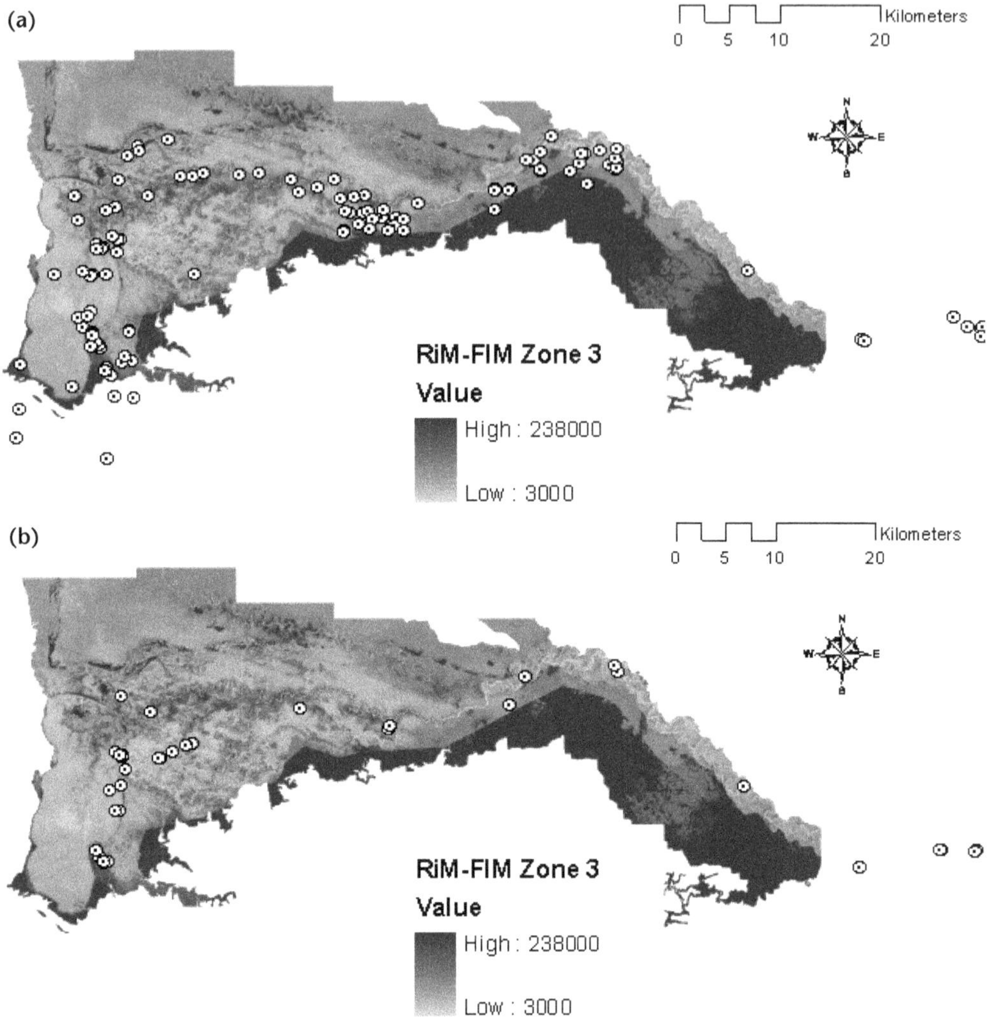

Figure 19.2: Distributional representation of species in different functional groups in Barmah-Millewa Forest (natives and exotics) overlain on CTF values of the Murray floodplain in GIS. (a) Terrestrial dry. (b) Amphibious fluctuation responders (floating leaves).

seasonality of plant growth and variability of sampling effort over a long period. Absence of a weed may result from lack of efficient dispersal, competitive exclusion or lack of comprehensive survey data. Unfortunately, point-location data cannot discriminate between these potential causes.

The current distribution of weeds at regional scales is a product of a series of historical events, including dispersal processes and land-management practices. While the RiM-FIM model is a useful tool for mapping distributional data of exotic species at the scale of the entire River Murray, the CTF value is a poor predictor of weed distribution at that scale, due to lack of a quantitative hydrological parameter relevant to plant life-cycles (e.g. flood depth and duration required for germination). Other hydrological gradients may be important in structuring species distribution boundaries, such as the depth, duration or frequency of flooding, which require further modelling.

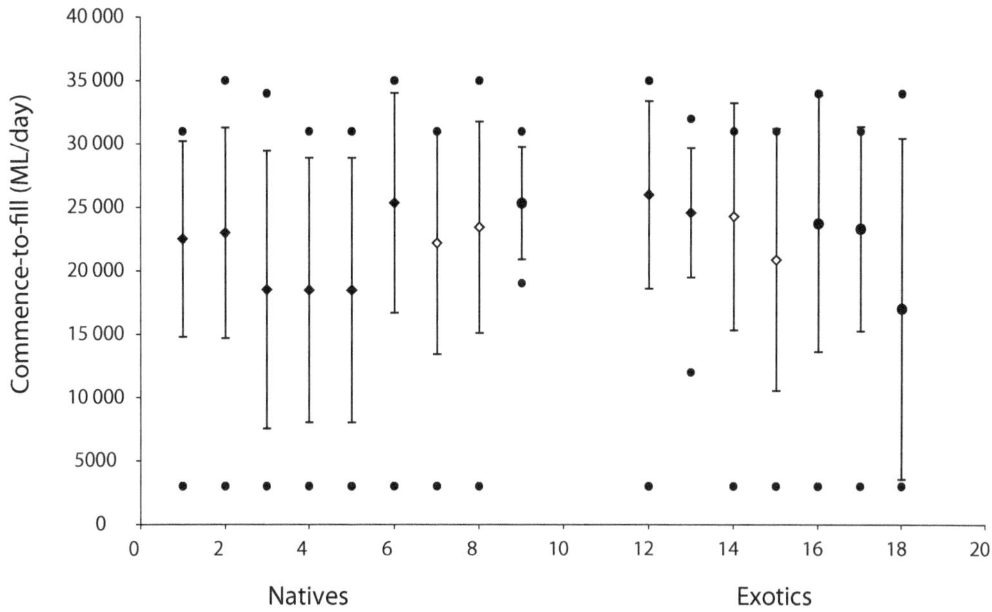

Figure 19.3: Mean (◆), 95% confidence intervals, maximum (•) and minimum (•) CTF values for exotic and native species in different functional groups (♦ = terrestrial dry, ◇ = terrestrial damp, • = amphibious fluctuation tolerators) for Zone 9.

To model trends in the diversity and abundance of exotic species over time, the small-scale data set for Barmah Forest is more effective because it contains information on species abundance as well as geographical location.

Distribution of exotic plant species over time related to flooding characteristics at Barmah Forest

The data set of Ward (2007) was analysed using ordination (ter Braak and Smilauer 2002), a technique which orders objects that are characterised by multiple values (e.g. plot species composition) so that similar objects are near each other and dissimilar objects are further from each other. Changes in vegetative community composition at Barmah-Millewa Forest were analysed over the study period. Overall variation of all plots within and between monitoring dates was assessed using Detrended Correspondence Analysis (CANOCO Package, ter Braak and Smilauer 2002), which identifies the main environmental gradients structuring community composition. Results indicate that plant communities at Barmah Forest have altered in similarity between the two observation periods (Fig. 19.4a). In spring 2006/07, segregation of vegetative communities occurred along a gradient of water availability. Drier communities were associated with lower axis 1 scores and wetter communities were represented by higher axis 1 scores, correlated with increasing open water. High cover values of exotics are associated with lower axis 1 scores and occur mainly at drier sites (Fig. 19.4b). Axis 1 clearly relates to a flooding gradient and the abundance of amphibious species is correlated with high axis 1 scores. Terrestrial dry species are correlated with low axis 1 scores.

Changes in the proportions of plant functional groups at Barmah Forest were also examined over time. Results indicate that exotic species occur most commonly in functional groups characterised by species preferring drier conditions. The proportion of exotic species in these

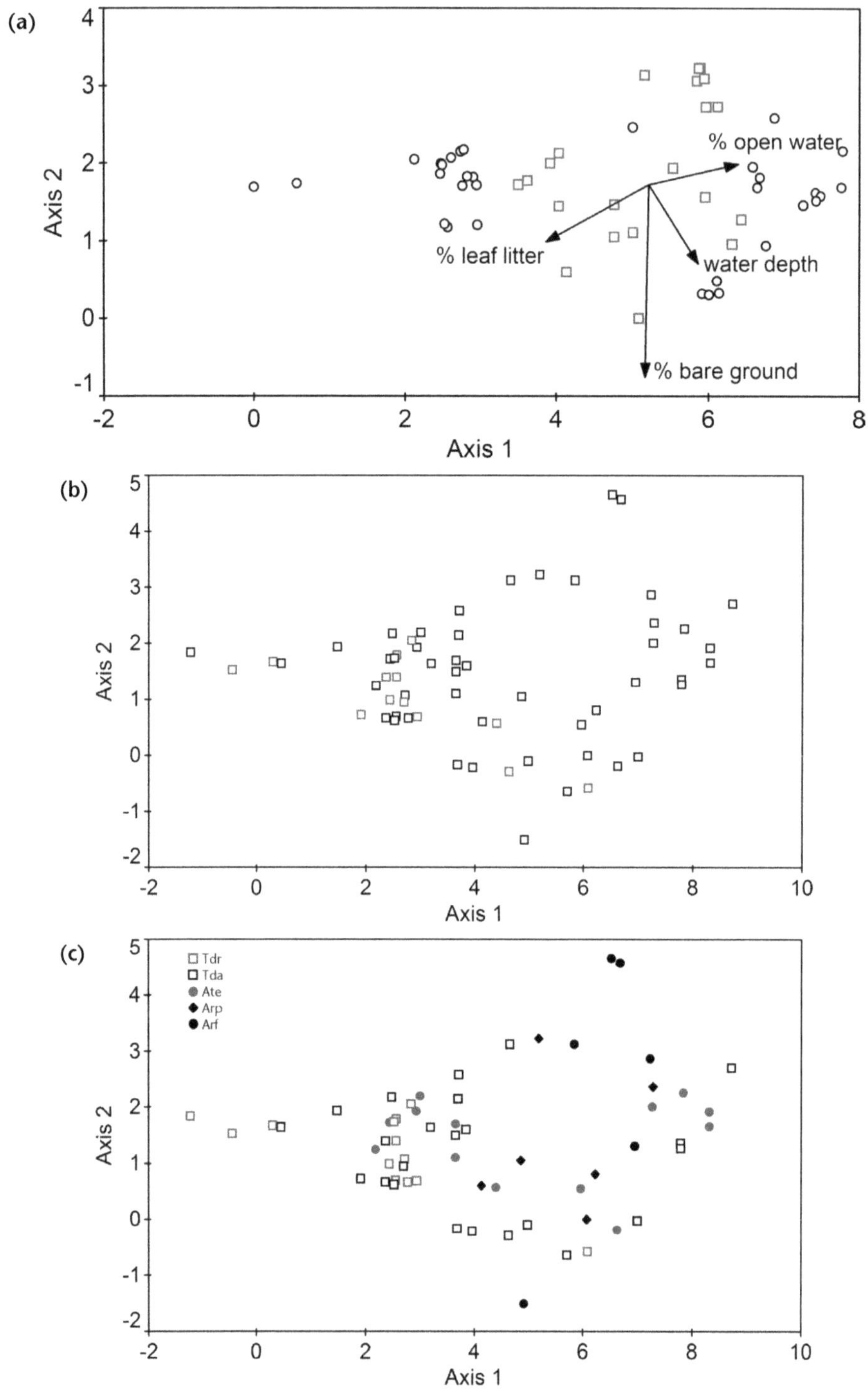

Figure 19.4: (a) Quadrats plotted by their scores on the first two axes from the DCA analysis during spring in Barmah-Millewa Forest. Species abundance data are average mid-point cover estimates of flora per 20 x 20 m quadrat. Quadrats coloured black represent site data collected in 1993/94 and those in grey represent data collected in 2006/07. (b) In the species ordination exotics are coloured grey and natives are coloured black. (c) Species ordination classified by functional group.

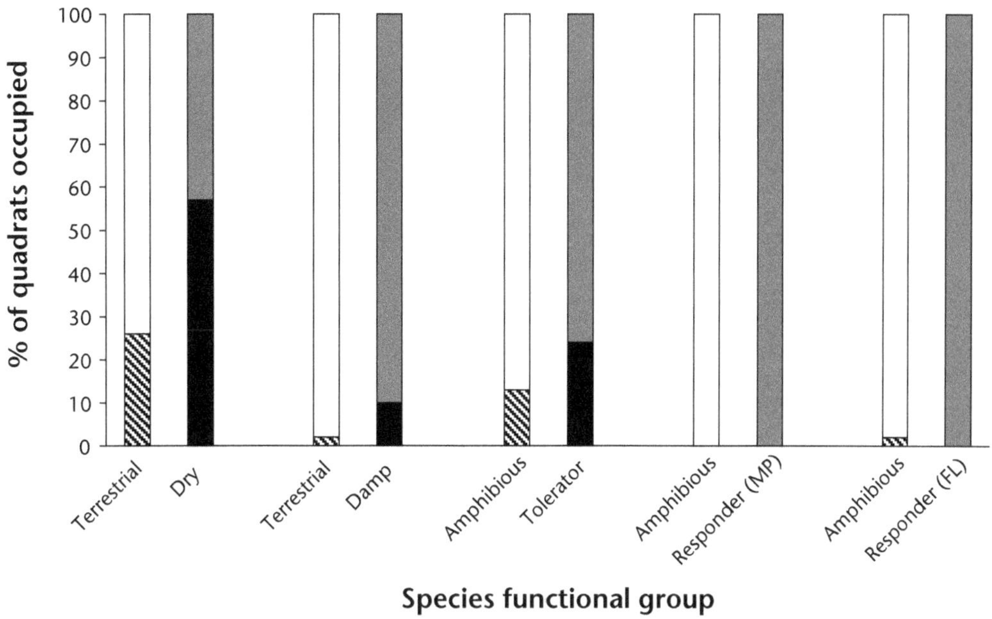

Figure 19.5: Mean changes in abundance (proportional number of quadrats occupied) for five plant functional groups in Barmah Forest between 1993/94 (striped/white) and 2006/07 (black/grey). Exotic species are coloured black/striped and native species are grey/white.

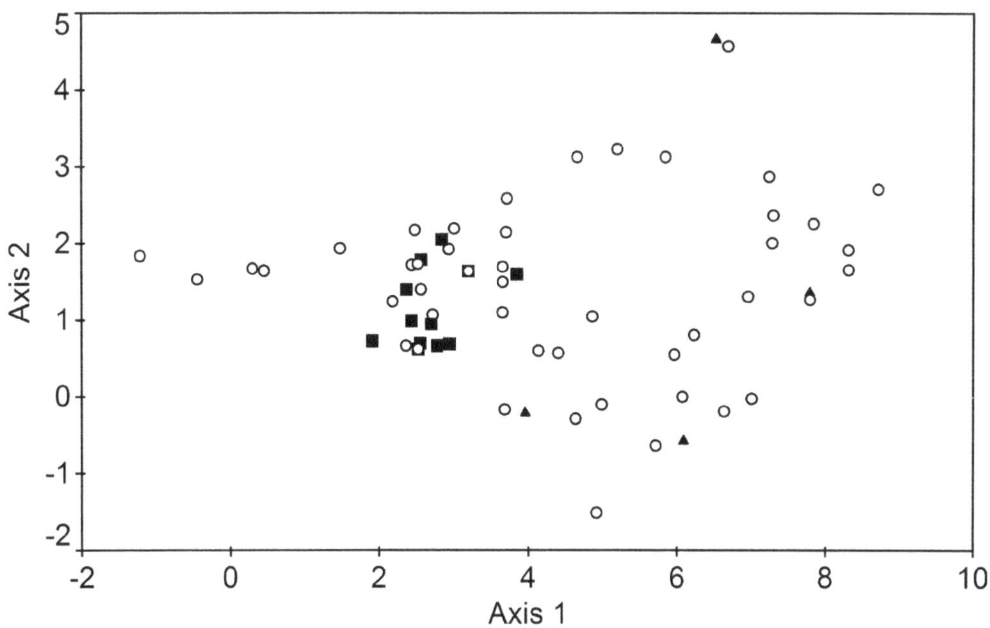

Figure 19.6: Species ordination during spring at Barmah Forest using data from 1993/94 and 2006/07 with species classified as annual (square), perennial (circle) and facultatively perennial (triangle).

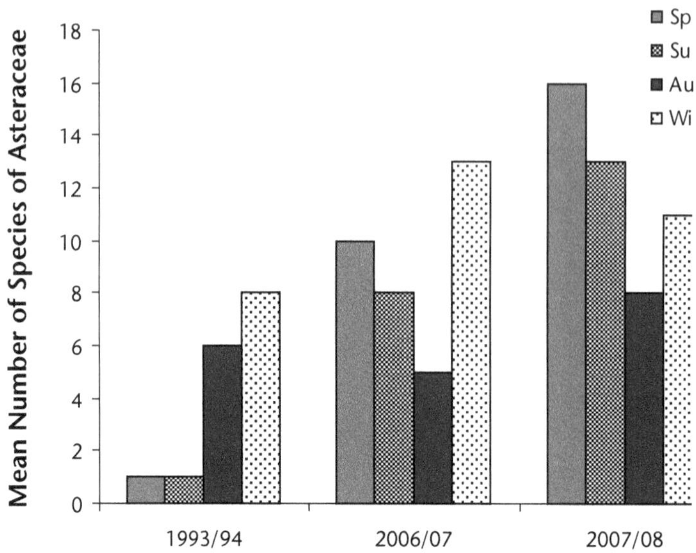

Figure 19.7: Increasing diversity of species of Asteraceae at Barmah Forest from 1993 to 2008.

groups increased between 1993/94 and 2006/07 (Fig. 19.5). The proportion of exotic species classified as hydrological responders decreased between the two monitoring dates.

Species occurring in spring were categorised as annual or perennial, revealing a higher number of annual species that were exotics (Fig. 19.6, cf Fig. 19.4b). Further classification to family level showed that the Asteraceae were the largest proportional representation of annual exotics. The number of recorded species belonging to this family were counted at each of the survey dates within each season, indicating a clear increase in the Asteraceae over time (Fig. 19.7).

Overall patterns in native and exotic species richness were examined to determine whether environmental conditions or biotic resistance determine weed infestation. In Barmah Forest, native species richness is positively correlated with exotic species richness, suggesting that at this scale invasion processes are less regulated by competition with native plants than by periodic inundation events.

Conclusions

Results demonstrate a strong trend towards the development of a drier flora over time. The main management implication is that attempting to control invasive plants at site scale is likely to be most productive if environmental water allocations are increased, since fewer exotic invaders occur in the amphibious functional groups. In general there has been an increase in the number of exotic annuals within the floodplain, which usually adopt herbaceous growth forms and tend to grow rapidly, many completing their life-cycle between disturbance events and producing a substantial long-lived seedbank. To summarise, our data indicate that the risk of plant invasions on floodplains following flood events is such that natives and exotics have an equal opportunity to recruit but the risk of invasion by terrestrial dry species is higher if the interflood period is large. The converse may also be true, as demonstrated by a recent study (Alexander *et al.* 2008) which conducted environmental watering on black box depressions, resulting in rapid sequential declines of terrestrial invasive species following flooding.

The different representation of exotic and native functional groups currently occurring within River Murray floodplains has future consequences for the supply of plant reproductive units and ultimately the regeneration of communities. Long-distance dispersal of buoyant and non-buoyant seed and vegetative fragments is known to occur during high-flow events (Boedeltje *et al.* 2004), leading to potential increases in exotic species abundance at sites of high hydrological connectivity. Incorporating spatial quantification of flood frequency into the modelling process could increase information at data-sparse sites with limited information on exotic species composition due to infrequent surveying. Additionally, incorporating information on weed population demography, or fluctuations in the size of populations over time, could help identify which weed populations are increasing and decreasing within the landscape in response to fluctuations in hydrological parameters.

Ultimately, alterations in community composition derive from land-use legacy and the challenge is to predict future species assemblages in a system where water resources are insufficient to recover natural flood regimes. The functional groupings of species presented here provide an indicator of underlying flood regimes which may be used to formulate spatially targeted management strategies to buffer floodplain vegetation communities from weed invasion during drought periods. The species classification techniques used would be transferable to the northern Murray-Darling Basin, despite species composition of exotic and native communities in northern catchments differing from those in the south. RiM-FIM proved an essential tool for mapping distributional data of exotic species against hydrological variables, and a similar model would be required to address the problem in northern regions of the basin. The construction of more accurate ecological response models for invasive floodplain weeds would require mapping the current ranges of exotic species against a broader range of hydrological parameters, such as the depth, duration or frequency of flooding. This would provide a valuable perspective on ameliorating the impacts of declining water availability in floodplains vulnerable to fluctuations in water availability.

References

Alexander P, Nielsen DL and Nias D (2008). Response of wetland plant communities to inundation within floodplain landscapes. *Ecological Management and Restoration* **9**, 187–195.

Bickford S, Penton D and West J (2006). The River Murray plant species database. Available on CD. CSIRO Plant Industry.

Boedeltje G, Bakker JP, Ten Brinke A, Van Groenendal JM and Soesbergen M (2004). Dispersal phenology of hydrochorous plants in relation to discharge, seed release time and buoyancy of seeds: the flood pulse concept supported. *Journal of Ecology* **92**, 786–796.

Boutin C and Keddy PA (1993). A functional classification of wetland plants. *Journal of Vegetation Science* **4**, 591–600.

Bren LJ (1992). Tree invasion of an intermittent wetland in relation to changes in the flooding frequency of the River Murray, Australia. *Australian Journal of Ecology* **17**, 395–408.

Brock MA and Casanova MT (1997). Plant life at the edges of wetlands: ecological responses to wetting and drying patterns. In *Frontiers in Ecology: Building the Links.* (Eds N Klomp and I Lunt) pp. 181–192. Elsevier, Oxford.

CSIRO (2008). Water availability in the Murray. Report to the Australian Government from the CSIRO Murray-Darling Basin Sustainable Yields Project. CSIRO, Australia.

Fridley JD, Stachowicz JJ, Naeem S, Seabloom EW, Smith MD, Stohlgren TJ, Tilman D and Von Holle B (2007). The invasion paradox: reconciling pattern and process in species invasions. *Ecology* **88**, 3–17.

Grigulis K, Lavorel S, Davies ID, Dossantos A, Lloret F and Vila M (2005). Landscape scale positive feedbacks between fire and expansion of the large tussock grass, *Ampelodesmos mauritanica* in Catalan shrublands. *Global Change Biology* **11**, 1042–1053.

Keddy PA (1984). Plant zonation on lakeshores in Nova Scotia: a test of the resource specialization hypothesis. *Journal of Ecology* **72**, 797–807.

Keith DA, Holman L, Rodoreda S, Lemmon J and Bedward M (2007). Plant functional types can predict decade-scale changes in fire-prone vegetation. *Journal of Ecology* **95**, 1324–1337.

Kingsford RT (2000). Ecological impacts of dams, water diversions and river management on floodplain wetlands in Australia. *Austral Ecology* **25**, 109–127.

Lytle DA and Poff NL (2004). Adaptation to natural flow regimes. *Trends in Ecology and Evolution* **19**, 94–100.

McIntyre S, Lavorel S and Tremont RM (1995). Plant life-history attributes: their relationship to disturbance response in herbaceous vegetation. *Journal of Ecology* **83**, 31–44.

Nilsson C and Svedmark M (2002). Basic principles and ecological consequences of changing water regimes: riparian plant communities. *Environmental Management* **30**, 468–480.

Overton IC, McEwan K and Sherrah J (2006). The River Murray floodplain inundation model: Hume Dam to lower lakes. Technical report 2006. CSIRO Water for a Healthy Country, Canberra.

Poff NL, Allan JD, Bain MB, Karr JR, Prestegaard KL, Richter BD and Stromberg JC (1997). The natural flow regime: a paradigm for river conservation and restoration. *BioScience* **47**, 769–784.

Pysek P and Richardson DM (2007) Traits associated with invasiveness in alien plants: where do we stand? In *Biological Invasions, Ecological Studies.* (Ed. W Nentwig) pp. 98–111. Springer-Verlag, Berlin.

Richardson DM, Holmes PM, Esler KJ, Galatowitsch SM, Stromberg JC, Kirkman SP, Pysek P and Hobbs RJ (2007). Riparian vegetation: degradation, alien plant invasions and restoration prospects. *Diversity and Distributions* **13**, 126–139.

Silvertown JW, Dodd ME, Gowing DJG and Mountford JO (1999). Hydrologically defined niches reveal a basis for species richness in plant communities. *Nature* **400**, 61–63.

ter Braak CJF and Šmilauer P (2002). CANOCO reference manual and CanoDraw for Windows user's guide: software for canonical community ordination (Version 4.5). Microcomputer Power, Ithaca, New York.

Tickner DP, Angold PG, Gurnell AM and Mountford JO (2001). Riparian plant invasions: hydrogeomorphological control and ecological impacts. *Progress in Physical Geography* **25**, 22–52.

van Eck WHJM, van de Steeg HM, Blom CWPM and de Kroon H (2004). Is tolerance to summer flooding correlated with distribution patterns in river floodplains? A comparative study of 20 terrestrial grassland species. *Oikos* **107**, 393–405.

Walker KF and Thoms MC (1993) Environmental effects of flow regulation on the Lower River Murray, Australia. *Regulated Rivers: Research and Management* **8**, 103–119.

Ward PA (2007). Monitoring understorey vegetation response to flooding in Barmah Forest: 2006/7. Final report. Consultant report prepared as part of the Living Murray project MD815 for the Murray-Darling Basin Commission, Canberra.

Weiher E, van der Werf A, Thompson K, Roderick M, Garnier E and Eriksson O (1999). Challenging Theophrastus: a common core list of plant traits for functional ecology. *Journal of Vegetation Science* **10**, 609–620.

Westoby M (1998). A leaf-height-seed (LHS) plant ecology strategy scheme. *Plant and Soil* **199**, 213–227.

Chapter 20

Ecosystem response modelling for the Barmah-Millewa Forest: the interface between science and management

Keith Ward and Matthew Colloff

Introduction

The basis for natural resource management and the application of environmental water allocations require a sound understanding of hydro-ecological processes. These include geomorphological events such as erosion and sedimentation, water chemistry including blackwater events and factors associated with changes in salinity, pH, dissolved oxygen and organic matter, as well as ecological responses of the biotic community and the interactions of its components. This environmental knowledge base interfaces with the constraints and challenges imposed by environmental policy objectives at a variety of jurisdictional scales, resources available to address those objectives and the omnipresent politics of the day. Management strategies at various tiers of government have a very broad range of aims and objectives, which are often competing.

Application of natural resource management is usually undertaken with an inadequate knowledge and understanding of the hydro-ecological processes being managed, although research and monitoring projects are often targeted and commissioned in an attempt to fill such voids in knowledge. In recent years, there have been concerted efforts by various agencies to galvanise this information into models to test impacts of hydrological change on components of ecosystems that are usually represented by a number of focal species. This chapter presents a brief history of the strategies, research and models, and their natural resource management applications at the Barmah-Millewa Forest.

The science–management interface represents a balance between systems knowledge and its implementation. In reality, it is much more intricate, though a detailed treatment is beyond the scope of this chapter. Many aspects have been explored by Siebentritt (2007), focusing on knowledge exchange and decision-making regarding environmental water applications to wetlands. In essence, the interaction between natural resource scientists and managers is a relationship based on the opportunity for mutual learning between people with very different aims, motivations and drivers but with a shared sense of purpose – the conservation of and improvement in ecological health of our major environmental assets. The science–management interface works well where the common purpose can be realised without compromising individual goals, and where those goals are mutually understood. It works less well where scientists assume their research products will be of value to managers just because the science is good, or where managers assume that scientists can and will address any question, regardless of its complexity (Cullen 1990). One of the purposes of this chapter, therefore, is to explore the characteristics of good-quality science which is useful to natural resource managers.

Historical perspective

The history of managing flood events in the Barmah-Millewa Forest is that drastic changes in perceptions of the nature, value and impact of floods have driven the management responses. Changes in environmental drivers, historical effects of management events and changes in societal values and the importance of certain forest extractive uses,have had a major influence on these perceptions.

Floods were originally regarded as nuisance flows because they had adverse impacts on important economic forest activities such as logging, grazing and apiary, as well as on recreational pursuits. The first management response to flooding on the Murray was river regulation and the construction of the Hume Dam, commenced in 1919 (completed in 1936) following the extensive floods of 1917, the first major inundation event following pioneer settlement (see Bren *et al.* 1988 for map of flood extent).

The expansion of horticulture along the Murray after World War II led to rising demand for irrigation water. Management efforts became focused on mitigating the effects of unseasonal flooding, mostly caused by rainfall-rejection events whereby orders of summer irrigation water were cancelled following rainfall, leading to overbank floods (Chong and Ladson 2003).

Environmental flow allocations were first introduced on the Murray and northern Victorian wetlands in 1979 with the Murray Flora and Fauna Bulk Entitlement to Victoria of 27.6 GL, following completion of the Dartmouth Dam. By 1989 there was increased impetus from the Murray-Darling Basin Commission for flood management to achieve requirements for the health of native vegetation, resulting in The Living Murray program in 2002. In 1993 the Murray-Darling Basin Ministerial Council authorised an annual environmental water entitlement to Barmah-Millewa Forest of 100 GL (MDBC 2000). The first author joined the Victorian Department of Conservation and Environment in 1990 to undertake hydro-ecological research and develop the first water management strategy for Barmah Forest. Environmental water allocations have been used to enhance natural flows at Barmah-Millewa Forest in 1998 and 2000/01 (Maunsell McIntyre 1999, 2001, summarised by Abel *et al.* 2006), 2002 (Abuzar and Ward 2003; Ward *et al.* 2002) and 2005/06 (Abuzar and Ward 2006; Ward and O'Connor 2006).

The forest last received flooding in 2005/06 (the same year the asset management plan for Barmah-Millewa was published; MDBC 2005). The Victorian Government recently implemented its environmental watering program, focusing on survival of key wetland species and communities (DSE 2009). Managers have thus moved from managing water on wetlands and floodplains to attempting to manage undesirable ecological responses to the lack of flows.

Models of flow–inundation area relationships

The hydrograph forms the basic data set for management of flows and for development of hydrological models. This may seem obvious, but there were almost 100 years between the commencement of data collection from gauging stations above Barmah-Millewa Forest and the development of the first models relating volume of flow to area inundated. The flow–inundation area relationship for Barmah Forest (but not Millewa) of Bren and Gibbs (1986) and Bren *et al.* (1988) formed the basis for water management for some years. The model was derived from an analysis of 34 maps compiled between 1963 and 1984 by foresters, who scored

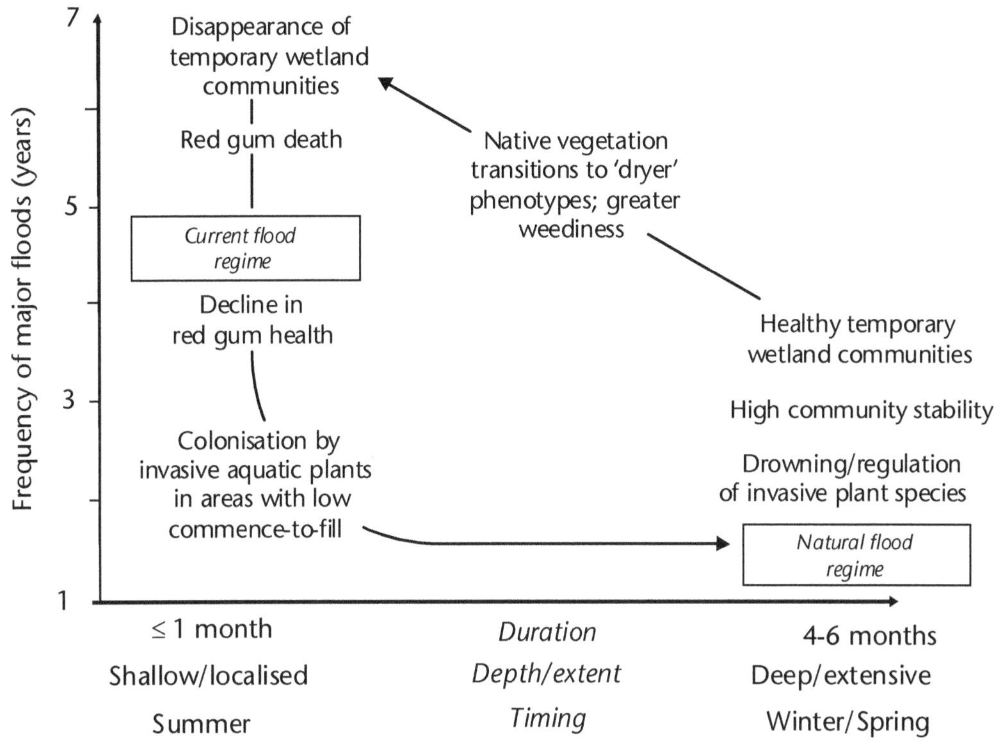

Figure 20.1: Conceptual model of the effects of alteration of flood regimes on responses of native vegetation in Barmah Forest.

flood occurrence on a grid of over 1200 0.5 × 0.5 km cells (see map by Bren 1991). Flood frequency indices were generated for each cell and the resulting map overlain with the vegetation map of Chesterfield (1986) to generate relationships between flood frequency and vegetation class. Finally, a predictive equation of the relationship between area inundated and maximum daily flow at Tocumwal was developed (see Bren *et al.* 1987 for details). The model is reasonably reliable for floods with low volumes – at 17 000 ML d^{-1}, the model predicted 28% inundation versus 25% actual (Maunsell McIntyre 1999) – but is less accurate for higher flows (68 500 ML d^{-1} predicted 96% inundation whereas a 35 000 ML d^{-1} event in 2005/06 flooded 57% of Barmah Forest). At 92 200 ML d^{-1} the forest is effectively fully flooded, so the consistency of 94% predicted versus 85% actual is unremarkable.

Overton (2005) and Overton *et al.* (2006) used satellite imagery to collate the flood history of Barmah-Millewa Forest as part of the River Murray Flood Inundation Model (RiM-FIM). Using image interpolation, the flood extent was derived from every 1000 ML d^{-1} flow at Tocumwal. In this model, for which simulations can be run from a desktop computer, the relationship between area inundated and peak discharge (Tocumwal) for the Barmah-Millewa Forest is much higher than that of Bren (1988) for Barmah Forest alone, partly because of different flood history data sets used for each model and the changes in flood patterns and extent due to regulators and barriers installed since 1988. Bren's relationship is derived from the maximum extent of inundation following a natural flood event, whereas RiM-FIM is the initial flood extent (effectively commence-to-fill).

Subsequently, Water Technology (2005) used LiDAR digital elevation data to construct a hydrodynamic model of flooding for the Barmah-Millewa Forest using MIKE FLOOD software. This model is better able to predict the effects of regulators, levee banks and channels under different configurations, including the addition of new ones or removal of old ones. The model was calibrated against the flood maps produced by Abuzar and Ward (2003). The most recent iteration of the MIKE FLOOD model includes a refined LiDAR digital elevation model, and overlays of vegetation communities on maps of predicted flood extent (Fig. 20.2/colour plate, p. 318). It requires quite high computing capacity (though can be run on a desktop) and specialist expertise to run scenarios.

Linking flood models to vegetation and ecology

From the commencement of attempts to model flood extent derived from river flows, the underlying driver has been to relate flood extent to frequency of inundation and thence to distinct vegetation communities. Every plant has its hydrological niche, and every plant community represents a cluster of species with shared overlapping water requirements. It follows that when those requirements change, so do the plant communities. This is precisely what has been happening for some time at Barmah-Millewa Forest, with invasion of giant rush and red gums and shrinkage of the area of Moira grass plains. Vegetation transitions were first documented by Chesterfield (1986) then modelled by Bren (1992). A conceptual model of ecological responses of vegetation communities to changes in flooding regimes is shown in Figure 20.1.

Vegetation communities tend to form the basic units of management (Tables 20.1 and 20.2), partly because they tend to be distinctive and definable, reflecting the envelope of environmental variables that prevail within the area they occupy, but also because they represent the structural elements within the landscape that provide habitat for other organisms, fulfil certain ecosystem functions and have attributes relating to their relative occurrence (commonness or rarity) that deem them worthy of protection and conservation.

Mapping of vegetation communities within Barmah Forest derived four structural classes (rushland, grassland, open forest and woodland), containing one dominant community each in the first two classes (giant rush and Moira grass), seven understorey communities, characterised by dominant grasses or sedges, in open forest and two in woodland (Chesterfield1986). Detailed mapping by Frood resulted in a much higher diversity of understorey communities and structural classes (Frood and Ward 2001). The final vegetation map contains a higher number of categories than optimal for practical use in redelineating water management areas. Consequently, it has been used as the basis for integration into 20 Victorian ecological vegetation community classes. These maps have been used within the latest version of the MIKE FLOOD model produced by Water Technology (Fig. 20.2.)

Abuzar and Ward (2003, 2006) used satellite imagery to map the inundation extent from the October–December 2002 and October–November 2005 flood periods. They mapped vegetation growth using a normalised difference vegetation index. The distribution of surface flooding and fresh vegetation growth was mapped for three periods between October and December 2002 and for two periods between October and November 2005. The area of most actively growing vegetation corresponded well with areas that had been flooded. This plant functional growth response approach using remote sensing shows considerable promise for long-term monitoring of ecological effects of flooding. A similar approach, at larger spatial scale, was used by Cunningham *et al.* (2008) for assessment of red gum health along the River Murray valley.

The Murray Flow Assessment Tool (MFAT) is a model for predicting the effects of different flow scenarios of the River Murray system on the physical condition of habitat for native fish, waterbirds and vegetation communities of floodplains and wetlands, as well as for a threatening process – algal growth (Young *et al.* 2003; MDBC 2004). The ecological components of the

Table 20.1: Summary of knowledge on environmental water requirements, Barmah-Millewa Forest

Watering target	Discharge required (ML d^{-1})	Duration (months)	Frequency of event	Max. interval (yr)	Area inundated (000 ha)
Low-lying wetlands	11 000	1	9:10	1	1
Wetlands and Moira grass plains; trigger for bird breeding	18 000	5	8:10	2	7–10
Wetlands and red gum forest	30 000	3	5:10	3	17
Red gum woodland	60 000	1	3:10	3	24

Timing of diversion from Murray River: Aug.–Nov. (+/-1 mo.). Volume returned = c. 95%.
a: 50% bankfull (BF).
b: all BF.

model are based on preference curves, which are indexed ecological responses over time of particular biotic groups to flood variables (timing, duration of inundation, depth and its variability, duration of interflood periods). The ecological models have been parameterised based on expert opinion (Ward 1991, 1994; Roberts and Marston 2000). The vegetation communities relevant to Barmah-Millewa Forest that are contained in MFAT include river red gum (*Eucalyptus camaldulensis*) forest and woodland, black box (*E. largiflorens*) woodland, rushlands dominated by cumbungi (*Typha* spp.), common reed (*Phragmites australis*) and giant rush (*Juncus ingens*) as well as Moira grass (*Pseudoraphis spinescens*) plains.

Water management areas

Water management areas, or units, represent the integration of vegetation classes and flood characteristics. Some 11 areas were delineated by Maunsell (1992) and refined by Ward *et al.* (1994), who identified management objectives for each. Definition and refinement of water management areas represents an exercise in adaptive management, including trade-offs regarding conflicts of water allocation, prioritisation of management objectives and prediction of the likelihood of achieving ecological objectives.

A better alignment between vegetation communities and water management areas and more precise temporal and spatial allocation of water to vegetation communities were recommended by Abel *et al.* (2006). These authors considered that the need to make trade-offs within the forest and allocate scarce water effectively required water management units to be managed autonomously to some extent. Harvey (2006) attempted a classification of vegetation to allow delineation of water management units by allocating the species of Frood's map units to a hydrological functional group classification (Brock and Casanova 1997; Casanova and Brock 2000). The resulting functional community map is quite complex, and was not interfaced with hydrological models. This approach would have allowed refinement of functional units based on flooding.

Vegetation mapping using Victorian ecological vegetation classes is now precise enough for defining water management areas. However, the vegetation communities at Barmah-Millewa are characterised by differential rates of change and transition, caused by invasion of giant rush and red gums onto Moira grass plains due to underlying changes in plant hydrological niches. Thus there is a chicken-and-egg paradox. To improve the management of vegetation communities using environmental flows, the requirement is for better alignment of vegetation communities with water management areas, yet the lack of environmental flows is the underlying driver of vegetation transition and change. The pressing issue for managers is not so much the definition and delineation of water management areas, as greater certainty of

Table 20.2: Vegetation structural types of Chesterfield (1986) ordered according to flood frequency (modified after Bren 2005)

Vegetation structural type	Estimated % of area occupied (1986)	Mean flood frequency (22 yr data)	Flood frequency class
Giant rush	1.8	1.26	Very high
Moira grass plains and red gum with Moira grass understorey	23.4	1.33	High
Red gum forest with sedge, wallaby or warrego grass	57.7	1.57	Moderate
Red gum with spike rush, and black box woodland	17.1	2.75	Low

whether future environmental water allocations will be available in the face of climate change, water-sharing and trading rules, and the complexity of priorities and demands for water management in the Murray-Darling Basin. These uncertainties are partly environmental but are also driven by political, economic and institutional factors, largely beyond the control of managers at the asset site scale. The Commonwealth Water Act of 2007 required the development of an environmental watering plan for the Murray-Darling Basin, which is in progress.

Modelling threatening processes

A statistical ecophysiological model of blackwater events has been applied to Barmah-Millewa and other wetlands downstream (Howitt *et al.* 2007). Warm months with large accumulations of leaf litter and low flows are triggered as risk and can signal the need for increased volumes of water with flushing flows.

MFAT contains an algal growth model (Young *et al.* 2003; MDBC 2004) which enables simulation of blue-green algal population increase in weir pools under different flow scenarios, water quality and weather conditions. The output is an index relating to algal toxin risk.

Effects of flow regime on fauna, including rare and threatened species

There is relatively little information on how flow regime affects populations of the rare and threatened species listed in Appendices 4 and 5 of the Barmah Forest Ramsar plan (DSE 2003). The first step in improving this is to find the species in the field, and build up a knowledge base of their ecology and flow requirements from repeated monitoring, systematic observations and accurate recording of locality data. This information could be included in a floodplain inundation model.

Data on birds in Table 20.3 has been generalised from expert knowledge in MFAT (Young *et al.* 2003). Data from wetlands within Barmah-Millewa Forest were used for MFAT modelling. These sites ought to be used for subsequent monitoring, so that the model can be tested and improved. The models need to be made spatially explicit for the management of Barmah-Millewa Forest.

Long-term monitoring at Barmah-Millewa Forest

The basis of determining long-term monitoring of ecological responses to floods rests with standardised, repeated-measures sampling designs for monitoring. Long-term data sets include those for understorey vegetation at permanent transects – monitored since 1991 (Ward 1991, 1994, 2007, 2008, 2009; Ch. 19, this volume), as well as data on waterbirds (Table 20.3) and frogs (Ward 2001, 2002, 2003, 2004, 2006).

Long-term vegetation transitions, originally modelled by Bren (1992) from aerial photographs, are in the process of being remodelled and updated with new data. Giant rush monitoring will be included with the new model (Maxwell in prep.). Future monitoring and research projects will help validate existing and new models.

Knowledge requirements for better management

Water management is based on correlations between patterns of inundation and the distribution and observed responses of species and ecological communities. As modelling has started to provide a better systems understanding, a stronger adaptive management program can be pursued, including experimental manipulations. But there remain some significant knowledge gaps.

Some gaps relate to improving a systems understanding of flooding, including the modelling of timing, depth, duration and frequency, rates of change and duration of dry periods. There is a need to complete a detailed hydrological analysis of the Barmah-Millewa floodplain, including characterisation of natural, historical, current and future flow regimes.

Some gaps relate to physical and operational issues of delivering water to specific points and at particular times. These include a better understanding of the role of distributary channels through Barmah-Millewa to maintain floodplain functions, as well as the identification of opportunities and methods for flooding higher terrain using embankments, pumping and different regulator combinations. There is also a need for better river flow–floodplain inundation correlations, including how the relationships vary with the operation of different regulators and the effects of antecedent conditions. Water loss to the floodplain (or return volumes to the river) also requires quantification, for environmental water allocation accounting purposes.

Other knowledge gaps relate to ecological responses. These include the identification of differences in biotic response to cooler-season flooding versus warmer-season flooding. An important knowledge need is better support for management of trade-offs between water allocations to areas of relatively low conservation significance versus sites of high ecological value. This necessitates the inclusion of scenarios of rotational watering, as well as clarifying the trigger points for management intervention. The largest deficit remains a clear understanding of how the responses of one component of the biota interact with another. The interaction of species and communities is one of the defining characteristics of ecosystems, because all species are linked through processes such as feeding and predation, competition, mutualism, nutrient cycling, resource use and habitat preference. Almost all ecological response modelling focuses only on the prediction of growth, or population increase, or survival, or recruitment, of single species or groups. This is not how ecosystems work. Such an approach is likely to provide only a limited predictive framework.

Conclusions

Water management is largely based on the natural paradigm (the assumption that patterns of vegetation communities are direct responses to the natural flow regime, and that any change towards a more natural flow regime will in the long term result in a more natural vegetation pattern). This is unsustainable in a regulated river system, especially with such largely modified flow regimes and the manifestation of drier conditions from climate change. Re-evaluation of current ecological objectives is warranted. The major driver for systems understanding at

Table 20.3: Publications relating to waterbird ecological objectives and required hydrological conditions

Publication	Area	Evidence	Value to managers
Crome (1986)	Murray-Darling Basin	*Observations*. Observational design with measurement of environmental variables.	*Moderate*
Briggs and Maher (1985), Briggs and Thornton (1999), Briggs *et al.* (1985, 1997)	Mid Murrumbidgee	*Observations*. Observational design with measurement of flood variables (duration, depth, extent).	*High*: targets wetland managers
Leslie (2001)	Barmah-Millewa Forest	*Scientific commentary*: Study of hydrology–waterbird breeding relationships. Calibrated modelling.	*High*: adaptive management in action
Young *et al.* (2003)	Murray River	*Conceptual models*. MFAT: synthesis of waterbird habitat requirements into ecological response models.	*Moderate*: not spatially explicit at asset scale
O'Connor and Ward (2003)	Barmah Forest	*Observations*. Waterbird breeding response to environmental water allocation.	*High*: adaptive management in action
Webster (2004)	Barmah Forest	*Observations*. Waterbird breeding response to annual variations in flow.	*Moderate*

Modified after Abel *et al.* (2006).

Barmah-Millewa has been adaptive management – learning by doing. Lessons are learnt from water management at other sites as well as within Barmah-Millewa Forest. One lesson is that it is easier to protect and maintain an ecological asset than attempt to restore or replace it once it is damaged or gone.

Natural resource managers are highly aware of competing and conflicting issues. Community expectations vary and community stakeholders are often vocal, requiring responses varying from information provision via documentation and consultation to active engagement in management. Strong research and monitoring data are paramount not only for supporting natural resource management directions and activities, but also for educating an often strongly opinionated community which otherwise bases its information on long-held beliefs, on large-scale extrapolation from small-scale observations or on hearsay.

We flagged the issue of the characteristics of science that is useful to natural resource managers. One feature, exemplified by the work of Bren (1992) and Chesterfield (1986), is that the precision of model predictions may at times be less important than their capacity to provide a systems-based understanding. The value to managers of this work was not so much the ability to predict accurately how many hectares of forest would be flooded, but in relating flood frequency to the occurrence of key vegetation communities. This provided a basis for collation of data on water requirements for these communities (Table 20.1). Each row in the table represents an hypothesis of vegetation water requirements based on a combination of scientific prediction and observation, adaptive management and experience, as well as informed estimation. The availability of new high-precision hydrodynamic and flood inundation models has begun to replace the earlier less-precise flow–inundation–vegetation relationships which previously formed the basis of management.

As a result of observation, monitoring and an improved systems understanding, the ecological objectives and options for water delivery have been refined and adapted. The hydro-ecological management of Barmah-Millewa Forest has become more systems-based and adaptive as the knowledge base improves. Only part of this knowledge base is driven by science.

Acknowledgements

Matthew Colloff acknowledges funding support from the CSIRO Water for a Healthy Country National Research Flagship. We thank Ian Overton and Tanya Doody (CSIRO Land and Water, Adelaide) for reviewing the manuscript.

References

Abel N, Roberts J, Reid J, Overton I, O'Connell D, Harvey J and Bickford S (2006). Barmah Forest: a review of its values, management objectives, and knowledge base. Report to the Goulburn Broken Catchment Management Authority, Shepparton. CSIRO Sustainable Ecosystems, Canberra.

Abuzar M and Ward KA (2003). Flood and vegetation mapping in the Barmah-Millewa Forests during October–December 2002 using remote sensing technology. Report prepared for the Barmah-Millewa Forum. NSW Dept of Primary Industries, Tatura.

Abuzar M and Ward KA (2006). Flood and vegetation mapping in the Barmah-Millewa Forests during October–November 2006 using remote sensing technology. Report to the NSW Dept of Primary Industries, Tatura, and Goulburn Broken Catchment Management Authority, Shepparton.

Bren L (1988). Flooding characteristics of a riparian red gum forest. *Australian Forestry* **51**, 57–62.

Bren L (1991). Modelling the influence of River Murray management on the Barmah river red gum forests. *Australian Forestry* **54**, 9–15.

Bren L (1992). Tree invasion of an intermittent wetland in relation to changes in the flooding frequency of the River Murray, Australia. *Australian Journal of Ecology* **17**, 395–408.

Bren L (2005). The changing hydrology of the Barmah-Millewa Forests and its effects on vegetation. *Proceedings of the Royal Society of Victoria* **117**, 61–76.

Bren L and Gibbs N (1986). Relationships between flood frequency, vegetation and topography in a river red gum forest. *Australian Forest Research* **16**, 357–370.

Bren L, O'Neill IC and Gibbs NL (1987). Flooding in the Barmah Forest and its relation to flow in the Murray-Edward river system. *Australian Forest Research* **17**, 127–144.

Bren LJ, O'Neill IC and Gibbs NL (1988). Use of map analysis to elucidate flooding in an Australian riparian red gum forest. *Water Resources Research* **24**, 1152–1162.

Briggs SV and Maher MT (1985). Limnological studies of waterfowl habitat in south-western New South Wales. II. Aquatic macrophyte productivity. *Australian Journal of Marine and Freshwater Research* **36**, 707–715.

Briggs SV and Thornton SA (1999). Management of water regimes in river red gum *Eucalyptus camaldulensis* wetlands for waterbird breeding. *Australian Zoologist* **31**, 187–197.

Briggs SV, Maher MT and Carpenter SM (1985). Limnological studies of waterfowl habitat in south-western New South Wales. I. Water chemistry. *Australian Journal of Marine and Freshwater Research* **36**, 59–67.

Briggs SV, Thornton SA and Lawler WG (1997). Relationships between hydrological control of river red gum wetlands and waterbird breeding. *Emu* **97**, 31–42.

Brock MA and Casanova MT (1997). Plant life at the edges of wetlands: ecological responses to wetting and drying patterns. In *Frontiers in Ecology: Building the Links*. (Eds N Klomp and I Lunt) pp. 181–192. Elsevier, Oxford.

Casanova MT and Brock MA (2000). How do depth, duration and frequency of flooding influence the establishment of wetland communities? *Plant Ecology* **147**, 237–250.

Chesterfield EA (1986). Changes in the vegetation of the river red gum forest at Barmah, Victoria. *Australian Forestry* **49(1),** 4–15.

Chong J and Ladson AR (2003). Analysis and management of unseasonal flooding in the Barmah-Millewa Forest, Australia. *River Research and Applications* **19**, 161–180.

Crome FHJ (1986). Australian waterfowl do not necessarily breed on a rising water level. *Australian Wildlife Research* **13**, 461–480.

Cullen P (1990). The turbulent boundary between water science and water management. *Freshwater Biology* **24**, 201–209.

Cunningham S, MacNally R, White M, Read J, Baker PJ, Thomson J and Griffoen P (2008). Mapping the current condition of river red gum (*Eucalyptus camaldulensis* Dehnh.) stands along the Victorian Murray River floodplain. Report to the Northern Victorian Catchment Management Authorities and the Dept of Sustainability and Environment. Australian Centre for Biodiversity, School of Biological Sciences, Monash University, Melbourne.

DSE (2003). Barmah Forest Ramsar Site strategic management plan. Victorian Dept of Sustainability and Environment, Melbourne.

DSE (2009). Environmental watering in Victoria 2007/08. Victorian Dept of Sustainability and Environment, Melbourne.

Frood D and Ward K (2001). Mapping of the understorey vegetation in Barmah Forest, Victoria. Dept of Natural Resources and Environment, Tatura.

Harvey J (2006). Chapter 9 Appendix: exploring alternative ways of classifying and mapping vegetation in Barmah Forest to enable the delineation of water management units. In Barmah Forest: a review of its values, management objectives, and knowledge base. (Eds N Abel, J Roberts, J Reid, I Overton, D O'Connell, J Harvey and S Bickford) pp. 183–187. Report to the Goulburn Broken Catchment Management Authority. CSIRO Sustainable Ecosystems, Canberra.

Howitt JA, Baldwin DS, Rees GN and Williams JL (2007). Modelling blackwater: predicting water quality during flooding of lowland river forests. *Ecological Modelling* **203**, 229–242.

Leslie DJ (2001). Effect of river management on colonially nesting waterbirds in the Barmah-Millewa Forest, south-eastern Australia. *Regulated Rivers: Research and Management* **17**, 21–36.

Maxwell R (In prep.). Mapping of historic vegetation of the Barmah-Millewa Forest. Dept of Primary Industries, Tatura.

Maunsell Pty Ltd (1992). Barmah-Millewa Forests water management plan. Final report. Murray-Darling Basin Commission, Canberra.

Maunsell McIntyre (1999). Report on Barmah-Millewa flood of October 1998, and the first use of Barmah-Millewa Forest allocation. Maunsell McIntyre Pty Ltd, Sydney.

Maunsell McIntyre (2001). Report on Barmah-Millewa Forest flood of Spring 2000 and the second use of Barmah-Millewa Forest environmental water allocation, Spring–Summer 2000/2001. Maunsell McIntyre Pty Ltd, Sydney.

MDBC (2000). The Barmah-Millewa Forests water management strategy. Murray-Darling Basin Commission, Canberra.

MDBC (2004). Inside MFAT. Available at www.mdbc.gov.au/subs/information.mfat/index.htm. Accessed 10 April 2009.

MDBC (2005). The Barmah-Millewa Forest Asset Environmental Management Plan 2005/2006. MDBC Publication No. 31/05. Murray-Darling Basin Commission, Canberra.

O'Connor PG and Ward KA (2003). Waterbird monitoring in Barmah Forest, 2002–2003. Victorian Dept of Sustainability and Environment, Tatura.

Overton IC (2005). Modelling floodplain inundation on a regulated river: integrating GIS, remote sensing and hydrological models. *River Research and Applications* **21**, 991–1001.

Overton IC, McEwan K, Gabrovsek C and Sherrah J (2006). The River Murray flood inundation model – RiM-FIM. Technical report. CSIRO Land and Water, Canberra.

Roberts J and Marston F (2000). Water regime of wetland and floodplain plants in the Murray-Darling Basin: a sourcebook of ecological knowledge. Technical report 30-00. CSIRO Land and Water, Canberra.

Siebentritt M (2007). Watering wetlands: impediments and challenges to the transfer of knowledge between wetland managers and scientists. Land and Water Australia, Canberra.

Stokes KE, Ward KA, Ward PA and Colloff MJ (2010). Modelling invasive plants in relation to flooding and drying: implications for ecosystem functions. In *Ecological Response Modelling in the Murray-Darling Basin.* (Eds N Saintilan and I Overton) pp. 333–344. CSIRO Publishing, Melbourne.

Ward KA (1991). Investigation of the flood requirements of the Moira grass plains in Barmah Forest, Victoria. Unpublished report. Victorian Dept of Conservation and Environment.

Ward KA (1994). Flood requirements of the wetland flora in Barmah Forest, Victoria. Unpublished report. Victorian Dept of Conservation and Environment.

Ward PA (2001). Monitoring frog response to flooding in Barmah Forest, 2000–01. Final report to Barmah-Millewa Forum for the Murray-Darling Basin Commission, Canberra.

Ward PA (2002). Monitoring frog response to flooding in Barmah-Millewa Forest, 2001–02. Final report to the Barmah-Millewa Forum for the Murray-Darling Basin Commission, Canberra.

Ward PA (2003). Monitoring frog response to flooding in Barmah-Millewa Forest, 2002–03. Final report to the Barmah-Millewa Forum for the Murray-Darling Basin Commission, Canberra.

Ward PA (2004). Monitoring frog response to flooding in Barmah-Millewa Forest, 2003–04. Final report to the Barmah-Millewa Forum for the Murray-Darling Basin Commission, Canberra.

Ward PA (2006). Monitoring frog response to flooding in Barmah-Millewa Forest, 2005–06. Final report to the Murray-Darling Basin Commission, Canberra.

Ward PA (2007). Monitoring understorey vegetation response to flooding in Barmah Forest, 2006–07. Final report prepared as part of The Living Murray project MD815 for the Murray-Darling Basin Commission, Canberra.

Ward PA (2008). Monitoring understorey vegetation response to flooding in Barmah-Millewa Forest, 2007–08. Final report prepared as part of The Living Murray Condition Monitoring Program for the Barmah-Millewa Icon Site, managed by the Dept of Sustainability and Environment on behalf of the Murray-Darling Basin Commission, Canberra.

Ward PA (In prep.). Monitoring understorey vegetation response to flooding in Barmah-Millewa Forest, 2008–09. Final report prepared as part of The Living Murray Condition Monitoring Program for the Barmah-Millewa Icon Site, managed by the Dept of Sustainability and Environment on behalf of the Murray-Darling Basin Authority, Canberra.

Ward KA and O'Connor PG (2006). Adaptive monitoring of an environmental water allocation in the Barmah wetlands, 2005–06. Goulburn Broken Catchment Management Authority, Shepparton, and Dept of Sustainability and Environment, Tatura.

Ward KA, Leitch C, Lloyd LN and Atkins BA (1994). Interim water management strategy for Barmah Forest, Victoria. Integrated Watering Strategy Report No. 12. Floodplain Ecology Group, Victorian Dept of Conservation and Natural Resources, Shepparton.

Ward KA, O'Connor PG and Caldwell M (2002). Survey for colonially nesting waterbirds in Barmah Forest, November/December 2002. Goulburn Broken Catchment Management Authority, Shepparton, and Dept of Primary Industries, Tatura.

Water Technology (2005). Barmah-Millewa Forest hydrodynamic model. Report prepared for the Goulburn Broken Catchment Management Authority. Report No. J129/R01.

Webster (2004). The effect of flooding on waterbirds within Barmah-Millewa Forest, 1999–2003. Ecosurveys Pty Ltd, Deniliquin.

Young WJ, Scott AC, Cuddy SM and Rennie BA (2003). Murray Flow Assessment Tool: a technical description. Client report for the CRC for Freshwater Ecology. CSIRO Land and Water, Canberra.

Chapter 21

Ecosystem response modelling in the Chowilla Floodplain, Lindsay and Wallpolla Islands icon site

Ian Overton and Tanya Doody

Introduction

Native riparian vegetation communities on the lower River Murray floodplain in a semi-arid region of Australia are suffering severe health declines due to changes in the surface water and groundwater systems. It is estimated that over 60% of riparian tree vegetation in wetlands and floodplains within the Murray-Darling Basin are in poor health or dead (Cunningham *et al.* 2007). The native vegetation, predominantly eucalyptus open forest and woodlands, are dependent upon a combination of surface flooding and groundwater. Eighty years of river regulation (installation of locks and weirs), water extraction and a severe long-term drought of over seven years have reduced the available surface water. Reduced flooding and low river levels have caused a decrease in flow and lateral groundwater recharge from fresh river water. In the lower River Murray, saline groundwater from the underlying regional aquifer has risen as a result of locking, land clearing and irrigation adjacent to the river. Increased soil salinity, driven by a lack of flooding and rising saline groundwater, has caused a decline in tree health, leading to transitions of large areas of riparian vegetation to rangeland species.

Various management strategies have been investigated for their potential to improve vegetation health for the Chowilla floodplain, to maintain or improve 70% of red gums and 45% of black box to meet ecological targets of the Living Murray Initiative (MDBC 2006). Management strategies include installation of a salt interception scheme, groundwater lowering, water pumping to water key areas and installation of a large regulator to elevate river levels and provide vital surface flows to floodplain areas. In order to understand the potential ecosystem response to changes in surface water and groundwater resources induced from these management strategies, particularly the response of vegetation, a tree health response model has been developed to enable provision of scientific information to aid decision-makers. The model is based on soil water availability, which is driven by surface water inundation, freshwater recharge and soil salinisation. Salinisation is modelled from surface water–groundwater interactions using a spatial and temporal model of salt accumulation from groundwater depth, groundwater salinity, soil type and flooding frequency. The derived soil water availability index model (WINDS) is used to infer vegetation health for the icon site and has been calibrated for the Chowilla and Lindsay-Wallpolla floodplain areas. The modelling work indicates there is a severe risk to the floodplain vegetation from current unnatural flow regimes and

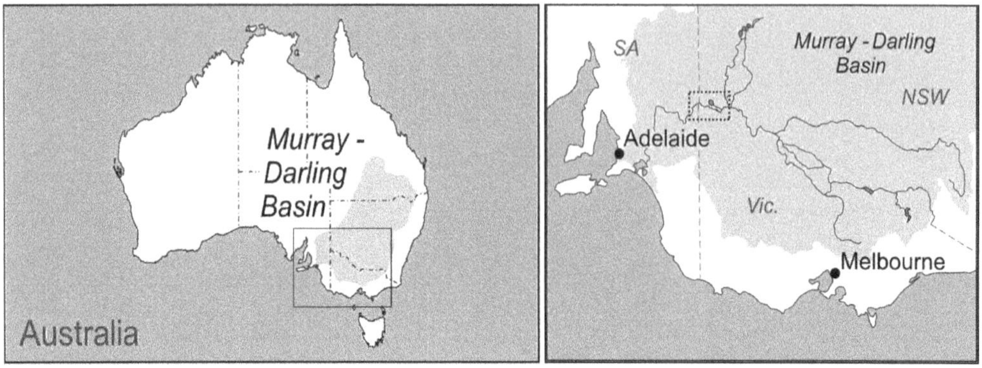

Figure 21.1: Location of Chowilla floodplain, Lindsay and Wallpolla Islands icon site within the Murray-Darling Basin.

elevated groundwater levels. Investigations indicate that increased surface flows via installation of a regulator is currently the preferred management option.

Modelling vegetation response on the Chowilla icon site

Site description

The Murray-Darling Basin Commission's Living Murray program identified a number of icon sites along the River Murray which are of high ecological value and require preservation (MDBC 2005). In the lower River Murray, a large floodplain area between Mildura (in Victoria) and Renmark (in South Australia) has been designated as the Chowilla floodplain, Lindsay and Wallpolla Islands icon site (Fig. 21.1).

The icon site is located on the borders of South Australia, NSW and Victoria, situated along the River Murray between Mildura in Victoria (above Lock 10) to approximately 50 km east of Renmark (downstream of Lock 6). The climate is semi-arid with an annual rainfall of about 260 mm yr^{-1} and a potential evaporation of approximately 2000 mm yr^{-1}.

The icon site shown in Figure 21.2 (colour plate, p. 319) comprises a wide floodplain and a network of anabranch streams which include Wallpolla Creek, Lindsay River (which bypasses Lock 7) and Chowilla Creek (which bypasses Lock 6). The Chowilla floodplain alone is the largest remaining natural riverine forest in South Australia and thus has high ecological significance. It forms part of the Riverland Ramsar wetland of international significance. The floodplain supports a high diversity of terrestrial and aquatic habitats, including populations of rare, endangered and nationally threatened species. It also contains many sites of European and indigenous cultural and heritage significance.

Although the region has experienced a drought for the last seven years, with the lowest river flows since river regulation began in the 1920s, the floodplain still retains much of its natural character and attributes even though the icon site is experiencing a significant decline in health.

WINDS model

The WINDS model evolved from the concept of a moving salt front model developed by Slavich *et al.* (1996). The moving salt front model depicts the movement of a salt front upwards during drought and salinisation, and downwards during leaching by floods. It combines limiting vegetation characteristics, soil hydraulic properties and flood history to develop a salinity index indicative of vegetation health (Fig. 21.3/colour plate, p. 320). Overton *et al.*

(2006a) applied this method within a GIS of the floodplain to create the WINDS tree health model. The WINDS model requires spatial and non-spatial data sets of vegetation type, groundwater depth, groundwater salinity, soil type, flooding frequency, rainfall and the presence of alternate water sources. The model has been independently reviewed and found to be fit for purpose (Costelloe and Western 2007). The acquisition of the spatial layers for the Chowilla and Lindsay-Wallpolla floodplain modelling project are outlined below.

Spatial WINDS model parameters and site characteristics

Vegetation type

Overton *et al.* (2005) developed a vegetation map for the Chowilla floodplain from South Australian vegetation floristic and health mapping (DEH 2005) and from NSW (DSNR 2003) vegetation community and health mapping of Chowilla (Overton and Jolly 2004). A vegetation map for the Lindsay-Wallpolla floodplain was created from surveys undertaken in NSW and from Victorian ecological vegetation community class mapping (Overton and Doody 2008b). In general, the terrestrial habitats include forest, woodland, shrubland, herbfield and grassland areas, which are distributed across the floodplain according to soil, salinity and hydrological gradients (MDBC 2006). Although the floodplain supports a diverse range of vegetative communities (O'Malley and Sheldon 1990; DSNR 2003; Overton and Jolly 2004; Smith and Kenny 2005), black box (*Eucalyptus largiflorens* F.Muell.) woodland is the dominant vegetation class (Overton *et al.* 2005), followed by river red gum (*E. camaldulensis*), lignum (*Muehlenbeckia cunninghamii*) and annual grasses. Black box is located on higher areas of the floodplain, as these trees are more drought-tolerant. Red gums occur along stream and creek courses (Dawes *et al.* 1998). The health of the tree species was recorded in 1993 (Taylor *et al.* 1996) and 2003 (DEH 2005).

Soil type

Soil maps in relation to associated soil parameters for the icon site were generated by Overton and Jolly (2004) and Overton and Doody (2008b) from surveys undertaken by the Australian Bureau of Rural Sciences using geophysical imagery at Lindsay and Wallpolla Islands and from a soil survey for Chowilla, conducted by Hollingsworth (1990) and CSIRO. In general, the floodplain consists of a range of soil types from sandy ridges to heavy clay depositional areas. A complex arrangement of soils exists due to changing depositional environments as the river meanders. Much of the floodplain is overlain by Coonambidgal clay, which acts as a semi-confining layer to the aquifer underneath and forms the floodplain surface. Coonambidgal clay is a low-permeability alluvial clay up to 5 m thick (Jolly *et al.* 1994), and its presence or absence is important to the rate of salinisation, leaching of soil salt stores during floods and recharge of soil water stores.

Flooding

In the past, the floodplain was entirely flooded approximately once every 13 years (Sharley and Huggan 1995) and 50% of the floodplain was inundated once every two years. A complex pattern of flooding is driven by a variation in elevation across the floodplain and the River Murray Floodplain Inundation Model (RiM-FIM; Overton *et al.* 1999) has been used to map flood inundation patterns from increasing river flows, for input into the WINDS model. More recently, hydrodynamic models such as MODFLOW, which predict water movement across the floodplain, have been developed, producing layers which can be input into the WINDS model.

Groundwater depth

A groundwater depth spatial layer for the Chowilla floodplain was generated from MODFLOW modelling (Yan *et al.* 2004), while groundwater depth was determined from density-corrected groundwater level contours at Lindsay-Wallpolla. Elevation derived from LiDAR imagery was subtracted from the groundwater layer to provide a layer with depth to groundwater (m) from the surface.

As a result of river regulation, irrigation and land clearing, groundwater depths are less than 4 m under most of the floodplain (Overton *et al.* 2005). Jolly *et al.* (1993) showed that groundwater shallower than an average value of 4 m will rise due to evaporative capillary rise, with 4 m considered a critical depth for salt accumulation, although this depth will vary with different soil types.

Groundwater salinity

Groundwater salinity was formerly modelled from interpolated piezometer readings for Chowilla, however, Munday *et al.* (2008) used airborne electromagnetic imagery (AEM) to provide a continuous layer of groundwater salinity. AEM data were not available for the Lindsay-Wallpolla floodplain, so the groundwater salinity layer was generated from groundwater salinity contours.

There is a large variation in groundwater salinity across the icon site, with low salinities close to the river where hydrostatic pressure forces recharge of the adjoining aquifer (the flushed zone) and salinity ranges from 35 to 85 dS m^{-1} in other areas (Noyce and Nicolson 1993). Groundwater salinity can be lower in areas where the surface clay layer is absent, as these areas act as recharge zones that allow freshening of the groundwater. In areas overlain by the surface clay layer, flooding results in only limited freshening of the groundwater due to the high impermeability of the clay.

Water source

To model vegetation health, it is necessary to consider areas of lateral recharge from losing and gaining creeks and freshwater lenses. These are identified by mapping salt concentration in the soil profile for lateral creek recharge and freshwater lenses. Identifying deeper flushed zones that recharge the groundwater, requires the identification of salt concentration in upper parts of the saturated zone. AEM data were used to map lateral recharge and freshwater lenses using a density slice at 2–4 m. The flushed zone and deep freshwater lenses were mapped using an AEM slice at 6–8 m (Munday *et al.* 2008), creating a spatial layer which identifies areas where additional water sources (other than rainfall and flooding) are available to the vegetation.

Model outputs

At the completion of each model run, two layers are output. The first layer predicts the soil salinity for the icon site in dS m^{-1} for the model end date. From this layer, vegetation health is inferred from determination of soil water availability, producing a WINDS health layer. This layer represents areas of poor to good health (broken down into three to five classes) with WINDS index values of 1 to –2. Values approaching 1 represent healthy vegetation.

Floodplain processes

Salt accumulation

Accumulation of salt in floodplain soil occurs due to capillary rise leading to discharge of shallow groundwater to the atmosphere by evapotranspiration processes such as soil surface evaporation and/or transpiration. Salts will remain in the upper soil layers, either at the soil surface or within the root zone.

A number of factors affect the extent of soil salinisation, including groundwater salinity, groundwater depth and soil texture. Accumulation rates within floodplain soils have increased significantly since river regulation and the impact of salt accumulation is exacerbated by the reduction in flood frequency and duration. Weir pool operation has caused river water to recharge adjacent floodplain aquifers, creating local freshwater zones which are beneficial to vegetation health (i.e. upstream of the locks). However, elevated weir pool levels limit groundwater discharge from floodplain aquifers to the river, causing groundwater levels to rise across parts of the floodplain.

Red gum and black box vegetation existing in semi-arid climates rely on periodic flooding to supply fresh water and leach salt from the soil profile in a predominantly saline groundwater environment. A number of studies (Hollingsworth 1990; Jolly *et al.* 1993; Mensforth *et al.* 1994; Thorburn 1993) indicate that soil salinity is a major factor affecting the health of floodplain trees.

Vegetation response to groundwater, surface water and salinisation

Vegetation health on the Chowilla floodplain, Lindsay and Wallpolla Islands icon site is directly related to a number of different floodplain processes. Some areas predominantly experience high levels of salinisation (Fig. 21.4a/colour plate, p. 321) and other areas decline as a result of moisture stress (Fig. 21.4b/colour plate, p. 321).

The different floodplain processes have been captured conceptually. Figure 21.5 (colour plate, p. 322) represents a floodplain cross-section influenced by flow regulation and reduced flooding frequency. River regulation in the form of locking creates an elevated river level caused by water ponding, which reverses the natural flow of floodplain groundwater. As a result, lateral bank recharge occurs, with low-salinity river water creating a flushed zone (Jolly and Walker 1995). Water may also discharge into nearby creeks and streams (gaining systems). While stream banks are recharged with fresh water, creating an environment conducive to good vegetation health, lowlying bank species become permanently inundated and die as a result of anoxic soil conditions (Fig. 21.5).

Beyond the flushed zone, salt accumulates in the unsaturated soil zone where groundwater is less than 4 m. Red gum and black box survive by extracting groundwater between flood periods, however, they are sensitive to salinities greater than 30 dS m^{-1} and 55 dS m^{-1} respectively (Overton and Jolly 2004). When groundwater salinity levels are high, these two species preferentially extract water from the upper soil profile and thus exacerbate salt accumulation, which leads to health decline and tree death from drought stress (Overton *et al.* 2006a). Rainfall is very low at the icon site and the presence of an impermeable clay layer prevents soil recharge in most areas. Rainfall therefore evaporates or exits the floodplain as run-off to streams.

Freshwater creeks can lose water to the saturated zone when the watertable is below the water level of the creek. Such losing creeks may be situated beyond the flushed zone (Fig. 21.6a/colour plate, p. 323). Creek banks in these areas may recharge and provide fresh water to the bank vegetation, however, outside the recharge zone vegetation health again begins to decline as salinity increases. In instances where saline groundwater levels are higher than creek levels, a gaining creek situation occurs and saline water enters the creek. This causes the health of fringing vegetation to decline (Fig. 21.6b/colour plate, p. 323). These two situations highlight manipulation of creek levels as a potential management option for preserving the health of fringing creek vegetation.

A second conceptual model (Fig. 21.7/colour plate, p. 324) illustrates localised recharge in the event of an overbank flood, in areas where the clay layer is thin or absent. These floodplain areas are generally sandy areas away from the main river, such as old meanders or dunes, and are characterised by vegetation in good health during drought conditions. A lens of fresh water sits above groundwater of higher salinity due to density differences caused by the salt, the lens providing a low-salinity water source which is refreshed by flooding events and by rainfall. The lens may become diminished after extended periods of low rainfall and no flooding. In the event of an overbank flood (highlighted by the dashed line in Fig. 21.7), localised recharge occurs through the sandy soils as floodwater infiltrates and replenishes the freshwater lens. Minimal recharge (and therefore salt leaching) occurs through the soil surface after a river has broken its banks, in areas where the surface is overlain by the thick impermeable clay layer.

Modelling vegetation response

The WINDS model is run in a raster GIS using 30 × 30 m grid cells over the whole floodplain. A WINDS value is calculated by determining the groundwater discharge rate from the depth of the watertable and soil hydraulic properties. Soil salinity is then calculated for each required five-year period using the discharge rate, the number of days dry and flooded and the soil parameters. The result is a predicted soil salinity map for the floodplain (Fig. 21.8/colour plate, p. 324), which can be used to assess the impacts of management options on salt accumulation. Soil salinity can also be compared with the salinity tolerances of trees and other vegetation types (Overton *et al.* 2006a).

A final step in the WINDS model is to calculate the soil water availability for each of the five-year periods with reference to the specific vegetation's salinity tolerance. The cumulative soil water availability of the last 15 years provides an indication of the plant growth conditions affecting tree health status. A WINDS index value between –2 (dead) and 1 (good health) is produced for each grid cell and, since black box and red gum have different salinity tolerances, the model can predict the health of these two species. The result is a tree health map which can be classified into three to five classes, ranging from dead to good by comparing WINDS values to actual health. Comparison of WINDS modelled vegetation health for the Chowilla floodplain has been tested against field-assessed data, producing a 76% spatial match between modelled data and field-assessed data (Overton and Doody 2008b). This gives confidence in the model results for this area. Predicted vegetation health for the Chowilla and Lindsay-Wallpolla icon site is shown in Figure 21.9 (colour plate, p. 325).

The WINDS model has been used to assess the tree health response on the Chowilla floodplain to a proposed environmental regulator (Overton and Doody 2008a), as well as a range of other management options that vary in degree of infrastructure required (Overton *et al.* 2006). Future predictions of vegetation health for the Chowilla floodplain were made for 2037, using the last 15 years of flow records repeated twice to provide a prediction of flow for the next 30 years. The last five years of flow have been particularly low and using such low flows will predict very poor future health outcomes in 2037. Therefore, the last 15 years of data have been repeated twice in an effort to model more realistic future flow conditions.

Modelling results

Soil and tree health response to a proposed environmental regulator on Chowilla Creek

An environmental regulator across Chowilla Creek in the western part of the floodplain is being considered to ameliorate the severe decline in tree health. Chowilla Creek bypasses Lock 6 (Fig. 21.2) and as a result there is a 3 m fall from its off-take from the River Murray upstream of the lock, to where it re-enters the River Murray downstream of the lock. If an environmental regulator is placed across the creek at the same height as the lock, flood extents equivalent to a natural flood of 65 000 ML d^{-1} in the River Murray (a one-in-three year flood naturally inundating approximately 30% of the floodplain) can be created using only 5000 ML d^{-1} in the River Murray (approximately baseflow). This will allow substantial areas of the floodplain to be inundated under relatively low-flow conditions. It will also produce high creek water levels

Table 21.1: Vegetation health in 2037 under do-nothing scenario and operation of the regulator

		Dead (%)	Very stressed (%)	Stressed (%)	Okay (%)	Healthy (%)
Red gum	Do-nothing	19	6	11	16	48
	Regulator	6	3	6	9	76
Black box	Do-nothing	0	34	30	23	13
	Regulator	0	24	27	19	30

during operation, which will increase lateral bank recharge and improve conditions for riparian zones where 80% of red gum trees occur. Groundwater levels are predicted to rise during the operation of the environmental regulator, however, this occurs naturally during flood periods and the environmental regulator is planned to operate only for a three-month duration once every three years.

Soil salinity response

The long-term predictions for salt accumulation under a do-nothing scenario in 2037 show an increase in the area of the floodplain that falls into the high-salinity category, if the 15-year flow history prior to 2007 were repeated into the future (Fig. 21.10a/colour plate, p. 326). The largest area of salt accumulation increase occurs in the eastern portion of the floodplain, with an increase also evident in other areas.

Under regulator operation, large areas of the western part of the floodplain where the regulator inundates show a reduction of soil salinity up to 50 dS m^{-1} (Fig. 21.10b/colour plate, p. 326) in 2037. The environmental regulator is therefore predicted to have a beneficial impact on soil salinity by increasing the amount of salt leaching.

Vegetation response

Future tree health predictions for 2037 under a do-nothing scenario are shown in Figure 21.11a (colour plate, p. 327), with 36% of red gums and 64% of black box either dead or stressed. Under a scenario of regulator operation, improvements to tree health are evident, especially in the western part of the floodplain (Fig. 21.11b/colour plate, p. 327), with substantial improvement in understorey health as a result of the increase in flooding frequency. Only 15% of red gum and 51% of black box are classed as dead or stressed. Red gum and black box classified as healthy, increasing from 48% to 76% and from 13% to 30% respectively (Table 21.1).

Impact of watering on floodplain vegetation

A series of watering trials have been undertaken on the Chowilla floodplain and the Bookpurnong floodplain downstream of Renmark. The extent and effectiveness of wetland watering on the fringing vegetation has been linked to the extent of lateral recharge of the surrounding groundwater, which is influenced by bank and aquifer hydraulic conductivity. The effectiveness of the watering trials was found to be variable (Holland *et al.* 2008). However, both projects clearly indicated that the trials had the capacity to impede further decline in vegetation health. The overall benefit of artificial watering across the site was reflected in an 80% and 47% reduction in plant water stress in red gums and black box respectively (Aldridge *et al.* 2006; Holland *et al.* 2008). The results provide reasonable justification for installation and operation of a regulator on the Chowilla floodplain to halt further tree decline and to improve health.

Following from single water applications, it has become obvious that repeated watering over time is required to maintain the initial vegetation response and to ensure the tree vegetation remains unstressed (Aldridge *et al.* 2006). Repeated use of the regulator is therefore required to maintain water availability. A groundwater-lowering scheme could increase the benefit of the regulator, by drawing freshened water via lateral recharge further out from creeks and wetlands by reversing the groundwater gradient.

Modelling response of other management options

Additional modelling was undertaken by Overton *et al.* (2006b) to identify other management techniques for the Chowilla floodplain, which might result in improved environmental outcomes over a 30-year period. Model results for each management option are presented in Table 21.2. They include a salt interception scheme of 32 bores positioned across the Chowilla floodplain, the temporary raising of the Lock 6 weir level to raise the anabranch creek levels, a smaller version of the environmental regulator located further upstream, earthworks to create surface flow channels to increase flooding in lowlying areas, pumping water from the River Murray and anabranch creeks into ephemeral wetlands, and combinations of these options. The most applicable techniques to meet the ecological objectives were determined by measuring the predicted improvement in the health of red gum and black box communities within the floodplain area, and by assessing the infrastructure and operating costs of the management options.

The results indicate that the option of installing a large regulator at the downstream end of the floodplain (described above) maintains or improves 66% and 33% of red gum and black box health respectively. These values are close to the Living Murray targets of maintaining or improving 70% of red gums and 45% of black box. The combined option of a salt interception scheme and a new regulator increases the health response marginally, from 66% to 67% for red gums and from 33% to 37% for black box (Table 21.1). This indicates that the greatest benefit in terms of improving floodplain vegetation health in relation to the vegetation targets would come from a combined approach, even though that approach is extremely expensive and the added improvement minimal. The combined approach offers further benefits as, potentially,

Table 21.2: Vegetation health in 2037 under do-nothing scenario and other scenarios

Management option outcomes in 2037	Red gum (2533 ha) WINDS maintained or improved (%)	Red gum WINDS declined (%)	Black box (3445 ha) WINDS maintained or improved (%)	Black box WINDS declined (%)
Do-nothing	33	67	6	4
Salt interception scheme (SIS)	38	62	13	7
Lock 6 weir raising	38	62	8	92
Small environmental regulator	61	39	28	72
Large environmental regulator	66	34	33	67
Surface water channels	37	63	7	93
Pumping into wetlands	35	65	8	92
SIS and weir raising	42	58	14	86
SIS and small regulator	63	37	33	67
SIS and large regulator	67	33	37	63
Weir raising, channels and pumping	41	59	9	91

new and larger freshwater lenses will result from lowering of saline groundwater in conjunction with surface water application, providing an essential water source for future hyper-drought periods. Although the improvement in health is minimal with this combined approach, it is suggested that it is more beneficial for long-term vegetation health. Freshwater lenses may be created or replenished over deeper saline groundwater. Groundwater lowering to below 4 m will reduce the rate of salt accumulation in areas where leaching rates are low due to the overlying impermeable clay layer. Flow management options alone are not enough to improve the overall floodplain health, as some important floodplain areas lie beyond the zone of influence that can be reached with surface water management options (MDBC 2006). In addition, lowering of the saline water alone is not sufficient to reduce plant water stress, as floodplain trees require a source of low salinity (flood/fresh groundwater) to reduce plant water stress (Holland *et al.* 2008). Therefore, the combined option is preferred. However, as groundwater management schemes are expensive a thorough cost-benefit analysis is required.

Sensitivity analysis of WINDS model

A sensitivity analysis of the WINDS model was undertaken where model parameters were altered by various degrees of magnitude. The parameter changes were used to test the shift from a base case scenario (in percentage shift in health class). Table 21.3 indicates the parameter layers used for each scenario and their degree of magnitude change from the base case scenario. Scenario 1 is composed of parameter layers which cause the WINDS health value to fall below that of the base case WINDS value, with consideration of different starting salinities, the presence of recharge zones (water areas) and a 20% error in the flood inundation extent. In scenario 2, each layer causes the WINDS health value to become more positive than the base case WINDS value, in the above analysis. Scenarios 3 and 4 represent a similar analysis, but in relation to the salinity layer which is output from the WINDS model.

The combination of layers which represent each scenario were input into the WINDS model and the output results were compared to the base case WINDS and salinity values (Table 21.4) to determine where the main sensitivities lie. Table 21.4 indicates that altering parameters (as shown in Table 21.3) to produce layers which individually lead to a decrease in WINDS values from the base case, leads to a base case matching of 71.55% to 61.80% for scenarios 1.1–1.4, with the majority of values decreasing by one health class where matching does not occur. For example, a pixel may fall into the 'good' health class but combined errors in the various layers mean those values are likely to fall into the 'ok' health class. Alternatively, 'ok' trees might be modelled as 'poor'. The best matching in this scenario is seen when initial groundwater salinity is 18 mS cm^{-1} and alternative sources of water are available around permanent watercourses. The layers used in this scenario have very little impact on the result of the output salinity layer, with 99.11% and 97.48% matching to base case.

Substantially less matching is achieved when layers which lead to an increase in WINDS health values are combined (scenario 2), with 37.49% and 35.63% matching to base case. In this instance, most pixels which do not match the base case are assigned to one health class higher than the base case (i.e. from 'poor' to 'ok', or 'ok' to 'good'), thus overestimating modelled vegetation health. Matching of the output salinity layer with the base case is only 20% in this case.

Scenarios 3 and 4 are similar to 1 and 2, but involve a comparison of the salinity layer produced from the model run with values in mS cm^{-1}, with base case salinity. The input parameter layers are a different combination from those used in scenarios 1.1–1.4. Poor matching exists when parameter layers are used which cause the salinity values to be less than that of the base case (37%). Good matching occurs when parameter layers are used which cause the salinity values to be more than that of the base case (~62%).

Table 21.3: Parameter adjustment for cumulative WINDS sensitivity analysis compared to the base case

Scenario	Drought health effect	Soil Ks, Qd, Qs	Groundwater depth Zw	Groundwater salinity Cg	Initial soil salinity Cs	Water areas	Magnitude of flood to cause inundation extent
1.0: base case	–0.3				5 mS/cm around permanent watercourses, 18 mS/cm across rest of floodplain	Alternate sources of water available around permanent watercourses	
1.1	–0.5	+50%	–0.5 m	+10 mS/cm	18 mS/cm	As per base case	+20%
1.2	–0.5	+50%	–0.5 m	+10 mS/cm	35 mS/cm	As per base case	+20%
1.3	–0.5	+50%	–0.5 m	+10 mS/cm	18 mS/cm	Alternate sources of water available across whole floodplain	+20%
1.4	–0.5	+50%	–0.5 m	+10 mS/cm	35 mS/cm	Alternate sources of water available across whole floodplain	+20%
2.0	–0.1	–50%	+0.5 m	–10m S/cm	0 mS/cm	As per base case	–20%
2.1	–0.1	–50%	+0.5 m	–10m S/cm	0 mS/cm	No alternate sources of water available across floodplain	–20%
3.0	–0.3	+50%	+0.5 m	–10 mS/cm	0 mS/cm	As per base case	–20%
4.0	–0.3	–50%	–0.5 m	+10 mS/cm	18 mS/cm	As per base case	+20%
41	–0.3	–50%	–0.5 m	+10 mS/cm	35 mS/cm	As per base case	+20%

Ks = hydraulic conductivity, Qd = unsaturated soil porosity, Qs = saturated soil porosity

Table 21.4: % matching of each model run with the base case and % of WINDS value shift up or down one or two health classes

Model name	WINDS	WINDS	WINDS	WINDS	WINDS	Salinity	Salinity
	–2 class shift (%)	–1 class shift (%)	0 (%)	+1 class shift (%)	+2 class shift (%)	Higher or same (%)	Lower (%)
1.1	2.37	25.30	71.55	0.77	0.01	97.48	2.52
1.2	2.87	29.61	66.94	0.58	0.01	99.11	0.89
1.3	2.86	33.13	63.55	0.44	0.01	97.48	2.52
1.4	3.55	34.20	61.80	0.44	0.01	99.11	0.89
2.0	0.62	5.71	37.49	54.64	1.54	20.05	79.95
2.1	0.62	5.65	35.63	56.27	1.83	20.05	79.95
3.0	1.49	8.99	49.52	39.33	0.66	23.26	76.74
4.0	3.75	32.08	63.71	0.45	0.01	99.35	0.65
4.1	4.54	33.82	61.22	0.41	0.01	99.48	0.52

Overall, the analysis indicates that an approximate sensitivity of ~±55% from the base case for WINDS-derived tree health and an approximate sensitivity of ~±40% from the base case for WINDS-derived salinity values. These sensitivities must be acknowledged when WINDS results are relied upon to inform management decisions.

Active floodplain concept

Red gum communities are establishing in lowlying area of wetlands and ephemeral creeks. In these habitats they have often been considered as weeds (Fig. 21.12a and b/colour plates, p. 328). If higher-elevation areas with river red gums can no longer be maintained via sufficient flooding frequency, then lower-elevation–higher-flooding frequency areas may need to be established for red gum recruitment.

A floodplain that inundates across a range of flooding frequencies up to approximately once in 13 years (the natural return period for most of the floodplain) can be defined as the natural 'active floodplain'. An active floodplain could be considered one that has a full range of healthy ecosystem functions with native floodplain flora. This active floodplain range is defined by the extent of a 50 000–70 000 ML d^{-1} flood (currently a once in 13 year flood event). Figure 21.13 highlights the active floodplain concept for the Chowilla floodplain under the current flooding regime. Historically, the riparian floodplain vegetation established and thrived in its current locations as a result of reliable flooding frequency and extent. The distribution of red gum and black box trees on the floodplain in relation to flow size illustrates that most of the black box and some of the red gum situated at higher elevations are situated in the old active floodplain area and are likely to be suffering from water stress and salinisation. The boundary of the active floodplain is likely to keep shifting to the left if the current do-nothing scenario continues. The graph is supported by the percentages of healthy red gum and black box on the floodplain from survey and model results.

If the floodplain is allowed to settle into the new active floodplain area under a once in 13 year return period, we will see a healthy floodplain ecosystem in the areas that are blue in Figure 21.14a (colour plate, p. 329). By comparison, the environmental regulator will support a much larger area of active floodplain (Fig. 21.14b/colour plate, p. 329).

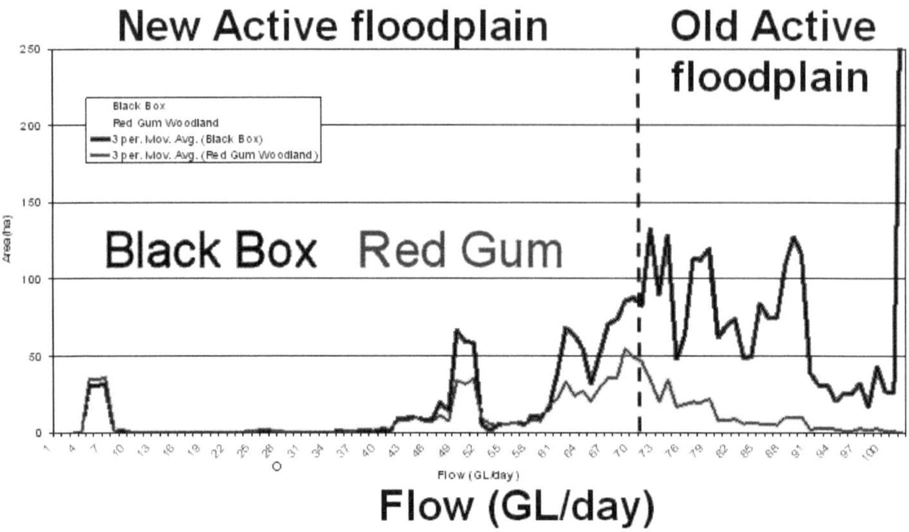

Figure 21.13: Distribution ('000s ha) of black box and red gum on the Chowilla floodplain in relation to Murray River flows. The active floodplain concept suggests the trees on the right of the dotted line are under threat from water stress and possible salinisation.

Future prospects for the icon site

The icon site is currently severely degraded. There are long-term prospects of drier and warmer conditions under climate change, reduced river flows from increased upstream water usage and increased salinisation from shallow saline groundwater. These factors are likely to increase the degradation of the floodplain. Historical natural flow regimes are not achievable with the water-sharing requirements in the Murray-Darling Basin. In addition, there is a large amount of salt stored in the floodplain soils as a result of elevated saline groundwater, high evapotranspiration rates and lack of flooding. The flooding frequency needs to be higher than natural to combat this and leach salt from the soils.

Long-term floodplain management in the broader context of River Murray management will require definition of the assets that society values, the components and sections of the floodplain which are important and those which can be compromised. Once defined, it may be possible to set environmental targets through understanding environmental water requirements and the management options required to achieve them. The River Murray is a working river, which will require managers to engineer what is required through a process of adaptive management rather than a process of restoration through return to natural conditions.

Conclusions

The ecosystem response to changing surface water and groundwater across the Chowilla floodplain, Lindsay and Wallpolla Islands icon site has been modelled using a tree vegetation health prediction model (WINDS). The WINDS model predicts soil salinity, which is used to infer vegetation health. The WINDS model has been independently reported as fit for purpose (Costelloe and Western 2007) and the sensitivity analysis shows that the model produced similar results even with parameter changes that decrease WINDS tree health values. The model results for sensitivity scenarios that increase WINDS tree health values do show a deviation from the base case and must be considered when interpreting model predictions. The modelling presented in this chapter has shown a large benefit to trees on the Chowilla floodplain from the operation of an environmental regulator on Chowilla Creek. However, factors other than tree health need to be considered in the operational regime and in relation to the overall benefits and negatives of the regulator. Maintaining a critical flow velocity through the system to encourage fish migration will need to be managed in conjunction with increasing flood inundation extent. Brookes *et al.* (2006) identified a number of potential risks from regulator operation, including cyanobacterial blooms, black-water events (deoxygenation caused by microbial degradation of natural organic matter), invasion by weeds, reduced lotic or flowing water habitats, interrupted fish passage, decrease in large-bodied native fish populations and increase in carp numbers. Conversely, environmental benefits of regulator operation include improved connectivity between riverine and floodplain habitats, improved soil condition, improvement of existing vegetation, development of new floodplain and wetland plant communities, and enhanced regional biodiversity (Brookes *et al.* 2006).

Modelling presented supports regulator installation as the best management option currently being considered. Its benefit would be supported by installation of a groundwater interception scheme to lower groundwater levels, increase lateral recharge and provide opportunities for development of freshwater lenses which will prevent health decline through future hyper-drought situations. A groundwater interception scheme will also assist in combating salt loads generated by flood inundation.

The Living Murray targets for trees specify that 70% of the existing area of red gums and 45% of existing black box area are required to be improved or maintained. If the current do-nothing scenario were to continue, there would only be decline in vegetation health of both species. Under operation of the proposed regulator, the WINDS model results indicate that 66% of red gum vegetation and 33% of black box vegetation would be maintained or improved. Although the black box Living Murray target is not met, the vegetation health results of 76% river red gum healthy and 30% of black box healthy – compared to the do-nothing scenario of 48% and 13% for red gum and black box respectively – are a vast improvement over doing nothing.

Acknowledgments

The authors would like to acknowledge funding from the Murray-Darling Basin Commission's Living Murray program and the South Australian Department of Water, Land and Biodiversity Conservation. They would like to thank Ian Jolly (CSIRO) and Peter Slavich (NSW Department of Primary Industries) for input into the development of the WINDS model, Tony Herbert and Brad Hollis (SA Murray-Darling Basin Natural Resource Management Board) for advice and review, and Kate Holland (CSIRO) for the Bookpurnong field results.

References

Aldridge K, Deegan B, Nicol J, Burch M and Brookes J (2006). Ecological response to watering trials on Chowilla Floodplain, 2005–2006. SA Dept of Water, Land and Biodiversity Conservation.

Bacon PE, Stone C, Binns DL, Leslie DJ and Edwards DW (1993). Relationships between water availability and *Eucalyptus camaldulensis* growth in a riparian forest. *Journal of Hydrology* **150**, 541–561.

Brookes JD, Baldwin D, Ganf G, Walker K and Zampatti B (2006). Comments on the ecological case for flow regulator on Chowilla Creek. Report prepared for the SA Dept of Water, Land and Biodiversity Conservation. University of Adelaide.

Brookes JD, Burch M, Wallace T and Baldwin D (2007). Risk assessment of cyanobacteria and blackwater events in Chowilla floodplain. Report prepared for the SA Murray-Darling Basin Natural Resource Management Board. University of Adelaide.

Costelloe JF and Western AW (2007). Peer review of the proposed Chowilla groundwater salt interception scheme (with an ecological focus) and the WINDS model. Dept of Civil and Environmental Engineering, University of Melbourne.

Cunningham SC, MacNally R, White M, Read J, Baker PJ, Thomson J and Griffoen P (2007). Mapping the current condition of river red gum (*Eucalyptus camaldulensis* Dehnh.) stands along the Victorian Murray River floodplain. Report to the Northern Victorian Catchment Management Authorities and the Victorian Dept of Sustainability and Environment.

Dawes WR, Slavich PG, Hatton TJ and Walker GR (1998). Modelling water and salt movement on the Chowilla floodplain. In WAVES: an integrated energy and water balance model. (Eds L Zhang and WR Dawes). Technical report 31/98. CSIRO Land and Water, Canberra.

DEH (2005). Floristic vegetation and tree health mapping, River Murray floodplain, South Australia. Internal report. SA Dept for Environment and Heritage, Adelaide.

DSNR (2003). Kulcurna Crown Reserve natural resource inventory. Report prepared for the Chowilla Integrated Natural Resource Management Project by the NSW Dept of Sustainable Natural Resources.

Holland KL, Charles AH, Jolly ID, Overton IC, Gehrig S and Simmons CT (In press). Effectiveness of artificial watering of a semi-arid saline wetland for managing riparian vegetation health. *Hydrological Processes.*

Hollingsworth ID (1990). A reconnaissance soil survey of the Chowilla anabranch system of the River Murray in South Australia and New South Wales. Report prepared for Murray-Darling Basin Commission by the Loxton Research Centre, SA Dept of Agriculture, Adelaide.

Jolly ID and Walker GR (1995). Floodplain salinisation in the Chowilla anabranch and its implications on river salinity and native riparian forests. Murray-Darling Basin Workshop, Wagga Wagga.

Jolly ID, Walker GR and Thorburn PJ (1993). Salt accumulation in semi-arid floodplain soils with implications for forest health. *Journal of Hydrology* **150**, 589–614.

Jolly ID, Walker GR and Thorburn PJ (1994). Salt balances in semi-arid floodplain soils and consequences for riparian vegetation health. In *Proceedings of International Groundwater and Hydrology Conference. Water Down Under 1994.* Adelaide. Institution of Engineers Australia National Conference Publication No. 94/14, pp. 687–690.

Mensforth LS, Thorburn PJ, Tyerman SD and Walker GL (1994). Sources of water used by riparian *Eucalyptus camaldulensis* overlying highly saline groundwater. *Oecologia* **100**, 21–28.

Munday T, Overton IC and Fitzpatrick A (2008). Managing and protecting the local ecology and biodiversity of the Lower River Murray floodplains in south-east Australia using bio-physical data derived from airborne EM survey data. In *Proceedings of the 2nd International Salinity Forum.* April 2008.Adelaide.

MDBC (2005). The Living Murray Foundation report on the significant ecological assets targeted in the First Step Decision. Murray-Darling Basin Commission Publication No. 09/05.

MDBC (2006). The Chowilla Floodplain and Lindsay-Walpolla Icon Site environmental management plan 2006–2007. Murray-Darling Basin Commission Publication No. 33/06.

Noyce T and Nicolson K (1993). Chowilla flood and groundwater analysis. Information Systems Branch, SA Dept of Housing and Urban Development, Adelaide.

O'Malley C and Sheldon F (1990). Chowilla floodplain biological study. Nature Conservation Society of South Australia.

Overton IC and Doody TM (2008a). Summary of the groundwater and vegetation responses to the Chowilla regulator. Report for SA Murray-Darling Basin Natural Resource Management Board.

Overton IC and Doody TM (2008b). Salt accumulation and tree vegetation response to surface water–groundwater interactions in the mallee region of the lower River Murray. CSIRO Scientific Report prepared for the Mallee Catchment Management Authority, Victoria and the Lower Murray-Darling Catchment Management Authority, NSW.

Overton IC and Jolly ID (2004). Integrated studies of floodplain vegetation health, saline groundwater and flooding on the Chowilla floodplain, South Australia. Technical report No. 20/04. CSIRO Division of Land and Water, Canberra.

Overton IC, Walker GR and Jolly ID (1995). Modelling soil salinisation risk on a River Murray floodplain using a geographical information system, Chowilla, South Australia. In *AURISA '95. Proceedings of the Australasian Urban and Regional Information Systems Association*. Melbourne. pp. 197–210.

Overton IC, Newman B, Erdmann B, Sykora N and Slegers S (1999). Modelling floodplain inundation under natural and regulated flows in the lower River Murray. In *Proceedings of the 2nd Australian Stream Management Conference*. Adelaide, Australia.

Overton IC, Rutherford JC and Jolly ID (2005). Flood extent, groundwater recharge and vegetation response from the operation of a potential weir in Chowilla Creek, South Australia. Client report for the SA Dept of Water, Land and Biodiversity. CSIRO Division of Land and Water.

Overton IC, Jolly ID, Slavich PG, Lewis MM and Walker GR (2006a). Modelling vegetation health from the interaction of saline groundwater and flooding on the Chowilla floodplain, South Australia. *Australian Journal of Botany* **54**, 207–220.

Overton IC, Slarke S and Middlemiss H (2006b). Chowilla management options. Final report. Report to SA Dept of Water, Land and Biodiversity Conservation, Adelaide.

Overton IC, Munday TM and Fitzpatrick A (2008). Managing and protecting the local ecology and biodiversity of the Lower Murray River floodplains in SE Australia using biophysical data derived from airborne EM survey data. In *Proceedings of the 2nd International Salinity Forum*. April 2008. Adelaide, Australia.

Sharley T and Huggan C (1995). Chowilla resource management plan. Final report. Murray-Darling Basin Commission, Canberra.

Slavich PG, Walker GR and Jolly ID (1996). Vegetation response to modified flooding regimes and groundwater depth on a saline floodplain. In *Proceedings of Hydrology and Water Resources Symposium*. Hobart. Institution of Engineers Australia National Conference Publication No. 96/05, pp. 505–510.

Smith FM and Kenny SK (2005). Floristic vegetation and tree health mapping, River Murray floodplain, South Australia. Internal report, SA Dept for Environment and Heritage, Adelaide.

Taylor PJ, Walker GR, Hodgson G, Hatton TJ and Correll RL (1996). Testing of a GIS model of *Eucalyptus largiflorens* health on a semi-arid, saline floodplain. *Environmental Management* **20**, 553–564.

Thorburn PJ (1993). Groundwater uptake by eucalyptus forests on the Chowilla floodplain, South Australia: implications for groundwater hydrology and forest ecology and health. PhD thesis. Flinders University of South Australia, Adelaide.

Yan W, Howles SR and Marsden Z (2004). Chowilla floodplain numerical groundwater model. Report No. 2004/65. SA Dept of Water, Land and Biodiversity Conservation, Adelaide.

Chapter 22

Ecosystem response modelling of the Murray Mouth and Coorong lagoons

Rebecca E. Lester and Peter G. Fairweather

Introduction

The Coorong is the estuary and coastal lagoon complex for the Murray-Darling Basin, on the South Australian coast (Phillips and Muller 2006; Lester and Fairweather 2009a; Fig. 22.1/colour plate, p. 330). The Coorong is a long, narrow lagoonal system comprising two main lagoons – the North Lagoon and the South Lagoon – and an estuarine region around the Murray Mouth (Phillips and Muller 2006). The water sources into the Coorong include freshwater flows from the River Murray across the barrages, marine water as tidal input through the Murray Mouth, fresh to brackish inputs from the Upper South-east Drainage scheme via Salt Creek near the south-east end of the system, and more localised rainfall and groundwater inputs (Webster 2007).

The Coorong is part of the Coorong, Lower Lakes and Murray Mouth (CLLAMM) region that is listed under the Ramsar Convention on Wetlands of International Importance (DEH 2000) and is designated as one of six icon sites within the Murray-Darling Basin (Phillips and Muller 2006). The region supports more than 1% of the global population (one of the criteria for Ramsar listing) of nine bird species, including the curlew sandpiper (*Calidris ferruginea*), red-necked stint (*Calidris ruficollis*) and sharp-tailed sandpiper (*Calidris acuminata*) (Paton *et al.* 2009). The region also has significant cultural, economic and recreational values, with sizeable local tourism and commercial fishing industries, nearby agriculture and a local indigenous Australian community, the Ngarrindjeri nation. Many of the bird, plant and fish species are of significant cultural value to the Ngarrindjeri, adding to the region's importance (Phillips and Muller 2006). Much of the Coorong is designated as national park (DEH 2000).

The management of the CLLAMM region is shared among several government agencies at the state and federal levels. Several main management levers can be used to affect ecosystem condition in the region, including:

- environmental flows of fresh water, either from the Upper South-east Drainage via Salt Creek or from the River Murray;
- continual dredging of the Murray Mouth to maintain the connection to the ocean;
- barrage operations controlling the flow of water between the Lower Lakes and the Coorong, including the operation of fishways;
- various local engineering options, including pumping and dredging options to achieve a variety of objectives.

These, combined with climatic forcing factors, determine the ecological condition of the Coorong and the identified 'water benefits' of the region, such as a viable fishery based on fish recruitment, tourism, cultural values, sustained waterbird populations and an open Murray Mouth.

Hydrologically, the Coorong is complex in both space and time (Webster 2005, 2007). Water levels are largely controlled by mean sea level in the Southern Ocean, with water levels highest in winter and early spring. Without significant freshwater inflows, water levels drop in late spring and are lowest in summer, when the two lagoons can become disconnected. Over shorter time scales, tidal influences, local weather systems, wind-driven seiching and the degree of openness of the Murray Mouth also influence the water and salinity levels in the Coorong (Webster 2005). Near the Murray Mouth, salinity can range from near-fresh to marine depending on River Murray inflow levels. Salinity naturally increases from the Murray Mouth to the South Lagoon. In recent years, following a prolonged period without River Murray inflow, an extreme salinity gradient has developed in the system. Salinity up to six times that of seawater was found in the South Lagoon in 2008.

The aim of this research was to construct a single, system-wide response model for the ecosystem condition of the region, which was explicit in both space and time. Previously, knowledge of system functioning has relied on (often untested) conceptual models developed from expert opinion; the aim was to create a model from the available data. The modelling was focused on providing direction as to the best long-term strategy for managing the system, rather than short-term response to recent events. This chapter briefly outlines the development of the ecosystem response model designed to meet these criteria, and introduces the ecosystem states of the Coorong as described by that model. We then give an example of the model' s application, by describing changes in ecosystem condition between 2000 and 2007.

Developing an ecosystem response model for the Coorong

Ecological modelling is a relatively new scientific discipline, with ongoing development of new methods of representing and modelling multivariate ecological data. Numerous options exist for any modelling problem (see Sutherland 2006 for a review of some possible approaches). Any complex modelling task, such as ecosystem response modelling, is likely to involve a combination of several different modelling techniques.

Attempts to model the response of key performance indicators of particular floral or faunal species are often based on restrictive assumptions dictating the data sets that are included and/ or the form that the relationships take. One of the major impediments to detailed ecosystem modelling is having only heterogeneous incomplete data sets that are patchy in space and time. In developing an ecosystem model for the Coorong, we aimed to address these challenges. As is often the case, the data set on which the modelling could be based is disparate in nature and collection methods, includes qualitative, semi-quantitative and quantitative data, is patchy in both space and time, and does not necessarily cover all parameters of potential interest. To identify the techniques most likely to be of use, we reviewed a variety of the best candidates (Lester and Fairweather 2008). The review concluded that a statistical model would be most appropriate, combining several methodologies including classification and regression tree analyses, state-and-transition modelling and Bayesian belief networks.

The model approach is based on an adaptation of the theory of alternative stable states (Petraitis and Dudgeon 2004, 2005) existing within an ecosystem. This theory contends that there are ecosystems for which more than one stable biological community is possible in the one place for a given set of environmental conditions, and that the state at any given time is path-dependent (Petraitis and Dudgeon 2004). Path dependency implies a memory within the system, or that the current state is a result of the states that previously existed in that location. A detailed account of how the alternative stable state theory has been modified to be relevant in this instance is given in Lester and Fairweather (2009a). Briefly, the data available for the Coorong do not allow us to determine the stability, or otherwise, of any identified states. Also, the continued decline of the system means that the assumption of equilibrium is not met. Therefore, we have extended the theory beyond equilibrium conditions to include transient ecosystem states that may appear during the decline or recovery of an ecosystem. This removes the emphasis on stability, and extends the theory's applicability to estuaries in decline worldwide. It explicitly models the various stages of decline, rather than simply capturing the end-points of the transition.

Data included in the model

Significant effort was expended in compiling as complete a data set as possible for the Coorong from the available literature, data collected by CLLAMMecology Research Cluster researchers (both prior to and during CLLAMMecology, a Research Cluster funded by the CSIRO Collaboration Fund that aimed to improve overall ecological knowledge of the Coorong, Lower Lakes and Murray Mouth region; www.csiro.au/partnerships/CLLAMMecologyCluster), monitoring data collected by state or federal agencies relevant to the region, and outputs from other models constructed for the Coorong. These data were divided into two sets: a biological data set and an environmental data set. The biological data set included abundance data for bird and benthic fauna, plant coverage and abundance of propagules for the dominant macrophyte species, *Ruppia tuberosa*, and catch-per-unit effort data for the local commercial fishery. The environmental data set included modelled water levels, depths and salinities along the length of the Coorong, daily tidal range, flows over the barrages, meteorological data and measured water quality parameters, including the concentration of nutrients, pH, alkalinity and turbidity. A list of the variables included in the model, the provenance of the data and references to sampling methods are given in Lester and Fairweather (2009a) for both the environmental and biological data sets.

Because of the varied provenance of the data sets, we have reported values in the units in which they were originally measured and not attempted to convert them to more standard values. For example, salinity was measured in grams per litre rather than practical salinity units. The conversion between the two is not straightforward (see Poisson and Gadhoumi 1993 for the formula to calculate practical salinity), so grams per litre has been retained.

For the environmental data set, a range of parameters was calculated for each variable, including maxima, minima, ranges, variability, moving averages (three years) and lagged effects (one and two years), particularly for flow and depth characteristics. This variety of parameters was based on the understanding that the mean conditions now are not always the best predictors of ecosystem state or condition (Gaines and Denny 1993). In constructing the data set, we divided the Coorong into a series of 12 study sites that corresponded with the field sites used during data collection for CLLAMMecology (see Lester and Fairweather 2009a for more detail). Most variables were collected between 1999 and 2007, so that nine-year period was used as the basis for the model.

Model development

An in-depth description of the development of the ecological response model is beyond the scope of this chapter, and is presented in Lester and Fairweather (2009a). We have, however, given an overview of the modelling technique. The biological and environmental data sets were treated separately and analysed alternately. This provided a series of checks during model construction, to ensure that results that were relevant for each data set reflected the properties of the other, and that the biological and environmental conditions were linked as expected.

First, we clustered the biological data set. The analysis used a series of cases that were delineated by year (nine years) and site (12 sites), for a total sample size of 80 (minus some cases that had missing data). The initial identified clusters were designated as preliminary states. Next, we used the environmental data set to identify the physical characteristics that were driving membership in those preliminary states. Classification and regression tree (CART) analyses (De'ath and Fabricius 2000), undertaken using the CART Pro 6.0 software package (Steinberg and Golovnya 2007), identified the parameters that could best differentiate between the preliminary states and the quantitative thresholds for each. CART analysis explains the variation in a univariate response variable by repeated binary splitting based on the best predictor among a range of explanatory variables. It has very few assumptions and is able to detect non-linear high-order relationships within data sets (De'ath and Fabricius 2000). In this way, the variables that best describe the variance in the response variable are identified from the long list of potential explanatory variables. Where explanatory variables were significantly correlated, one was removed and the model rerun. The stability of the CART model was tested using a bagging algorithm (Jones *et al.* 2006), a bootstrapping technique that draws independent subsamples from the data set, performs a CART analysis on each and produces an averaged 'committee of trees' as an output. This indicates the stability of the original tree and the relative strength of its predictions (Steinberg and Golovnya 2007). Where a committee of 100 regression trees did not explain significantly more of the variance than the initial tree and significant predictors were similar, the initial tree was accepted as a stable model.

The next step involved state membership from CART being tested using the biological data set. ANOSIM analyses in PRIMER 6 software (Clarke and Gorley 2006) tested that the states created from the environmental data set differed biologically from one another. ANOSIM analysis tests the dissimilarity between sample groups based on randomisation and permutations of similarity matrices (Clarke and Gorley 2006). Various subsets of the data were tested to account for missing data in some variables for some cases. Those states that did support statistically distinct biota were confirmed as ecosystem states. The non-significant states were merged, after carefully considering the statistical power of each test.

This model was used to classify new cases into the identified ecosystem states using the environmental data set. The new cases were ones for which missing data precluded their initial inclusion in cluster analyses. For each state, the biological data of the new cases were compared to those of the original cases to ensure that the new cases did not differ significantly from the biota of the original states. This was also done using ANOSIM analyses, and provided a check that the model was correctly predicting the biological community based on environmental conditions. Finally, the ecosystem states identified in the model were characterised based on their biological and environmental attributes. This was done using SIMPER procedures in PRIMER 6, calculating the mean conditions for parameters known to be of interest to managers, and determining the abundances of key species from a commercial, conservation or ecological perspective. SIMPER identifies the variables that contribute most to the dissimilarity between sample groups (Clarke and Gorley 2006), allowing the identification of indicator species for each of the identified states.

Describing long- and short-term trends

The initial model, based on annual time-step data at 12 sites along the Coorong, identified three distinct ecosystem states within the Coorong and explained 86% of the variance. The significant predictors were mean annual salinity and average water level from the previous year. However, it did not adequately identify the recent (since 2005) decline in condition along the length of the Coorong because of insufficient cases that were representative of that decline. To better characterise these more-degraded states, we constructed a second model at a shorter, quarterly time-step for the same 12 sites from 2005 to 2007. The biological data set varied slightly from that for the longer-term, annual analyses. As *Ruppia tuberosa* distributions were only mapped annually, those variables were excluded from the short-term analyses. Migratory species of birds were also excluded, because their presence or absence throughout the year is seasonal and not necessarily related to the environmental condition of the region. In addition to the commercial fish-catch data, data from fishery-independent fish surveys were included. These surveys targeted a wider range of species, including those too small to be commercially fished but with the potential to be key regional species (e.g. small-mouthed hardyhead, *Atherinosoma microstoma*). Finally, benthic invertebrate sampling for juvenile invertebrates was included. The environmental data set included the same set of parameters, although values were recalculated to be relevant to the shorter time-step.

This short-term model identified six distinct ecosystem states within the Coorong and explained 66% of the variance within the data set. It was expected that this proportion would be lower than for the long-term model – no dummy variable was included for season, so any seasonal variability would not be represented in the model and thus would constitute extra unexplained variance. The explanatory variables in this model included average daily tidal range, the maximum number of days since flow over the barrages, the average quarterly water level, the average quarterly salinity from the previous year, and the average quarterly water levels and depths from the previous year.

As the goal was to create a single ecosystem response model for the Coorong that was capable of predicting the longer-term trends in ecosystem condition as well as the deterioration associated with drought conditions, we combined the two models. Using ANOSIM analyses, we identified which short-term states adequately represented their long-term counterparts. One of the three long-term states was biologically distinct from any of the short-term states; the other two were not statistically different from several of the short-term states. An environmental data set was constructed using the cases from the short-term states, as well as the one distinct long-term state. This created a combined model by undertaking a new CART analysis on cases from both individual models (the long-term and short-term models). The new combinations of short- and long-term states resulted in the identification of a distinct extra state, resulting in eight final ecosystem states (Lester and Fairweather in press, 2009). Once the model had been finalised, all short- and long-term cases were reclassified based on the final ecosystem states identified for the region. The characterisation of each state was undertaken separately for the long-term and short-term data sets.

Ecosystem states of the Coorong

The model for the Coorong identified eight distinct ecosystem states. The environmental parameters that differentiated among the various states were a combination of water quality, quantity and flow variables (Lester and Fairweather 2009a). They included the average daily tidal range, maximum number of days since flow across the barrages, average water level and salinity at the location and average depth of water in the previous year. The appearance of average tidal range as the first split variable effectively divided the Coorong into two basins, with four states possible within each. The northern basin existed around the Murray Mouth estuary and down to the northern part of the North Lagoon, to about Noonameena (Fig. 22.1), where the states were essentially marine in character. The southern basin included the southernmost part of the North Lagoon and the entire South Lagoon and had four hypersaline states.

We gave each of the eight states a name, for ease of interpretation. The names were based on the environmental conditions of each state, and the biota supported by each. The trend of declining biotic richness and the variables for which thresholds were significant (e.g. length of time since flow over the barrages) led us to believe that the states represent a continuum from a healthy ecosystem to a degraded ecosystem in each basin. We named the states accordingly. However, these names do not imply that a single state (e.g. the 'healthy' state) should, or even could, necessarily exist in each region. One characteristic of the Coorong is that it has always supported a range of biota. The ideal condition for the system as a whole is therefore likely to include a mixture of states. In addition, the definition of a state in this modelling means that they may be transitional states, that will exist for a period before transitioning into another state, or be stable over long periods. Within the available data set, it is not possible to identify which states fall into each category. Finally, despite several states being labelled 'unhealthy' and 'degraded', this does not mean that no biota exists. As described below, even degraded states continue to support a range of biota, which is available as food and habitat resources for other organisms.

The northern basin consisted of four states, including those named Estuarine/Marine, Marine, Unhealthy Marine and Degraded Marine. These states had greater tidal ranges than the four of the southern basin: Healthy, Average, Unhealthy and Degraded Hypersaline. The long-term biological and environmental characteristics of each state are shown in Tables 22.1 and 22.2, respectively, while the short-term characteristics are shown in Tables 22.3 and 22.4. For five of the eight states, the long-term differences in multivariate environmental parameters are shown in Figure 22.2. Only five states are shown because missing values exclude a sixth state. Two states did not appear within the long-term analyses. These five, however, had significant differences in environmental parameters (ANOSIM Global R = 0.536 p = 0.001). This difference between the environmental parameters of each state supports the notion that the model is describing distinct ecosystem states. The environmental characteristics of all states were quite distinct, with very little overlap. The biological community supported by each of the five states also varied (ANOSIM Global R = 0.402, p = 0.001), based on the macrophyte, bird and commercial fish abundances found.

Figure 22.2: nMDS ordination plot of long-term environmental characteristics for Coorong ecosystem states. The figure is based on a Euclidean similarity matrix of normalised data for modelled salinity, depth and water levels, flow characteristics, meteorological and water quality data for 12 sites between 1999 and 2007 (minus cases with missing values, n = 55). A maximum of 25 runs were used.

Table 22.1: Long-term biological characteristics of the eight Coorong ecosystem states

	Marine states				Hypersaline states			
Species	Estuarine/ Marine	Marine*	Unhealthy Marine	Degraded Marine†	Healthy Hypersaline	Average Hypersaline	Unhealthy Hypersaline	Degraded Hypersaline
Macrophyte (% cores containing *Ruppia*)								
Ruppia tuberosa	0.02	0.00	0.21	NA	NA	51.97	22.69	NA
Fish (average CPUE for net days)								
Yellow-eyed mullet	6.08	6.62	8.47	9.45	NA	3.90	0.00	NA
Bony bream[a]	0.70	0.03	1.28	0.00	NA	0.53	0.00	NA
Mulloway[b]	1.02	0.69	0.89	0.11	NA	0.13	0.00	NA
Greenback flounder[c]	0.40	0.07	0.32	0.52	NA	0.09	0.00	NA
Bronze whaler shark[d]	0.22	0.41	0.35	0.00	NA	0.00	0.00	NA
Australian salmon[e]	0.84	1.06	1.23	0.00	NA	0.03	0.00	NA
Birds (average abundance)								
Australian shelduck[f]	383	0	286	13 922	NA	1023	905	NA
Red-necked avocet[g]	160	18	61	1260	NA	506	613	NA
Hoary-headed grebe[h]	147	0	265	497	NA	686	904	NA
Curlew sandpiper	283	21	711	238	NA	288	5	NA
Chestnut teal[i]	1047	3	655	5593	NA	1341	373	NA
Banded stilt	529	0	159	15 487	NA	2418	20 270	NA

Biota included are based on the results of SIMPER analyses outlining the species that drove the similarities within states and the differences between states for the long-term analysis. Several SIMPER analyses were undertaken on various subsets of data due to missing values. The species included are those that most commonly and significantly differentiated between pairs of ecosystem states.

* While the Marine state has few of the listed bird species , it supported more abundant populations of other species.

† Caution should be used in interpreting these results, because only one case of the Degraded Marine state exists. Figures are indicative only.

NA = No sites were categorised as Healthy Hypersaline or Degraded Hypersaline in the long-term analyses.

a = *Nematolosa erebi*, b = *Argyrosomus japonicus*, c = *Rhombosolea tapirina*, d = *Carcharhinus brachyurus*, e = *Arripis truttaceus*, f = *Tadorna tadornoides*, g = *Recurvirostra novaehollandiae*, h = *Poliocephalus poliocephalus* and i = *Anas castaneta*.

Table 22.2: Long-term environmental characteristics of the eight Coorong ecosystem states

	Marine states				Hypersaline states			
Parameter	**Estuarine/ Marine**	**Marine**	**Unhealthy Marine**	**Degraded Marine†**	**Healthy Hypersaline**	**Average Hypersaline**	**Unhealthy Hypersaline**	**Degraded Hypersaline**
Flow								
Maximum number of days since flow	127	376	376	376	NA	152	376	NA
Water quantity								
Average daily tidal range (m)	0.11	0.38	0.19	0.00	NA	0.03	0.03	NA
Variance of average water level	0.03	0.02	0.02	0.02	NA	0.08	0.03	NA
Water quality								
Maximum salinity (g L^{-1})	57.56	48.11	61.64	88.24	NA	123.28	176.47	NA
Average [total phosphate] (mg L^{-1})	0.14	0.05	0.12	0.35	NA	0.27	0.50	NA
Average [ammonia] (mg L^{-1})	0.04	0.05	0.05	0.04	NA	0.20	0.30	NA
Average [Chlorophyll b] (μg L^{-1})	2.38	0.45	2.33	4.20	NA	10.39	7.11	NA
Average turbidity (NTU)	14.59	1.68	11.14	35.00	NA	17.55	23.53	NA
Average [TKN] (mg L^{-1})	1.85	0.61	1.22	6.57	NA	5.61	6.90	NA
Meteorological*								
Maximum daily minimum temperature (°C)	24.06	28.75	28.73	29.00	NA	24.11	28.75	NA
Average annual rainfall (mm yr^{-1})	467.46	300.95	301.35	295.80	NA	476.10	300.95	NA

Parameters are based on the results of SIMPER analyses outlining the variables that drove the similarities within states and the differences between states for the long-term analysis. Several SIMPER analyses were undertaken on various subsets of data due to missing values. Prameters are those that most commonly and significantly differentiated between pairs of ecosystem states.

* Meteorological parameters were calculated from data collected at Meningie weather station, approximately 10 km away.

† Caution should be used in interpreting these results, as only one case of the Degraded Marine state exists. Figures are indicative only.

NA = No sites were categorised as Healthy Hypersaline or Degraded Hypersaline in the long-term analyses.

Table 22.3: Short-term biological characteristics of the eight Coorong ecosystem states

	Marine states				Hypersaline states			
Species	**Estuarine/ Marine**	**Marine**	**Unhealthy Marine**	**Degraded Marine†**	**Healthy Hypersaline**	**Average Hypersaline**	**Unhealthy Hypersaline**	**Degraded Hypersaline**
Fish (average abundance)								
Small-mouthed hardyhead	ND	132	963	708	ND	ND	406.40	80
Yellow-eyed mullet	ND	477	212	171	ND	ND	0.00	0
Australian salmon	ND	118	49	15	ND	ND	0.00	0
Sandy sprat[a]	ND	294	156	49	ND	ND	0.00	0
Birds (average abundance)								
Chestnut teal	36	10	37	16	96	92	71	36
Hoary-headed grebe	31	11	1	0	295	199	416	3
Banded stilt	6	0	1	54	406	868	1361	1894
Red-necked stint	62	152	78	86	23	499	119	211
Grey teal[b]	106	113	67	35	292	157	176	0
Whiskered tern[c]	34	58	72	1545	2	75	600	3
Invertebrates (average abundance)*								
Capitella worms	8084	3619	10 775	ND	0	0	ND	ND
Arthritica bivalves	2439	1561	4500	ND	0	0	ND	ND
Simplesetia worms	1264	1327	3424	ND	0	0	ND	ND
Insect larvae	319	336	177	ND	740	34	ND	ND

Biota included are based on the results of SIMPER analyses outlining the species that drove the similarities within states and the differences between states for the short-term analysis. Several SIMPER analyses were undertaken on various subsets of data due to missing values. The species included are those that most commonly and significantly differentiate between two ecosystem states.

† Caution should be used in interpreting these results, as only one case of the Degraded Marine state exists. Figures are indicative only.

* All invertebrate abundances reported here are for juveniles.

a = *Hyperlophus vittatus*, b = *Anas gracilis* and c = *Chlidonias hybridus*.

ND = indicates that no data were available for that species for that particular state.

Table 22.4: Short-term environmental characteristics of the eight Coorong ecosystem states

	Marine states				Hypersaline states			
Variables	**Estuarine/ Marine**	**Marine**	**Unhealthy Marine**	**Degraded Marine***	**Healthy Hypersaline**	**Average Hypersaline**	**Unhealthy Hypersaline**	**Degraded Hypersaline**
Flow								
Maximum number of days since flow	141	540	536	652	2	166	559	505
Average flow volume from previous year (GL)	47.70	55.56	57.48	0.00	56.89	45.81	85.48	0.00
Water quantity								
Average maximum water level (m AHD)	0.34	0.39	0.22	ND	0.42	0.15	−0.03	−0.25
Maximum water level (m AHD)	0.69	0.78	0.56	ND	0.58	0.43	0.33	−0.07
Minimum water level (m AHD)	−0.12	−0.27	−0.25	ND	0.21	−0.11	−0.35	−0.40
Average minimum water level (m AHD)	0.13	−0.01	0.02	ND	0.38	0.12	−0.08	−0.29
Average water level (m AHD)	0.22	0.15	0.12	0.04	0.40	0.13	0.04	−0.25
Average daily tidal range (m)	0.21	0.39	0.20	ND	0.03	0.02	0.03	0.03
Change in water level over quarter (m)	0.81	1.05	0.81	ND	0.38	0.53	0.68	0.32
Water quality								
Average maximum salinity (g L^{-1})	49.82	44.12	52.11	71.60	113.03	123.73	128.87	178.87
Average minimum salinity (g L^{-1})	32.05	34.72	39.45	55.20	99.63	95.22	97.39	120.53
Average salinity (g L^{-1})	41.46	37.98	45.04	64.62	105.94	110.78	112.56	151.37

Variables included are based on the results of SIMPER analyses outlining the factors that drove the similarities within states and the differences between states for the short-term analysis. Several SIMPER analyses were undertaken on various subsets of data due to missing values. The parameters included are those that most commonly and significantly differentiated between pairs of ecosystem states.

* Caution should be used in interpreting these results, as only one case for the Degraded Marine state exists. Figures are indicative only.

ND = indicates that no data were available for that variable for that particular state.

Describing the ecosystem states

The Estuarine/Marine state was characterised by lower salinities and shorter periods without flow than other states. The state supported large numbers of marine and estuarine fish. Bird fauna was dominated by piscivores, although waterfowl and migratory waders were also present. *Ruppia tuberosa* was present but with limited cover, and invertebrates were diverse.

The Marine state had low salinities and high variability in water levels. Biologically, the state supported a wide range of marine and estuarine fish species, but fewer piscivores than the Estuarine/Marine state. There were greater numbers of waterfowl and a more diverse polychaete fauna.

The Unhealthy Marine state had longer periods without freshwater flow than the Estuarine/Marine or Marine states. *Ruppia tuberosa* was common, as were piscivores and yellow-eyed mullet *Aldrichetta forsteri*. Polychaetes such as capitellids and *Simplesetia aequisetis* were found in high abundances.

The Degraded Marine state was represented only twice in the data set, so was difficult to characterise. It had higher salinities and lower water levels than other marine basin states. Few fish species were found and chironomid larvae were the only invertebrate taxon recorded.

The Healthy Hypersaline state had the highest freshwater flows of any state, although it tended to have higher salinities than the marine basin states. It had low variability in water levels, but high depths and nutrient concentrations. The state supported high numbers of waders and waterfowl, and chironomid larvae were abundant.

The Average Hypersaline state also had frequent freshwater flows and high nutrient concentrations, with a slightly higher average salinity than the Healthy Hypersaline state. Few fish species or piscivorous birds were characteristic of this state but *Ruppia tuberosa* was common, as were waterfowl and waders. Chironomid larvae and amphipods were abundant.

The Unhealthy Hypersaline state had lower water levels and higher salinities than the Average Hypersaline state. Banded stilt *Cladorhynchus leucocephalus* was characteristic of the state, which had a sparse invertebrate fauna, although chironomid larvae were still common.

The Degraded Hypersaline state had the highest salinities and lowest water level of all states. The length of time without freshwater inputs was only longer for the Degraded Marine state. Small-mouthed hardyhead was one of the few fish species present and there were few invertebrates. Banded stilt were common, with other species of waders and waterfowl also present.

Additional detail for each ecosystem state is given in Lester and Fairweather (2009b). Details are also given regarding other states that have been hypothesised to exist (e.g. a more estuarine state) but are outside the experience of the data set used here. The possible existence of other states that fall outside the bounds of the data set is important to keep in mind when interpreting these results with a view to management of the system.

Mapping the ecosystem states in space and time

One of the unique features of the Coorong is the range of conditions that it supports at any one time along its 110 km length. This was one of the reasons for its listing on the Ramsar Convention, and its status as a Murray-Darling Basin icon site (MDBC 2006). It is important that our modelling reflect that mix. This section demonstrates changes in the spatial and temporal mix of ecosystem states and describes the observed changes in ecosystem response that have occurred over the last 10 years.

State of the Coorong in 2000

Near the beginning of our time series, during 2000, two ecosystem states were observed along the length of the Coorong (Fig. 22.1a, p. 330). The Murray Mouth region and the North Lagoon existed in the Estuarine/Marine state (characterised by relatively low average salinities, recent freshwater flow over the barrages, high average water depths and high variability of water levels throughout the year). The South Lagoon, however, was then in the Average Hypersaline state, with the characteristic environmental conditions including higher average salinities than found in the northern-region states, low average depths, but recent freshwater flows over the barrages. Water quality parameters indicated high nutrient concentrations and high phytoplankton abundances.

State of the Coorong in 2007

In addition to varying along the length of the Coorong, the environmental conditions and corresponding mix of ecosystem states have changed through time. As the drought has continued, and worsened, the previous states have gradually been replaced by more degraded states. A lack of fresh water has led to a loss of estuarine conditions from the system. With seawater now the only source of water replacing evaporative losses, the salt loads within the system have increased, with the South Lagoon becoming extremely hypersaline (up to a measured maximum of 220 g L^{-1} in 2007). The changing environmental conditions have resulted in changes in the biological conditions, the original ecosystem states being replaced with new ones.

The mix of states observed in 2007 illustrates these changing conditions (Fig. 22.1f, p. 330). Three ecosystem states existed along the length of the Coorong, all of which were different from the ones that were present in 2000. The northern part of the Coorong, including the Murray Mouth region, was predominantly in the Unhealthy Marine state (seven sites), with two sites existing in a Marine state. The Marine state had the lowest average salinity of any state (37.4 g L^{-1}), but had low average water levels and a long time since flow had passed over the barrages. The water quality reflected the marine inputs through the Murray Mouth, with low values of both nutrient concentrations and turbidity. The Unhealthy Marine state was characterised by a relatively long time since receiving freshwater input over the barrages, but high variability in water levels throughout the year. Compared with the Marine state, salinity was high, but nutrient concentrations and turbidity remained relatively low.

Future state of the Coorong

The change in the distribution of states, from more natural states with a higher diversity to more degraded states, can be seen for the 12 sites studied within CLLAMMecology (Fig. 22.1). The progressive change in the distribution of ecosystem states is shown in Figure 22.3 (colour plate, p. 331; see below for explanation of the 114-year sequence). It is clear that the system as a whole has deteriorated in average condition over the nine years investigated. Should current conditions continue, we would expect the system to experience further declines in condition. There are a large number of species (including estuarine fish species and the macrophyte, *Ruppia tuberosa*) that require lower salinities and higher water levels to successfully recruit and are currently declining rapidly, or at best ageing without reproducing in the system. We would expect these species to disappear from the system under long-term low flows of fresh water, along with other species that rely on them for food or shelter.

Adapting our ecosystem state model for detailed scenario analyses

The next step in the ecological response modelling of the Coorong was to develop and analyse a series of scenarios to determine the ecological impact of each (Lester *et al.* 2009). The scenarios included a combination of climate change possibilities and management options, to see how these interacted and what the possible futures of the Coorong could be like. This was done using the ecosystem state model (Lester and Fairweather 2009b) in sequence after the hydrodynamic model that has been produced for the region (Webster 2007), the latter providing the necessary input data for predicting ecosystem responses. This will provide managers with detailed information on the consequences of various management options under varying degrees of climate change and allow them to include scientifically rigorous assessments in the political process of decision-making for the system.

The predictive ecosystem model was used to run a number of scenarios to inform the management of the Coorong (Lester *et al.* 2009). The results of these scenario analyses have been designed to assist managers in identifying, and justifying, the most appropriate management options for the Coorong, based on the best available evidence. The scenarios were various climate change scenarios, including the identified median and dry future climates as defined in the Murray-Darling Basin Sustainable Yield project (CSIRO 2008), various management scenarios including different flow regimes to be delivered across the barrages, and an exploration of the effects of several management interventions including the benefit currently derived from dredging the Murray Mouth and the effects of additional water delivery through Salt Creek via the Upper South-east Drainage Scheme (Lester *et al.* 2009). Figure 22.3 presents the results of the baseline scenario, which simulated the effect of the historic climate from 1895 to 2008, with current extraction levels for the entire run and average flows through the drainage scheme, but no dredging (see Lester *et al.* 2009 for further details and additional scenarios). The sequence of states occurring in the Coorong through the 114-year model illustrate the uniqueness of the current drought situation: no other drought had a comparable effect on ecosystem states.

Defining ecosystem states in the manner outlined here provides a series of decision rules that can be used for detailed scenario modelling. While the initial data set included an extensive list of environmental parameters of potential interest, the analytical process identified a much smaller subset with the strongest relationships to the biota upon which to focus future modelling. Identified thresholds between states provide a set of cut-offs (from the CART analyses) arranged in a logic tree (Lester and Fairweather 2009a, b) that collectively determine the state of a given site at a given time. This definition of transitions dramatically simplifies the task of modelling future condition because very few variables need to be simulated. By measuring water levels, flow over the barrages and salinities, we can define the eight ecosystem states for the Coorong. In addition, the results of such simulations are more interpretable than some other modelling techniques. Each state is well characterised environmentally and biologically, so transitions between states can be interpreted as a significant leap between biological communities and/or environmental conditions. This can be easily applied to management of complex ecosystems because a suite of conditions, as represented by an ecosystem state, can be defined as desirable or otherwise for each location at each point in time and the scenarios that provide the most appropriate mix of states can be readily identified.

The ecosystem state model described here has been used as the basis of a state-and-transition model to determine the mix of states present in the Coorong over time. The defined ecosystem states, and the thresholds between them, were the initial set of states and transitions included in that model. An extension of this approach would be the use of a Bayesian belief network framework. This would allow us to include hypothesised states as alternatives and to constrain allowable transitions based on adjacency and pathways. In addition to including the well-defined ecosystem states, hypothesised states may include a truly estuarine state (Lester and Fairweather 2009b). Transitions could then be constrained by how likely they were to occur given a site's current state (pathway constraint) and the state of neighbouring sites in space or time (proximity constraint). Including these states and transitions would maximise the predictive capacity of the model, allowing it to make reasonable predictions for thresholds that fall outside the range of conditions modelled here but within known or suspected historic limits. Additional thresholds could be based on the available evidence, including anecdotal and piecemeal evidence, as well as the expert opinion of researchers. The use of a Bayesian belief network would enable the model to be continually updated as new data become available (Marcot *et al.* 2006).

We could include additional thresholds between states, where relevant. These may include the conditions under which characteristic taxa occur, based on experimental data, and piecemeal evidence that could not be included in the original modelling. Finally, we would be able to account for the spatial and temporal limitations of transitions between states. Sites are more likely to transition into states that are present in neighbouring sites, and are more likely to remain in states present in the previous time-step than to randomly switch to another state. This model could form the basis of the predictive model used for scenario analyses (Lester *et al.* 2009), and additional information and limitations could be incorporated where justifiable.

Applying the ecosystem state model to other systems

The Coorong was an ideal candidate upon which to test the utility of ecosystem state modelling. Its long narrow form with a distinct environmental gradient made the derivation of a set of ecosystem states relatively straightforward. There is good reason to believe that the technique will be useful in other situations which may initially seem less suitable.

The initial alternative stable state theory was originally developed in systems that do not necessarily have such pronounced longitudinal gradients, e.g. freshwater lakes (Ibelings *et al.* 2007) and benthic marine communities (Petraitis and Dudgeon 2004). The changes that we have made to alternative stable state theory (i.e. removing the emphasis on stability and including transitional states) extends the applicability of the original theory, rather than constraining it (Lester and Fairweather 2009b). Ecosystem state theory is applicable to ecosystems that are in decline or recovery as well as those in equilibrium, and there is no requirement to demonstrate the stability of each identified state. Therefore, there should be no theoretical barrier to applying the theory to a wide range of systems, including those that do not have the obvious gradient of conditions as in the Coorong.

One constraint on the applicability of ecosystem state theory to other systems is the availability of data in the target system. While the techniques we have described are designed to maximise the benefit of patchy, disparate data sources, developing a robust set of ecosystem states does require a solid basis of biological and environmental data. A time series of abundance data for a range of locations within the system is ideal. The more taxa that are represented in the database, the more the model will truly represent the ecosystem rather than focus on the charismatic birds, fish or vegetation that are often included in management plans. Because of the need to cluster the biological data, this is the section of modelling that is least tolerant of missing data (because the cluster analyses are based on Bray-Curtis similarity matrices that are undefined for missing data; Clarke and Gorley 2006).

The environmental data set from which thresholds are identified can be drawn from data that are patchy in space and time. CART analysis can include variables with some missing data and can include qualitative and quantitative parameters (De'ath and Fabricius 2000). A time series of variables for various locations within the system is ideal, but it is also possible to include measures for which data are incomplete or limited in spatial or temporal coverage. Where hydrodynamic (or other) deterministic models are available and well-calibrated (Webster 2007), it is possible to include their output as explanatory parameters, in addition to measured data.

The development of ecosystem state models should be possible for a wide range of aquatic systems. Many systems, such as lakes, wetlands and estuaries, have a tendency to experience stepped changes in condition (Folke *et al.* 2004); for example, where wetting and drying cycles are present or where ecosystem-engineering species exist, whose presence/absence dramatically changes the character of a system (Folke *et al.* 2004). The basis for modelling ecosystem states is the link between biotic assemblages and thresholds in environmental conditions. This link occurs in the vast majority of aquatic systems so the approach should be applicable in each, but it needs to be made explicit and be empirically based. At this stage in its development, it is certainly worth attempting to apply ecosystem state modelling elsewhere, if only to demonstrate its generality or otherwise. Should such modelling be as successful elsewhere as in the Coorong, it may prove to be a significant step forward in our ability to manage environmental assets at an ecosystem scale.

Conclusions

The modelling approach used here addresses challenges relating to our disparate data sources that cover various spatial and temporal domains and are collected in different ways. This enables the use of as much data as possible relating to ecosystem structure and environmental condition, making the model more relevant at an ecosystem scale. The inclusive approach to data means that the modelling is not limited by predetermined data types or by the disciplinary preferences of modellers. Alternating analysis between environmental and biological data sets provides corroboration of incremental results and increases the overall robustness of model predictions.

This approach of defining ecosystem states has significant implications for the research and management of other estuaries or, indeed, other ecosystem types such as floodplain wetlands. The multivariate data input and exploratory nature of the technique has the advantage of greatly simplifying the task of defining ecosystem condition and may revolutionise the setting of management targets. It allows management of a system at an ecosystem scale – aiming for a mix of appropriate ecosystem states, rather than using surrogate variables or indicator species that are often arbitrarily chosen or defined and may lack adequate testing. Our ecosystem states are combinations determined by the data in hand rather than by preconceptions about co-occurring focal variables. By managing for a set of states, there is significantly more flexibility in the definitions of acceptable change, allowing for redundancy in an ecosystem (where multiple species could perform the same function) to be recognised and acknowledged in the management strategy. A deliberately multivariate approach discourages the temptation to search for a singular 'magic bullet' solution to ecological problems by recognising the interconnectedness of environmental and biotic conditions and thus the complexity of ecosystems as a whole. Finally, the approach is flexible enough to provide the basis for simple indices indicating overall condition and likely future condition, as well as the more detailed scenario analyses that are the next step in our work on the Coorong.

Acknowledgments

This research was part of the CLLAMMecology Research Cluster supported by the CSIRO Flagship Collaboration Fund. It represents a collaboration between CSIRO, the University of Adelaide, Flinders University and SARDI Aquatic Sciences. Thanks to the members of the Cluster, Gene Likens, Peter Petraitis and the other participants at the three CLLAMM Futures workshops for their constructive input into this modelling approach.

Invaluable access to data and assistance in interpretation of those data were provided by David Paton and Daniel Rogers from the University of Adelaide, Sabine Dittmann and Alec Rolston from Flinders University, Qifeng Ye and Craig Noell from SARDI Aquatic Sciences, Joseph Davis from the Murray-Darling Basin Authority, and the Australian Wader Study

Group. Scenario analyses were undertaken in collaboration with Ian Webster from CSIRO Land and Water. The generosity of these contributors in sharing their valuable data sets is gratefully acknowledged; the foresight of these scientists in collecting these data sets is exemplary. Funding bodies contributing to the original collection of the data include the South Australian Department for Environment and Heritage, Earthwatch and the Fisheries Research and Development Corporation. Additional data were supplied by the South Australian Department for Environment and Heritage, Primary Industries and Resources South Australia and the Australian Bureau of Meteorology.

We also gratefully acknowledge the research assistance provided by Rebecca Langley and Stephanie Duong, and assistance with map-making from Craig Noell at SARDI Aquatic Sciences. Thanks also go to the participants in Ecological Response Modelling in the Murray-Darling Basin conference in Sydney, 5–6 November 2008, for their helpful comments and questions.

Finally, Rod Oliver, Mike Geddes, Sebastien Lamontagne and Nicky Grigg provided helpful comments that improved this manuscript.

References

Clarke KR and Gorley RN (2006). PRIMER v6: User manual/tutorial. PRIMER-E Ltd, Plymouth.

CSIRO (2008). Water availability in the Murray. Report to the Australian Government from CSIRO Murray-Darling Basin Sustainable Yield Project. CSIRO, Canberra.

De'ath G and Fabricius KE (2000). Classification and regression trees: a powerful yet simple technique for ecological data analysis. *Ecology* **81**, 3178–3192.

DEH (2000). Coorong and Lakes Alexandrina and Albert Ramsar management plan. SA Dept for Environment and Heritage, Adelaide.

Folke C, Carpenter S, Walker B, Scheffer M, Elmqvist T, Gunderson L and Holling CS (2004). Regime shifts, resilience, and biodiversity in ecosystem management. *Annual Review of Ecology, Evolution and Systematics* **35**, 557–581.

Gaines SD and Denny MW (1993). The largest, smallest, highest, lowest, longest, and shortest: extremes in ecology. *Ecology* **74**, 1677–1692.

Ibelings BW, Portielje R, Lammens EHRR, Noordhuis R, van den Berg MS, Joosse W and Meijer ML (2007). Resilience of alternative stable states during the recovery of shallow lakes from eutrophication: Lake Veluwe as a case study. *Ecosystems* **10**, 4–16.

Jones KB, Neale AC, Wade TG, Cross CL, Wickham JD, Nash MS, Edmonds CM, Riitters KH, O'Neill RV, Smith ER and van Remortel RD (2006). Multiscale relationships between landscape characteristics and nitrogen concentrations in streams. In *Scaling and Uncertainty Analysis in Ecology.* (Eds J Wu, KB Jones, H Li and OL Loucks) pp. 205–224. Springer, Dordrecht.

Lester RE and Fairweather PG (2008). Review of modelling alternatives for CLLAMM futures. CSIRO Water for a Healthy Country National Research Flagship, Adelaide.

Lester RE and Fairweather PG (2009a). Modelling future conditions in the degraded semi-arid estuary of Australia's largest river using ecosystem states. *Estuarine, Coastal and Shelf Science* **85**, 1–11.

Lester RE and Fairweather PG (2009b). Ecosystem states of the Coorong: an ecosystem response model: method development and sensitivity analysis. CSIRO Water for a Healthy Country, Adelaide.

Lester RE, Webster IT, Fairweather PG and Langley RA (2009). Predicting the future ecological condition of the Coorong: the effect of management actions and climate change scenarios. CSIRO Water for a Healthy Country, Adelaide.

Marcot BG, Steventon JD, Sutherland GD and McCann RK (2006). Guidelines for developing and updating Bayesian belief networks applied to ecological modeling and conservation. *Canadian Journal of Forest Research* **36**, 3063–3074.

MDBC (2006). The Lower Lakes, Coorong and Murray Mouth Icon Site environmental management plan 2006–2007. Murray-Darling Basin Commission, Canberra.

Paton DC, Rogers DJ, Hill BM, Bailey CP and Ziembicki M (2009). Temporal changes to spatially stratified waterbird communities of the Coorong, South Australia: implications for the management of heterogeneous wetlands. *Animal Conservation* **12**, 408–417.

Petraitis PS and Dudgeon SR (2004). Detection of alternative stable states in marine communities. *Journal of Experimental Marine Biology and Ecology* **300**, 343–371.

Petraitis PS and Dudgeon SR (2005). Divergent succession and implications for alternative states on rocky intertidal shores. *Journal of Experimental Marine Biology and Ecology* **326**, 14–26.

Phillips B and Muller K (2006). Ecological character description: Coorong, Lakes Alexandrina and Albert, wetland of international importance. SA Dept for Environment and Heritage, Adelaide.

Poisson A and Gadhoumi MH (1993). An extension of the Practical Salinity Scale 1978 and the Equation of State 1980 to high salinities. *Deep-Sea Research Part I – Oceanographic Research Papers* **40**, 1689–1698.

Steinberg D and Golovnya M (2007). CART 6.0 user's guide. Salford Systems, San Diego.

Sutherland GD (2006). Predicting the ecological consequences of environmental change: a review of the methods. *Journal of Applied Ecology* **43**, 599–616.

Webster IT (2005). An overview of the hydrodynamics of the Coorong and Murray Mouth. CSIRO Water for a Healthy Country National Research Flagship, Canberra.

Webster IT (2007). Hydrodynamic modelling of the Coorong. CSIRO Water for a Healthy Country National Research Flagship, Canberra.

Chapter 23

Ecosystem response modelling in the Goulburn River: how much water is too much?

Peter Cottingham, Michael Stewardson, Jane Roberts, Rod Oliver, David Crook, Terry Hillman and Ian Rutherfurd

Introduction

The expansion of water markets and resultant intervalley trade has led to an increase in water demand in the Sunraysia district in north-west Victoria over the last five years. To meet this demand, intervalley transfers (IVTs) will be needed, most likely between January and March using the lower Goulburn River and River Murray as carriers. Within Victoria, there has been concern that the delivery of IVT flows well above those which would occur normally under low-flow summer and autumn conditions will adversely affect the environmental values of the lower Goulburn River.

The lower Goulburn River has a long history of river regulation. Lake Eildon, the second-largest storage in Victoria, dams the headwaters of the Goulburn River and supplies water across northern Victoria by a network of channels, storages and natural watercourses (Fig. 23.1). This water is released from Lake Eildon and diverted from the Goulburn River at Goulburn Weir. As a direct consequence, mean annual discharge in the Goulburn River downstream of Goulburn Weir has been severely reduced. Typically, more than 50% of annual inflows are diverted from the river. With increasing trade of water downstream to Sunraysia, Eildon irrigation releases are passing Goulburn Weir and flow to the River Murray, augmenting summer flows. In 2003, an environmental flow study (Cottingham *et al.* 2003) recommended that minimum summer–autumn flows in the lower Goulburn River should be increased from approximately 250 ML d^{-1} to >610 ML d^{-1} to maintain habitat for native fish. This recommendation would still have summer and autumn flows at the lower end of their natural range. No upper bounds were put on the recommendation. At the time, the need for frequent or increasingly large IVTs to be delivered to growth areas further west via the Goulburn River was not identified.

As the likelihood of IVTs became apparent, it also became apparent that the environmental consequences of IVTs would need to be assessed within a short time-frame; contemporary rules for intervalley water trade were focused on capacity and delivery and on commercial and contractual obligations only. Accordingly, the scientific panel that developed the environmental flow recommendations in 2003 was reconvened in 2006–07 to consider the ecological implications of IVTs for the lower Goulburn River, based on existing knowledge. It soon became apparent to the scientific panel that there was limited scientific understanding of the effect of augmented summer flows on river ecology and very little quantitative research to draw on for evaluating the effects of different volumes of IVTs in summer. The panel therefore developed a decision support tool based on modelled ecological and geomorphological responses that included both ecosystem processes as well as the habitat requirements for key biota. The decision support tool could then be used to consider the ecosystem implications of new water delivery scenarios, without the need to reconvene the scientific panel.

This chapter describes the approach adopted by the scientific panel in the more recent study, and the basis of the decision support tool and the ecosystem modelling that supports it. Two case histories (concerning riverbank vegetation and in-channel primary production) are presented to illustrate the distinctive features of this work.

Approach

Philosophically, our approach is one of hydraulic habitat conservation (Hillman 2009). This approach seeks to provide a hydraulic environment that will support the ecosystem (biota and processes) in a healthy condition, rather than seeking to manage individual components of the ecosystem. Our modelling aims to ensure that all key components of the hydraulic environment are considered appropriately. The environmental modelling that underpins the decision support tool (Cottingham *et al.* 2007) is a series of separate conceptual flow–response models (for geomorphology, instream primary production, macro-invertebrates, riverbank vegetation and native fish), developed individually but evaluated collectively. For example, the flow–response models cover geomorphic processes such as rates of rise and fall or the persistence of water levels at particular stage heights, and ecosystem processes such as the effects of retention time for primary production and area of hydraulic habitat for macro-invertebrates. The IVTs are considered using the same two reaches and representative sites below Goulburn Weir as in the environmental flow study (Fig. 23.1): Goulburn Weir to Shepparton (the Murchison reach) and Shepparton to the River Murray (the Wyuna reach).

The first step was to establish a means of evaluating the various IVT scenarios. Five guiding principles were developed, covering the management context and ecological aspirations:

- IVTs must, where possible, be consistent with the intentions of previous environmental flow recommendations and considerations;
- IVTs must not lead to a decline in ecosystem condition, structure or function, and should be used to improve conditions if possible;
- IVTs must not be to the detriment of existing regional, state or national programs or strategies for river protection and rehabilitation;
- IVTs will be considered in terms of their potential impacts on components of the natural flow regime that are considered important for ecosystem function and the protection or rehabilitation of assets and values;
- IVTs will be considered in terms of their potential for missing components of a natural flow regime, from the perspective of the total flow regime (including intra- and interannual variability), and in terms of alleviating potential negative ecosystem impacts due to the inversion of the natural seasonal pattern of flow.

The second step was to review previous environmental flow recommendations and identify the potential for achieving ecological benefits through IVTs, as well as any disbenefits or risks. As part of this, the flow-related ecological objectives for riverine attributes (river geomorphology, riverine productivity, aquatic and riverbank vegetation, macro-invertebrates and native fish) were considered in relation to changes to the flow regime as a result of IVTs. Using ecology–flow responses as a conceptual basis made it possible to identify the critical flow

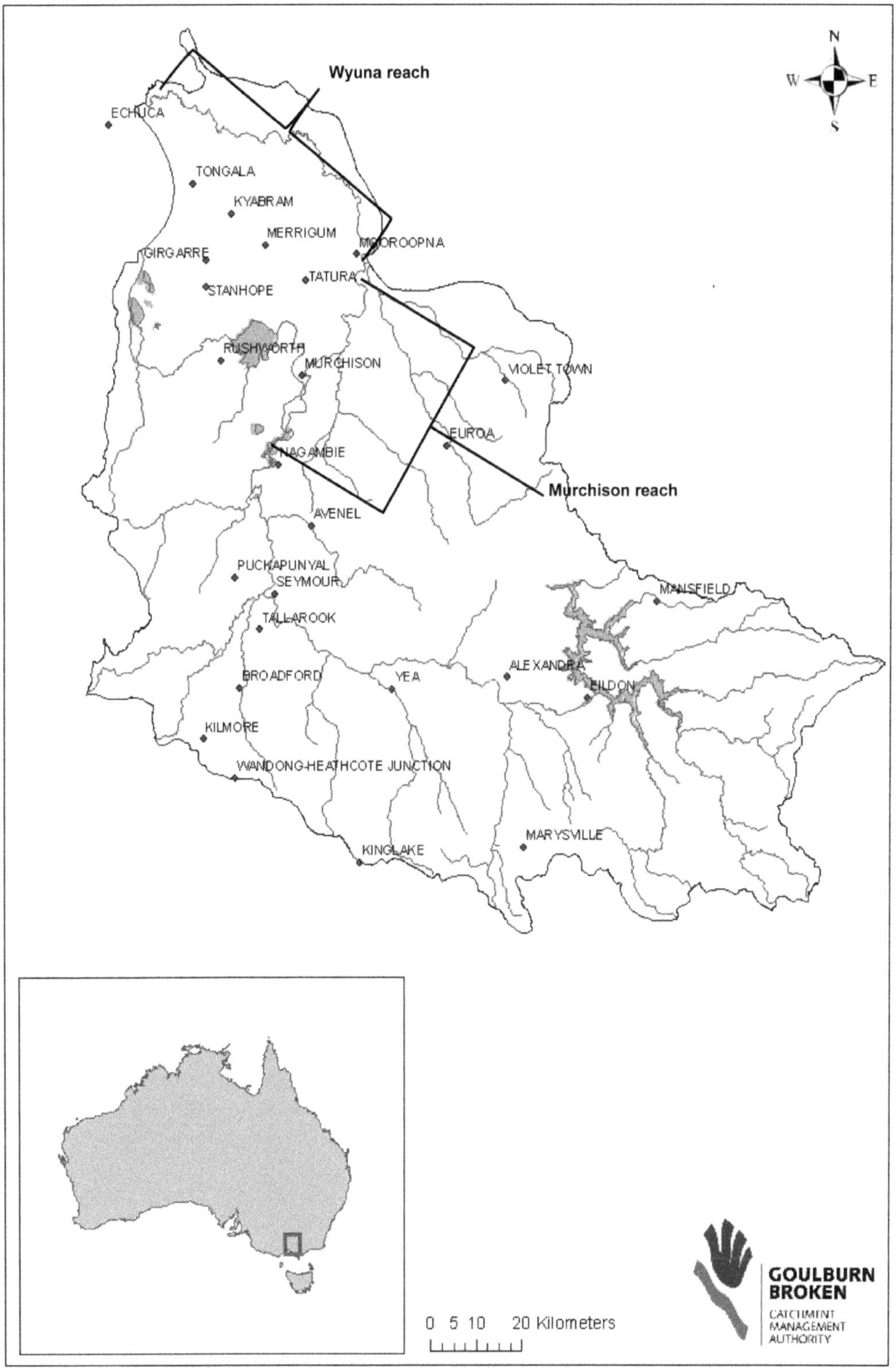

Figure 23.1: Goulburn-Murray Water region, including the two study reaches.

drivers (or stressors) affecting ecosystem processes or the amount and timing of available habitat. For example, flow drivers/stressors considered for the two riverine attributes of river-bank vegetation and in-channel primary production were:

- mean residence time;
- euphotic depth in relation to mean depth;
- shear stress;
- variability in water level fluctuations;
- periods of inundation at various stage heights;
- rates of rise and fall.

The third step was to define risks and associated thresholds (critical flows) in terms that could be modelled. This was done by setting two limits or thresholds. The first limit was one considered by the scientific panel to represent the threshold between low and moderate risk in terms of achieving flow-related ecosystem objectives. The second represented the threshold beyond which the flow elements were considered to pose a high risk to achieving ecosystem objectives. The decision support tool was used to generate statistical summaries of flow driver/stressors based on a 25-year series of daily flow (1975–2000) corresponding to three flow management scenarios: unregulated or modelled natural flows, historic (recorded) flows and projected IVT flows (Cottingham *et al.* 2007).

Quantitative flow analysis

Link between objectives and flow stressors and related elements

A flow stressor is a specific characteristic of flow identified in the flow–ecology conceptual models (Table 23.1) as being directly related to an ecological response. The majority are defined hydraulically rather than by flow volume. Many relate to the duration and/or frequency of spells above (or below) hydraulically defined flow thresholds and refer to a specific time-frame, usually a single year or season, or frequency of occurrence at longer time-scales. There is no single or exclusive relationship between a flow stressor and an ecological objective: an ecosystem objective can be affected by a single or multiple flow stressors.

Each flow driver/stressor is expressed quantitatively using one or more flow elements (flow metrics and units). For example, the flow stressor specified as the 'proportion of time when mean bed shear stress is low, leading to deposition of fine sediment' actually has three flow elements corresponding to three different shear stress thresholds below which sedimentation might occur (1 N m^2, 2 N m^2 and 3 N m^2). Using flow–shear stress relationships, it is possible to define the discharge and duration that would avoid excessive rates of sedimentation. Flow elements are calculated for one or more ecologically defined time periods, referred to in the modelling as seasons (Table 23.1). Most cases are for a calendar season (e.g. December–February for summer) but a few are for the summer–autumn irrigation season (i.e. December to April) or for the whole year.

Thresholds for the flow elements were set corresponding to the likelihood of achieving the ecological outcome, based on scientific evidence or the experience of scientific panel members. In most cases, two thresholds could be specified for each percentile: one corresponded to the threshold for a high likelihood of achieving the ecological objectives (assuming other environmental constraints were addressed) and the other corresponded to a high likelihood the ecological objective will not be met. Thresholds could be expressed as an upper and/or lower limit for the flow element.

This procedure was completely dependent on conceptual models of the effect of flow alteration on stream ecology (Cottingham *et al.* 2007). For this reason, we devoted considerable effort, in individual research and group discussions, to developing these models. We used them to establish ecological objectives for any environmental flow release. They also provided the basis for defining the flow stressors (and their quantitative expression as flow elements). We frequently referred back to the conceptual models, which were fundamental to communicating our findings with the client and other stakeholders. Some aspects of our models have transferability to other rivers, but there will be some river-specific issues that demand expertise and time for developing conceptual models in any environmental flow study.

Decision Support Tool

The majority of environmental flow studies in Victoria (Cottingham *et al.* 2003, 2008; LREFSP 2002; SKM 2006) are required to make recommendations based on specified flow components (e.g. cease-to-flow, minimum flows, flow fresh, bankfull flows, overbank flows) and single recommendations for each flow component are designed to meet multiple ecological objectives. Minimum flows are expressed as a single minimum flow threshold and flow-pulse recommendations are expressed as a single flow magnitude, duration and frequency for a particular season. A limitation of this approach is that it is not possible to assess the implication of non-compliance with the flow recommendations for each ecological objective. To overcome this problem, we developed individual environmental-flow recommendations for each ecological objective. Each recommendation may be expressed as thresholds for multiple flow stressors, each flow stressor may be quantified by multiple flow elements and each flow element may be calculated for one or more season. It is difficult, however, to express recommendations in a simple table within a report. This apparent complexity can be daunting for practitioners used to the simpler results produced in previous studies.

To help environmental flow managers apply our recommendations, we developed a spreadsheet decision support tool (using Microsoft Excel) for each study reach. The tool allows users to assess compliance of a proposed flow-management scenario with our environmental-flow recommendations. Users enter a modelled daily-flow series, representative of the scenario under consideration. The tool assesses compliance against each recommendation and reports compliance in a single table arranged by flow elements (rows) and environmental flow objects (columns). Coloured cells (cf traffic lights) in the table identify non-compliance with our environmental flow recommendations (Fig. 23.2/colour plate, p. 332).

Case studies: conceptual approach and identification of flow drivers/stressors

The following sections on riverbank vegetation and primary production are presented as two case studies to highlight the approach used to evaluate potential risks from IVTs.

Riverbank vegetation

The in-channel vegetation of the Goulburn River downstream of Murchison has three distinctive characteristics. First is the extent of terrestrial vegetation down the riverbank: terrestrial grasses, typically the tussock-forming *Poa labillardieri*, occur down the face of the bank to within a metre (visually estimated) of the summer water level. Second is the presence of terrestrial vegetation within the channel: saplings and young trees of river red gums *Eucalyptus camaldulensis* and silver wattles *Acacia dealbata* are common on the mostly sandy instream bars and benches. Third is the general lack of amphibious plants and hence overall low structural diversity: herbs such as *Lythrum salicaria* only occur as scattered individuals at or near the summer waterline.

As stated above, flows in the lower Goulburn River have been severely reduced through diversions at Goulburn Weir since its completion in 1891 and have been stabilised by regulation.

Table 23.1: Relationships between flow-related objectives and flow stressors

Ecological value	Attribute	Ecological objective	Seasons*	Stressor mechanism
Source of food for fish and invertebrates and influence on river nutrient and chemical conditions	Planktonic algae	Production rates, biomass levels and community composition more resembling unimpacted sites and dynamic diverse food webs	Su, Sp	Increased channel retention due to reduced water velocity and/or hydraulic retention zones allows accumulation of biomass if growth rates exceed loss rates
			Su, Sp	Proportion of time planktonic algae spend in the euphotic zone determines whether net production is possible
			Sp	Proportion of time planktonic algae spend in the euphotic zone multiplied by mean surface irradiance determines the relative level of production
			Su, Sp	Water depth influences the rate of deposition of planktonic algae (it takes longer for settling in deeper water)
Source of food for fish and invertebrates, habitat and influence on river nutrient and chemical conditions	Periphytic algae	Production rates, biomass levels and community composition more resembling unimpacted sites and dynamic diverse food webs	Su, Sp	Benthic production is restricted to wetted perimeter within the euphotic zone (where light penetrates to the channel bed and banks)
			Su, Sp	High velocities influencing biofilm stability. Area of colonisation determined by extent of light zone – use euphotic depth, but limited by velocity
			Sp	Establishment of biofilm requires that the wetted surface remains wet and within the euphotic depth for some time. Drying and submersion below the euphotic depth will adversely affect biofilm
Contributes to primary production, habitat for macro-invertebrates and native fish	Macrophytes	Production rates, biomass levels and community composition more resembling unimpacted sites and dynamic diverse food webs	Su, Au, Wi, Sp	Benthic production is restricted to wetted perimeter within the euphotic zone (i.e. where light penetrates to the channel bed and banks)
			Su, Au, Sp	High velocities influencing biofilm stability. Area of colonisation determined by extent of light zone – use euphotic depth, but limited by velocity
			Su, Au, Sp	Establishment of aquatic macrophytes requires that the wetted surface remains wet and within the euphotic depth for some time. Drying and submersion below the euphotic depth will adversely affect macrophytes

Ecological value	Attribute	Ecological objective	Seasons*	Stressor mechanism
Natural gradient of native terrestrial vegetation up the riverbanks	Terrestrial bank vegetation	Maintain native terrestrial cover at top of banks and reduce cover of terrestrial vegetation in bank areas influenced by flow regulation	Dec–Apr	Duration of submergence (inundation) has potential to drown terrestrial vegetation, due to carbon and oxygen starvation. Critical values for duration tolerance expected to vary between seasons, being much longer in cool (autumn–winter) than in warm growing (spring–summer) season
Diverse and resilient aquatic macro-invertebrate fauna that are an integral part of food webs	Macro-invertebrates	Provision of conditions suitable for aquatic vegetation, which provides habitat for macro-invertebrates	Su	Slow shallow velocities required for establishment of aquatic vegetation
			Su, Au, Wi, Sp	Short-term flow fluctuations can adversely affect aquatic vegetation growing along the channel margins
	Macro-invertebrates	Submersion of snag habitat within the euphotic zone to provide habitat and food source for macro-invertebrates	Su, Au, Wi, Sp	Quantity and variety of snags dependant on volume (possibly modified by biodiversity and productivity of snag biofilm - depth and variability of light climate).
			Su, Au	High shear stresses can lead to biofilm instability
			Su, Au, Wi, Sp	Loss of pools
			Dec–Apr	Reduced flow results in drying of large woody debris
	Macro-invertebrates	Provision of slackwater habitat favourable for planktonic production (food source) and habitat for macro-invertebrates	Su	Increased flow velocity and rapid rates of rise and fall affect availability of shallow slackwater habitat for macro-invertebrates
			Su, Au, Wi, Sp	Daily fall in stage

Ecological value	Attribute	Ecological objective	Seasons*	Stressor mechanism
	Macro-invertebrates	Entrainment of litter packs available as food/habitat source for macro-invertebrates	Su, Au, Wi, Sp	Shear stress required to disrupt (refresh) biofilms and entrain organic matter
			Su, Au, Sp	Shear stress required to disrupt (refresh) biofilms and entrain organic matter
			Su, Au, Wi, Sp	Overbank events may entrain organic matter
	Macro-invertebrates	Maintenance of water quality suitable for macro-invertebrates	Su, Au, Wi, Sp	Temperature, nutrients and salinity assumed not significant, pollution effects (toxicants) not known. Sediment deposition noted and known to remove susceptible taxa
Diversity of native species, naturally self-reproducing populations of native fish, threatened and iconic native species	Native fish	Suitable in-channel habitat for all life stages	Su, Au, Wi, Sp	Slow shallow habitat required for larvae/juvenile recruitment and adult habitat for small-bodied fish
			Su, Au, Wi, Sp	Deep water habitat for large-bodied fish
		Cues for adult migration during spawning season	Su, Sp	Flow variation required as a cue for migration and spawning
		Suitable off-channel habitat for all life stages	Sp	Inundation of floodplain required by some species and for transport of nutrients and organic matter to drive food webs
Natural channel form and dynamics	Geomorphology	Avoid notching	Dec–Apr	Long duration of stable flow followed by rapid drawdown. Impact likely to be exacerbated by loss of bankside vegetation
	Geomorphology	Avoid slumping	Su	Excessive rates of fall in river level
	Geomorphology	Maintain pool depth	Su	Unseasonal events that fill pools with sediment but do not flush them
	Geomorphology	Maintain natural rates of geomorphic disturbance	Dec–Apr	High-velocity discharge increases disturbance of sand substrates and aquatic macrophytes

* Seasons: Su = summer, Au = autumn, Wi = winter, Sp = spring.
From Cottingham et al. (2007).

For example, a flow event with a 10-year ARI (average recurrence interval) under the current flow regime is equivalent to two-year ARI under natural conditions (Cottingham *et al.* 2003). This represents a considerable reduction in riverbank inundation and a considerable increase in the time that the riverbank remains dry. In addition, targets for minimum passing flows for the downstream end of Wyuna Reach resulted in flows of 250–400 ML d^{-1} instead of 500–5000 ML d^{-1} for 70% of the time (Cottingham *et al.* 2003), indicating considerable loss of flow variability. Given this level of flow reduction and stabilisation, these vegetation patterns were interpreted as signs of terrestrialisation, encroachment and niche compression, and attributed to the relatively long history of regulation of the Lower Goulburn River.

The flow tolerance of riverbank vegetation in Australia has been largely ignored, except for the investigations by Lowe (2002) on a sedge and a stoloniferous herb. There is no information on tolerances of species such as *Poa labillardieri,* which is common along the banks of the Goulburn River. In contrast, in Europe there has been a surge of interest in the effects of submergence on terrestrial plants, focusing on inundation tolerance (Mommer *et al.* 2004, 2005, Voesenek *et al.* 2006) and how this differs between seasons (van Eck *et al.* 2004, 2005). As might be expected, the European studies show that species vary enormously in tolerance (Mommer *et al.* 2006, van Eck *et al.* 2004, 2006) but, despite this variability, the overall trends are clear. All terrestrial species tested survived longer, sometimes much longer, when submerged in winter. Species that were recognised *a priori* as flood-tolerant had a median survival (LT_{50}) of 95–100 days submergence in winter compared with 38–55 days in summer. Moreover, flood-tolerant plants survived longer in summer (LT_{50} = 38–55 days) than terrestrial species recognised as flood-sensitive (LT_{50} = 5–10 days). Partitioning terrestrial species into these two functional groups – flood-sensitive versus flood-tolerant – is a useful distinction for situations required to address summer flows.

The distinction and European survival estimates were applied to the dominant tussock grass on the riverbanks of the lower Goulburn. *Poa labillardieri* was considered to be a flood-sensitive terrestrial plant. The flow conditions of interest were the duration of stage rises in the summer months and whether these lasted or were expected to last long enough to adversely affect the target species. Median survival was set to 15 days, rather than 5–10 days, a slightly conservative adjustment to allow for the buffering effects of its dense tussock habit. The basis of the analysis became the occurrence and vertical distribution up the bank of flow events expected to stress (if isolated) or exclude (if persistent or repeated) flood-sensitive terrestrial plants, referred to here as excluding events.

First, and partly by way of trialling the concept of excluding events, the occurrence of 50 cm-deep events lasting 15 days or longer in the December–March period under modelled natural flows was compared with their occurrence under regulated (historical) flows, for four bands up the bank (Fig. 23.3). This depth was chosen as a tussock would be submerged, and the bands were defined as quartiles of the natural flow regime for the gauge at Murchison, just upstream of the site considered representative of the reach. The marked decrease in the frequency of excluding events over the lower part of the bank under regulated conditions relative to natural (Fig. 23.3) strengthens the interpretation of riverbank terrestrialisation as a historically recent process. Clearly there is an area towards the top of the bank where excluding events are rare even under natural flow conditions and hence where, in general terms, terrestrialisation is not an issue. The location of this area is defined rather crudely when using 25th percentile bands, as the top 25th percentile refers to the entire bank above 4 m in elevation, equivalent to 2–4 m depending on the reach.

Ecological objectives were originally proposed for riverbank vegetation (Cottingham *et al.* 2007) with the intention that, in combination, these would return some diversity in functional

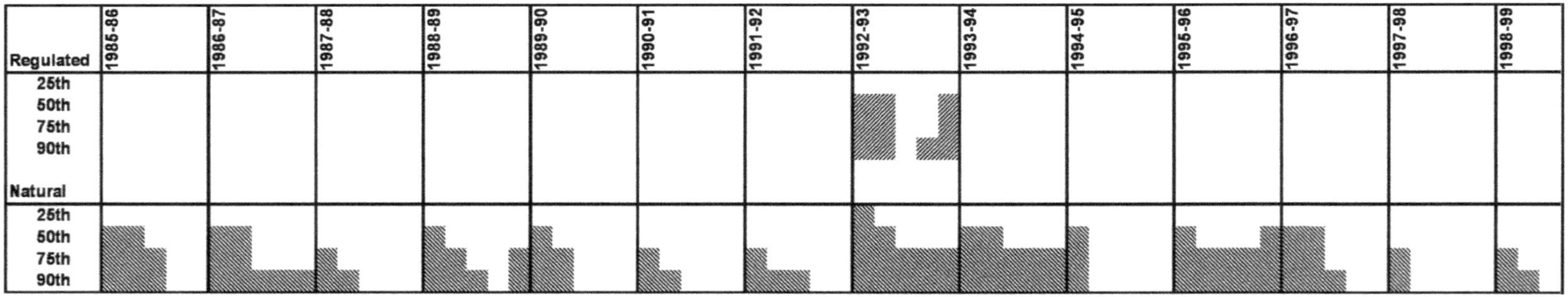

Figure 23.3: Riverbank as a habitat for flood-sensitive plants. Occurrence of flow conditions in the five months from December to March inclusive, as indicated by shading, for each year in a 14-year period expected to cause inundation stress to tussock grasses on the banks of the Goulburn River, under regulated and modelled natural flow regimes. The bank is presented as four zones, defined as 90th, 75th, 50th and 25th percentile of natural flows, and corresponding to 0.6 m, 1 m, 1.8 m and 4 m stages on the gauge at Murchison. Inundation is to a depth of 50 cm (minimum) to allow for complete submergence of the grass.

groups (terrestrial, amphibious, aquatic) towards that assumed for pre-regulated conditions. Two objectives that linked terrestrial vegetation to flow were:

- maintain persistent cover over the upper part of the bank (equivalent to natural flow percentiles where bank inundation has same/similar duration under natural as historic) – i.e. protect those areas of the bank unaffected by regulation;
- reduce the cover of terrestrial vegetation over the lower part of the bank – i.e. move towards assumed natural, for those parts of bank falling below threshold (threshold is flow percentile where duration under natural regime and historic regulated regime are approximately equal).

For the scenario testing, the definition of an excluding event was simplified to just be duration, but the bank was partitioned into nine elevation bands corresponding to deciles of the natural flow regime (10th percentile to 90th percentile). As stated above, the upper and lower parts of the riverbank had differing objectives. For the lower part, where terrestrialisation was deemed to have occurred, the loss of tussock grass cover would be considered a reversal of the terrestrialisation process, thus the occurrence of excluding events was not considered a threat. For the upper part, the presence of *Poa labillardieri* needed to be protected and the occurrence of excluding events would be a threat. The distinction between these two parts of the riverbank was defined as the flow at which the duration of inundation events was the same for natural as for regulated flows – 6600 ML d^{-1}, corresponding to the 30th percentile. Risks were defined in precautionary terms as nine out of 10 years (the 90th percentile, relative to the 'recommended' definition) and were different for each part of the bank. For the upper part, the risk to the persistence of *Poa labillardieri* increases if an event duration exceeds the recommended 15 days, whereas for the lower part the risks increase if event duration falls below the recommended 40 days. The recommended durations differ because of their differing purposes (protect versus decrease) and the confidence of achieving them.

The flow series for the most recent IVTs (2005/06) was then entered into the decision support tool. The results (not shown) indicated that flows to inundate the 40th, 50th and 60th percentile stage heights were of insufficient duration to limit the encroachment of terrestrial vegetation down the riverbank. This suggests that, if carefully managed to avoid undue rates of rise and fall, IVTs could be used to limit the encroachment of terrestrial vegetation and increase the habitat suitable for plants in the amphibious functional group. For example, flows could be held above the 40th–60th percentile stage height for longer than 20 days over the summer period to meet downstream demand. This would reduce the high risk of encroachment by flood-sensitive species.

Riverine (in-channel) primary production

The contributions of phytoplankton, biofilms and submerged macrophytes to riverine primary production are affected by flow rates. In particular, changes in water depth and water velocity affect the habitats that determine whether primary producers can establish photosynthetically active communities (Biggs *et al.* 1998; Madsen *et al.* 2001; Oliver and Lorenz 2007). Light energy declines exponentially with depth, confining the growth of photosynthetic organisms to well-illuminated surface layers (Reynolds 1984; Blanch *et al.* 1998). Few underwater light measurements are available for the Goulburn River (Webb and Chan 2004), so a data set from the Murray and Murrumbidgee Rivers (Oliver, unpubl.) was used to relate vertical attenuation coefficient (k_d) for photosynthetically active radiation to monitored nephelometric turbidity.

Phytoplankton

Models of daily phytoplankton photosynthesis and respiration indicate that zero net production occurs when the ratio of the euphotic depth (z_{eu}, the depth of 1% light penetration) to the mean depth of the water column (z_{mean}) is of the order 0.3 (Talling 1957; Reynolds 1984; Oliver *et al.* 1999). This indicates a shift from light sufficiency, where photosynthesis exceeds respiration, to conditions of light limitation where growth is curtailed. Applying this ratio to the Goulburn River indicates that the water column is well-illuminated most of the time (Fig. 23.4). Although this index was used to delineate the proportion of time suitable for growth, it does not capture the continuous nature of the phytoplankton response to light where gross primary production (*GPP*) reduces progressively as light availability declines. Empirical models were applied to describe these responses.

Studies in the River Murray (Oliver and Merrick 2006) show that *GPP* is a linear function of the mean irradiance of the water column, provided the illuminated zone is within the water depth ($z_{eu}/z_{mean} \leq 1$). Assuming similar conditions in the Goulburn River, and a typical daily summer average incident irradiance of 900 $\mu E\ m^{-2}\ s^{-1}$ (Oliver, unpubl. data), the influence on *GPP* of light changes caused by the alterations in depth resulting from changing flow rates was estimated and expressed as a maximum potential specific daily growth rate for periods when the z_{eu}/z_{mean} ratio ≤ 1 (Fig. 23.4). These calculations show a decline in *GPP* at Murchison as flow increases, and at Wyuna there is a substantial effect leading to a halving of *GPP*. Calculations for net phytoplankton production (*NP*) were similarly based on an empirical model derived from studies on the River Murray. Although the responses are similar to those of *GPP*, relative changes in net production are larger and become negative at Wyuna as flows increase. Mean net production rates for selected periods were used to assess alterations in growth conditions as a function of flow rates.

Change in water depth also influences phytoplankton loss through sedimentation, with the concentration of phytoplankton declining exponentially over time as a function of sinking rate and the mean depth of the water layer (Reynolds and Wiseman 1982). The daily sinking loss rate, calculated as a fractional decline but expressed as a positive specific growth rate for ease of comparison with *NP*, is related to flow rate through the resulting depth changes. The effect is demonstrated for the dominant diatom *Aulacoseira granulata* (Fig. 23.4), which has a high sinking rate of 0.95 $m\ d^{-1}$ (Sherman *et al.* 1998). At Murchison, net growth rates are similar to sinking rates while at Wyuna sinking rates significantly exceed the expected production rates. This changing balance between positive and negative phytoplankton production along rivers has been widely reported (Reynolds and Descy 1996; Wehr and Descy 1998; Bahnwart *et al.* 1999; Leland 2003). A moderate sinking rate of 0.4 $m\ s^{-1}$ was applied to characterise mean loss rates of less dense phytoplankton (Reynolds 1984).

In the River Murray, phytoplankton biomass increased exponentially as flow declined below 0.2 $m\ s^{-1}$ (Oliver and Lorenz 2007). Similar observations in other large rivers reflect the inability of phytoplankton to establish large populations when retention times (the inverse of velocity) are short (Bowles and Quennell 1971; Reynolds 1988). Discharges of 875 $ML\ d^{-1}$ at Murchison and 1100 $ML\ d^{-1}$ at Wyuna generate velocities of 0.2 $m\ s^{-1}$ and at higher values phytoplankton populations are likely to remain sparse, even with sufficient light. At lower flow rates there is an increasing chance of thermal stratification, especially during summer, that improves growth conditions for phytoplankton such as buoyant cyanobacteria. In south-east Australia this generally occurs below average cross-sectional velocities of ~0.05 $m\ s^{-1}$ (Bormans and Webster 1997; Mitrovic *et al.* 2003), so problems of cyanobacterial surface blooms are most likely below flows of 80 $ML\ d^{-1}$ at Murchison and 34 $ML\ d^{-1}$ at Wyuna. These critical water velocity levels were used to indicate the effects of flow velocity on phytoplankton biomass.

Periphyton, biofilms and macrophytes

The potential habitat suitable for attached primary producers is the area of sufficiently illuminated submerged sediment. The micro-algae in periphyton and in biofilms have light requirements similar to that of the phytoplankton (Reynolds 1984) and the 1% light-penetration

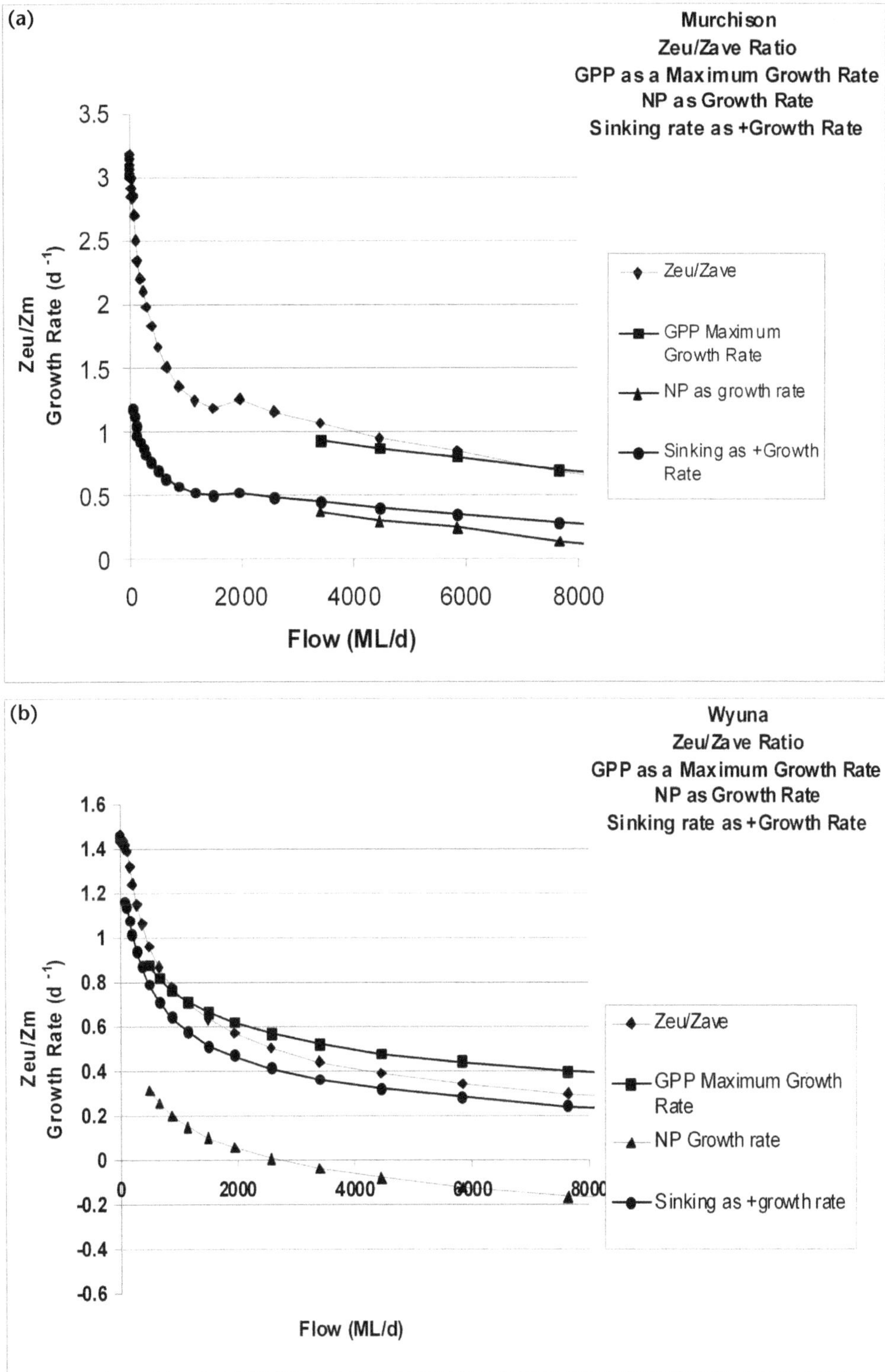

Figure 23.4: Characteristics of light and phytoplankton responses to flow at (a) Murchison and (b) Wyuna.

depth was used to estimate the limits of their depth distribution. Zero net production for charophytes and angiosperms occurs in the range 3.5–47.3 μmol m^{-2} s^{-1} (Blanch *et al.* 1998), which is 0.5–5% of typical incident summer irradiances. The 1% light penetration level was chosen to delineate the area of suitable light habitat for macrophytes. Average channel cross-sections and typical turbidity values were used to calculate the area of illuminated benthos at different discharges for each of the two river reaches (Cottingham *et al.* 2007).

The distribution of submerged macrophytes and biofilms are also influenced by water velocity. In river reaches with steady flow conditions and velocities <0.2 m s^{-1}, up to 75% of the sediment can be covered with macrophytes. Less than 10% is covered where velocities exceed 0.9 m s^{-1} (Biggs 1996). Madsen *et al.* (2001) found photosynthetic rates increased with flow of 0–0.1 m s^{-1}, while above this macrophyte biomass was negatively correlated with flows up to ~1 m s^{-1}, when they were no longer present. Water velocities also impinge on the structure and function of biofilms. Chlorophyll content and ash-free dry mass increase as velocities decrease below 0.3 m s^{-1} (Jowett and Biggs 1997). Mucilaginous diatom/cyanobacterial mats show maximum biomass at near-bed velocities of ~0.2 m s^{-1}, whereas the maximum of filamentous green algal biofilms occurs at near-bed velocities <0.2 m s^{-1} (Biggs *et al.* 1998). At faster flow rates the biomass of biofilms declines exponentially to minimal levels at velocities >0.4 m s^{-1} (Biggs *et al.* 1998; Ryder *et al.* 2006). These flow influences were included by curtailing illuminated areas by velocities of 0.2, 0.3 and 0.4 m s^{-1} to indicate suitable areas for biofilm occurrence, and 0.2 m s^{-1} and 0.9 m s^{-1} to estimate growth areas for macrophytes.

Oscillations in water depth influence the structure and composition of attached primary producers (Sheldon and Walker 1997; Burns and Walker 2000). The maximum net primary productivity of biofilms developing on redgum blocks occurred after 29 days of submersion then fell rapidly (Ryder 2004). Blocks undergoing high-frequency oscillations (five days submerged and nine days emersed) developed a heterotrophic biofilm after 75 days with negative *NP*, while those exposed to a moderate cycle (11 days submerged and 21 days emersed) had an autotrophic biofilm with higher *NP* per algal biomass. These effects will be important at the water's edge, especially under fluctuating flows. The effect of changing water levels was assessed from the proportion of time that various heights on the bank were in the euphotic zone. In the case of biofilms, 14 days or longer was considered sufficient time for their establishment and was set as the minimum wet period. A similar analysis was used for submerged macrophytes, except that typical establishment periods are four to six weeks (Larned *et al.* 2006) so a minimum inundation period of 42 days was used to indicate the minimum wet period for establishment. This analysis is similar to that described above for terrestrial river-bank vegetation.

The indices selected for phytoplankton, epiphytic algae, biofilms and submerged macrophytes attempted to clarify the potential for supporting in-channel biomass levels and productivity. Measures of the indices when sites were unimpacted by flow regulation were used as reference conditions for comparison with current values and values under IVT flow conditions. In general, pre-regulation values for particular seasons were used as recommended targets, while moderate risk values were considered to be ±20% of these. Values were only considered relevant if they fell within the critical bounds outlined above.

In summary, the following objectives were set for in-channel primary production;

- planktonic algae:
  - biomass levels resembling sites unimpacted by flow regulation;
  - productivity resembling sites unimpacted by flow regulation;
  - community composition resembling sites unimpacted by flow regulation.
- periphytic algae, biofilms and submerged macrophytes:
  - biomass levels resembling sites unimpacted by flow regulation;

- productivity resembling sites unimpacted by flow regulation;
- community composition resembling sites unimpacted by flow regulation.

Flow stressors for in-stream production were:

- planktonic algae:
  - mean residence time (hr/km);
  - mean ratio of sinking velocity (0.40 m s^{-1}) to mean water depth (assess sinking losses from the water column of average plankton);
  - mean ratio of fall velocity (0.95 m s^{-1}) to mean water depth (assess sinking losses of the diatom *Aulacoseira granulata*);
  - proportion of time when euphotic depth is less than 0.3 times the mean depth (assess when in-channel production exceeds respiration, i.e. *NP* is negative);
  - mean ratio of euphotic depth to mean water depth (assess light-limited *GPP* and *NP* in the channel).
- periphytic algae, biofilms and submerged macrophytes:
  - mean illuminated area of benthos (m^2 per m length of channel) (assess relative area of benthic production);
  - mean illuminated area of benthos with velocity <0.2 m s^{-1}, <0.3 m s^{-1} and <0.4 m s^{-1} (m^2 per m length of channel) (assess velocity conditions suitable for benthic production);
  - proportion of time the bank has been within the euphotic zone for at least 14 days, at selected stages above the bed corresponding with various exceedence flows in the natural regime.

Based on the primary production conceptual model presented in Cottingham *et al.* (2007), we identified the three flow stressors affecting the contribution of epiphytic, biofilms and submerged macrophyte primary production. These are discussed below, along with their corresponding flow elements, seasons and thresholds.

Mean illuminated area of the benthos

The illuminated area of the benthos is calculated as the area of the channel bed and banks with a water depth less than euphotic depth. There was a concern that the illuminated area might stray outside the natural range of values under the effects of flow regulation. For this reason, upper and lower bounds were placed on the mean area of illuminated benthos (high- and low-risk thresholds were used). More specifically, the upper threshold was placed on the 90th percentile year (it was accepted that in one out of 10 years mean illuminated area could exceed this value). Similarly, a lower threshold was placed on the 10th percentile year. The thresholds were based on the natural range of values for the flow element. To ensure a high likelihood of achieving the ecological objective, the 10th and 90th percentile values modelled for the pre-regulation flow regime were chosen as the lower and upper thresholds.

Mean illuminated area of the benthos in slackwater areas

The second flow stressor is the illuminated area within slackwater zones. This second flow stressor is very similar to the first. The literature suggests that slackwater should be defined by an upper velocity of 0.2–0.4 m s^{-1} (Cottingham *et al.* 2007). Given this range, we decided to use flow elements defined by velocity thresholds of 0.2 m s^{-1} and 0.4 m s^{-1}. Upper and lower thresholds were set for summer and spring values using the same approach as the first flow stressor.

Proportion of time the bank remains within the euphotic zone for at least 14 days

Establishment of biofilm requires that the wetted surface remains wet and within the euphotic depth. Drying or submersion below the euphotic depth will adversely affect biofilm production. Devising a flow stressor for this flow dependency is relatively complicated because it deals with the effect of temporal sequences of flow on persistence of inundation at different locations across the channel. There is unlikely to be a single duration of inundation to establish biofilm production; a period of 14 days was considered an appropriate minimum threshold to define this period (Cottingham *et al.* 2007).

To calculate flow elements, we divided the channel bed and banks into regions inundated at different discharges, as for the bank vegetation flow stressor (i.e. stages corresponding to the 10th, 20th … 90th flow percentiles). The flow element was calculated as the proportion of the season when the bank (at the various flow stages) was persistently in the euphotic zone. Only periods of illumination longer than two weeks contributed to this flow element. Nine flow elements were calculated for the nine different flow levels considered.

Decision support tool results

Results based on the flow series for the most recent IVTs (2005/06) (not shown) indicated high risk related to factors such as prolonged periods in a season where shear stress was insufficient to renew biofilms or when benthic and macrophyte production would be unduly favoured over phytoplankton production, compared with the natural flow regime. Rating curves for each flow stressor contained in the support tool (e.g. flow versus shear stress, flow versus wetted area in the euphotic zone) and thresholds for the proportion of time (in a season) that stage height remains at a particular level, could be used to identify the appropriate stage height period of inundation required to reduce the risks associated with the IVTs, in a similar fashion to that described for bank vegetation.

Conclusions

The decision support tool was developed and used to evaluate the potential risks to ecological objectives should the lower Goulburn be used as a carrier for IVTs. The general approach and the tool developed by the scientific panel have since proved to have a wider application. For example, it was used to consider the effects of periodic IVTs moving down the Campaspe River downstream of Campaspe Weir (Cottingham *et al.* 2008). This was possible because the principles were expressed in general terms rather than being specific to the lower Goulburn River, and the ecology–flow relationships were based on available scientific knowledge.

The decision support tool was developed so that any future flow scenario could be considered in terms of potential risks to stated ecosystem objectives. It can be used to identify adjustments to a flow scenario that could reduce potential risks, independent of the scientific panel. However, it is important to remember that the decision support tool does not predict the future condition or functioning of the river. The user inspects the display and decides whether to proceed, based on personal assessment of how many, or which, high or moderate risk stressors to address.

Acknowledgments

This project was funded by the Goulburn-Broken Catchment Management Authority. The assistance and support of Geoff Earl and Wayne Tennant was greatly appreciated.

References

Bahnwart M, Hübener T and Schubert H (1999). Downstream changes in phytoplankton composition and biomass in a lowland river-lake system (Warnow River, Germany). *Hydrobiology* **391**, 99–111.

Biggs BJF (1996). Hydraulic habitat of plants in streams. *Regulated Rivers: Research and Management* **12**, 131–144.

Biggs BJF, Goring DG and Nikora VI (1998). Subsidy and stress responses of stream periphyton to gradients in water velocity as a function of community growth form. *Journal of Phycology* **34**, 598–607.

Blanch SJ, Ganf G and Walker K (1998). Growth and recruitment in *Vallisneria americana* as related to average irradiance in the water column. *Aquatic Biology* **61**, 181–205.

Bormans M and Webster I (1997). A mixing criterion for turbid rivers. *Environmental Modelling and Software* **12**, 329–333.

Bowles B and Quennell S (1971). Some quantitative algal studies of the River Thames. *Journal of Society of Water Treatment and Examination* **20**, 35–51.

Burns A and Walker KF (2000). Biofilms as food for decapods (Atyidae, Palaemonidae) in the River Murray, South Australia. *Hydrobiologia* **437**, 83–90.

Cottingham P, Stewardson M, Crook D, Hillman T, Roberts J and Rutherfurd I (2003). Environmental flow recommendations for the Goulburn River below Lake Eildon. Technical Report 01/2003. CRC for Freshwater Ecology and CRC for Catchment Hydrology, Canberra.

Cottingham P, Stewardson M, Crook D, Hillman T, Oliver R, Roberts J and Rutherfurd I (2007). Evaluation of summer inter-valley water transfers from the Goulburn River. Report to the Goulburn Broken Catchment Management Authority. Peter Cottingham and Assoc.

Cottingham P, Horne A, Crook D, Hillman T, Roberts J, Sharpe A and Stewardson M (2008). Campaspe summer flows project: assessment of potential summer releases along the lower Campaspe River. Report to the North Central Catchment Management Authority. Peter Cottingham and Assoc.

Hillman TJ (2009). The policy challenge of matching environmental water to ecological need. In *Policy and Strategic Behaviour in Water Resource Management.* (Eds A Dinar and J Albiac) pp. 109–124. Earthscan Publishing, London.

Jowett I and Biggs BJF (1997). Flood and velocity effects on periphyton and silt accumulation in two New Zealand rivers. *New Zealand Journal of Marine and Freshwater Research* **31**, 287–300.

Larned ST, Suren AM, Flanagan M, Biggs BJF and Riis T (2006). Macrophytes in urban stream rehabilitation: establishment, ecological effects, and public perception. *Restoration Ecology* **14(3)**, 429–440.

Leland HV (2003). The influence of water depth and flow regime on phytoplankton biomass and community structure in a shallow, lowland river. *Hydrobiologia* **506-509**, 247–255.

LREFSP (2002). Environmental flow determination of the Loddon River Catchment. Final report. Unpublished report by the Loddon River Environmental Flows Scientific Panel to the North Central Catchment Management Authority and Dept of Natural Resources and Environment.

Lowe B (2002). Impacts of altered flow regime on riverbank vegetation in a lowland floodplain river. PhD thesis. Charles Sturt University, Wagga Wagga.

Madsen JD, Chambers PA, James WF, Koch EW and Westlake DF (2001). The interaction between water movement, sediment dynamics and submersed macrophytes. *Hydrobiologia* **444**, 71–84.

Mitrovic S, Oliver R, Rees C, Bowling L and Buckney R (2003). Critical flow velocities for the growth and dominance of *Anabaena circinalis* in some turbid freshwater rivers. *Freshwater Biology* **48**, 164–174.

Mommer L, Pedersen O and Visser EJW (2004). Acclimation of a terrestrial plant to submergence facilitates gas exchange under water. *Plant, Cell and Environment* **27**, 1281–1287.

Mommer L, de Kroon H, Pierik R, Bogerman GM and Visser EJW (2005). A functional comparison of acclimation to shade and submergence in two terrestrial plant species. *New Phytologist* **167,** 197–206.

Mommer L, Lenssen JPM, Huber H, Visser EJW and de Kroon H (2006). Ecophysiological determinants of plant performance under flooding: a comparative study of seven plant families. *Journal of Ecology* **94,** 1117–1129.

Oliver RL and Lorenz Z (2007). Murray River metabolism: quantifying the food supplies that support riverine food webs. CSIRO Water for a Healthy Country National Research Flagship, Canberra.

Oliver RL and Merrick CJ (2006). Partitioning of river metabolism identifies phytoplankton as a major contributor in the regulated Murray River (Australia). *Freshwater Biology* **51**, 1131–1148.

Oliver R, Hart B, Olley J, Grace M, Rees G and Caitcheon G (1999). The Darling River: algal growth and the cycling and sources of nutrients. CRCFE and CSIRO report to the Murray-Darling Basin Commission, Canberra.

Reynolds CS (1984). *The Ecology of Freshwater Phytoplankton.* Cambridge University Press, Cambridge.

Reynolds C (1988). Functional morphology and the adaptive strategies of freshwater phytoplankton. In *Growth and Survival Strategies of Freshwater Phytoplankton.* (Ed. CD Sandgren) pp. 388–433. Cambridge University Press, Cambridge.

Reynolds C and Descy JP (1996). The production, biomass and structure of phytoplankton in large rivers. *Archives of Hydrobiology* (Suppl.) **113**, 161–187.

Reynolds C and Wiseman S (1982). Sinking losses of phytoplankton in enclosed limnetic systems. *Journal of Plankton Research* **4**, 489–522.

Ryder DS (2004). Response of biofilm metabolism to water level variability in a regulated floodplain river. *Journal of the North American Benthological Society* **23**, 214–223.

Ryder DS, Watts RJ, Nye E and Burns A (2006). Can flow velocity regulate epixylic biofilm structure in a regulated floodplain river? *Marine and Freshwater Research* **57**, 29–36.

Sheldon F and Walker KF (1997). Changes in biofilms, induced by flow regulation, may explain the extinction of gastropods in the lower River Murray. *Hydrobiologia* **347**, 97–108.

Sherman BS, Webster IT, Jones GJ and Oliver RL (1998). Transition between *Aulacoseira* and *Anabaena* dominance in a turbid river weir pool. *Limnology and Oceanography* **43**, 1902–1915.

SKM (2006). Campaspe River environmental flows assessment. SKM report to the North Central Catchment Management Authority.

Talling JF (1957). The phytoplankton population as a compound photo-synthetic system. *New Phytologist* **56**, 133–149.

van Eck WHM, van de Steeg HM, Blom CWPM and de Kroon H (2004). Is tolerance to summer flooding correlated with distribution patterns in river floodplains? A comparative study of 20 terrestrial grassland species. *Oikos* **107**, 393–405.

van Eck WHJM, Lenssen JPM, Rengelink RHJ, Blom CWPM and de Kroon H (2005). Water temperature instead of acclimation stage and oxygen concentration determines response to winter floods. *Aquatic Botany* **81**, 253–264.

van Eck WHJM, Lenssen JPM, van de Steeg HM, Blom CWPM and de Kroon H (2006). Seasonal dependent effects of flooding on plant species survival and zonation: a comparative study of 10 terrestrial grassland species. *Hydrobiologia* **565**, 59–69.

Voesenek LAC, Colmer TD, Pierik R, Millenaar FF and Peeters AJM (2006). How plants cope with submergence. *New Phytologist* **170**, 213–226.

Webb JA and Chan TU (2004). Ecological risk associated with irrigation systems in the Goulburn-Broken catchment. Phase II: Priority risk – blue-green algal blooms. Water Studies Centre, Monash University, Melbourne.

Wehr JD and Descy J (1998). Use of phytoplankton in large river management. *Journal of Phycology* **34**, 741–749.

Conclusions: lessons from ecosystem response modelling and future directions for the Murray-Darling Basin

Neil Saintilan and Ian Overton

Introduction

The chapters in this book have identified the broad areas of scientific endeavour associated with ecosystem modelling.

- **Inventory and health assessment.** Chapters have identified the need for improved inventory work including the health and extent of water-dependent ecosystems in the Murray-Darling Basin and their hydrology to be used as baselines to assess changes in ecosystem health. This includes a need for improved assessment of 'health' versus condition.
- **Data shortage.** Many authors have highlighted the lack of long-term monitoring projects that can demonstrate the link between changes in hydrology and changes in ecosystems. Empirical and scientifically robust measurements of ecosystem change are needed across multiple spatial and temporal scales in all areas of ecosystem science.
- **Ecosystem responses.** Ecosystem managers require strong links between hydrology and ecology with rules that can guide the management of flows and water levels. Ecohydrology responses are most easily combined when they are in the form of prescribed hydrology variables predicted to create a variable or threshold response in particular components of the ecosystem. Relationships are complicated by other factors such as salinity, water quality, land management and complex interactions between organisms.
- **Complexity of ecosystem science.** Challenges include the non-linearity of ecosystems, the transferability of responses to other river systems or wetlands, the lack of data on many components of the ecosystem and the potential unpredictability of ecosystems. There is a need for improved complexity of models, to include multiple spatial and temporal scales. Many ecosystem functions such as bird breeding links to wetland inundation need to be considered at a basin scale rather than in individual wetlands. Temporal responses include hydrological requirements across the time of bird nesting, fledging and feeding of young for successful breeding to occur.
- **Whole of ecosystem functions.** A method of identifying the risk of ecosystem changes going to new ecosystem states was presented for the Coorong and Murray Mouth region. Understanding ecosystem changes requires an understanding not just of individual

component responses to changes in hydrology but of the whole ecosystem function. Ecosystem function requires ecological and geomorphological states to function across the biogeochemical, energy and food web cycles. Measurable evidence in changes in ecosystem function are limited in scientific research.

- **Economic and social costs and benefits.** One role of healthy ecosystems is to provide ecosystem services that can be measured with economic methods. This allows changes to hydrology to be valued in a dollar sense for the loss of these services. The cost of environmental flow releases needs to be considered in terms of the cost of water and the value from improved ecosystem health. Ecosystem response models can be integrated with models of economic and social costs and benefits. Ganf *et al.* (Ch. 18, this volume) proposed a method for the integration of environmental, economic and social models for water allocation and prioritisation of environmental flows.

Marsh and Cuddy (Ch. 1, this volume) describe a classification of ecosystem response models and provide examples of models that have been developed within the basin. They identify that ecosystem response models are important tools for predicting impacts of management actions on the natural environment. They explain that models have had a varying degree of success in assisting management decisions, as these usually require multiple ecological outcomes using multiple ecosystem models that tend to be in incompatible forms. The complexity of the task of integrating ecosystem response models that focus on ecosystem components such as species and life-stages should not be underestimated. Chapter 1 presents the Murray Flows Assessment Tool as an example of an integrated model. Kingsford *et al.* (Ch. 2, this volume) describe the range of time scales over which organisms respond to environmental flows, and stress the importance of spatially explicit modelling which integrates the responses of key suites of organisms. This modelling suggests indicators which can be monitored in an adaptive management context. Overton *et al.* (Ch. 16, this volume) and Overton and Doody (Ch. 21, this volume) provide examples of spatially explicit models that have been developed in the River Murray.

Ecosystem response models can be described by their basic taxonomic components Ch. 1, this volume). These include the ecosystem component being modelled, the drivers of the change or the scenarios, the abstraction method used and the computational form. Model abstraction, or the conceptual approach, usually takes one of the following forms – habitat, primary production, mass balance or community models such as condition, presence/absence, aggregated population, size or age structure (Ch. 1).

The development and validation of ecosystem models was the focus of Stewardson and Webb (Ch. 3, this volume). They argue that improved processes of model development, including conceptual models, causal pathways and hierarchical models, will improve model assessment from sparse data. The analysis of ecosystem responses is limited by the amount of empirical data that can be used to show causal relationships between ecological outcomes and flow regimes. Potential data across the basin was recently identified and, despite the hundreds of data sets on biological condition and abundance collected as part of Commonwealth and state monitoring assessments, intervention and event monitoring and academic research studies, the amount of data suited to analysing broad basin-scale ecosystem responses is limited (Overton *et al.* 2008).

Quantitative ecosystem response models are being developed in many parts of the Murray-Darling Basin to better inform decisions on environmental water allocation. The Narran Ecosystem Project (2004–07) developed a linked hydrodynamic–hydrological model of the Narran Lakes, which provided a basis for modelling the ecological impact of alterations in river flows (Thoms *et al.* 2007). The Australian National Water Commission, through the Water Smart

Australia program, is funding the application of this approach in other wetlands of the northern basin, including the Macquarie Marshes, the Gwydir wetlands and the Lowbidgee wetlands (Saintilan *et al.*, Ch. 4, this volume), including the development of decision support systems.

A whole of ecosystem model is required to assess water requirements for key assets. Lester and Fairweather (Ch. 22, this volume) provide an example of a statistical model of the Coorong ecosystem at the mouth of the Murray-Darling Basin that predicts the ecosystem state under various environmental flow scenarios.

This book has detailed ecological responses of primary production, aquatic vegetation, fish, birds and amphibians to water modelled at a range of temporal scales at many locations across the basin. The models describe aquatic species condition and abundance, ecosystem function or changes of state resulting from altered hydrological regimes and colonisation by invasive species, and in many cases explore the uptake of this information into water and reserve management systems. This concluding chapter synthesises the insights of these studies.

Northern (Darling) basin synthesis

The management of wetland mosaics with environmental water is a complex interdisciplinary task. These systems are large, and significant wetland assets are distributed across a network of anastomising channels, flood-runners and floodplains. The optimisation of environmental water requires an understanding not only of the water requirements of these assets but an appreciation of optimal delivery strategies through a hydrologically complex system.

In the northern Murray-Darling Basin, rivers flow through semi-arid and arid landscapes where rainfall is increasingly variable with distance from the eastern highlands. Balcombe *et al.* (Ch. 5, this volume) describe the challenges posed by flow variability to fish populations, which find refuge in pools and waterholes during dry periods. Fish biomass responses to flow in these situations could be predicted on the basis of waterhole habitat variables in the Warrego River, but not in the Moonie River. The result challenges the assumption that flow–ecology relationships generated for one river or wetland can be transferred to another.

Wilson *et al.* (Ch. 7, this volume) describe the difficulty in developing ecosystem response models of the relationship between fish assemblage composition and flow in the Gwydir wetlands. However, flow variability during the spawning period did appear to have a positive influence on the growth rate of Australian smelt and bony bream. Flow variability may also be important in decreasing competitive interactions between native and juvenile fish. Kelleway *et al.* (Ch. 6, this volume) describe a shift in the trophic position of silver perch compared to introduced carp during an environmental flow in the Gwydir. Both studies suggest new indicators for monitoring the success of environmental flows, but point to hydrological variability as a characteristic of semi-arid rivers vital to the preservation of biological diversity in these systems.

The application of remote-sensing techniques to vegetation extent and condition assessment has allowed the modelling of vegetation–flow responses at broad spatial and temporal scales. This is a particularly powerful approach if wetland inundation characteristics are modelled over similar scales. Thomas *et al.* (Ch. 9, this volume) show how the Landsat archive can be used to reconstruct a spatially explicit history of inundation frequency over large inland wetland complexes. They applied this technique to the Macquarie Marshes, the Gwydir wetlands and the Lowbidgee wetlands in the southern basin. The Macquarie Marshes case study demonstrates that, at the landscape scale, interdecadal trends in wetland vegetation can be linked to changing inundation frequency. This spatially explicit analysis provides quantitative support to previously anecdotal observations of wetland decline in response to flow alteration, and suggests a framework for assessing vegetation change in an adaptive monitoring context.

Decision support systems are one means of integrating the ecological and hydrological models required to optimise the delivery of environmental water. Several such systems have been applied in recent years to manage environmental water in the Murray-Darling Basin, including the Murray Flow Assessment Tool (Young *et al.* 2003) and the Water Allocation decision support system (Letcher 2005). In this volume, Merritt *et al.* (Ch. 8) outline the IBIS decision support system which, as applied to the Gwydir wetlands, integrates a water balance model of the Gwydir wetlands (Powell *et al.* 2008) with models of vegetation and fish response to flow. The developers present Bayesian networks as a useful tool in dealing with the uncertainty inherent in ecological prediction. Options are compared in terms of the probability of occurrence of various ecological outcomes. The approach underscores the importance of good models populating the decision support system. We are often constrained by a lack of clear relationships between flow and response for many species, in particular the emergent relationships of predation and competition between species.

Another degree of complexity is added to the task of water planning in the northern basin by the threat of climate change. The CSIRO Sustainable Yields program (CSIRO 2008) has modelled potential hydrological impacts of climate change scenarios for major river catchments within NSW. On the basis of a thorough review of the water requirements of biota inhabiting the Macquarie Marshes, Rogers *et al.* (Ch. 10, this volume; 2008) have modelled the effects of the forecast decline in inundation frequency and the shift towards more constant low-flow conditions. Some species will benefit from these changes, but most of the species for which the Macquarie Marshes are famous will decline.

The basin will not be uniformly affected by climate change, though. Models forecasting regional climate change impacts become less certain in the northern basin as the relative influence of tropical storm patterns becomes stronger. There are several large-scale drivers of interannual and interdecadal variability in rainfall over the northern basin – the El Niño Southern Oscillation, the Interdecadal Pacific Oscillation and the Madden-Julien Oscillation (Thoms *et al.* 2007) – and their links with global warming are poorly understood. Given the prognosis of continuing decline in the availability of water in the south, the northern basin wetlands could be the more sustainable in the long term provided water allocations are appropriately balanced between environmental and extractive needs.

Southern (Murray) basin synthesis

The southern basin is dominated by large, highly regulated lowland rivers such as the River Murray, Murrumbidgee, Lachlan, Goulburn and other southern tributaries of the River Murray. The high degree of regulation since the 1920s has led to a large decline in aquatic ecosystem condition and a reduction in ecosystem function. The southern basin also contains the River Murray estuary and the Coorong region that is facing ecological collapse.

The floodplain of the Lachlan River contains one of the most extensive networks of arid inland wetlands in the Murray-Darling Basin. Challenges of low flows and groundwater-dependent ecosystems are described by Driver *et al.* (Ch. 12, this volume), who summarise previous and ongoing studies to develop a whole-of-catchment flow–ecology model.

To date, most environmental flows have targeted the triumvirate of vegetation, birds and fish, though recently the decline in the southern bell frog populations in their Lowbidgee stronghold has prompted targeted environmental releases in support of frog breeding. Wassens (Ch. 14, this volume) has provided a well-developed framework for the prioritisation of wetlands for inundation, with considerations including the proximity of wetlands to drought refugia, likely exposure to predation, and suitability of habitat and transfer losses. The model would be best served by extensive survey prior to application. The study highlights the need both for good models (a good understanding of the relationships between populations, habitats and flow) and good antecedent data to apply the best frameworks.

Reduced flow variability within rivers has longitudinal effects on a diverse range of plants and animals within the river channel and its immediate environs. Long periods of high and stable discharge on the upstream sections of the Murrumbidgee River has resulted in detrimental effects on biofilms, riverbank plant understoreys and the larvae of fish and invertebrates (Watts, Ch. 13, this volume). Weir pools have permanently inundated previously ephemeral wetlands, leading to reduced floristic diversity, while downstream reductions in inundation frequency and duration have had detrimental effects on floodplain wetlands (Wen 2008; Ch. 13, this volume; Wen *et al.*, Ch. 15, this volume).

Wen *et al.* (Ch. 15, this volume) demonstrate how CART analysis can be used to model critical thresholds relevant to the condition and type of floodplain wetland vegetation community. This work related antecedent inundation history at Yanga National Park in the Lowbidgee to the condition categories identified and mapped by McCosker (2008). Not only can survival thresholds be identified, but so can the flood timing and duration required to maintain river red gum in better condition.

Ward and Colloff (Ch. 20, this volume) identify that the hydrograph forms the basic data set for management of flows but, despite much of the identification of floodplain and wetland ecosystem responses being linked to floodplain inundation, it was almost 100 years between commencement of data collection from gauging stations above the Barmah-Millewa Forest and the development of the first models relating volume of flow to area inundated. The development of the River Murray Floodplain Inundation Model (RiM-FIM) (Overton *et al.*, Ch. 16, this volume) provided a consistent assessment of floodplain inundation and wetland connectivity for the whole of the River Murray. Overton *et al.* used the RiM-FIM to establish a link between floodplain vegetation and flooding regimes. This flooding regime was further linked to invasive plant distributions along the River Murray (Stokes *et al.*, Ch. 19, this volume).

The importance of floodplain inundation was also identified by Oliver and Lorenz (Ch. 17, this volume) in the link between in-stream primary productivity and exchange of carbon with floodplains. Ganf *et al.* (Ch. 18, this volume) showed the link between aquatic macrophytes and wetland connectivity.

Ecosystem response modelling at the Living Murray icon sites has provided valuable input into environmental flow management and the development of infrastructure plans. The Barmah-Millewa Forest (Ch. 20, this volume), the Chowilla floodplain and Lindsay-Wallpolla Islands (Overton and Doody, Ch. 21, this volume) and the Coorong (Lester and Fairweather, Ch. 22, this volume) show different approaches to ecosystem response modelling driven by different management drivers and available data.

The integration of a range of ecosystem responses using expert panel assessment and numerical models was provided by Cottingham *et al.* (Ch. 23, this volume) for the Goulburn River in Victoria.

Conclusion

John Kenneth Galbraith once remarked that the primary function of economic forecasting was to make astrology look respectable. Some would propose a similar role for ecological forecasting. In projecting the relationship between flow and ecological response, there are many sources of potential error. The behaviour of water over the nearly flat floodplain surface can vary between nearly identical hydrological inputs, depending on a host of antecedent conditions. The response of vegetation can depend on the cumulative effects of several years of flooding history, local-scale changes in topography and soil type, and interactions with groundwater (Overton and Doody, Ch. 21, this volume). Birds may or may not utilise a flooded wetland, depending on regional population trends and little-known triggers, and the relationships between fish and flow are still poorly characterised.

In proposing quantitative relationships between flow and wetland response, are we setting ourselves up for failure? Not if we consider our models as hypotheses to be refuted on the basis of better information and better models. As long as there is a commitment to ongoing monitoring there is an opportunity to improve our understanding of the relationships between flow and ecological response, and the temporal and spatial scales over which these should be measured. Environmental watering events, both singularly and collectively, become opportunities to test hypotheses, and so the nexus between hypothesis and observation finds a corollary in adaptive environmental management.

The past nine years (2000–09) have seen us perform an on-ground experiment on a vast scale, pertaining to the limits of wetland resilience to drought. The effect of an unusually prolonged sequence of below-average rainfall has been exacerbated by the diversion of flows which, under natural conditions, would have sustained core wetland areas and other refugia. Our vegetation mapping has confirmed the decline of habitats over this period, with widespread dieback in vegetation documented in many of the most important wetlands in the Murray-Darling Basin (Saintilan *et al.*, Ch. 4, this volume; Thomas *et al.*, Ch. 9, this volume; Overton and Doody, Ch. 21, this volume; Overton *et al.*, Ch. 16, this volume). The real outcome of this experiment will only be known when flooding returns to these systems. Only then will the fate of the seedbanks be known, and the resilience of populations which transcend individual wetlands. What emerges may be very different from the wetlands which entered this long dormancy nearly a decade ago.

Our goal for these rivers and wetlands is therefore not conservation but restoration. This necessity is reflected in the names given to the government-funded programs supporting ecosystem response modelling in the northern basin – the Wetland Recovery Plan and the Rivers Environmental Restoration Program. Successful restoration will depend not only on the restoration of hydrological conditions suitable for the maintenance of wetland health. It is equally important to maintain the mechanisms of recruitment that allow the emergence of new wetlands when hydrological conditions permit. This will involve taking particular care to preserve or even to create critical aquatic refuges during prolonged droughts. The role of bird migrations in reseeding wetlands needs to be appreciated, and the maintenance of regional and international populations requires the constant vigilance of governments, and of non-government organisations.

The components of our river and wetland ecosystems must therefore be managed at a range of temporal and spatial scales. For several species of birds utilising the wetlands of the Murray-Darling Basin these scales will be at least continental. Vegetation habitats are best managed at the scale of the wetland mosaic, remembering the inherently dynamic relationship between habitat distribution and geomorphic change. One of the primary tasks of the new Murray-Darling Basin Authority in the development of the basin plan should be the setting of overarching ecological objectives for the basin, within which wetland-specific ecological goals should be couched. While individual wetland systems will have their own water allocations and their own environmental management objectives, the development of a basin plan provides an opportunity to apply the Commonwealth's adaptive environmental water account to the pursuit of broader objectives set at the scale of the basin, and to prioritise watering of ecosystem assets that contribute to the ecological character of the basin. As the summer 2008 event in the Narran Lakes illustrated, there is a role for the Commonwealth in investing opportunistically in critical events at critical times for the betterment of basin-wide bird populations.

Whatever form basin-wide ecological models take, they should build upon models being developed for critical aquatic ecosystems within the basin. These models have concentrated on iconic sites, often Ramsar-listed wetlands. Setting objectives at broader spatial scales is presently impeded by an insufficient understanding of the distribution and attributes of the myriad of wetlands, and the dangers of extrapolating flow–response models from specific, well-studied situations to other locations (Balcombe *et al.*, Ch. 5, this volume). Improved data-sharing and a co-ordinated investment in research are preconditions to the responsible management of the river and wetland estate in the long term.

Ecosystem models have a range of functions including the communication of system understanding. Stewardson and Webb (Ch. 3, this volume) and Cottingham *et al.* (Ch. 23) argue strongly for the use of conceptual models to assist in this communication, not only for non-scientific audiences, but also for environmental scientists to fully identify the causal links within the system and to improve the efficiency of the modelling process. Cogle *et al.* (Ch. 11, this volume) identify a role of ecosystem models to communicate the confidence and uncertainty in the model outcomes.

This volume has focused on the development and implementation of ecosystem response models within the Murray-Darling Basin to inform the management of environmental water releases. Although ecosystem response models are valuable, it must be recognised that they are only one of a number of tools and criteria that water managers use when making environmental watering decisions (Cogle *et al.*, Ch. 11, this volume; Ward and Colloff, Ch. 20, this volume).

References

Balcombe SR, Huey JA, Lobegeiger JS, Marshall JC, Arthington AH, Davis LC, Sternberg D and Thoms MC (2010). Comparing fish biomass models based on biophysical factors in two northern Murray-Darling Basin rivers: a cautionary tale. In *Ecological Response Modelling in the Murray-Darling Basin.* (Eds N Saintilan and I Overton) pp. 67–84. CSIRO Publishing, Melbourne.

Bowen S and Simpson S (2008). Changes in extent and condition of the vegetation communities of the Macquarie Marshes floodplain 1991–2008. Draft Adaptive Environmental Management Plan. Water for the Future, Canberra.

Bowen S, Simpson S and McCosker R (2008). Changes in extent and condition of floodplain wetland vegetation communities in the Gwydir wetland and floodplain. NSW Dept of Environment and Climate Change, Sydney.

CSIRO (2008). Water availability in the Murray-Darling Basin. Report to the Australian Government. CSIRO Water for a Healthy Country, Canberra.

Kelleway J, Mazumder D, Wilson G and Kobayashi T (2010). Using isotopic techniques to assess trophic structure in northern Murray-Darling Basin wetlands. In *Ecological Response Modelling in the Murray-Darling Basin.* (Eds N Saintilan and I Overton) pp. 85–102. CSIRO Publishing, Melbourne.

Letcher RA (2005). Implementation of a water allocation decision support system in the Namoi and Gwydir valleys. In *MODSIM05: International Congress on Modelling and Simulation.* December 2005. pp. 1456–1552. Modelling and Simulation Society of Australia and New Zealand.

McCosker R (2008). Yanga vegetation mapping: historical community extent and condition. Technical report to NSW Dept of Environment and Climate Change, Sydney.

MDBC (2008a). Murray-Darling Basin Rivers: ecosystem health check, 2004–2007, sustainable rivers audit report 1. Murray-Darling Basin Commission, Canberra.

MDBC (2008b). The Living Murray: Icon Site condition report, October 2008. Murray-Darling Basin Commission, Canberra.

Merritt W, Powell S, Pollino C and Jakeman T (2010). IBIS: a decision support tool for managers of environmental flows into wetlands. In *Ecological Response Modelling in the Murray-Darling Basin.* (Eds N Saintilan and I Overton) pp. 119–136. CSIRO Publishing, Melbourne.

Overton IC, Doody TM and Cuddy SM (Eds) (2008). Ecological outcomes of flow regimes: hypotheses, conceptual models and data availability. Report to the National Water Commission. CSIRO Water for a Healthy Country, Adelaide.

Powell SJ, Letcher RA and Croke BFW (2008). Modelling floodplain inundation for environmental flows: Gwydir wetlands, Australia. *Ecological Modelling* **211**, 350–362.

Rogers K, Ralph T and Imgraben S (2010). Water requirements of biota, geomorphology and climate change in the Macquarie Marshes. In *Ecological Response Modelling in the Murray-Darling Basin.* (Eds N Saintilan and I Overton) pp. 151–170. CSIRO Publishing, Melbourne.

Rogers K *et al.* (2008). Macquarie Marshes and climate change. Report to Dept of Environment and Climate Change.

Saintilan N, Ling J and Maher M (2010). Science and management challenges for the wetlands of the northern Murray-Darling Basin. In *Ecological Response Modelling in the Murray-Darling Basin.* (Eds N Saintilan and I Overton) pp. 55–66. CSIRO Publishing, Melbourne.

Thoms M, Suter P, Roberts J, Koehn J, Jones G, Hillman T and Close A (2000). Report of the River Murray Scientific Panel on environmental flows: River Murray–Dartmouth to Wellington and the Lower Darling River. Murray-Darling Basin Commission, Canberra.

Thoms M, Capon S, James C, Padgham M and Rayburg S (2007). The Narran Ecosystem Project: the response of a terminal wetland system to variable wetting and drying. Final report to the Murray-Darling Basin Commission. MDBC Publication No. 40/08.

Walker KF, Sheldon F and Puckeridge JT (1995). A perspective on dryland rRiver ecosystems. *Regulated Rivers: Research and Management* **11(1)**, 85–104.

Wassens S (2010). Flooding regimes for frogs in lowland rivers of the Murray-Darling Basin. In *Ecological Response Modelling in the Murray-Darling Basin.* (Eds N Saintilan and I Overton) pp. 213–228. CSIRO Publishing, Melbourne.

Watts RJ (2010). Ecosystem response modelling in the Murrumbidgee River. In *Ecological Response Modelling in the Murray-Darling Basin.* (Eds N Saintilan and I Overton) pp. 197–212. CSIRO Publishing, Melbourne.

Wen L (2008). Reconstruction of natural flow in a regulated system, the Murrumbidgee River, Australia, using time series analysis. *Journal of Hydrology* **164**, 216–226.

Wen L, Saintilan N, Rogers K and Ling J (2010). Linking river red gum condition to hydrological change at Yanga National Park. In *Ecological Response Modelling in the Murray-Darling Basin.* (Eds N Saintilan and I Overton) pp. 229–242. CSIRO Publishing, Melbourne.

Wilson GG, Spencer JA and Heagney EC (2010). Responses of fish and waterbirds to flow variability in the Gwydir wetlands. In *Ecological Response Modelling in the Murray-Darling Basin.* (Eds N Saintilan and I Overton) pp. 103–118. CSIRO Publishing, Melbourne.

Young WJ, Scott AC, Cuddy SM and Rennie BA (2003). Murray Flow Assessment Tool: a technical description. Client report. CSIRO Land and Water, Canberra.

Index

www.ingramcontent.com/pod-product-compliance
Lightning Source LLC
LaVergne TN
LVHW070846210926
848482LV00016B/878
9780643096134